THE ELECTRON CAPTURE DETECTOR AND THE STUDY OF REACTIONS WITH THERMAL ELECTRONS

THE ELECTRON CAPTURE DETECTOR AND THE STUDY OF REACTIONS WITH THERMAL ELECTRONS

E. C. M. CHEN
E. S. D. CHEN

WILEY-INTERSCIENCE
A JOHN WILEY & SONS, INC., PUBLICATION

Published by John Wiley & Sons, Inc., Hoboken, New Jersey.
Published simultaneously in Canada.

Library of Congress Cataloging-in-Publication Data:

Chen, E. C. M.
The electron capture detector and the study of reactions with thermal electrons / E. C. M. Chen, E. S. D. Chen.
p. cm.
Includes bibliographical references and index.

ISBN 0-471-32622-4 (Cloth)

1. Gas chromatography. 2. Electrons–Capture. I. Chen, E. S. D. II. Title.
QD79.C45 C49 2004
543′.85–dc22 2003023101

Printed in the United States of America

10 9 8 7 6 5 4 3 2 1

This book is dedicated to Professor Wayne E. Wenthworth who made it possible for the authors to become scientists. Without him the work in this book could not have been accomplished.

We also recognize the contributions of Dr. James. E. Lovelock, the inventor of the electron capture detector.

CONTENTS

FOREWORD **xiii**

PREFACE **xv**

1. Scope and History of the Electron **1**

1.1 General Objectives and Organization 1
1.2 General Scope 2
1.3 History of the Electron 4
References 6

2. Definitions, Nomenclature, Reactions, and Equations **8**

2.1 Introduction 8
2.2 Definition of Kinetic and Energetic Terms 8
2.3 Additional Gas Phase Ionic Reactions 15
2.4 Electron Affinities from Solution Data 16
2.5 Semi-Empirical Calculations of Energetic Quantities 17
2.6 Herschbach Ionic Morse Potential Energy Curves 18
2.7 Summary 19
References 20

3. Thermal Electron Reactions at the University of Houston **22**

3.1 General Introduction 22
3.2 The First Half-Century, 1900 to 1950 23
3.3 Fundamental Discovery, 1950 to 1960 25
3.4 General Accomplishments, 1960 to 1970 27
3.4.1 Introduction 27
3.4.2 The Wentworth Group 28
3.4.3 Stable Negative-Ion Formation 28
3.4.4 Dissociative Thermal Electron Attachment 33
3.4.5 Nonlinear Least Squares 35
3.5 Milestones in the Wentworth Laboratory and Complementary Methods, 1970 to 1980 37
3.6 Negative-Ion Mass Spectrometry and Morse Potential Energy Curves, 1980 to 1990 40
3.7 Experimental and Theoretical Milestones, 1990 to 2000 41

3.8 Summary of Contributions at the University of Houston 42
References 43

4. Theoretical Basis of the Experimental Tools 47

4.1 Introduction 47
4.2 The Kinetic Model of the ECD and NIMS 47
4.3 Nondissociative Electron Capture 50
4.4 Dissociative Electron Attachment 59
4.5 Electron Affinities and Half-Wave Reduction Potentials 64
4.6 Electron Affinities and Ionization Potentials of Aromatic Hydrocarbons 66
4.7 Electron Affinities and Charge Transfer Complex Energies 67
4.8 Summary 71
References 73

5. Experimental Procedures and Data Reduction 75

5.1 Introduction 75
5.2 Experimental ECD and NICI Procedures 76
5.3 Reduction of ECD Data to Fundamental Properties 85
5.3.1 Introduction 85
5.3.2 Acetophenone and Benzaldehyde 86
5.3.3 Benzanthracene, Benz[a]pyrene, and 1-Naphthaldehyde 87
5.3.4 Carbon Disulfide 89
5.3.5 Nitromethane 90
5.3.6 Consolidation of Electron Affinities for Molecular Oxygen 91
5.4 Reduction of Negative-Ion Mass Spectral Data 93
5.5 Precision and Accuracy 96
5.6 Evaluation of Experimental Results 97
5.7 Summary 101
References 101

6. Complementary Experimental and Theoretical Procedures 103

6.1 Introduction 103
6.2 Equilibrium Methods for Determining Electron Affinities 105
6.3 Photon Techniques 110
6.4 Thermal Charge Transfer Methods 116
6.5 Electron and Particle Beam Techniques 121
6.6 Condensed Phase Measurements of Electron Affinities 124
6.7 Complementary Theoretical Calculations 125
6.7.1 Atomic Electron Affinities 126
6.7.2 Polyatomic Molecules 128
6.8 Rate Constants for Attachment, Detachment, and Recombination 132

6.9 Summary 134
References 134

7. **Consolidating Experimental, Theoretical, and Empirical Data** **139**

7.1 Introduction 139
7.2 Semi-Empirical Quantum Mechanical Calculations 140
7.3 Morse Potential Energy Curves 150
7.3.1 Classification of Negative-Ion Morse Potentials 151
7.3.2 The Negative-Ion States of H_2 153
7.3.3 The Negative-Ion States of I_2 156
7.3.4 The Negative-Ion States of Benzene and Naphthalene 157
7.4 Empirical Correlations 161
7.5 Summary 165
References 166

8. **Selection, Assignment, and Correlations of Atomic Electron Affinities** **168**

8.1 Introduction 168
8.2 Evaluation of Atomic Electron Affinities 169
8.3 Mulliken Electronegativities 178
8.4 Electron Affinities of Atomic Clusters 184
8.5 Summary 189
References 190

9. **Diatomic and Triatomic Molecules and Sulfur Fluorides** **193**

9.1 Introduction 193
9.2 Diatomic Molecules 194
9.2.1 Electron Affinities and Periodic Trends of Homonuclear Diatomic Molecules 194
9.2.2 Electron Affinities and Morse Potential Energy Curves: Group VII Diatomic Molecules and Anions 197
9.2.3 Electron Affinities and Morse Potential Energy Curves: Group VI Diatomic Molecules and Anions 205
9.2.4 Electron Affinities and Morse Potential Energy Curves: Group IA and IB Homonuclear Diatomic Molecules and Anions 209
9.2.5 Electron Affinities and Morse Potential Energy Curves: NO and NO(−) 214
9.3 Triatomic Molecules and Anions 216
9.4 Electron Affinities and Morse Potential Energy Curves: Sulfur Fluorides and Anions 224
9.5 Summary 229
References 229

10. Negative Ions of Organic Molecules **234**

10.1 Introduction 234
10.2 Electron Affinities and Potential Energy Curves for Nitrobenzene and Nitromethane 235
10.3 Electron Affinities Determined Using the Magnetron, Alkali Metal Beam, Photon, and Collisional Ionization Methods 238
10.3.1 Electron Affinities Determined Using the Magnetron Method 238
10.3.2 Electron Affinities Determined Using the AMB Method 240
10.3.3 Electron Affinities Determined Using Photon Methods 241
10.3.4 Electron Affinities Determined Using Collisional Ionization Methods 243
10.4 Electron Affinities Determined Using the ECD, NIMS, and TCT Methods 244
10.4.1 Electron Affinities of Aromatic Hydrocarbons by the ECD Method 244
10.4.2 Electron Affinities of Organic Carbonyl Compounds by the ECD Method 246
10.4.3 Electron Affinities of Organic Nitro Compounds the ECD and TCT Methods 253
10.5 Electron Affinities of Charge Transfer Complex Acceptors 257
10.6 Substituent Effect 261
10.7 Summary 263
References 263

11. Thermal Electrons and Environmental Pollutants **266**

11.1 Introduction 266
11.2 Alkyl Halides 267
11.2.1 Morse Potential Energy Curves 267
11.2.2 Experimental Activation Energies 269
11.2.3 Alkyl Fluorocompounds 272
11.2.4 Electron Affinities of the Alkyl Halides 274
11.3 Aromatic Halides 276
11.3.1 Electron Affinities of Fluoro- and Chlorobenzenes 276
11.3.2 Electron Affinities from Reduction Potentials and CURES-EC 283
11.3.3 Negative-Ion Mass Spectra and Electron Affinities 284
11.4 Negative-Ion Mass Spectrometry 287
11.5 Calculation of the ECD and NIMS Temperature Dependence 291

11.6 Summary 293
References 293

12. Biologically Significant Molecules **296**

12.1 Introduction 296
12.2 Electron Affinities of Purines and Pyrimidines 299
12.2.1 Predictions of Electron Affinities 299
12.2.2 Electron Affinities from Reduction Potentials 300
12.2.3 Gas Phase Measurements of Electron Affinities 302
12.2.4 Theoretical Electron Affinities 305
12.3 Electron Affinities of Biological Molecules from Reduction Potentials 307
12.4 Gas Phase Acidities of Nucleic Acids 310
12.5 Morse Potential Energy Curves for Thymine and Cytosine 311
12.6 Gas Phase Acidities and Electron Affinities of the Amino Acids 315
12.7 The Calculation of the ECD and NIMS Temperature Dependence 316
12.8 Electron Affinities of AT AU and GC 318
12.9 Radiation Damage in DNA 320
12.10 Summary 326
References 327

APPENDICES **329**

I Glossary of Terms, Acronyms, and Symbols **331**

II Structures of Organic Molecules **336**

III General Least Squares **339**

IV Tables of Evaluated Electron Affinities **349**

Table A1.1 Atoms 349
Table A1.2 Main Group Homonuclear Diatomic Molecules 351
References 352
Table A2.1 and A2.2 CH Molecules 355
References 357
Table A2.3 and A2.4 CHX Molecules 357
References 359
Table A3.1 and A3.2 CHNX Molecules 360
References 361
Table A4.1 and A4.2 CHO Molecules 362
Table A4.3 and A4.4 CHOX Molecules 366

References 369
Table A5.1 and A5.2 CHON Molecules 370
Table A5.3 and A5.4 CHONX Molecules 375
References 376
Table A6.1 Bergman Dewar set 377
Table A6.2 Values Different from NIST Values (from Tables A2.1 to A5.4) 378
Table A6.3 Unpublished or Updated Gas Phase Values not in NIST Tables 380
Table A6.4 Values for Adenine, Guanine, Cytosine, Uracil, Thymine, and Their Hydrates 382
Table A6.5 Values for Charge Transfer Complex Acceptors not in NIST Tables 382
Table A6.6 Values for Chlorinated Hydrocarbons from Reduction Potentials and CURES-EC 383
Table A6.7 Values for Biological Compounds from Reduction Potentials 383

AUTHOR INDEX **387**

SUBJECT INDEX **395**

FOREWORD

As the title indicates, the principal focus of this text is to describe the Electron Capture Detector (ECD), both in terms of its fundamental principles and its numerous applications to various analytical disciplines. Consequently, this text will be of extreme importance to analytical chemists who may need the ECD in obtaining a satisfactory analysis. It is extremely useful where sensitivity is of major concern, such as for environmental pollutants and biologically important analyses. Special attention should be given to the Appendices where there are tables giving a comprehensive summary of the electron affinities of various classes of compounds. The listings are given in order of both the electron affinity value and the compound molecular weight. This is very convenient to find the electron affinity of a compound that has been studied previously. Also these tables are useful in estimating the electron affinity of a compound not previously investigated. These tables are of immense value to an analytical chemist who would like to know the sensitivity of compounds to the ECD. The authors have been extremely thorough in their literature search and evaluation to obtain the most reliable, self consistent electron affinities.

The analytical chemist will find Chapters 4 and 5 most beneficial to a fundamental understanding of the basic principles of the ECD. The ECD is somewhat simple to operate to an experienced analyst, but the process of electron capture can be quite complex. The authors have done an excellent job in presenting these principles in a very understandable fashion. With the knowledge from these chapters the analytical chemist can use the ECD to obtain fundamental properties associated with thermal attachment to molecules.

In order to support the fundamental parameters obtainable from the ECD, the authors have carried out a thorough investigation of the literature for complementary techniques for the study of electron attachment to molecules. This is a monumental task and the authors should be proud of their accomplishments in this regard. To my knowledge this has never been attempted before; this information is of value both to the understanding of the ECD and to the complementary techniques. To assist in these correlations between the ECD and complementary techniques, the authors have carried out appropriate theoretical calculations to assist in the interpretation. It is now understood that a molecule can have multiple negative ion states and the theoretical calculations can be useful in defining these different states. Frequently, different negative ion states are involved in different types of studies, and an understanding of these states is necessary in order to correlate the results. This text is also of extreme value to investigators involved in the study of electron attachment to molecules by alternative methods.

I compliment the authors on a job well done. I hope that the text will accomplish its objective giving the analytical chemist the information to operate the ECD optimally, based upon fundamental principles, and to use the results in the evaluation of parameters associated with the thermal attachment to molecules.

Professor of Chemistry
University of Houston

W. E. WENTWORTH

PREFACE

More than a century ago Thompson determined the mass-to-charge ratio of the electron and established its fundamental nature. It remains the only one of the subatomic particles that has not been subdivided. Simultaneously, Tswett initiated the study of modern chromatography. Fifty years later Lovelock observed that the reaction of molecules with thermal electrons greatly perturbed ionization currents generated by radioactivity in air. This led to the electron capture detector (ECD) and inextricably bound chromatography and the reactions of thermal electrons with molecules.

Since its introduction, the ECD has become the most selective and sensitive technique for determining the concentration of compounds such as explosives, polychlorinated biphenyls, pesticides, and freons. It is less recognized as an unsurpassed probe of the kinetics and thermodynamics of the reactions of thermal electrons with molecules. In the 1960s the kinetic model for this system was proposed and verified. The pulse-sampling ECD operates under conditions of atmospheric pressure, low fields, with low-energy electrons and in the presence of complex organic molecules. A capillary column provides an ultra-pure test sample of a region where thermal electrons are present. The concentration of the thermal electrons can be measured without disturbing the thermal energy distribution of the electrons. The change in the electron concentration in the presence of the test species is the response of the detector. The measurement of the temperature dependence of this response leads to the fundamental thermodynamic and kinetic properties of molecules: the electron affinities of molecules and radicals, the bond dissociation energies of molecules and negative ions, the partition function ratios for the anion and neutral, and the pre-exponential terms and activation energies of the rate constants for attachment, detachment, and negative-ion dissociation.

In 1972 Wentworth, Chen, and Steelhammer set out to write a monograph entitled *Negative Ions Reaction and Formation in the Gas Phase,* so scientists could plan future research using the ECD. At the time few fundamental properties of thermal electron reactions had been measured. Now many molecular electron affinities and rate constants for thermal electron attachment have been measured. Currently electron affinities and bond dissociation energies can be verified using theoretical SCF calculations on desktop computers. It is especially timely to review the techniques for studying reactions of thermal electrons with molecules and to evaluate the results.

This book is based on the reactions of thermal electrons with molecules. The ECD, negative-ion chemical ionization (NICI) mass spectrometry, and polarographic reduction in aprotic solvents methods are used to determine the kinetic and thermodynamic parameters of these reactions. The chromatograph gives a small pure sample of the molecule. The temperature dependence of the response of the ECD and NIMS is measured to determine fundamental properties. The ECD measurements are verified and extended by correlations with half-wave reduction potentials in aprotic solvents, absorption spectra of aromatic hydrocarbons and donor acceptor complexes, electronegativities, and simple molecular orbital theory.

Chemistry is a central science, but there are many diverse areas of study within it. Since the quantities determined using the ECD are important to all areas of chemistry, this book should be of interest to a variety of scientists, including environmental, analytical, physical, organic, and inorganic chemists. An untapped application is the study of thermal electron reactions with biological molecules. This is especially important since the equipment for the application of these tools is relatively inexpensive and now commercially available.

The present book was partially inspired by a conversation the authors had with Dr. Alan Marchand. When reminded that 1997 was the 100th anniversary of the "discovery of the electron," he laughed out loud. He noted that it was a bit presumptuous of modern scientists to state that the electron had been discovered. According to Marchand it is and always has been a fundamental particle, but indeed it remains the only member of the triad—the electron, proton, and neutron—that has not been further divided. Once we established that this groundwork had already been established, an interesting discussion of the ECD and its use in determining the fundamental properties of the reaction of thermal electrons with molecules evolved. Professor Marchand was unaware of the efforts in this field and believed that the only method of measuring electron affinities was powerful and expensive mass spectrometers. From this conversation came the impetus for our book.

CHAPTER 1

Scope and History of the Electron

1.1 GENERAL OBJECTIVES AND ORGANIZATION

> The electron theory holds that the various atoms of matter have the power, in different degrees, of gaining or losing electrons; and it assumes that electrons form a natural part of all atoms, revolving around them or in them at enormous velocities.
>
> —Robert Sanderson Mulliken,
> Newburyport High School, Salutatorian
> Newburyport, Massachusetts, 1913

The electron is a universal quantity, as noted in the high-school address of renowned chemist Mulliken [1]. It is a particle, a wave, and an integral part of all matter. We now know that the natural phenomenon observed with marvel by ancient man, fire, sunlight, lightning, and the modern things we take for granted, such as electricity, radio, television, and nuclear power, all involve electrons.

The primary objective of this book is to present a concise description of the experimental and theoretical procedures developed to determine the fundamental energetic and kinetic properties of electron molecule reactions using the electron capture detector (ECD), negative ion mass spectrometry (NIMS), negative ion Morse potential energy curves, and semi-empirical quantum mechanical calculations. Professor Wayne E. Wentworth, inspired by his collaboration with James E. Lovelock, the inventor of the ECD, led the original work. In the past fifteen years the authors have evaluated and assigned electron affinities, obtained the maximum information from the ECD data, calculated and categorized negative ion potential energy curves by combining diverse data, and applied these techniques to biological molecules.

This chapter covers the specific objectives of the book, discusses its importance to various areas of chemistry, and provides a brief history of the characterization of the electron. The fundamental definitions, nomenclature, symbols, and equations are then summarized in Chapter 2. Chapter 3 gives a chronological history of the events leading up to this book in the Wentworth laboratory, including summaries of

The Electron Capture Detector and the Study of Reactions with Thermal Electrons
by E. C. M. Chen and E. S. D. Chen
ISBN 0-471-32622-4

material only published in dissertations. This is especially important now that it is possible to access or order dissertations online [2]. In Chapter 4 the theoretical background for the tools used to obtain new information are presented. Chapter 5 and Appendix III describe the experimental and data reduction procedures. Procedures that have been used to obtain complementary data are discussed in Chapter 6. Chapter 7 examines the procedures for consolidating diverse data. The next five chapters present representative evaluations of data for atoms, homonuclear diatomic molecules, small inorganic molecules, organic molecules, environmental pollutants, and biologically significant molecules. A more complete list of evaluated molecular electron affinities is given in the appendices.

1.2 GENERAL SCOPE

In the first half of the twentieth century, positive-ion molecule reactions and the interaction of hyperthermal electrons with molecules were emphasized. Some thermal electron molecule reactions in flames and electron swarms were investigated [3]. Prior to 1950 only the electron affinities of hydrogen and the halogen atoms had been measured. A 1953 review on electron affinities noted

> The determination of further electron affinities is not an easy matter; it is possible that the electronic equilibrium method could be extended to a few more elements, but at the temperatures involved, molecules and radicals would be decomposed. The only reasonable hope of estimating the electron affinities of radicals would seem to lie in a study of the appearance potential of negative ions, and the determination of their kinetic energies, although it must be borne in mind that a careful search of the mass spectrum of methane has failed to reveal the existence of a CH_3(-)ion. [4]

Fortunately, this dire prediction was not realized.

In the late 1960s several major advances were made in the study of thermal electron reactions. These were based on the ECD, the extension of the "magnetron" method of studying electron molecule reactions to determine equilibrium constants for electron molecule reactions, and the invention of high-pressure thermal electron negative-ion sources for mass spectrometry [5–7]. Electron swarms were also used to determine rate constants for thermal electron reactions [8, 9]. The electron affinities of molecules were measured using electron and alkali metal beams [10, 11]. Relative electron affinities were obtained from the direction of the reaction of a negative ion with a molecule [12, 13]. Other major advances were photodetachment and photoelectron spectroscopy [14–17].

This book is based on the study of reactions of thermal electrons with molecules using the ECD, negative ion chemical ionization (NICI) mass spectrometry in the gas phase and polarographic reduction in aprotic solvents [18]. Only the complementary studies related to our research are considered here.

Until recently, there were few theoretical methods available to test experimental results. With the developments in computer software and hardware, it is now easy to compare theory and experiment. The use of multiconfiguration configuration interaction (MCCI), in a post-self-consistent field calculation improves the agreement to chemical accuracy. The systematic use of MCCI has been assigned the acronym CURES-EC. It appropriately offers a cure for the electron correlation problem in semi-empirical calculations. The CURES-EC method has been successfully applied to over 200 electron affinities, including those for purines and pyrimidines [19, 20].

The physical properties determined using the ECD are important to different areas of chemistry. Analytical chemistry deals with "how much and what" are involved in a chemical reaction. Expressed differently, it establishes what we refer to as the QQQ: quantitation, qualitative identification, and the quality of the results. The determination of the electron affinities of the chlorinated biphenyls, dioxins, and phenols and the prediction of the response of the ECD and NIMS are important to qualitative and quantitative analyses of environmental pollutants [21]. Polarographic reduction in solutions likewise gives accurate and precise qualitative and quantitative results. The quality of the analyses is expressed by the random and systematic uncertainties in the reported values. These are obtained from the same principle of weighted least squares used to obtain information from ECD data. Wentworth has described the application of the general least-squares procedure to chemical problems [22, 23].

The experimental and theoretical methods of measuring or calculating the kinetic and thermodynamic properties of reactions of thermal electrons are important to physical chemistry. The ECD method will be compared to other methods. The CURES-EC procedure can be applied to other energetic quantities. We present a method to consolidate diverse data into "pseudo-two-dimensional" potential energy curves for these reactions [9, 10].

The biochemical applications involve the electronic nature of the components of DNA and proteins, especially the charge distributions, electron affinities, and gas phase acidities of purines and pyrimidines and amino acids. The role of electron reactions in diverse areas such as cancer, electron conduction, and sequence recognition all depend on fundamental energetic properties such as electron affinities and solution energies. We explain nonadiabatic experimental data from radiation chemistry by excited anionic states of biological molecules [24].

Inorganic chemistry concerns molecules of all the atoms. The electron affinities of atoms, small molecules, and radicals and their relationship with the Periodic Table, electronegativities of elements, Morse curves of diatomic anions, and the energies of ion molecule reactions and bond energies are inorganic problems we have considered. Ionic radii can be estimated using potential energy curves.

The reactions of organic molecules in solution are related to gas phase electron affinities and electronegativities. Anions are often intermediates in such reactions. The electron conduction of polymers is related to the electron affinities of the components. The theoretical calculations of electron affinities of aromatic hydrocarbons and the effect of substitution on electrons affinities and gas phase acidities are important to organic chemistry. Pseudo-two-dimensional Morse potentials have been used to represent the dissociation of organic molecules and their anions [18].

A major objective of this book is to evaluate the reported values of molecular electron affinities and their errors and to assign them to specific states. Prior to 1970 the magnetron and ECD methods were used to measure the majority of gas phase molecular electron affinities. An extensive compilation of unevaluated experimental, empirical, and theoretical electron affinities of atoms, molecules, and radicals was published before 1990 [9]. The electron affinities measured in the gas phase are now available on the Internet but have not been evaluated [26]. The molecular E_a in this list is defined and evaluated in Appendix IV. Values that are significantly lower than the selected values will be assigned to excited states. Semi-empirical calculations and the CURES-EC technique support these assignments. Unpublished electron affinities and updated electron affinities from charge transfer complex data and half-wave reduction potentials are given in Appendix IV.

1.3 HISTORY OF THE ELECTRON

The first recorded study of the electron dates back to 600 B.C when a Greek philosopher, Thales of Miletus, observed that amber, when rubbed with a cloth, attracted bits of feathers and the pith of plants. In 1600 William Gilbert, an English physician, characterized static electricity and used the Latin word for amber, *electrum,* to characterize substances that behave like amber. In 1646 Sir Thomas Browne, another English physician, first used the word "electricity," deriving it from the Greek word *elektron* for amber. In the early to middle 1700s an English scientist Stephen Gray and a French scientist Charles-François Du Fay identified two types of electricity of opposite signs so that those with like signs repelled each other, while those of opposite signs attracted each other. In the mid-1700s Benjamin Franklin used the terms "positive" and "negative" to describe the two types of electricity. In 1752 he performed his famous kite experiment, which led to the discovery of the lightning rod, the first invention to protect man from the hazards of electricity.

The late 1700s witnessed the work of the French physicist Charles-Augustin de Coulomb, and the Italian physicists Luigi Galvani and Alessandro Volta; their research led to development of the battery and the first source of a steady current of electricity. In the next half-century the studies of the Danish physicist Hans Christian Ørsted, French physicist André-Marie Ampère, German school teacher Georg Simon Ohm, and German physicist Thomas Johann Seebeck related electricity to magnetism and heat. These concepts were then put into practice by Michael Faraday, the English physicist, and Joseph Henry, an American contemporary. Faraday recognized the fundamental nature of the electron and its relationship to electrochemistry. The events surrounding "the discovery of the electron" began in 1833, when Faraday postulated the following two laws of electrolysis:

1. The chemical power of a current of electricity exists in direct proportion to the absolute quantity of electricity that passes through it.
2. The equivalent weight of bodies is simply those quantities that contain equal quantities of electricity.

Scottish physicist James Clerk Maxwell, who developed the theory of electricity and magnetism in the late 1800s, proposed that Faraday's laws required that "one molecule" of positive and negative electricity is involved in electrolysis. Irish physicist George Johnstone Stoney believed in this "molecule of electricity" and set out to measure the definite quantity of electricity, called e1. He suggested that if this unit of electricity was adopted, it would represent a very important step in our study of molecular phenomena. In 1891 Stoney referred to these charges as "electrons." The controversy of the day then revolved around whether these electrons were waves or particles.

The relationships to electromagnetic waves postulated by the German physicist Heinrich Rudoph Hertz led to the work of the English physicist Sir Joseph John Thomson in 1897, which is often linked to the actual discovery of the electron [26]. The measurement of the e/m and m of the corpuscles called electrons by Thompson settled this controversy. Electrons were at least particles, but other studies suggested that they were also electromagnetic radiation. Thompson described his conclusions as follows:

> If, in the intense electric field in the neighbourhood of the cathode, the molecules of the gas are dissociated and are split up, not into the ordinary chemical atoms, but into these primordial atoms, which we shall for brevity call corpuscles; and if these corpuscles are charged with electricity and projected from the cathode by the electric field, they would behave exactly like the cathode rays. They would evidently give a value of (e/m) which is independent of the nature of the gas and its pressure, for the carriers are the same whatever the gas may be.... Thus on this view we have in the cathode rays matter in a new state, a state in which the subdivision of matter is carried very much further than in the ordinary gaseous state: a state in which all matter—that is, matter derived from different sources—is of one and the same kind; this matter being the substance from which all the chemical elements are built up. [27]

In the next decade the physicists Townsend, Charles Wilson, Sir John Planck, Ernest Rutherford, Johannes Geiger, Max Begeman, and finally Robert Millikan estimated the charge on an electron, the quantity e1, and by inference, its mass. Millikan deserves the greatest credit because his measurements proved that e1 is a fundamental and invariant unit of charge. Subsequently, experiments by Planck introduced the two fundamental quantum constants, h and k, now known as Planck's constant and Boltzman's constant, respectively. Planck recognized that k is the gas constant for one molecule that gives the value of Avogardro's number N_0.

This led to the important discoveries of Albert Einstein and the quantum revolution at the heart of modern chemistry. Modern chemical pioneers Gilbert Newton Lewis, Linus Pauling, and Mulliken elucidated the fundamental concepts of the electron as the heart of chemical bonding [1, 28, 29]. The role of the electron in modern society is much too extensive to chronicle. Radio, television, nuclear weapons and power are but a few applications. Indeed, the ultimate importance of the electron is easily realized if we return to the idea that all matter contains electrons.

Methods of studying the reactions of molecules with thermal electrons have not been concisely described. This book is written with the hope that it will encourage others to use these techniques to add to the knowledge of the reactions of thermal electrons as chemical reagents. Indeed, we explore the role of the electron as the zeroth member of the Periodic Table and use the ECD as the focal point of these investigations.

REFERENCES

1. Mulliken, R. S. *Selected Papers of R. S. Mulliken*, edited by D. A. Ramsay and J. Hinze. Chicago, IL: The University of Chicago Press, **1975**, p. 15.
2. Available at http://www.umi.com/hp/Products/DisExpress.html.
3. Massey, H. S. W. *Negative Ions*. New York: Cambridge University Press, **1976**.
4. Pritchard, H. O. *Chem. Rev.* **1953**, 52, 529.
5. Lovelock, J. E. "The Electron Capture Detector, a Personal Odyssey," in *The Electron Capture Detector*, edited by A. Zlatkis and C. Poole. New York: Elsevier, **1981**, p. 1.
6. Page, F. M. and Goode, G. C. *Negative Ions and the Magnetron*. New York: Wiley-Interscience, **1969**.
7. Von Ardenne, M.; Steinfelder, K.; and Tummler, R. *Angew. Chem.* **1961**, 3, 135.
8. Pack J. L. and Phelps, V. *Phys. Rev. Lett.* **1961**, 6, 111.
9. Christodoulides, A. A.; McCorkle, D. L.; and Christophorou, L. G. "Electron Affinities of Atoms, Molecules and Radicals," in *Electron-Molecule Interactions and Their Applications*. New York: Academic Press, **1984**.
10. Herschbach, D. R. *Adv. Chem. Phys.* **1966**, 10, 250.
11. Kleyn, A. W. and Moutinho, A. M. C. *J. Phys. B.* **2001**, 34, R1.
12. Rains, R. J.; Moore, H. W.; and McIver, R. T. *J. Chem. Phys.* **1978**, 68, 3309.
13. Kebarle, P. and Chowhury, S. *Chem. Rev.* **1987**, 87, 513.
14. Branscomb, L. M. In *Photodetachment in Atomic and Molecular Processes*, edited by D. R. Bates, New York: Academic Press, **1962**, p. 100.
15. Mead, R. D.; Stevens A. E.; and Lineberger, W. C. "Photodetachment in Negative Ion Beams," in *Gas Phase Ion Chemistry*, Vol. 3, edited by M. T. Bowers. New York: Academic Press, **1984**, p. 214.
16. Drzaic, P. S.; Marks, J.; and Brauman, J. I. "Electron Photodetachment in Gas Phase Ions," in *Gas Phase Ion Chemistry*, Vol. 3, edited by M. T. Bowers. New York: Academic Press, **1984**, p. 168.
17. Reinstra-Kiracofe, J. C.; Tschumper, G. S.; Schaefer, H. F.; Nandi, S.; and Ellison, G. B. *Chem. Rev.* **2002**, 102, 231.
18. Streitweiser, A. *Molecular Orbital Theory for Organic Chemists*. New York: Wiley, **1961**.
19. Dewar, M. J. S. *The Molecular Orbital Theory of Organic Chemistry*. New York: McGraw-Hill, **1969**.
20. Chen, E. S. D.; Chen, E. C. M.; Sane, N.; Talley, L.; Kozanecki, N.; and Shultze, S. *J. Chem. Phys.* **1999**, 110, 9319.
21. Stemmler, E. A. and Hites, R. A. *Electron Capture Negative Ion Mass Spectra*. New York: VCH, **1988**.

22. Wentworth, W. E. *J. Chem. Ed.* **1965**, 42, 96.

23. Wentworth, W. E. *J. Chem. Ed.* **1965**, 42, 162.

24. Szent-Györgi, A. *Introduction to a Submolecular Biology.* New York: Academic Press, **1960**.

25. National Institute of Standards and Technology (NIST). *Chemistry WebBook*, **2003**. Available at http://webbook.nist.gov.

26. The names and events given in this brief chronological history were searched on the Internet. Thales of Miletus electron gives the derivation of the word "electron," whereas J. Thomson electron offers a complete slideshow on the "discovery of the electron."

27. Thompson, J. J. *Philosoph. Mag.*, **1897**, 44, 293, and Thompson, J. and Thompson, G. P. *Conduction of Electrons Through Gases*, Dover reprint, New York: Vol. 1, **1928**, Vol. 2, **1933**.

28. Lewis, G. N. *Valence and the Structure of Atoms and Molecules.* New York: Chemical Catalog Co. **1923**.

29. Pauling, L. *The Nature of the Chemical Bond*, 3rd ed. Ithaca, NY: Cornell University Press, **1960**.

CHAPTER 2

Definitions, Nomenclature, Reactions, and Equations

2.1 INTRODUCTION

The objectives of this chapter are:

1. To define the generally accepted terms applicable to thermal electron reactions. This nomenclature is used specifically in our work.
2. To classify reactions of thermal electrons and negative ions.
3. To describe experiments used to study these reactions.
4. To present the equations relating half-wave reduction potentials and charge transfer absorption spectra to electron affinities.
5. To describe a semi-empirical procedure for quantum mechanical calculations of electron affinities.
6. To present fundamental relationships for the calculation of Morse potentials.
7. To classify negative-ion Morse potentials based on fundamental data.

2.2 DEFINITION OF KINETIC AND ENERGY TERMS

Two energy terms applicable to thermal electron reactions referenced to the ground state of the neutral are the electron affinity E_a and the vertical electron affinity VE_a. E_a is the difference in energy between the most stable state of the neutral and a specific state of a negative ion. VE_a is the difference in energy between the anion and neutral species in the geometry of the neutral. The anion can be temporary and relax to either the ground state, an excited valence-state anion, or dissociate.

Each state will have an E_a and a VE_a. By convention E_a are positive when the anion is more stable than the neutral. It was once believed that only one bound

The Electron Capture Detector and the Study of Reactions with Thermal Electrons
by E. C. M. Chen and E. S. D. Chen
ISBN 0-471-32622-4

anion state could exist. However, multiple bound states for anions have been observed. The ground-state or adiabatic E_a, AE_a, is the difference in energy between the anion and the neutral when each exists in its most stable state. It is the largest E_a by definition. Since electrons are attracted to all elements, even helium, by ionic polarization forces, the AE_a is always positive. The AE_a should be greater than the bond energy of He_2, where only van der Waal attractions are present. Short-range or valence-state electron affinities can be either negative or positive. When all the valence-state electron affinities, or simply E_a, are negative, the balance between the short-range repulsion and long-range attractions determines the AE_a. Molecules with a large dipole moment can form a stable dipole bound (DB) state in the same geometry as the neutral. These DBEA are generally less than 200 meV. All molecules will be bound to electrons by polarization forces leading to a polarization electron affinity PE_a. These are smaller than DBEA.

The energy required to eject an electron from the negative ion in its most stable configuration is the photodetachment energy E_{pd}. The reorganization energy E_{rr} is equal to the vertical electron affinity minus electron affinity for the same state. The AE_a for benzene is a small positive value. The excited state VE_a is -1.13 eV and the E_a is -0.78 eV, so E_{rr} is -0.35 eV. For naphthalene the VE_a is -0.20 eV and the AE_a is 0.16 eV, yielding an E_{rr} of -0.36 eV. In diatomic molecules there is only one geometrical parameter, the internuclear distance. In polyatomic molecules the energies refer to the completely optimized geometries. These quantities are defined for all states.

The electron affinities and reorganization energies are illustrated for F_2 in Figure 2.1, where semi-empirical Morse potentials for the neutral and negative ions are presented. The properties of the ground-state negative ion have been measured. The vertical electron affinity has been obtained from the solution charge transfer energy E_{CT}. The activation energy for thermal electron attachment has been measured using the electron capture detector (ECD). The dissociation energies of the excited states are obtained through the isoelectronic principle by analogy to the rare gas positive ions. The other parameters for the excited states are obtained from the absorption spectra, or E(abs), of the anion and electron impact (EI) spectra for the formation of the fluorine ion by monoenergetic electrons. The procedures for calculating these types of curves will be discussed in Chapter 7 [1].

Nondissociative electron attachment and detachment are general terms that refer to the process whereby an electron $e(-)$ reacts with a molecule AB to form the molecular negative ion $AB(-)$ and the reverse process whereby the electron is removed. It is also designated as the molecular anion formation:

$$e(-) + AB \underset{k_{-1}}{\overset{k_1}{\Leftrightarrow}} AB(-) \qquad \text{(2.1 and 2.2)}$$

For thermal electrons nondissociative reactions will take place in the typical temperature range of the ECD when the quantity $D_{AB} - E_a(B)$ is greater than about 1.5 eV, where D_{AB} is the bond dissociation energy and $E_a(B)$ the electron affinity of the dissociated species. This quantity is defined as minus the energy for

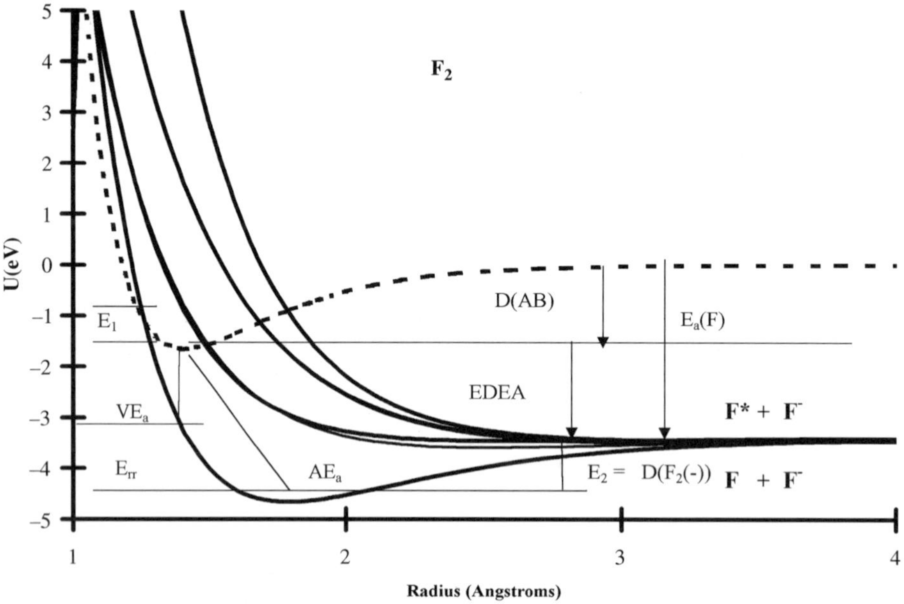

Figure 2.1 Morse potential energy curves for the neutral and negative-ion states of F_2. The vertical electron affinity VE_a, adiabatic electron affinity AE_a, activation energy for thermal electron attachment E_1, $E_{rr} = AE_a - VE_a$, EDEA $= E_a$(F) − D(FF), and dissociation energy of the anion E_2 are shown.

dissociative electron attachment, or −EDEA. Aromatic hydrocarbons, such as anthracene, undergo nondissociative thermal electron attachment reactions since −EDEA = 4.5 − 1.5 eV = 3.0 eV. This can be seen in Figure 2.2, where the neutral and negative ion curves are shown in pseudo-two-dimensional Morse potential energy curves. The activation energy to form the ground state is greater than that which forms the first excited state. The rearrangement energy for the ground state is smaller than that for the excited states [2, 3].

The word "attachment" implies that the molecule holds the electron loosely. This is only true for the initial interaction since the added valence-state electron generally resides in the lowest unoccupied molecular orbital (LUMO) of *AB*. The first step is the loose attachment of the electron to the molecule at a large distance and the second step stabilization to the ground state. The overall rate constant for the formation of the ground state consists of two terms, one of which is dependent on the specific bath gas and the potential relaxation modes of the initial anion. The rate constants for the forward and reverse process are k_1 and k_{-1} [$k_1 = A_1 T^{-1/2} \exp(-E_1/RT)$ and $k_{-1} = A_{-1} T \exp(-E_{-1}/RT)$] with activation energies E_1 and E_{-1}. The ratio k_1/k_{-1} is the equilibrium constant K_{eq} for the thermal electron attachment reaction and the reaction energy at 0 K is $E_a = -E_1 + E_{-1}$. Thus, it is possible to determine the electron affinity E_a of the molecule using the ECD. In some experiments, those involving k_1, the rate constants for the forward reaction

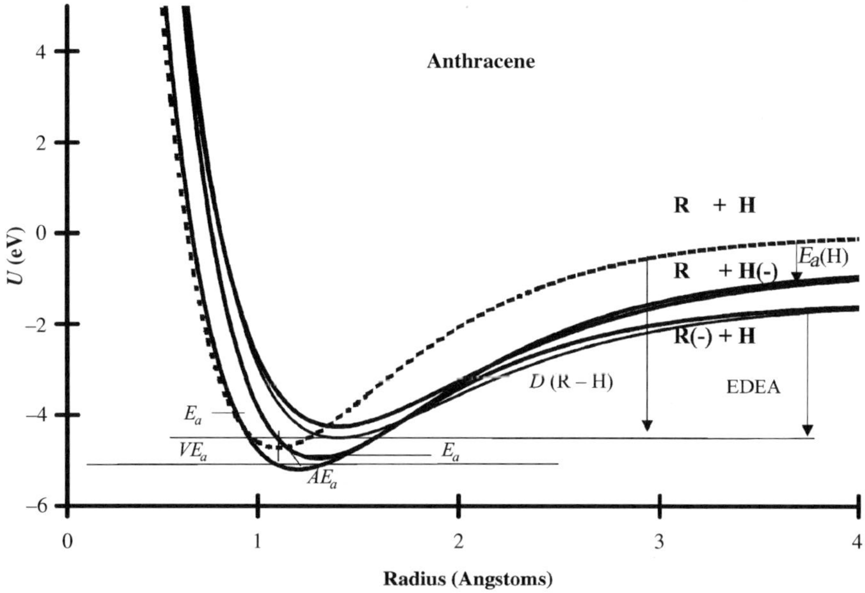

Figure 2.2 Morse potential energy curves for the neutral and negative-ion states of anthracene. The vertical electron affinity VE_a, adiabatic electron affinity AE_a, and activation energy for thermal electron attachment E_1 are shown. The two E_a are 0.68 eV and 0.53 eV observed in ECD data. There will be nine other negative ion curves, yielding a total of thirteen anion curves, four each for the different C—H bonds and a polarization curve. Some of these will be accidentally degenerate.

alone can be determined. The activation energy for electron attachment to F_2 is shown in Figure 2.1.

In the event that the combination of the electron and molecule have energies greater than required to form a fragment negative ion $B(-)$, direct dissociation can take place:

$$e(-) + AB \xrightarrow{k_{12}} A + B(-) \tag{2.3}$$

The rate constant for this process is designated as k_{12} for unimolecular dissociation. The kinetic expression is the same as for direct attachment so that this can also be designated as k_1. Activation energies for this reaction with thermal electrons have been measured in the ECD and are as large as 0.5 eV to 0.8 eV. With electron beam studies the distribution of the dissociated ions is measured as a function of electron energy to yield energy-dependent cross-sections. In electron swarms cross-sections can be measured as a function of the average energy of electrons in an electric field. Thermal electron reactions with chlorinated methanes are examples of this type of reaction. This will occur when the bond dissociation energy is less than the E_a of the dissociating species by about 15 kcal/mole or 0.7 eV. In Figure 2.3 this

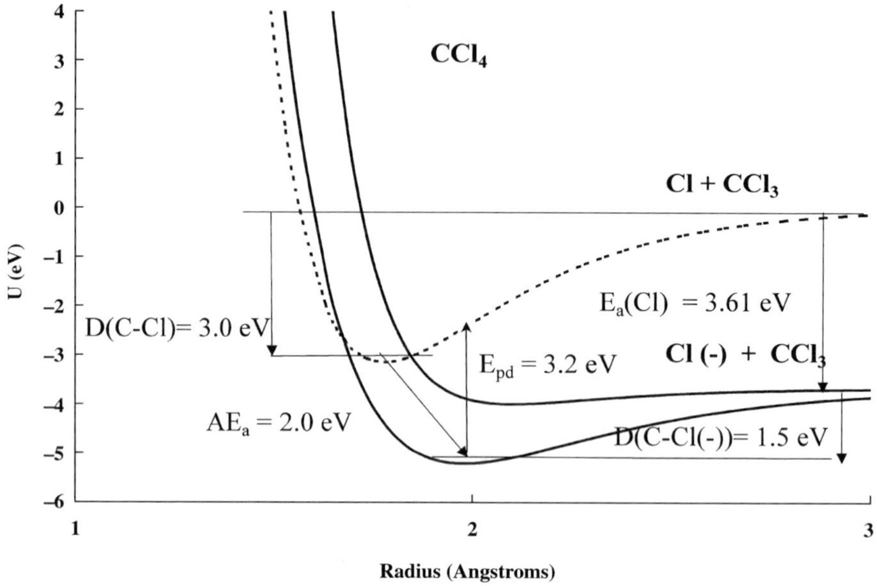

Figure 2.3 Morse potential energy curves for the neutral and negative-ion states of CCl_4. The new quantity illustrated in this figure is photodetachment energy. It is larger than AE_a and is the peak in the photodetachment spectrum. Thermal electron attachment is exothermic, that is, EDEA = a positive quantity. Two other states dissociating to $Cl + CCl_3(-)$ and the polarization curve are not shown.

is illustrated for CCl_4. The quantity EDEA is 0.61 eV. Two other curves that dissociate to the limit $CCl_3(-) + Cl$ are not shown. The activation energy for thermal electron attachment is about zero since the negative-ion curve crosses the neutral curve at the internuclear distance of the neutral [4].

Another mode of dissociation occurs via an intermediate molecular ion:

$$e(-) + AB \Leftrightarrow AB(-) \xrightarrow{k_2} A + B(-) \tag{2.4}$$

The rate constant for this dissociation is k_2: $k_2 = A_2T \exp(-E_2/RT)$. The pre-exponential term (A_2T) for these rate constants is one of the largest observed pseudo-Arrhenius terms. Activation energies as large as 40 kcal/mole have been measured in the ECD for fluoronitrobenzenes [5].

The major reactions for the permanent loss of electrons and negative ions recombine with positive ions. These reactions are given as

$$e(-) + P \xrightarrow{k'_D} \text{Neutrals} \tag{2.5}$$

and

$$AB(-) \quad \text{or} \quad B(-) + P \xrightarrow{k'_N} \text{Neutrals} \tag{2.6}$$

At atmospheric pressure and in the presence of an unchanging concentration of positive ions P, the bimolecular rate constants can be replaced by pseudo-unimolecular rate constants $k_D = k'_D[P]$ and $k_N = k'_N[P]$. These are approximately temperature-independent because recombination is exothermic since the energy is E_a − IP.

Other experimental procedures measure the energetics and kinetics of thermal electron reactions in the same manner as the ECD and NIMS. These are designated equilibrium methods. The direct capture magnetron method (MGN) and the individual determination of the individual rate constants k_1 and k_{-1} for thermal electron reactions are also equilibrium methods. The latter was first carried out in an electron swarm (ES) for O_2, but can be applied to any system to measure thermal electron reactions. These methods differ in how the electron and ion concentrations are generated and measured [6, 7].

The equilibrium methods are based on the measurement of the equilibrium constant K_{eq} for the thermal electron reactions since the energy of reaction at 0 K is $E_a = -E_1 + E_{-1}$. The ratio k_1/k_{-1} is the equilibrium constant

$$e^- + AB \underset{k_{-1}}{\overset{k_1}{\Leftrightarrow}} AB^- \qquad (2.1 \text{ and } 2.2)$$

$$K_{eq} = k_1/k_{-1} = [g(A(-))/\{g(A)g(e(-))\}]\exp(E_a/RT) \qquad (2.7)$$

The $g(A(-)), g(A)$, and $g(e(-))$ are the partition functions of the respective species:

$$K_{eq} = h^3/(2\pi m_e kT)^{3/2} Q_{an} \exp(E_a/RT) \qquad (2.8)$$

Taking logs and multiplying through by $T^{3/2}$, we obtain

$$\ln\ K_{eq}\ T^{3/2} = \ln[h^3/(2\pi m_e k)^{3/2}] + \ln[Q_{an}] + E_a/RT \qquad (2.9)$$

$$= 12.43 + \ln[Q_{an}] + E_a/RT \qquad (2.10)$$

The value of 12.43 is calculated from fundamental constants h, k, and m_e. Q_{an} is the ratio of the partition functions of the anion and neutral with the spin multiplicity of the anion canceling that of the electron.

The electron capture coefficient is the experimental molar response of the ECD normalized to a constant concentration of electrons. This concentration may change because of changes in impurities in the carrier gas. It is best to minimize the changes, but if it is assumed that the reactions of the impurities only reduce the concentration of electrons and do not react with the specific test molecule, the correction is simply a multiplication factor. The electron capture coefficient K_{ECD} is calculated from the concentrations of the electrons I_b in the absence of the test molecule, the concentration of the electrons in the presence of the test molecule I_e, and the concentration of the capturing species $[AB]$. The corresponding equation is

$$\frac{I_b^2 - I_e^2}{I_e I_b} = K_{\mathrm{ECD}}[AB] \qquad (2.11)$$

At the limit of low capture, $I_b = I_e$, giving

$$I_b^2 - I_e^2 = (I_b - I_e)(I_b + I_e) = 2I_b(I_b - I_e) \qquad (2.12)$$

and equation 2.11 becomes

$$\frac{I_b - I_e}{I_e} = K_{\mathrm{ECD}}[AB]/2 \qquad (2.13)$$

The electron capture coefficient K_{ECD} is related to the fundamental rate constants defined in reactions 2.1 to 2.6, as will be seen in the kinetic model of the ECD and NIMS. The affinity of molecules for electrons and electron capture ability have been loosely used to describe the response of the ECD. These refer to the sensitivity of a molecule in the ECD regardless of the process. Molecules can have a high response for electrons in the ECD without having a high electron affinity. Likewise, some molecules with high electron affinities such as benzoquinone have a low ECD response. The upper limit to the response of the ECD response is based on the DeBroglie wavelength of the electron. The maximum value of A_1 for the reactions with thermal electrons is the DeBroglie A_1 value (DeBA). The value of ln(DeBA) is about 36 at 400 K [4].

The measurement of the equilibrium constant as a function of temperature leads to an absolute value of the electron affinity. The magnetron technique was the first to obtain accurate electron affinities of organic molecules. Some of the MGN values for quinones are for excited states [4, 7]. Of these techniques, the ECD method has been applied to the greatest number of molecules. In cases where sufficient data exist in the equilibrium region, such as for acetophenone, the electron affinities can be measured to a precision of 2 meV. Both the ground and excited states of hexafluorobenzene and carbon disulfide have been observed in a single ECD experiment. The measurement of the temperature dependence of ions by NIMS can lead to accurate and precise E_a. The mere observation of a negative ion is evidence of the positive electron affinity of a molecule. This has not been utilized extensively, but there is a large body of unanalyzed published data [8]. The addition of an electron monochrometer source to a GC/NIMS makes this observation possible at thermal energies [9].

One of the early confirmations of the ECD model was based on the agreement of the ECD measurements for the rate constants and activation energies for thermal electron attachment to SF_6 and C_7F_{14} with the values obtained using the microwave method. In the interim other methods have been developed. Indeed, the determination of the rate constants as a function of both electron energy and temperature has been achieved [10–15].

2.3 ADDITIONAL GAS PHASE IONIC REACTIONS

The first experimental observation of negative ions in the gas phase used the parabola mass spectrometer described by J. J. Thompson [16]. Thus, the halogens sulfur, hydrogen, and oxygen form negative ions, whereas nitrogen and the rare gases do not. Few definitive properties of negative ions were determined before 1950. The only electron affinities in the 1953 review article by H. O. Pritchard [17] matching the current "best" values were identified for F, Cl, Br, I, and CN.

In the second half of the twentieth century four experimental techniques were developed to measure E_a: the equilibrium, the photon, the beam, and the thermal charge transfer methods [4, 18–29]. In addition to the above reactions observed in the ECD and NIMS, negative ions can be formed in other reactions and complementary energetics and kinetics determined. The ECD and NIMS results have been integrated and compared with data obtained from other studies [4]:

$$AB(-) + CD \underset{k_{-\mathrm{ET}}}{\overset{k_{\mathrm{ET}}}{\Leftrightarrow}} AB + CD^{-}(-) \qquad \text{Electron transfer} \qquad (2.14)$$

$$AB + e(-) \quad \text{or} \quad M \quad \text{or} \quad h\nu \underset{k_{12}}{\overset{k_{IP}}{\rightarrow}} P(+) + B(-) \qquad \text{Ion pair formation} \qquad (2.15)$$

$$e(-)(\text{hyper}) + AB \rightarrow A + B(-) \qquad \text{Electron impact} \qquad (2.16)$$

$$AB(-) + h\nu \rightarrow AB + e(-) \qquad \text{Photodetachment} \qquad (2.17)$$

$$A(-) + BH \rightarrow AH + B(-) \qquad \text{Proton transfer} \qquad (2.18)$$

$$AB(-) + L \underset{k_{-L}}{\overset{k_L}{\Leftrightarrow}} AB(-) \bullet L \qquad \text{Ligand clustering} \qquad (2.19)$$

Another pertinent energy quantity is the gas phase acidity (GPA) of a molecule. It is also called the deprotonation energy of a molecule. It is the energy for the reaction

$$AH \rightarrow H^{+} + A^{-} \qquad (2.20)$$

The deprotonation energy is related to the A—H bond dissociation energy, the electron affinity of the radical, and the ionization potential of the hydrogen atom as

$$\mathrm{GPA} = \mathrm{IP(H)} + E_a(\mathrm{A}) - D(\mathrm{A{-}H}) \qquad (2.21)$$

The temperature dependence of reactions 2.14 and 2.18 yields relative energetics. If the reactant negative ion is a radical, the hydrogen transfer reaction results in relative gas phase acidities. If the reactant ion is a molecule, the relative molecular electron affinities are obtained. If we assume similar entropy changes, the ratio of the ions and that of the concentration of the neutral reactants give a value of the equilibrium constant and hence a measure of the relative acidity or electron affinity. In some cases only the direction of the reaction is observed to yield limiting electron affinities or acidities. In order to obtain an absolute value, the value of the reference compound must be known. A better procedure is to determine the temperature dependence of the equilibrium constant in order to remove the assumptions concerning the entropy terms. The equilibrium TCT methods have been used to determine the electron affinities of many organic molecules. When relative acidities are obtained, the corresponding bond dissociation energy of the X—H bond must be known to obtain the electron affinity of the radical [21–23].

Photodetachment (PD) and photoelectron spectroscopy (PES) utilize reaction 2.17 [24–27]. In the PD method the threshold photon energy for the destruction of the anions is related to the electron affinity of the molecule or radical. In the case of the PES method the electrons generated by the interaction of photons at a fixed frequency are energy-resolved and the electron affinity determined from the threshold. The PD and complementary photoabsorption procedure led to accurate electron affinities of the atoms. In the case of the PD of molecules the onset is equal to E_a, whereas the peak is E_{pd}. This is illustrated in Figure 2.3 for carbon tetrachloride. The PES method has been applied to a few large molecules. The determination of the threshold is complicated by the change in geometry in forming the anion limiting the precision. In PES the state of the negative ion must be identified. The two states for the anthracene anion shown in Figure 2.2 will yield two PES peaks.

Another method used alkali metal beams (AMB) to give a negative ion and a positive ion (equation 2.15). The threshold is the sum of the ionization potential of the alkali metal and the electron affinity of the molecule. A related procedure is the determination of the threshold for endothermic charge transfer (EnCT) from one anion to a neutral to form a second anion. The threshold is equal to the difference in the electron affinities of the species so that the electron affinity of the reactant must be known. These procedures yielded accurate electron affinities of the halogen molecules [18–19, 28, 29].

Electron beam (EB) and electron transmission (ET) experiments measure the vertical electron affinities of molecules. In the EB experiments the ions formed as a function of electron energy are measured. The ion distributions give an indication of a potential energy surface in the Franck Condon region of the neutral. Very good mono-energetic electrons can be obtained. The temperature dependence of the zero energy cross-section can be related to the activation energy for thermal electron attachment. In ET spectroscopy the transmitted current is measured and resonances indicated by a change in the derivative of the electron current as a function of energy. This complements electron impact studies since some transitions do not lead to the formation of negative ions [9, 14, 15]. This VE_a is illustrated in Figure 2.2 for anthracene [3].

2.4 ELECTRON AFFINITIES FROM SOLUTION DATA

Relative electron affinities of organic molecules can be obtained from half-wave reduction potentials in aprotic solvents. The electron affinities are related to the half-wave reduction potentials by

$$E_a(\text{eV}) = E_{\text{ref}} - (-\Delta\Delta G) + E_{1/2} = 4.71 - (-\Delta\Delta G) + E_{1/2} \tag{2.22}$$

where $-\Delta\Delta G$ is the solution energy difference for the reaction and 4.71 is the reference energy for an SCE electrode. The $-\Delta\Delta G$ term is a function of the solvent and the specific counter ion, but is approximately constant for aprotic solvents and similar molecules [30–32]. Because there were no experimental electron affinities of organic molecules before the 1960s, it was assumed that the $-\Delta\Delta G$ values were the same for different molecules. In the case of even the aromatic hydrocarbons, the $-\Delta\Delta G$ are a function of the charge distribution on the anion. Thus with experimental E_a and experimental $E_{1/2}$ values, it was possible to evaluate and classify the $-\Delta\Delta G$ values. With the ability to calculate charge distributions in the anions using quantum mechanical procedures, the value of $-\Delta\Delta G$ can be estimated to obtain more precise E_a. This is one of the major advances of the past decade and will be described in more detail in Chapter 4.

According to the Mulliken theory of charge transfer complexes, the vertical electron affinity (VE_a) of an acceptor and the vertical ionization potential (VIP) of a donor are related to the energy of maximum absorption of the complex (E_{CT}) by the following equation:

$$E_{CT} = \text{VIP} - VE_a - C_2 + C_1/(\text{VIP} - VE_a - C_2) \tag{2.23}$$

Therefore, if a series of complexes are studied in which there is one donor and multiple acceptors, the constants C_2 and C_1 can be determined. In order to use this equation, however, there must be experimental values of VE_a and VIP. If the data are calibrated to adiabatic values, then E_{CT} is related to AE_a for the unknown species. This type of analysis was conducted in 1975 with the data available at the time [33].

2.5 SEMI-EMPIRICAL CALCULATIONS OF ENERGETIC QUANTITIES

In 1995 the electron affinities of organic molecules calculated using the standard MINDO/3 or AM1 semi-empirical self-consistent field calculations agreed with experiment for molecules containing only CH and O to about 0.1 eV. However, the calculated values differed from experiment for molecules containing N and the halogens by as much as 1 eV.

The agreement with experiment has been improved by using multiconfiguration, configuration interaction (MCCI) with different numbers of orbitals to minimize the difference between experimental and calculated values. This procedure is called "CURES-EC." The geometry of both the ion and neutral are optimized and the energy calculated from the difference in the energy of the valence electrons after the inclusion of MCCI. The acronym CURES-EC stands for from "configuration interaction or unrestricted orbitals to relate experimental quantities to self-consistent field values by estimating electron correlation." The acronym is appropriate since it is a post-self-consistent field approach to cure the electron correlation problem in semi-empirical calculation. It has been used to obtain ionization potentials, gas phase acidities, and bond dissociation energies. With the general availability of modern desktop computers and software to carry out theoretical calculations, it is possible for anyone to extend the calculations described in this book [31].

2.6 HERSCHBACH IONIC MORSE POTENTIAL ENERGY CURVES

In 1963 negative-ion Morse parameters for the ground-state anions of Br_2 and I_2 were obtained by estimating D, r_e, and ν from the VE_a from charge transfer spectra and the properties of the excited states of the neutral [34]. This followed a principle postulated by Mulliken, "If for an electronic state, one knows only r_e, D, and υ, one cannot easily do better in constructing its U(r) curve than to use Morse's function" [35]. This is valid even if all the values must be estimated. Morse potential energy curves consolidate diverse data for anions [1]. The Morse potentials for the neutral as referenced to zero energy at infinite separation and the ionic Morse curves are given by

$$U(X_2) = -2D_e(X_2)\exp(-\beta(r - r_e)) + 2D_e(X_2)\exp(-2\beta(r - r_e)) \tag{2.22}$$

$$\begin{aligned} U(X_2(-)) = {} & -2k_A D_e(X_2)\exp(-k_B\beta(r - r_e)) + k_R D_e(X_2) \\ & \times \exp(-2k_B\beta(r - r_e)) - E_a(X) + E(X^*) \end{aligned} \tag{2.23}$$

where $D_e(X_2)$ is the spectroscopic dissociation energy, r is the internuclear separation, $r_e = r$ at the minimum of $U(X_2)$, $E(X^*)$ is the excitation energy, $\beta = \nu_e(2\pi^2\mu/D_e[X_2])^{1/2}$, μ is the reduced mass, and k_A, k_B, and k_R are dimensionless constants.

In 1966 Herschbach classified ionic Morse potential energy curves (HIMPEC) [34]. He identified nine types of $XY(-)$ potential energy curves, based on molecular anion formation or dissociation in a vertical transition, the signs of the vertical E_a, VE_a, and EDEA $= E_a(X) - BDE_{XY}$. The original classifications included impossible combinations. For example in Figure 2.4, if the EDEA is negative, then for dissociation to take place, the curves must cross on the front side and VE_a must be negative, making BII+ and CII+ impossible. Likewise if the EDEA is positive, then the vertical process will always lead to dissociation, making AI-impossible. In addition, the two different types of dissociative curves, B and C, which are completely repulsive, are eliminated since all excited state curves will have a minimum at long ranges due to at least polarization attractions. Adding the sign of the E_a as a classifier eliminated these gaps in the HIMPEC. This gives $2^3 = 8$ possible curves, described in a symmetrical notation: $M(m)$ and $D(m)$, $m = 0$ to 3, where m is the number of positive metrics. With the $r(e-)$-XY separation as a third dimension, eight subclasses $Mc(m)$ and $Dc(m)$ can be defined based on the crossing of the polarization and covalent curves to yield molecular ions or dissociation [36].

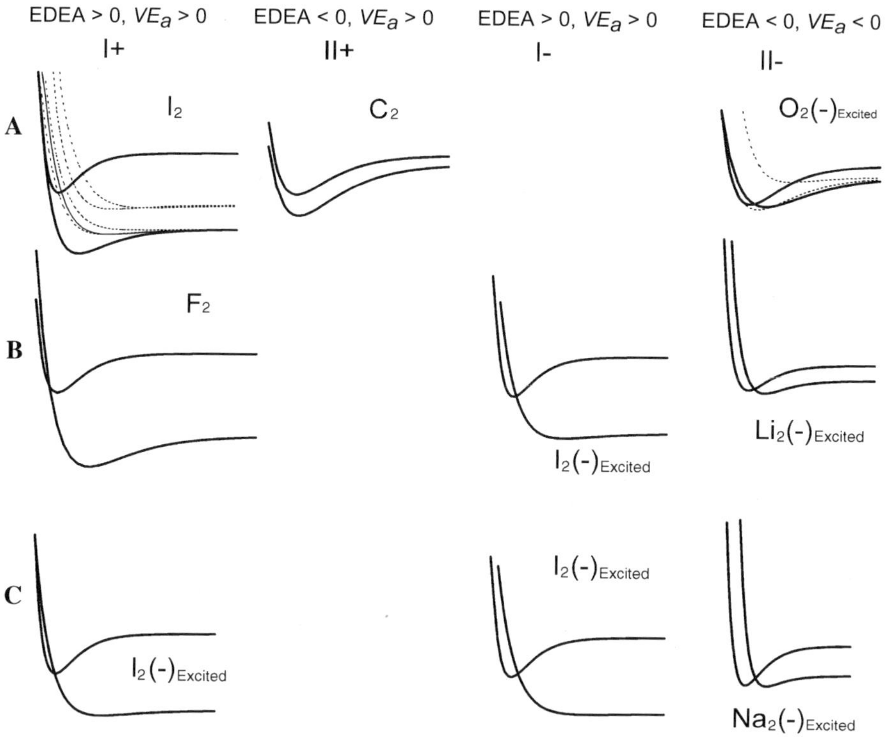

Figure 2.4 Classification of negative-ion Morse potential energy curves originally presented by Herschbach in 1966. The curves are calculated from actual data. The empty spaces are impossible combinations. These classifications have been modified so that they are symmetrical and all combinations are possible [34, 36].

2.7 SUMMARY

The reactions for the formation and destruction of negative ions have been listed. The specific techniques used to obtain energetic quantities have been described. The ECD and NIMS reactions have been identified. A brief discussion of the alternative methods of measuring electron affinities then follows. The classification of molecules to obtain electron affinities from reduction potentials and the relationship between charge transfer complex energies and electron affinities were introduced. The use of quantum mechanical calculations to obtain electron affinities, bond dissociation energies, and gas phase acidities was noted. We described and updated the general classification of types of Morse potential energy curves. A list of acronyms, symbols, and abbreviations is given in Appendix I.

REFERENCES

1. Chen, E. C. M. and Wentworth, W. E. *J. Phys. Chem.* **1985**, 84, 4099.
2. Chen, E. S. D. and Chen, E. C. M. *J. Chromatogr. A* **2002**, 952, 173.
3. Jordan, K. D. and Burrow, P. D. *Chem. Rev.* **1987**, 87, 557.
4. Chen, E. C. M. and Wentworth, W. E. *Mol. Cryst. Liq. Cryst.* **1989**, 171, 271.
5. Chen, E. S. D.; Chen, E. C. M.; and Wentworth, W. E. *J. Phys. Chem.* **1991**, 95, 520.
6. Pack J. L. and Phelps, V. *Phys. Rev. Lett.* **1961**, 6, 111.
7. Page, F. M. and Goode, G. C. *Negative Ions and the Magnetron.* New York: Wiley-Interscience, **1969**.
8. Stemmler, E. A. and Hites, R. A. *Electron Capture Negative Ion Mass Spectra.* New York: VCH, **1988**.
9. Laramee, J. A.; Mazurkiewicz, P.; Berkout, V.; and Deinzer, M. L. *Discrete Electron Capture Negative Ion Mass Spectrometry in Encyclopedia of Analytical Chemistry.* New York: Wiley, **2000**.
10. Chen, E. C. M.; George, R. D.; and Wentworth, W. E. *J. Chem. Phys.* **1968**, 49, 1973.
11. Mahan, B. H. and Young, C. E. *J. Chem. Phys.* **1966**, 44, 2192.
12. Crompton, R. W. and Hadad, G. N. *Aust. J. Phys.* **1983**, 36, 15.
13. Christophorou, L. G.; McCorkle, D. L.; and Carter, J. G. *J. Chem. Phys.* **1971**, 54, 253.
14. Smith, D. and Spanel, P. I. *Adv. At. Mol. Opt. Phys.* **1994**, 32, 307.
15. Spanel, P.; Matejcik, S.; and Smith, D. *J. Phys. B* **1995**, 28, 2941.
16. Thompson, J. J. *Philosoph. Mag.*, **1912**, 24, 205, and Thompson, J. J. and Thompson, G. P. *Conduction of Electrons Through Gases*, Dover reprint, New York, Vol. 1, **1928**, Vol. 2, **1933**.
17. Pritchard, H. O. *Chem. Rev.* **1953**, 52, 529.
18. Massey, H. S. W. *Negative Ions.* New York: Cambridge University Press, **1976**.
19. Christodoulides, A. A.; McCorkle, D. L.; and Christophorou, L. G. "Electron Affinities of Atoms, Molecules and Radicals," in *Electron-Molecule Interactions and Their Applications.* New York: Academic Press, **1984**.
20. National Institute of Standards and Technology (NIST). *Chemistry WebBook*, **2003**. Available at http://webbook.nist.gov.
21. Von Ardenne, M.; Steinfelder, K.; and Tummler, R. *Angew. Chem.* **1961**, 73, 135.
22. Rains, R. J.; Moore, H. W.; and McIver, R. T. *J. Chem. Phys.* **1978**, 68, 3309.
23. Kebarle, P. and Chowhury, S. *Chem. Rev.* **1987**, 87, 513.
24. Bartmess, J. E. and McIver, R. T. *The Gas Phase Acidity Scale*, Vol. 2, edited by M. T. Bowers. New York: Academic Press, **1979**, p. 88.
25. Mead, R. D.; Stevens A. E.; and Lineberger, W. C. "Photodetachment in Negative Ion Beams," in *Gas Phase Ion Chemistry*, Vol. 3, edited by M. T. Bowers. New York: Academic Press, **1984**, p. 214.
26. Drzaic, P. S.; Marks, J.; and Brauman, J. I. "Electron Photodetachment in Gas Phase Ions" in *Gas Phase Ion Chemistry*, Vol. 3, edited by M. T. Bowers. New York: Academic Press, **1984**, p. 168.

27. Reinstra-Kiracofe, J. C.; Tschumper, G. S.; Schaefer, H. F.; Nandi, N.; and Ellison, G. B. *Chem. Rev.* **2002**, 102, 231.

28. Kleyn, A. W. and Moutinho, A. M. C. *J. Phys. B* **2000**, 34, R1.

29. Hughes, B. M.; Lifschitz, C.; and Tiernen, T. O. *J. Chem. Phys.* **1973**, 59, 3162.

30. Briegleb, G. *Angew. Chem. Internat. Edit.* **1964**, 3, 617.

31. Chen, E. S. D.; Chen, E. C. M.; Sane, N.; Talley, L.; Kozanecki, N.; and Shultze, S. *J. Chem. Phys.* **1999**, 110, 9319.

32. Streitweiser, A. S. *Molecular Orbital Theory for Organic Chemists*. New York: Wiley, **1961**.

33. Chen, E. C. M. and Wentworth, W. E. *J. Chem. Phys.* **1975**, 63, 3183.

34. Herschbach, D. R. *Adv. Chem. Phys.* **1966**, 10, 250.

35. Mulliken, R. S. *Selected Papers of R. S. Mulliken*, edited by D. A. Ramsay, and J. Hinze. Chicago Il: The University of Chicago Press, **1975**, p. 310.

36. Chen, E. S. D. and Chen, E. C. M. *J. Phys. Chem. A* **2002**, 106, 6665.

CHAPTER 3

Thermal Electron Reactions at the University of Houston

3.1 GENERAL INTRODUCTION

> If enough histories, written while the ideas are still fresh in the minds of the people concerned, are available for a variety of discoveries of inventions, it may eventually be possible to lay down some of the principles required to facilitate the obtaining of fruitful results in scientific research in general. Clearly also the background of knowledge at the time the advance was made will be best understood if the history is as recent as possible.
>
> —Archer J. P. Martin, Nobel lecture, 1951.

These words were the catalyst for a look into the history of the first seventy-five years of chromatography [1]. Their significance is emphasized when we recall that the ECD was invented by James E. Lovelock at the behest of Martin, once the Welch Professor at the University of Houston. Martin passed away in 2002 with little note of his contribution to chromatography. In their history of chromatography, L. S. Ettre and A. Zlatkis first gave a biographical sketch of Martin and then a narrative of his accomplishments.

In this chapter we present a personal retrospective on the decades of work at the W. E. Wentworth Laboratories and summarize the advances made there in the past fifteen years, which constitute the major focus of this book. We give our impression of the applications of the scientific method that Wentworth passed on to his students and colleagues.

In the scientific method problems are neatly stated, postulates made, data collected, and solutions obtained. Seldom is this path unidirectional. More often, it is a matter of luck and the right combination of individuals at the right time that leads to a solution. Science could not progress without the critical eyes of others. It is often the skeptic questioning the results of other investigators who spurs and

The Electron Capture Detector and the Study of Reactions with Thermal Electrons
by E. C. M. Chen and E. S. D. Chen
ISBN 0-471-32622-4

contributes to the original search for answers. New data and techniques compiled by those same skeptics often lead to new experiments confirming or modifying the initial answers. Here we will summarize the status of problems studied at the time and related work by others.

Wayne E. (Sonny) Wentworth was born in Rochester, Minnesota, on May 29, 1930. For two years (1948–1950) he attended Rochester Junior College, where he was elected to the Phi Theta Kappa honorary scholastic society. Over the next two years he studied at St. Olaf College in Northfield, Minnesota, completing the requirements for a B. S. in chemistry. He attended graduate school at Florida State University from 1952 to 1956, during which time he was awarded an Ethyl research scholarship. In June 1957 he received his Ph.D. in analytical-physical chemistry.

From September 1956 to September 1959 Wentworth was employed by the RCA Service Company at Patrick Air Force Base, Florida, as a mathematical analyst. From September 1959 to the present he has taught at the University of Houston, serving as assistant professor from 1959 to 1963, associate professor from 1963 to 1969, and a full professor since 1969. He is married to the former Elise Hughes of Georgia. They had four children and have three grandchildren.

Wentworth has been a member of the American Chemical Society, the American Solar Energy Society, and Alpha Chi Sigma. He has published over 125 papers, written three textbooks, and received numerous patents. One of the publications from his doctoral dissertation was selected as a pioneer paper in analytical chemistry. His papers on the chemical applications of nonlinear least squares became Citation Index Classics. His patent for pulse discharge ionization detectors was selected as one of the 100 most significant inventions of 1997. He retired after four decades of teaching and research, having about forty graduate students, half of whom received doctorates. The Robert A. Welch Foundation supported his efforts every year.

3.2 THE FIRST HALF-CENTURY, 1900 TO 1950

The pioneer developers of chromatography and ECD included engineers, physicians, mathematicians, and physical, organic, inorganic, and biological chemists. Here we present some of the recollections of M. S. Tswett, Martin, M. C. E. Golay, S. R. Lipsky, Ettre, Zlatkis, Wentworth, and Lovelock. The majority may be found in the history of chromatography and the monograph on ECD cited at this chapter's end [1, 2].

Tswett introduced chromatography in 1903. Ettre discussed its discovery, rejection, and reinvention in the 1930s. The reasons for the rejection were as follows:

1. The method was too unique.
2. The results contradicted some "established" scientists of the time. Tswett was an "outsider" who did not hesitate to criticize results that had been demonstrated to be wrong by chromatographic analysis.

3. Tswett's findings were correct.
4. Tswett published the results of much of his work in a dissertation written in Russian. However, one of the scientists he criticized had the document translated one year after its publication.

Ettre has commented, "Tswett, the underdog was right but was pushed aside by the establishment" [1]. Tswett was simply ahead of his time. A change of philosophy was needed for his technique and investigations to become fully appreciated. This occurred at the end of the 1920s when the exponential evolution of chromatography finally began.

In his recollections Martin has stated:

> Then I realized it was not necessary to move both liquids (in a counter current extraction device). If I just moved one of them the required conditions were fulfilled. I was able to devise a suitable apparatus the very next day. This became the partition chromatograph that we now know. This work was eventually published in 1941 where we noted that the mobile phase could just as well be a gas as a liquid. We also predicted that with such a system, very refined separations of various kinds of compounds would be possible. Although this paper was widely read by chemists in different fields, no one thought this prediction worth testing experimentally. [1]

As Lovelock remembers,

> The discovery of the ECD dates back to a sensitive anemometer based on ions generated from a radioactive source made from radium extracted from the luminous dials of discarded aircraft instruments. This device was very sensitive but its response was perturbed by cigarette smoke. This was a drawback in the device and to discover its source, other compounds, such as halocarbons were tested. However, in 1948, this was something to merely note for the future. Looking back I realized that the key to invention is need. We did not at that time need a device to detect low levels of chlorofluorocarbons and so the electron capture detector was in a sense prematurely invented. [2]

Lovelock often simply tried to measure the response of every volatile compound in the laboratory to characterize a detector. This method of inquiry was a Lovelock trademark later adopted by the Wentworth group. In the first half of the twentieth century few reactions of thermal electrons with molecules were studied. Some studies of electrons and ions in flames and electron swarms were conducted during this time. H. S. W. Massey described the work on negative ions in a series of monographs [3].

Prior to 1950 only the electron affinities of hydrogen and the halogens had been measured. The most accurate value of the electron affinity of the hydrogen atom was calculated [4]. The E_a of the halogens that agree with current values were the Born Haber cycle values. Very few molecular electron affinities had been measured up to that point [5].

3.3 FUNDAMENTAL DISCOVERY, 1950 TO 1960

World War I led to Tswett's premature death at age 41 and interrupted the development of chromatography. World War II postponed the further study of gas chromatography until the early 1950s [1]. It was not until 1960 that gas chromatography was demonstrated, capillary columns were developed, and ionization detectors including the electron capture detector (ECD) were invented and commercialized [6–8].

Martin described the process of streamlining gas chromatography for practical application as such: "So (to improve James' morale) I suggested that we go back to the prediction . . . to use a gas instead of a liquid as a mobile phase; I was sure this would work. We spent our first week waiting for the bands (of fatty acids) to come out of the gas chromatograph; in fact, they had all come out in the first few seconds" [1]. The need for the ECD soon arose after the practical application of gas chromatography.

Lovelock recalls,

> In 1956, my brief from Archer Martin was to invent a more sensitive detector for gas chromatography. When a low voltage was applied to an ionization detector so that only thermal electrons are collected, a 1 mg sample of an allegedly pure sample of methyl caproate gave a never ceasing range of off scale peaks. I shall never forget the of amazement on Tony James' face. There were two problems with the detector, it was too sensitive and it was erratic. When I tried a mixture made up in the solvent CCl_4 the current fell to zero and there remained.

Chromatography was not ready for such a sensitive and selective detector so it was put aside for a few more years [1, 2].

The opportunity to develop the ECD came in 1958 in the form of Lipsky's invitation to colleagues to visit his laboratory at Yale in New Haven, Connecticut. Lipsky described the development of the plane parallel detector like this:

> One day, Jim (Lovelock) poked a hole in one of the tritium foils and the plane parallel detector was ready to test. . . . A mixture of components containing different functional groups along with 2 or 3 hydrocarbons was prepared and injected onto a capillary column. Positive deflections were recorded for the hydrocarbons and a series of negative deflections for the ketones, aldehydes, alcohols and particularly the halogenated substances. The electron capture detector came into being with full force.

Lovelock noted the following: "During the time at Yale, the key to the cure of the electron capture detector's bad habits came from an encounter with Dr. McAfee of Bell Laboratories. He had developed a pulse method for observing electron attachment in a drift tube." This method was seized on to overcome the erratic behavior of the ECD by applying a brief pulse of a high potential to collect the electrons rather than by using a DC voltage that could also lead to ionization and/or the development of contact potentials and space charges [1, 2].

Lovelock began the description of his personal odyssey with the ECD as such:

> When I first came to Houston in 1958 and felt the touch of its warm humid air on my skin it seemed like a place where interesting things might happen. . . . I was met at the airport by Al Zlatkis and taken directly to his laboratory at the University of Houston. There we set it (the ECD) up and in no time were running samples of products from the Houston Petrochemical Industry; mixtures called "platformates" and "reformates" which I had never heard spoken of before. They gave the most glorious and unbelievably excellent chromatograms resolving components hitherto only suspected to be present. It was not pure science but it surely was one of those days when everything goes right technically and life as a scientist seems to be very worthwhile. [1]

In the 1950s the discovery that very high resolution could be attained using a capillary column [8] added to the need for a selective detector that would be able to identify some of the many resolved peaks. Lovelock turned to the ECD for an answer.

> It was a great pleasure to be able to share the observation that . . . the powerful combination of the capillary column and a sensitive ionization detector were able to make exquisitely perfect chromatograms. Such were their quality that when Zlatkis and I tried to publish one, the referee (a well known expert on capillary chromatography) complained that they were too good to be real and must have been faked. Fortunately we succeeded in convincing him otherwise.

At the University of Houston, R. S. Becker joined Lovelock and Zlatkis and suggested determining the ECD response of aromatic hydrocarbons. In 1959 Wentworth joined the faculty at the University of Houston.

It is generally assumed that closed-shell molecules do not interact strongly with each other. However, as early as 1909, it was observed that new intense absorption bands were observed when I_2 was dissolved in an aromatic hydrocarbon. By the mid-twentieth century the concept of nonbonded charge transfer complexes was postulated to explain the intense new absorption spectrum that arose when a closed-shell donor was added to a closed-shell acceptor. The theory of such complexes was formulated in terms of the electron affinity of the acceptor and the ionization potential of the donor and led to the development of new techniques for the determination of properties of such complexes.

The phenomenon was addressed quantum mechanically by Robert S. Mulliken. He related the strength of the complex to the ionization potential of the donor and the electron affinity of acceptor. This prompted the determination of half-wave reduction potentials of acceptor molecules as a measure of electron affinity. In 1953 Mulliken commented,

> All electron affinities are of course relatively small, 1–3 eV. . . . From chemical behavior and stability of molecular complexes, we may reach some conclusions about relative electron affinities of acceptor molecules. . . . Quantitative methods for determining molecular electron affinities are not very well developed, but there seem to be possibilities for the future. [9]

In 1953 Thomas J. Watson and F. N. C. Crick proposed the double helix for the structure of DNA [10]. Even after this was postulated, the problem of a more precise structure of DNA remained. Many studies were begun during this period of time to characterize the electronic structure of the components of DNA, experimentally from spectroscopic studies and theoretically from quantum mechanics. Indeed, there were questions as to whether the components of DNA were acceptors or donors of electrons [11].

In the area of electron affinities of organic molecules, other electrochemical measurements were made and compared with half-wave reduction potentials. Quantum mechanical calculations for aromatic hydrocarbons were carried out using self-consistent field calculations. Many advances were made in the determination of the acidity of organic molecules. The effect of substitution and replacement on electron affinities and bond dissociation energies was recognized. This work is summarized in Chapters 10 and 12. A. S. Streitweiser provides an excellent review of the role of anions in organic chemistry up to 1960 [12].

3.4 GENERAL ACCOMPLISHMENTS, 1960 TO 1970

3.4.1 Introduction

The use of the ECD in environmental chemistry commenced in the late 1950s. As noted by Lovelock,

> In spite of the difficulties of using these primitive detectors, the capable professionals (from the Shell group at Kent, Goodwin et al., and the US FDA, Watts and Klein) successfully and accurately applied the method to the important environmental problem of pesticide residue analysis. ... They lent veracity to the unprovable statements of that remarkable book by Rachael Carson. [1, 13]

Lovelock was acutely aware of the significance of the electron in biochemistry. As reported,

> In testing many compounds of biological significance, it was observed that many of the compounds which responded well in the ECD fell into two categories, those which are important components of the biological system of electron transport or were highly toxic sometimes as carcinogens. It was tempting to speculate that the electron is a fundamental particle of biology as well as of chemistry and physics. It was a challenging coincidence that each alternate acid of the Krebs cycle was one of the very few organic compounds that reacted vigorously with free electrons.... It is still unclear as to whether this association is real or coincidental but there is no doubt that a remarkably high proportion of electron absorbers are biologically active and it is this which has made it so important a device in environmental science.

These observations led Lovelock to propose the following hypothesis: "Electron transport in the living cell proceeds through an ordered sequence of reversible electron traps. The introduction of a foreign, especially irreversible trapping agents is harmful because of their ability to remove electrons from the normal sequence" [2, 14].

Albert Szent Gyorgi recognized the role of electron transport in biological chemistry and the need to extend biological investigations to the submolecular level. He commented as follows on the concept of collaboration between the biochemist and the physicist in studying the role of the electronic dimension in biology: "It is not necessary for the biologist to acquaint himself with the intricacies of wave mechanics. ... The physicist had better stay on his side of the fence than become perhaps a second rate biologist." On the three methods of measuring donor and acceptor strengths, he said,

> All three methods of measurement and expression, redox potentials, ionization potentials and electron affinities, as well as orbital energies have their merits and shortcomings and at present there is no universally applicable method available. What would be needed is to bring the three methods to a common denominator that has all of the merits and none of the shortcomings. What would serve our purpose best (of understanding charge transfer in biology) would be a knowledge of IP and EA of all of the molecules in biological media, but this information is not available.

These remarks clearly previewed our studies on half-wave reduction potentials, semi-empirical calculations, and the direct measurement of electron affinities of molecules in the gas phase [15].

3.4.2 The Wentworth Group

The stage was set for the development of equilibrium methods of measuring electron affinities of molecules from the temperature dependence of thermal electron reactions, and the ECD, swarm, and magnetron methods. The magnetron direct capture method was the first developed [16]. By this mechanism the equilibrium constant for thermal electron attachment is measured as a function of temperature. Since the reactions are carried out at high temperatures with a hot filament, other processes can take place. However, when the proper temperature dependence is observed, the method gives precise electron affinities of molecules from 1 eV to 3.3 eV. The swarm method was used to obtain an electron affinity of O_2 [17]. Wentworth and Becker suggested a method for the measurement of the electron affinities using the ECD [18, 19].

3.4.3 Stable Negative-Ion Formation

It was postulated that the response of the ECD could be related to a precisely defined energetic quantity, the electron affinity. This term is defined as the difference in energy between the neutral and the negative ion in their most stable forms. The term "electron affinity" had previously been used to qualitatively describe the ability of a molecule to respond in the ECD, regardless of the mechanism. The carrier gas was a material such as argon/methane, nitrogen, or hydrogen, with no affinity for free electrons. Some misunderstood this rather loose definition and took it to mean thermodynamic electron affinity. Early work had to dispel this notion.

By assuming that the electron capture coefficient is equal to the equilibrium constant for the reaction of thermal electrons with aromatic hydrocarbons, the electron affinity can be obtained by measuring the response as a function of temperature. The equilibrium constant K_{eq} is related to the electron affinity of the molecule by

$$\ln K_{eq}T^{3/2} = 12.43 + \ln[Q_{an}] + E_a/RT \tag{3.1}$$

In equation 3.1 the spin terms of the negative species have been canceled out. The quantity 12.43 is obtained from fundamental constants and the translation partition function of the electron. Q_{an} is the ratio of the remaining partition function of the anion to that of the neutral. If the partition function ratio for the anion and neutral are assumed to be the same, this term is zero. With one value of the equilibrium constant the electron affinity of the molecule can be estimated. The statistical mechanical expression for K_{eq} refers to the absolute zero of temperature so that no temperature correction to E_a is necessary. Unfortunately, there were no values for the equilibrium constants or electron affinities. Thus, the value of K_{eq} for one molecule, anthracene, was determined and the electron affinity of other aromatic hydrocarbons referenced to that value. If the partition function ratios are equal,

$$E_a(\text{unknown}) = E_a(\text{ref}) - RT\ln(K_{\text{ref}}/K_{un}) \tag{3.2}$$

The electron affinities of triphenylene, phenanthrene, chrysene, benzo[c]phenanthrene, anthracene, pyrene, and benz[a]anthracene were reported in 1962 utilizing equation 3.2 and ECD response factors. The E_a of anthracene used to obtain absolute values was 200 meV too low. At the time theoretical SCF E_a of anthracene was calculated to be some 200 meV higher. The ECD values obtained from the original ECD data are the same as the current "best" values within the experimental error using the higher value. In Figure 3.1 a plot of the data for these seven compounds is given with the slope for anthracene set to the then reported theoretical value of 0.63 eV [17–20]. The E_a are next plotted against the current "best" values in Figure 3.2. The standard deviation is 0.08 eV.

Two important methods for verifying the relative values of the electron affinities obtained from the ECD method were introduced in an article cautiously entitled, "Potential Method for the Determination of Electron Affinities of Molecules: Application to Some Aromatic Hydrocarbons," with comparisons to half-wave reduction potentials and SCF calculations [18, 21]. The relative ECD values agreed with the half-wave reduction potential order from two independent sets of measurements. From this correlation the relative values had an error of 10 to 15%, or for a value of 0.6 eV an absolute error of ± 0.1 eV, because the electron affinity is logarithmically related to the K value. The agreement was within the experimental and calculation error. It was suggested that electronic absorption spectra, ionization potentials (through the constant electronegativity concept), and

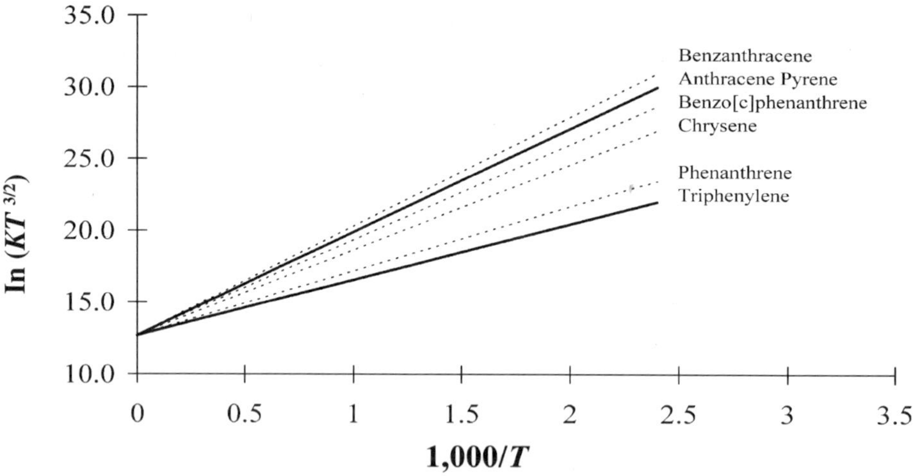

Figure 3.1 Linear plots of ECD data versus $1,000/T$. The compounds are assumed to have one α region. The intercept is determined by the E_a of anthracene calculated to be 0.63 eV in 1962. Using this value, the relative electron affinities of the other compounds are determined from the relative response of the ECD measured at one temperature [18–21].

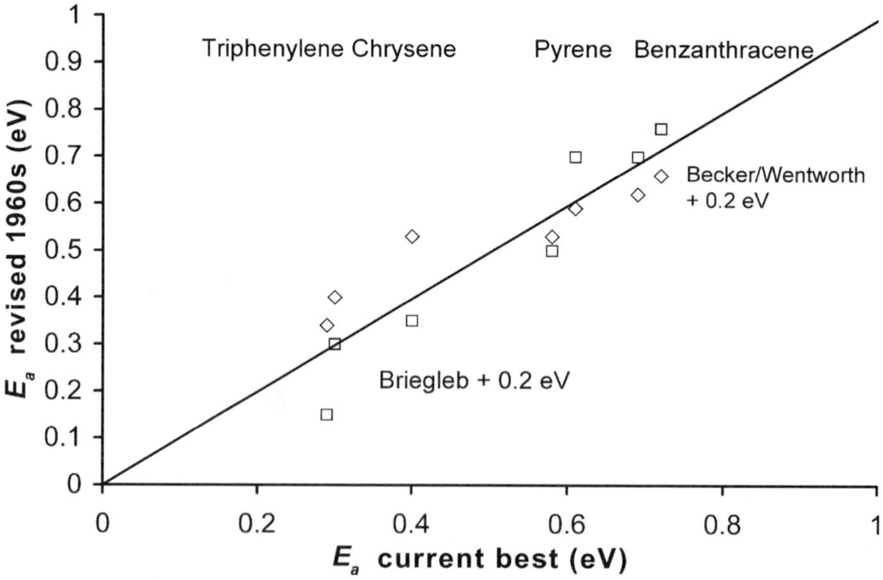

Figure 3.2 The electron affinities of triphenylene, phenanthrene, beno[c]phenanthrene, chrysene, pyrene, anthracene, and benz[a]anthracene given by Becker and Wentworth [18, 19] and by Briegleb [22] incremented by 0.2 eV are plotted versus the current "best" values. The value of 0.2 eV was established by obtaining a unit slope and zero intercept.

comparisons with Huckel theory might be fruitful areas of investigation. It was also proposed that solution energies could be determined from comparisons with half-wave reduction potentials. These ideas were successfully pursued in a second paper [19].

In 1964 G. Briegleb reviewed the electron affinities of organic molecules, including results from the ECD, half-wave reduction potentials, and charge transfer complex energies. The latter two methods yielded relative electron affinities that were in the right order for π charge transfer complex acceptors, but did not give accurate electron affinities. Because there were no absolute values available at the time, the values were about 1.4 eV too low. The AE_a for the aromatic hydrocarbons were also low by about 0.2 eV because they were scaled to ECD values. A plot of Briegleb's modified values for the aromatic hydrocarbons is given in Figure 3.2. The standard deviation is 0.08 eV. Briegleb also applied the ECD method to dissociative electron capture [22]. This created confusion in the literature that persisted as late as 1984, with the E_a of chlorobenzene listed as 0.40 eV. The current "best" value is 0.17(5) eV [23].

Lovelock's odyssey with the ECD continued. He described it as follows:

> In 1961, I came to Houston again. . . . This provided at last an opportunity to spend full time in discovering how the electron capture detector really worked. In this I was fortunate to have nearby both Dr. Zlatkis who by then had become a close friend and colleague and Dr. Wentworth and Dr. Chen of the University of Houston. This fruitful collaboration led to the first plausible kinetic model of the electron capture detector. [2]

To test the ECD hypothesis, E. C. M. Chen measured the temperature dependence of the molar response. This entailed a detailed study of all parameters associated with the pulse sampling method. For these molecules the most important reactions were postulated to be electron generation and collection, electron and ion recombination, electron attachment and detachment. It was discovered that the simple thermodynamic model was not applicable and that a kinetic model was necessary to explain the change in temperature dependence. If we assume a steady state exists, an expression can be obtained that relates the ECD molar response to kinetic rate constants for the above reactions [24, 25].

The electron capture coefficient for nondissociative electron attachment is

$$K = \frac{k_1(k_N)}{2\,k_D(k_N + k_{-1})} \tag{3.3}$$

When the kinetic expressions $k_N = A_N$, $k_D = A_D, k_1 = A_1\,T^{-1/2}\exp(-E_1/RT)$, and $k_{-1} = A_{-1}\;T\;\exp(-E_{-1}/RT)$ are substituted in equation 3.3, we get

$$K = \frac{A_N A_1\,T^{-1/2}\exp(-E_1/RT)}{2\,A_D[A_N + A_{-1}\,T\;\exp(-E_{-1}/RT)} \tag{3.4}$$

At low temperatures the recombination process predominates over the detachment process and K is relatively constant since $k_N \gg k_{-1}$:

$$K = k_1/2\,k_D \tag{3.5}$$

or

$$\ln KT^{3/2} = \ln(A_1/T) - \ln(2A_D) - E_1/RT \tag{3.6}$$

This has been designated the β temperature region. The activation energy for thermal electron attachment and the pre-exponential term for the rate constant for thermal electron attachment are obtained. At that time neither of these two quantities had been measured so comparisons could not be made. k_D was estimated from the electron concentration measured as a function of reaction time. The maximum value of k_1 was determined by the DeBroglie wavelength of the electron and stabilization to the ultimate ions by collisions at high pressures.

At high T where $k_N \ll k$, the expression for K becomes

$$K = k_1\,k_N/(2\,k_D k_{-1}) \tag{3.7}$$

or

$$\ln KT^{3/2} = 12.43 + \ln(A_N/2A_D) + \ln[Q_{an}] + E_a/RT \tag{3.8}$$

This has been designated the α region since it was the first observed temperature region. In this region the temperature dependence of the ratio $(A_N/2A_D)$ and its dependence on the positive-ion concentration will be minimized since both recombination terms will be affected in the same manner.

By 1967 the kinetic model for nondissociative thermal electron attachment and revised values for the electron affinities of 16 aromatic hydrocarbons and 7 aromatic carbonyl compounds were reported [24–26]. The ECD E_a values were correlated to theoretical calculations, electronegativities, spectroscopic data, and reduction potentials. The majority of these remain the most precise electron affinities for such compounds. Some values are assigned to excited states based on the multistate model of the ECD postulated in the 1990s [27, 28]. The electron affinities of atoms, molecules, and radicals were reviewed in 1966 [24]. The relative E_a of nitrobenzene, CS_2, and SO_2 were measured by the thermal charge transfer techniques and the E_a of O_2 by photodetachment [30–32].

The use of the ECD for physical measurements did not progress as rapidly as its use as an analytical method because of the lack of comparable information on the energetics and kinetics of thermal electron reactions. In the late 1960s and early 1970s many response factors for the ECD were measured, but the importance of temperature was often ignored. L. E. Lyons, G. C. Morris, and L. J. Warren confirmed the E_a for anthracene and pyrene by measuring temperature dependence in a static ECD using ethylene as a carrier in the late 1960s. In addition, an E_a for tetracene was obtained. For many years this value was considered the adiabatic electron affinity. With the two-state model and the determination of the electron affinity of tetracene by other techniques, this value can be assigned to an excited state [28, 33].

In 1965 we purchased a commercial chromatograph with an ECD. The first thing accomplished with the instrument was to obtain the temperature dependence for acetophenone. This work clearly supported the ECD model and formed the basis of the determination of the electron affinity of a series of additional substituted benzaldehydes acetophenones and benzonitrile. Lily Wang Ming, William Ristau, and John Wiley applied the kinetic model of the ECD to a series of aromatic aldehydes and ketones, some of which exhibited a change in geometry upon negative-ion formation [34, 35].

Wang and Charles Han calculated the electron affinities of aldehydes and ketones by using the parameterized Huckel theory. Eight parameters were used to calculate the electron affinities of 16 compounds with a deviation of only 0.05 eV. However, some of the data were not published until the 1970s [35]. By measuring relative electron capture coefficients and scaling to the acetophenone data, more precise electron affinities could be obtained. This was further support for the validity of the ECD model. M. J. S. Dewar reproduced the experimental electron affinities of aromatic hydrocarbons using the MINDO/3 method and calculated E_a from reduction potentials [36].

In 1967 B. H. Mahan and C. E. Young used a new microwave method to determine the rate constant for thermal electron attachment to molecules. These quantities were determined for SF_6 and C_7F_{14} using the ECD and agreed with the values reported using the microwave method at room temperature within the experimental error [37, 38]. In addition, the temperature dependence was determined so that activation energies were obtained. This was especially important in the case of strained molecules such as cyclooctatetrene [34].

3.4.4 Dissociative Thermal Electron Attachment

In 1964 a brief description of the ECD kinetic model was presented in *Nature*. This occurred in response to criticism of the use of ECD data to measure the affinity of biological molecules for free electrons. A new procedure for studying electron attachment in swarms and beams had been applied to chlorobenzene. Since the ECD response was originally referenced to that of chlorobenzene, critics emphasized the distinction between dissociative capture and nondissociative capture. They noted that dissociative capture can take place with thermal electrons. This was not disputed. It was realized that certain molecules could undergo dissociative electron capture and that the kinetic model would have to be expanded to include these types of compounds.

Lovelock commented on these exchanges as follows:

> The analyst is probably not too perturbed by the assertions of academic scientists that proves that the ECD detector does not work. I well recall the fierce attack on the

> method and its measurements by Stockdale and his colleagues in 1964. Their paper in *Nature* was a cry of sheer exasperation straight from the heart. It so well expressed the justifiable annoyance of professional chemical physicists who saw their neat and orderly pastures of work on ion molecule reactions trodden over by what seemed to be a mob of clumsy peasants. It is only fair to add that within a year the same authors confirmed the validity of measurements using the ECD.

This initial criticism led to an increase in the data for organic molecules in swarm and beam experiments and an understanding of the ECD kinetic model [39–41].

In the case of dissociative reactions, the ECD capture coefficient is given by

$$K = \frac{(k_1 + k_{12})(k_N + k_2)}{2\,k_D(k_N + k_{-1} + k_2)} \tag{3.9}$$

Roberta Tai Tung studied dissociative electron capture. The model was extended to unimolecular dissociative electron capture by recognizing that k_{12} and k_1 are the same if no molecular negative ion is formed. There is one β region where equation 3.6 applies. The positive slope is the activation energy for dissociative thermal electron attachment. The intercept is related to the DeBroglie wavelength of an electron. For endothermic reactions the activation energy will be at least equal to the energy for dissociative electron attachment. For exothermic reactions, such as for CCl_4, the activation energy will approach zero and the capture coefficient will be large. For the first time qualitative potential energy curves were used to describe the ECD mechanisms. It was noted that the prexponential was larger for the aromatic halogenated compounds [42–47]. Thus, a new mechanism had to be proposed. In this case there is a change in the direction of the temperature dependence in a transition from the α region to a new region called the γ region where $(k_2 > k_N < k_{-1})$, and the K value will be given by

$$K = \frac{k_1(k_2)}{2\,k_D(k_{-1})} \tag{3.10}$$

In this region the phenomenological activation energy will be equal to $D - E_a(X)$. Since the electron affinity of the halogen atoms is well established, this experimental property can be used to obtain the bond dissociation energy.

Tai Tung, Robert George, Joe Steelhammer, and Herman Keith also studied dissociative electron capture in the ECD [42–47]. The kinetic model was extended to three temperature regions, one of which presented a way to measure the electron affinities of acetate and the NO_2 radicals. Steelhammer recognized that dissociative thermal electron attachment could take place via a molecular ion intermediate. He also discovered a variation on that mechanism, whereby dissociation could occur through a single potential energy curve. In this case the experimental activation energy is equal to the quantity $D - E_a(B)$ so that if one is known, the other can be determined. In the instance of acetic anhydride the bond dissociation energy is well known so the electron affinity of the acetate radical is obtained [43].

Han refined the kinetic model and investigated the effect of flow rate on the model and the use of ethylene as a dopant in the ECD [46]. By 1969 the kinetic model of the ECD was firmly established, as summarized in one review [47]. The ECD, swarm, and beam data were consolidated into a consistent picture by applying the nonlinear least-squares procedure to the ECD data and by postulating that "two-dimensional" Morse potential energy curves could be used to represent the cross-section through a potential energy surface that then represents the negative ions. Steelhammer constructed pseudo-two-dimensional Morse potentials to describe these processes by relating the negative-ion curves to the properties of the neutral. These curves used the available data for dissociative capture from electron beam and swarm techniques [47].

Lovelock summarized the reasons for developing the ECD method as follows:

> In view of the many highly developed and accurate techniques already described, the introduction of a new method of measurement in electron-attachment studies would seem to need more than usual justification. The pulse-sampling method arose not from a perverse desire for novelty but as a simple and inexpensive qualitative detector for gas chromatography. Its later consideration as a potential method for electron attachment studies was made not because it was better able to perform the physical measurements, but because none of the previous methods are able to satisfy the severe chemical constraints imposed during the measurements of weakly volatile, highly polar, and possibly impure organic compounds, for example, steroids. The method is dynamic and designed for the observation of brief pulses of ultra pure, dilute vapor emerging from a chromatograph column. Static methods of observing electron attachment, although quite adequate with pure permanent gases are quite unsuited for this purpose. [25]

3.4.5 Nonlinear Least Squares

One of the major accomplishments of the Wentworth group has been the application of the nonlinear least-squares procedure to chemical problems. This has been generally recognized by the fact that two articles on the subject published in the *Journal of Chemical Education* in 1965 have become "classics." The following is taken from the March 31, 1986, Current Contents issue:

> Many equations in chemistry and related physical science are non-linear with respect to various parameters. The subject article describes how a least squares adjustment can be carried out rigorously using properly weighted observations. The procedure is illustrated by application to a kinetic rate expression and the Arrhenius equation. In 1986, these papers have been cited in over 340 and 115 publications respectively. [48, 49]

Wentworth was asked to write a comment about how this article came into being:

> While I was a graduate student in chemistry, I had several occasions to fit data to equations using the principle of least squares. However, I noticed that upon rearrangement

of the equation I would obtain a different least squares solution. This was puzzling to me and I could never get the problem resolved. Upon leaving graduate school in 1956, I decided to work as an applied mathematician in the "new" missile industry. Recall that digital computers were in their infancy at that time and all programming was carried out in machine language. It was during this employment as an applied mathematician that I was introduced to the generalized least squares adjustment. It became immediately apparent that there was only one least squares adjustment and the different least squares solutions that I had obtained in graduate school were the neglect of *weighting the residuals*. Once the appropriate weighting factors are used, there is only one solution. This assumes that the errors are sufficiently small that a Taylor's Series expansion including only the first order terms is satisfactory. This is the case when the parameters are defined with reasonable precision.

The generalized least squares adjustment was described thoroughly in a book written by W. E. Deming in 1943, well before the advent of digital computers [50]. Since the calculations required performing non-linear least squares adjustments are very tedious, it is understandable that the technique was not used extensively. In retrospect, I believe the reason for the attention given to my paper was a matter of timing. In the early to mid-1960s digital computers were finding their way into universities, and it was only natural that non-linear least squares would appeal to experimentalists who for years had struggled with less satisfactory approximation methods. My interest in writing the article was to present the subject at a level that the average experimental chemist could understand. Most chemists at that time did not have a background in matrix algebra and I tried to avoid the extensive use of matrices. The popularity of the article was apparent when 250 reprints were exhausted within two months and another 250 at the end of the year. Even today (1986), I get an occasional reprint request.

As graduate students in the Wentworth laboratory, we were all required to use nonlinear least squares initially, sometimes with a mechanical calculator whose best feature was that it could take precise square roots. Later, as we tested the ECD method, we dropped the least-squares procedure for the simpler determination of a slope through straight lines. This was generally correct, but as we learned when the dissociative mechanism of the ECD was included, there was a need to use the complete ECD equation. Hirsch studied the temperature dependence of the ECD and sought new correlations of the electron affinities and hydrogen bond strengths. In order to obtain the thermodynamic parameters for the complexes from the data, a nonlinear least-squares procedure to include data determined by other experiments was developed [51]. This procedure was applied to the ECD data for the multistate model.

The general least-squares procedures can now be implemented in spreadsheets programmed with macros. Adjustments once impossible are now trivial. The classification of molecules to obtain electron affinities from half-wave reduction potentials is an example of a linear least-squares adjustment. The determination of the adiabatic electron affinity for acetophenone is an example of a nonlinear two-parameter least-squares procedure. The nonlinear least-squares adjustment of ECD to the expanded kinetic model is one of the major advances of the 1990s.

In 1966 the relative electron affinities of charge transfer complex acceptors were calculated from spectral data and half-wave reduction potentials. Unfortunately, at the time, no accurate electron affinities of typical π charge transfer complex acceptors existed so one could obtain absolute electron affinities from either half-wave reduction potentials or charge transfer complexes. Thus, the magnetron E_a of 1.40 eV for the electron affinity of benzoquinone was selected. This is now known to be about 0.5 eV too low, making all the values low. This emphasizes the difference between the determination of relative electron affinities that depend on the absolute electron affinity of a reference compound and absolute ones from experimental measurements and fundamental constants.

3.5 MILESTONES IN THE WENTWORTH LABORATORY AND COMPLEMENTARY METHODS, 1970 TO 1980

Shen Nan Lin continued work on the relationship between the swarm and beam data and the ECD data, developing pseudo-two-dimensional potential energy curves for halogenated benzenes, including the pentafluorophenyl halides. He also used negative-ion mass spectrometry and plasma chromatography. He related the drift times in plasma chromatography to the size of the ions. The ECD data were not published because of the absence of confirmatory data and because some of the bond dissociation energies were "too low" based on the current best value. All the data obtained by other methods are in agreement with the ECD data. The results will be updated in this book [52].

Robert Freeman described his research problem as such:

> There is a continuing controversy concerning the mechanism by which a group of small molecules attach an electron, including oxygen, nitric oxide, nitrous oxide, and nitrogen dioxide. The study of the electron attachment to these molecules lends itself nicely to the ECD method. The purpose of this study is to investigate thermal electron attachment to these selected small molecules and to correlate the results with those obtained from other experiments that use complementary techniques in an attempt to elucidate the mechanism of attachment and to evaluate any molecular parameters such as electron affinities or activation energies. [53]

Theoretical potential energy curves were calculated for both NO and O_2 that included numerous excited states, only one of which had been observed experimentally. Freeman and independent investigators measured the E_a of what is now thought to be an excited state in the ECD [52, 53]. Freeman obtained unexplained results for electron attachment to both O_2 and NO that were included in the thesis, but were not understood until the rediscovery of a paper predicting the multiple negative-ion states of $O_2(-)$ by H. H. Michels [54]. The data for O_2 have been verified and the analysis published. A similar analysis for NO will be carried out in Chapter 9.

Freeman also demonstrated the effect of a change in geometry on the formation of the anion of N_2O and showed that the activation for thermal electron attachment was measured in the ECD. He used two-dimensional Morse potentials to obtain its first estimated molecular electron affinity [56]. He in addition utilized the kinetic model to illustrate that the response of the ECD is optimum at higher temperatures. He obtained ECD data for nitromethane and α-nitrotoluene [52]. Both the electron affinity of the molecule and radical were determined from these data since nondissociative and dissociative electron capture can take place. These data were not completely analyzed until recently and are discussed in Chapters 5 and 10.

Jorge Ayala determined the rate constants for thermal electron attachment to aliphatic halides and the halogen molecules to confirm values measured by other techniques. The electron affinities of the halogen molecules had been determined by endothermic charge transfer experiments [57–59]. In the case of the halogen molecules, the ECD results lead to the rate constant for thermal electron attachment rather than the electron affinity of the molecule. Two-dimensional Morse potentials for the anions were constructed based on these data. Freeman and Ayala searched for a nonradioactive source for the ECD. In 1975 the data on the electron affinities of atoms were summarized and correlations examined between these values and the position of the atoms in the Periodic Table [60]. A large number of the atomic electron affinities were measured by photoelectron spectroscopy [61]. A similar compilation of the electronegativities of elements was carried out. In this case some of the values were obtained from the work functions of salts [62]. These results will be updated in Chapter 8.

Between 1970 and 1975 a number of molecular electron affinities of charge transfer complex acceptors were determined using the alkali metal beam experiments. The magnetron studies were applied to many more molecules and the results summarized in one book in 1969 [16]. The electron affinities of charge transfer complex acceptors tetracyanoethylene and tetracyanoquinodimethane were measured. The E_a of p-benzoquinone was measured as 1.9 eV by alkali metal beam studies, as opposed to the magnetron value of 1.4 eV [63]. With these new data it was possible to determine the absolute electron affinities of organic molecules from half-wave reduction potential data and charge transfer spectra. The values were consistent with experiment for about a dozen E_a measured in the gas phase. The data are shown in Figure 3.3. The standard deviation of the zero intercept correlation line for both is ± 0.15 eV. The uncertainty in the magnetron values is the same. The uncertainty in the alkali metal beam value for p-benzoquinone is ± 0.3 eV. The study also suggested that the magnetron electron affinity of benzoquinone and anthraquinone could result from an excited state [64]. These values will be revisited in Chapter 10.

During the 1970s the ECD became firmly established as the most sensitive gas chromatographic detector for some compounds. The kinetic model was described in terms of a numerical solution of the differential equations. This was assisted by the development of the constant current mode of measuring the response and the development of Ni-63 sources for the detector. The purification of the carrier gas and the further development of capillary columns improved the operation of the ECD. In addition, chemical reactions were used to make derivatives with a greater sensitivity in the ECD. Other ion molecule reactions were used to improve the sensitivity of

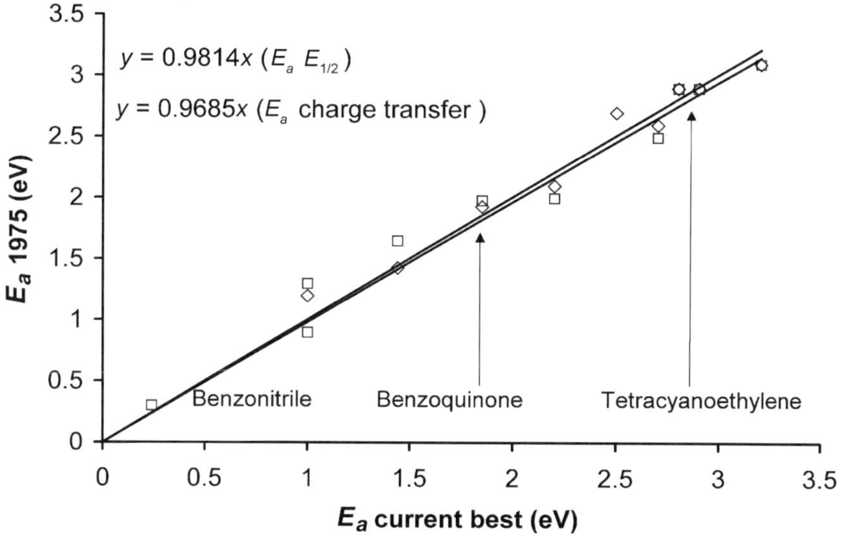

Figure 3.3 The E_a of benzonitrile, 1,2 dicyanoethylene, 1,2 dicyanobenzene, maleic anhydride, benzoquinone, s-tetracyanobenzene, fluoranil, chloranil, hexacyanobenzene, tetracyanoquinodimethane, tetracyanoethylene, and hexacyanobutadiene reported in 1975 [64] from reduction potentials and charge transfer absorption maximum versus the current "best" values. These are precision and accuracy plots. The slopes are determined with a fixed intercept of zero, so that the displacements are random uncertainties.

the detector. The diverse techniques used to improve the ECD for analytical purposes were summarized in one monograph [2]. A review chapter on the theory of the ECD showed how the temperature dependence and sensitivity of the detector to compounds could be predicted based on the chemical structure. The compilation of ECD response data can be used to obtain the physical properties of molecules, some of which have not been measured by any other technique. This will be addressed in Chapter 11 [2].

A comprehensive review of the use of negative-ion mass spectrometry was presented in 1973. It described its scope like this: "In this review, covering the literature since the last comprehensive review by Melton in 1960, an examination of current research areas involving gaseous negative ions and a discussion of the usefulness of the data in chemical analysis will be presented" [65]. In closing, the article stated, "The electron attachment method of Wentworth has not been included. . . . It is anticipated that the electron affinities of molecules and radicals may be measured with greater accuracy"[65]. Several new techniques for studying negative ions and measuring electron affinities in the gas phase were developed between 1970 and 1980. In the mid-1970s the photon techniques were applied to the measurement of the electron affinities of many atoms. For the electron affinities of molecules, these methods include the alkali metal beam, the endothermic charge transfer, photodetachment, photoelectron spectroscopy, thermal charge transfer using an ion cyclotron mass spectrometer, high-pressure mass spectrometry, and the flowing afterglow procedures. Another procedure developed for the study of the interaction of electrons with molecules during this time was electron transmission spectrometry. This was applied to aromatic and heterocyclic molecules and combined with half-wave reduction potentials to obtain electron affinities [66].

In the mid-1970s methods of measuring gas phase acidities using negative-ion mass spectrometry were developed. Two devices, the high-pressure negative-ion mass spectrometer and the ICR mass spectrometer, were used. In some cases the electron affinities of radicals had been measured using photodetachment or photoelectron spectroscopy methods so that the gas phase acidities could be confirmed from measured values of bond dissociation energies. The ECD electron affinities of the acetate radical and the NO_2 radical agree with the mass spectrometric values.

The theoretical calculation of the electron affinities of aromatic hydrocarbons was advanced by the development of the MINDO/3, MNDO, AM1, and PM3 semi-empirical techniques. These procedures gave the adiabatic electron affinities of molecules obtained from the ECD and from half-wave reduction potentials that agreed with the experimental values to within the experimental error. A different semi-empirical procedure yielded consistently lower values than the experimental values partially because they were adjusted to the lower values [67–69].

3.6 NEGATIVE-ION MASS SPECTROMETRY AND MORSE POTENTIAL ENERGY CURVES, 1980 TO 1990

In the decade of the 1980s, E. C. M. Chen continued to work with the Wentworth group and summarized the negative-ion states of the halogen molecules and determination of molecular electron affinities using the ECD. The work on the determination of the electron affinities of molecules and the ECD presented all the values understood at the time [70]. The negative-ion states of the halogens were based on the ECD data generated by Ayala, but incorporated concepts unknown to us at the time that had been postulated by D. R. Herschbach [71, 72]. Doctoral student C. K. Lee and master's student Nicholas Hernandez examined the response of the ECD at short reaction times. Tom Limero tested an experimental high-temperature tritium foil. Ela D'sa discovered a fourth temperature region in the ECD model [73–76].

In the period from 1980 to 1990 the atmospheric pressure ionization technique was applied to the study of negative ions. This led to the verification of the ECD kinetic model through NIMS studies. The electron affinities of C_6F_6 and SF_6 were determined based on this model. These values agreed with the results obtained from other gas phase techniques and support the assumption of an excited state in the TCT measurements. Limero, D'sa, and Lih Ren Shuie studied electron capture and obtained data for anion hydrates and other anion complexes using this equipment [76–80]. Other data were obtained for E_a's and rate constants for thermal electron attachment to larger molecules using swarm, beam, and AMB techniques. These were compared with ECD values when available [22, 81].

In 1983 the use of the ECD to obtain the absolute electron affinities of NO, O_2, CS_2, COS, N_2O CH_3NO_2, biacetyl, and benzophenone was described. The procedure for obtaining the electron affinities of NO_2 and the acetate radical determined using the ECD was summarized. These were the only ones with E_a measured by the ECD and some other gas phase technique. All these values remain valid today since some of the ECD values had lower limits and one of them, for CS_2, referred to an excited state [70, 81]. In 1989 the electron affinities of organic acceptors were summarized. For the first time variable solution energy differences were used to obtain absolute electron affinities from reduction potentials [81]. This was possible because of the determination of the relative E_a of organic acceptors using the mass spectrometric thermal charge transfer reactions, TCT. The first values were obtained in the late 1970s using an ICR mass spectrometer [82]. Later the high-pressure mass spectrometer (HPMS) was used [83]. In the early studies the ECD benzophenone value was used to anchor the scale. The E_a were subsequently anchored to the photoelectron spectroscopy E_a of SO_2.

3.7 EXPERIMENTAL AND THEORETICAL MILESTONES, 1990 TO 2000

In the decade of the 1990s the work in the Wentworth laboratories took several directions. The development of the PDECD and its application to the measurement of physical properties were studied by numerous students pursuing their Ph.D.'s and M. S. degrees [84–86]. The development of the ECD and the GC/MS method for determining electron affinities and ion energetics were extended [87–92]. Variable solution energy differences were applied to organic molecules and reduction potentials [93–95]. Morse potential energy curves were calculated for diatomic molecular anions and biochemical molecules. The Herschbach classification of ionic Morse potential energy curves was modified [96–102]. The kinetic model was extended to two states where x refers to an excited state:

$$K = \frac{k_{1x}(k_N + k_{2x})}{2(k_D)(k_{-1x} + k_N + k_{2x})} + \frac{k_1(k_{-1} + k_N + k_2)}{2(k_D)(k_{-1} + k_N + k_2)} \tag{3.11}$$

This expression is obtained by assuming equilibrium between the two states. If we assume nondissociative electron attachment, this equation can reproduce the temperature dependence for O_2, NO, CS_2, C_6F_5Cl, tetracene, and anthracene where excited state electron affinities have been measured in the gas phase [29, 103, 104]. The extension of the ECD model to two negative-ion states explains the structure in the data.

E. S. D. Chen was primarily responsible for the study of properties of biologically significant molecules. The E_a of the purines and pyrimidines were predicted using substitution and replacement effects and subsequently measured experimentally. Quantum mechanical calculations of the E_a supported these determinations. These studies led to a theory of electron transport through DNA [105–111]. These results will be considered in Chapter 12. The electron affinities of the carbon clusters were calculated using CURES-EC [112].

Recent reviews on alkali metal beam studies, theoretical and experimental determinations of electron affinities using photon methods, and atomic electron affinities and an Internet source for electron affinities all give compilations [113–117]. The evaluation of molecular electron affinities is a major objective of this book.

3.8 SUMMARY OF CONTRIBUTIONS AT THE UNIVERSITY OF HOUSTON

Wentworth began work on the ECD at the University of Houston based on the earlier work of Lovelock. Advances in chromatography by Martin inspired the practical invention of the ECD. The work of Zlatkis and Becker played an important part in its development. Wentworth and Becker then postulated the relationship between ECD molar response and molecular electron affinity. The kinetic model of the ECD was one focus of the Ph.D. dissertation of E. C. M. Chen.

Beginning in 1960 doctoral student E. C. M. Chen, master's students Hirsch, Tai, and Wang, and undergraduates Marie Cobb and Keith worked with the Wentworth group. By the end of the decade doctoral students Freeman and Lin had joined the group. Master's students, George, Steelhammer, Wiley, Han, and Ristau finished their work under Wentworth.

Steelhammer and Wiley began their research as undergraduate students. Steelhammer, Han, Ristau, and Wiley eventually obtained their Ph.D.'s at other universities. These students established the ECD model for obtaining fundamental information on the reactions of thermal electrons with molecules.

During the 1970s the work on the ECD was carried out by doctoral students: Freeman [52], Lin, [53] and Ayala [57]. In 1973 E. C. M. Chen began his academic career. He became a full professor at the University of Houston Clear Lake in 1981. Charles Batten soon joined the Wentworth laboratory. His expertise in mass spectrometry proved very valuable in the study of negative ions.

In the 1980s Shuie, D'sa, Limero, Gigi Bear, and Bernard White were the doctoral students at Wentworth. The latter two studied negative ions in flames. Master's student Hernandez examined the response of the ECD at short reaction times. Limero studied fluorobenzenes. D'sa discovered a fourth temperature region in the ECD for chloro and bromoethylenes. Limero, D'sa, and Shuie researched electron capture in the atmospheric pressure chemical ionization system and obtained data for anion complexes using such equipment. Batten became intimately involved with this work. Lee and R. Ranatunga, the doctoral students of Zlatkis, made additional contributions.

Professors Wiley, J. M. Robinson, and S. Ehdaie at the University of Texas in Permian Basin measured the reduction potentials of purines and pyrimidines. Wiley obtained NICI data for the chloroethylenes and purines and pyrimidines. The post-doctoral and doctoral students at the Wentworth laboratory in the 1990s were Gerard Gremaud, Huamin Cai, Janardhan Madabushi, J. Dojahn, and Kefu

Sun. Master's students included S. Watanesk, N. Helias, S. Mendonca, Lei Rao, Lei Zhang, Sui Shen, and Parmis Darvish. The work on bond dissociation energies, biological molecules, semi-empirical calculations, and the consolidation of data in Morse potentials was carried out by students at the University of Houston in Clear Lake: K. Albyns, L. Dussack, M. S. Milligan, P. Richardson, W. R. Reed, R. Swatlotski, J. Dojahn, W. Odegard, O. Zhang, R. George, N. Kozanecki, N. Sane, S. Carr, S. Schulze, E. S. D. Chen, and T. Brown.

REFERENCES

1. Ettre, L. S. and Zlatkis, A. *Seventy-five Years of Chromatography—A Historical Dialog.* New York: Elsevier, **1979**.
2. Lovelock, J. E. "The Electron Capture Detector, a Personal Odyssey," in *The Electron Capture Detector*, edited by A. Zlatkis and C. Poole. New York: Elsevier, **1981**, p. 1.
3. Massey, H. S.W. *Negative Ions*. New York: Cambridge University Press, **1976**.
4. Hylleras, E. *Z. Phys.* **1930**, 60, 624.
5. Pritchard, H. O. *Chem. Rev.* **1953**, 52, 529.
6. James, A. T. and Martin, A. J. P. *Biochem. J.* **1951**, 48, vii.
7. Lovelock, J. E. and Lipsky, S. R. *J. Amer. Chem. Soc.* **1960**, [19, 20] 82, 431.
8. Golay, M. J. E. *Anal. Chem* **1957**, 29, 928.
9. Mulliken, R. S. and Person, W. B. *Molecular Complexes, a Lecture and Reprint Volume*. New York: Wiley, **1969**.
10. Watson, J. D. and Crick, F. H. C. *Nature* **1953**, 171, 737.
11. Berthold, H.; Giessner-Prettre, G.; and Pullman, A. *Theor. Chem. Acta.* **1966**, 5, 53.
12. Streitweiser, A. S. *Molecular Orbital Theory for Organic Chemists*. New York: Wiley, **1961**.
13. Carson, R. *Silent Spring*. New York: Houghton Mifflin, **1962**.
14. Lovelock, J. E. *Nature* **1961**, 189, 720.
15. Szent-Györgi, A. *Introduction to a Submolecular Biology.* New York: Academic Press, **1960**.
16. Page, F. M. and Goode, G. C. *Negative Ions and the Magnetron*. New York: Wiley-Interscience, **1969**.
17. Pack J. L. and Phelps, V. *Phys. Rev. Lett.* **1961**, 6, 111.
18. Wentworth, W. E. and Becker, R. S. *J. Amer. Chem. Soc.* **1960**, 82, 431.
19. Becker, R.S. and Wentworth, W. E. *J. Amer. Chem. Soc.* **1962**, 84, 4263.
20. Lovelock, J. E.; Zlatkis, A.; and Becker, R. S. *Nature* **1962**, 193, 540.
21. Hoyland, J. R. and Goodman, L. *J. Chem. Phys.* **1962**, 36, 21.
22. Briegleb, G. *Angew. Chem. Internat. Edit.* **1964**, 3, 617.
23. Christodoulides, A. A.; McCorkle, D. L.; and Christophorou, L. G. "Electron Affinities of Atoms, Molecules and Radicals," in *Electron-Molecule Interactions and Their Applications*. New York: Academic Press, **1984**.
24. Chen, E. C. M. Ph.D. dissertation, University of Houston, **1966**.

25. Wentworth, W. E.; Chen, E. C. M.; and Lovelock, J. E. *J. Phys. Chem.* **1966**, 70, 445.
26. Chen, E. C. M. and Becker, R. S. *J. Chem. Phys.* **1966**, 45, 2403.
27. Wentworth, W. E. and Chen, E. C. M. *J. Phys. Chem.* **1967**, 71, 1929.
28. Chen, E. C. M. and Chen, E. S. D. *J. Chromatogr. A* **2002**, 952, 173.
29. Chen, E. C. M.; Carr, S. D.; Wentworth, W. E., and Chen, E. S. D. *J. Chromatogr. A* **1998**, 827, 91.
30. Branscomb, L. M. *Photodetachment in Atomic and Molecular Processes*, edited by D. R. Bates. New York: Academic Press, **1962**, p. 100.
31. Kraus, K.; Muller-Duysing, W.; and Neuert, H. *Z. Naturfor.* **1961**, 16A, 1385.
32. Henglein, A. and Muccini, G. A. *J. Chem. Phys.* **1959**, 31, 1426.
33. Lyons, L. E.; Morris, G. C.; and Warren, L. J. *J. Phys. Chem.* **1968**, 72, 3677.
34. Wentworth, W. E. and Ristau, W. *J. Phys. Chem.* **1969**, 73, 2126.
35. Wentworth, W. E.; Kao, L. W.; and Becker, R. S. *J. Phys. Chem.* **1975**, 79, 1161.
36. Dewar, M. J. S. *The Molecular Orbital Theory of Organic Chemistry*. New York: McGraw-Hill, **1969**.
37. Mahan, B. H. and Young, C. E. *J. Chem. Phys.* **1966**, 44, 2192.
38. Chen, E. C. M.; George, R. D.; and Wentworth, W. E. *J. Chem. Phys.* **1968**, 49, 1973.
39. Stockdale, J. A.; Hurst, G. S.; and Christophorou, L. G. *Nature* **1964**, 203, 1270.
40. Lovelock, J. E. *Nature* **1964**, 203, 1267.
41. Becker, R. S. and Wentworth, W. E. *Nature* **1964**, 203, 1268.
42. Wentworth, W. E.; Becker, R. S.; and Tung, R. *J. Phys. Chem.* **1967**, 71, 1652.
43. Wentworth, W. E.; Chen, E. C. M.; and Steelhammer, J. C. *J. Phys. Chem.* **1968**, 72, 2671.
44. Wentworth, W. E. and Chen, E. C. M. *J. Gas Chromatogr.* **1967**, 5, 17.
45. Wentworth, W. E; George, R.; and Keith, H. *J. Chem. Phys.* **1969**, 51, 1791.
46. Han, C. C. M. S. thesis, University of Houston, **1969**.
47. Wentworth, W. E. and Steelhammer, J. C. *Adv. Chem. Sers.* **1968**, 82, 75.
48. Wentworth, W. E. *J. Chem. Ed.* **1965**, 42, 96.
49. Wentworth, W. E. *J. Chem. Ed.* **1965**, 42, 162.
50. Deming, W. E. *The Statistical Adjustment of Experimental Data*. New York: Dover, **1964**.
51. Wentworth, W. E.; Hirsch, W.; and Chen E. C. M. *J. Phys. Chem.* **1967**, 71, 21.
52. Lin, S. N. Ph.D. dissertation, University of Houston, **1969**.
53. Freeman, R. R. Ph.D. dissertation, University of Houston, **1971**.
54. Van de Wiel, H. J. and Tommasen, H. *J. Chromatogr.*, **1972**, 71, 1.
55. Michels, H. H. *Adv. Chem. Phys.* **1981**, 45, 227.
56. Wentworth, W. E.; Chen, E.; and Freeman, R. *J. Chem. Phys.* **1971**, 55, 2075.
57 Ayala, J. A. Ph.D. dissertation, University of Houston, **1977**.
58. Ayala, J. A. Chen, E. C. M.; and Wentworth, W. E. *J. Phys. Chem.* **1981**, 85, 768.
59. Ayala, J. A. Chen, E. C. M.; and Wentworth, W. E. *J. Phys. Chem.* **1981**, 85, 3989.
60. Chen, E. C. M. and Wentworth, W. E. *J. Chem. Educ.* **1975**, 52, 486.
61. Hotop, H. and Lineberger, W. C. *J. Phys. Chem. Ref. Data* **1975**, 4, 539.
62. Chen, E. C. M.; Wentworth, W. E.; and Ayala, J. A. *J. Chem. Phys.* **1977**, 67, 2642.
63. Cooper, C. D.; Naff, W. T.; and Compton, R. N. *J. Chem. Phys.* **1975**, 63, 2752.

64. E. C. M. Chen and Wentworth, W. E. *J. Chem. Phys.* **1975**, 62, 3183.
65. Dillard, J. G. *Chem. Rev.* **1973**, 73, 700.
66. Nenner, I. and Schultz, G. J. *J. Chem. Phys.* **1975**, 62, 1747.
67. Dewar, M. J. S.; Hashmall, J. A.; and Trinajatic, N. *J. Amer. Chem. Soc.*, **1970**, 92, 5555.
68. Dewar, M. J. S. and Rzepa, H. S. *J. Amer. Chem. Soc.* **1978**, 100, 784.
69. Younkin, J. M.; Smith, L. J.; Compton, R. N. *Theor. Chem. Acta* **1976**, 41, 157.
70. Chen, E. C. M. and Wentworth, W. E. *J. Phys. Chem.* **1983**, 87, 45.
71. Chen, E. C. M. and Wentworth, W. E. *J. Phys. Chem.* **1985**, 89, 4099.
72. Herschbach, D. R. *Adv. Chem. Phys.* **1966**, 10, 250.
73. Zlatkis, A.; Lee, C. K.; Chen, E. C. M.; and Wentworth, W. E. *Anal. Chem.* **1983**, 54, 1596.
74. Hernandez, N.; Wentworth, W. E.; Limero, T.; and Chen, E. C. M. *J. Chromatogr.* **1984**, 312, 31.
75. Hernandez, N.; Wentworth, W. E.; Limero, T.; and Chen, E. C. M. *J. Phys. Chem.* **1984**, 88, 6181.
76. D'sa, E. D.; Wentworth, W. E.; Batten, C. F.; and Chen, E. C. M. *J. Phys. Chem.* **1988**, 92, 285.
77. Wentworth, W. E.; Limero, T.; and Chen, E. C. M. *J. Chem. Phys.* **1985**, 83, 6541.
78. Wentworth, W. E.; Limero, T.; and Chen, E. C. M. *J. Phys. Chem.* **1987**, 91, 214.
79. D'sa, E. D.; Wentworth, W. E.; Batten, C. F.; Shuie, L. R.; and Chen, E. C. M. *J. Chromatogr.* **1987**, 390, 249.
80. Wentworth, W. E.; Limero, T.; Batten, C. F.; and Chen, E. C. M.; *J. Chem. Phys.* **1988**, 88, 4711.
81. Chen, E. C. M. and Wentworth, W. E. *Mol. Cryst. Liq. Cryst.* **1989**, 171, 271.
82. Rains, R. J., Moore, H. W.; and McIver, R. T. *J. Chem. Phys.* **1978**, 68, 3309.
83. Kebarle, P. and Chowhury, S. *Chem. Rev.* **1987**, 87, 513.
84. Wentworth, W. E.; Huang, J.; Chen, E. C. M.; and Sterns, S. D. *Chromatographia* **1996**, 353, 43.
85. Wentworth, W. E.; Huang, J.; Chen, E. C. M.; Odegard, W.; and Sterns, S. D. *J. Chrom. Sci.* **1996**, 34, 368.
86. Wentworth, W. E.; Huang, J.; Sun, K.; Zhang, Y.; Rao, H.; Cai, H.; and Sterns, S. D. *J. Chromatogr. A* **1999**, 842, 229.
87. Chen, E. C. M.; Albyns, K.; Dussak, L.; and Wentworth, W. E. *J. Phys. Chem.* **1989**, 93, 6827.
88. D'sa, E. D.; Wentworth, W. E.; Batten, C. F.; Shuie, L. R.; and Chen, E. C. M. *Struc. Chem.* **1993**, 5, 250.
89. Chen, E. C. M.; Wiley, J. R.; Batten, C. F.; and Wentworth, W. E. *J. Phys. Chem.* **1994**, 98, 88.
90. Chen, E. C. M.; Chen, E. S. D.; Wiley, J. R.; Milligan, M. S.; and Wentworth, W. E. *J. Phys. Chem.* **1992**, 96, 2385.
91. Chen, E. C. M.; Wiley, J. R.; Batten, C. F.; and Wentworth, W. E. *J. Phys. Chem.* **1994**, 98, 88.
92. Wiley, J. R.; Chen, E. C. M.; and Wentworth, W. E. *J. Phys. Chem.* **1993**, 97, 1256.

93. Wiley, J. R.; Chen, E. C. M.; Chen, E. S. D.; Richardson, P.; Reed, W. R.; and Wentworth, W. E. *J. Electroanal. Chem. Interface.* **1991**, 307, 169.
94. Ruoff, R. S.; Kadish, K. M.; Boulas, P.; and Chen, E. C. M. *J. Phys. Chem.* **1995**, 99, 8843.
95. Chen, E. S. D., Chen, E. C. M.; Sane, N.; Talley, L.; Kozanecki, N.; and Shultze, S. *J. Chem. Phys.* **1999**, 110, 9319.
96. Dojahn, J.; Chen, E. C. M.; and Wentworth, W. E. *J. Phys. Chem.* **1996**, 100, 9649.
97. Dojahn, J.; Chen, E. C. M.; and Wentworth, W. E. *J. Phys. Chem. A* **1997**, 101, 3088.
98. Chen, E. S. and Chen, E. C. M. *Chem. Phys. Lett.* **1998**, 293, 491.
99. Chen, E. C. M.; Welk, N.; Chen, E. S.; and Wentworth, W. E. *J. Phys. Chem. A* **1999**, 103, 9072.
100. Chen, E. S.; Wentworth, W. E.; and Chen, E. C. M. *J. Mol. Struc.* **2002**, 606, 1.
101. Chen, E. S. and Chen, E. C. M. *J. Phys. Chem. A* **2002**, 106, 6655.
102. Chen, E. S. D. and Chen, E. C. M. *J. Phys. Chem. A* **2003**, 107, 169.
103. Chen, E. C. M.; George, R.; Carr, S. D.; Wentworth, W. E.; and Chen, E. S. D. *J. Chromatogr. A* **1998**, 811, 250.
104. Chen, E. C. M. and Chen, E. S. D. *J. Chromatogr. A* **2002**, 952, 173.
105. Chen, E. S. D.; Chen, E. C. M.; and Wentworth, W. E. *Biochem. Biophys. Res. Comm.* **1990**, 171, 97.
106. Wiley, J. R.; Robinson, J. M.; Ehdaie, S.; Chen, E. S. D.; Chen, E. C. M.; and Wentworth, W. E. *Biochem. Biophys. Res. Comm.* **1991**, 180, 841.
107. Zhang, O. and Chen, E. C. M. *Biochem. Biophys. Res. Comm.* **1995**, 217, 255.
108. Chen, E. S. D. and Chen, E. C. M. *Bioelectrochem. Bioenerget.* **1998**, 46, 15.
109. Chen, E. S. D.; Chen, E. C. M.; Sane, N.; and Schultz, S. *Bioelectrochem. Bioenerget.* **1999**, 48, 69.
110. Chen, E. C. M. and Chen, E. S. D. *J. Phys. Chem. B* **2000**, 104, 783.
111. Chen, E. S. D.; Chen, E. C. M.; and Wentworth, W. E. *Biochem. Biophys. Res. Comm.* **2001**, 289, 421.
112. Chen, E. C. M.; Schulze, S.; Welk, N.; Brown, T.; and Chen E. S. D. *The Chemist* **1999**, March/April, 13.
113. Kleyn, A. W. and Moutinho, A. M. C. *J. Phys. B* **2001**, 34, R1.
114. Reinstra-Kiracofe, J. C.; Tschumper, G. S.; Schaefer, H. F.; Nandi, S.; and Ellison, G. B. *Chem. Rev.* **2002**, 102, 231.
115. Nadeau, M-J.; Garwan, M. A.; Zhao, X-L.; and Litherland, A. E. *Nucl. Instr. Meth. Phys. Res. B* **1997**, 123, 521.
116. Andersen, T.; Haugen, H. K.; and Hotop, H. *J. Phys. Chem. Ref. Data* **1999**, 28, 1511
117. National Institute of Standards and Technology (NIST). *Chemistry WebBook*, **2003**. Available at http://webbook.nist.gov.

CHAPTER 4

Theoretical Basis of the Experimental Tools

4.1 INTRODUCTION

> Research is seeing what others have seen and thinking what nobody has thought.
> —Commonly attributed to Albert Syent Gyorgi

The equations used in our study of the thermodynamics and kinetics of thermal electron reactions using the ECD and NIMS are presented. The ECD and NIMS methods were developed in our laboratories. These are used to determine the rate constants, electron affinities, partition function ratios, and bond dissociation energies of molecules and energies for the formation of complexes of anions. The general kinetic model for the ECD and negative-ion mass spectrometry is presented. Molecules will be classified using example data.

The electron affinities of many of the molecules determined in the ECD or NIMS have been verified by half-wave reduction potentials and charge transfer complex data. These methods were developed in the 1960s but have been significantly improved. The relationship between the electronegativity and the electron affinities and ionization potentials for aromatic hydrocarbons can be used to support the E_a. The use of the ECD model and these techniques to estimate the electron affinities of aromatic hydrocarbons are illustrated for selected compounds. We will also describe the use of charge transfer complex data to obtain the electron affinities of acceptors.

4.2 THE KINETIC MODEL OF THE ECD AND NIMS

The reactions in the ECD and in NIMS were given in Chapter 2 as equations 2.1 to 2.6. They involve electron attachment and detachment, unimolecular dissociative

The Electron Capture Detector and the Study of Reactions with Thermal Electrons
by E. C. M. Chen and E. S. D. Chen
ISBN 0-471-32622-4

electron attachment, and sequential dissociation. These reactions are combined with a constant source of electrons to define the kinetic model. When the standard kinetic expressions are written for these reactions and steady state applied to the negative ions and electrons, an expression for the ECD response can be obtained. This is illustrated as follows:

$$R \overset{k_P}{\Rightarrow} e(-) + P \tag{4.1}$$

$$e(-) + P \overset{k'_D}{\rightarrow} \text{Neutrals} \tag{2.5}$$

$$AB^- + P \overset{k'_N}{\rightarrow} \text{Neutrals} \tag{2.6}$$

$$e(-) + AB \underset{k_{-1}}{\overset{k_1}{\Leftrightarrow}} AB(-) \tag{2.1 and 2.2}$$

$$AB(-) \overset{k_2}{\rightarrow} A + B(-) \tag{2.4}$$

$$B(-) + P \overset{k'_N}{\rightarrow} \text{Neutrals} \tag{2.6}$$

In the absence of *AB*

$$d[e]/dt = k_p - k'_D[P][e] = 0 \tag{4.2}$$

$$I_b = k_p/\{k'_D[P]\} \tag{4.3}$$

In the presence of *AB*

$$d[e]/dt = k_p - k'_D[P][e] - k_1[AB][e] + k_{-1}\{AB^-] = 0 \tag{4.4}$$

$$d[AB^-]/dt = k_1[AB][e] - (k_{-1} + k_2)[AB^-] - k'_N[P][AB^-] = 0 \tag{4.5}$$

$$d[B^-]/dt = k_2[AB^-] - k'_N[P][B^-] = 0 \tag{4.6}$$

Solving simultaneously and setting $k'_D[P] = k_D$ and $k'_N[P] = k_N$, we get

$$I_e = k_p(k_{-1} + k_N)/(k_D + k_1 k_N[AB]) \tag{4.7}$$

For low fractional capture

$$(I_b - I_e)/I_e = k_1(k_N + k_2)[AB]/2(k_{-1} + k_N + k_2)k_D \tag{4.8}$$

$$K_{\text{ECD}} = k_1(k_N + k_2)/2(k_{-1} + k_N + k_2)k_D \tag{4.9}$$

The values of some of the rate constants may be small because of energetic considerations. The original model considered only electron attachment to a single negative-ion state. With more than one state the ECD response is given by a sum of terms if we assume no interconversion of states. In the remainder of the book the subscript ECD will be dropped for convenience.

The response equation is

$$K = \sum \left[\frac{(k_{1i})(k_N + k_{2i})}{2(k_D)(k_{-i1} + k_N + k_{2i})} \right] \tag{4.10}$$

$$i = 1 \text{ to } n$$

For multiple states the expression is expanded. The value of k_1 can refer to direct dissociation, previously designated k_{12} or non-dissociative capture.

Four regions are observed in the nominal temperature range accessible to NIMS and ECD combined with chromatography. These are defined based on the relative values of the rate constants. From low (298 K) to high temperatures (600 K) these are:

1. The β region, where $(k_N \gg k_{-1} + k_2)$ and $K = k_1/2k_D$
2. The α region, where $(k_{-1} \gg k_N + k_2)$ and $K = [k_N/2k_D][k_1/k_{-1}]$
3. The γ region, where $(k_2 \gg k_N)$ and $(k_{-1} \gg k_2)$ and $K = [k_1 k_2/2k_D k_{-1}]$
4. The δ region, where $(k_2 \gg k_{-1} + k_N)$ and $K = k_1/2k_D$

Using these approximations and limiting values of equation 4.10, we obtain the fundamental kinetic and thermodynamic properties of the thermal electron reactions from ECD or NIMS data. In the α region the molecular electron affinity E_a and partition function ratio Q_{an} are measured. In the β region and δ region A_1 and E_1 are obtained, and in the γ region A_2 and E_2 are determined. Often the ECD gives data that overlap these regions so that nonlinear least-squares procedures are used. From equation 4.10 and the kinetic expressions an equation for least-squares analysis can be obtained:

$$k_N = A_N = \text{constant} \quad \text{(this can be estimated for a given system)}$$

$$k_D = A_D = \text{constant} \quad \text{(this can be estimated for a given system)}$$

$$k_1 = A_1 T^{-1/2} \exp(-E_1/RT)$$

$$k_{-1} = A_{-1} T \exp(-E_{-1}/RT)$$

$$k_2 = A_2 T \exp(-E_2/RT)$$

In the case of a single state, there are six parameters, two each for three rate constants k_1, k_{-1}, and k_2. The least-squares equation is

$$K = \frac{A_1 T^{-1/2} \exp(-E_1/RT)\{A_N + A_2 T \exp(-E_2/RT)\}}{2A_D[A_N + A_{-1} T \exp(-E_{-1}/RT) + A_2 T \exp(-E_2/RT)]} \tag{4.11}$$

The data can exhibit α, β, γ, and δ regions in a sufficiently large temperature range. In the event that two or more states are involved, there will be additional terms as indicated in equation 4.10. For two states there will be 12 parameters, two each for the six rate constants. We have not encountered a case in which all 12 parameters can be determined from ECD data. The most that we have determined is eight for the nondissociative two-state case (e.g., CS_2) or six for the dissociative single-state case (e.g., CH_3NO_2). We will now consider the limiting cases for the ECD model beginning with nondissociative electron capture [1].

4.3 NONDISSOCIATIVE ELECTRON CAPTURE

In the event that $[D(AB) - E_a(B) > 1.5 \text{ eV}]$, the value of k_2 will be small and the data can exhibit both an α and a β region. At most, four parameters will be determined. In some cases only a single slope and intercept are observed. We obtain the absolute electron affinity from the slope in the α region:

$$K = [k_N/2k_D][k_1/k_{-1}] = [k_N/2k_D][A_1/A_{-1}]T^{-3/2}\{\exp(E_a/RT)\} \tag{4.12}$$

$$\ln KT^{3/2} = \ln(A_N/2A_D) + \ln(A_1/A_{-1}) + E_a/RT \tag{4.13}$$

From equation 4.13 the slope in a plot of ln $KT^{3/2}$ versus $1{,}000/T$ is E_a/R. The intercept is ln $(A_N/2A_D) + \ln(A_1/A_{-1})$. The statistical mechanical expression for $k_1/k_{-1} = K_{eq}$ for the reaction of thermal electrons gives

$$(A_1/A_{-1}) = [g(A^-)/g(A)]h^3/(2\pi m_e k)^{3/2} \tag{4.14}$$

The g's are partition functions, k and h, the Boltzmann and Planck constants, and m_e, the electron mass.

From this equation $\ln([g(A^-)/g(A)]) = \ln(A_1/A_{-1}) - 12.43 - \ln(A_N/2A_D)$. The value of 12.43 is obtained from fundamental constants. This involves the ratio of $(A_N/2A_D)$ so that the concentration of the positive species and the temperature dependence of the intrinsic rate constants will cancel out and not affect the value of the slope. With an experimental intercept and $\ln(A_N/2A_D)$, a value for Q_{an}, the partition function ratio $Q_{an} = [g(A^-)/g(A)])$, can be calculated. The observed Q_{an} values range from 1 to 10^{-4}. The ECD and NIMS experiments are the primary source of Q_{an} in the literature. A value of $A_N/2A_D$ can be obtained by measuring the ECD temperature dependence of a compound with an accurate E_a and a unit partition function ratio such as acetophenone. A unit value of Q_{an} implies that the negative ion and neutral have the same partition functions except for spin multiplicity. Thus, equation 4.13 becomes

$$\ln KT^{3/2} = \ln(A_N/2A_D) + 12.43 + \ln(Q_{an}) + E_a/RT \tag{4.13b}$$

In Figure 4.1 a plot of ln $KT^{3/2}$ for acetophenone and several compounds that exhibit only an α region is given. Table 4.1 provides the values of E_a and Q_{an}. The data for Figure 4.1 were obtained using a parallel plate detector at the "long" reaction time of 1,000 μsec. The compounds are classified as Eql(1/1) since there is only one

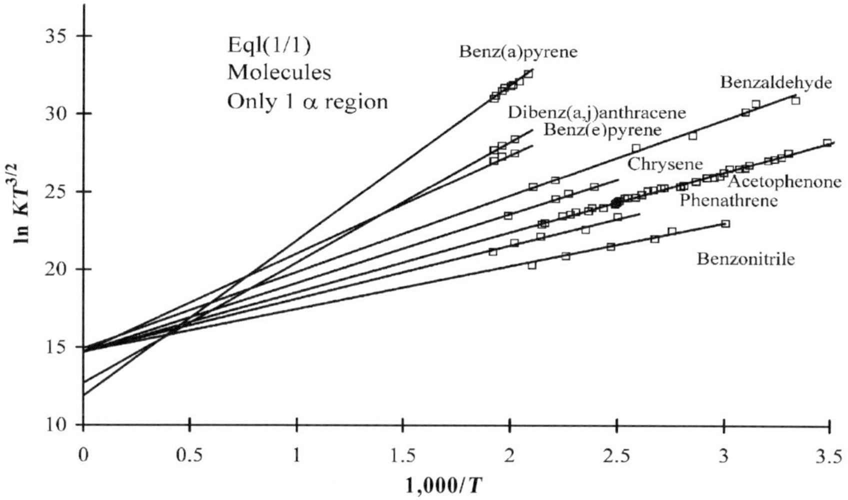

Figure 4.1 Linear plots of ECD data versus $1{,}000/T$. The compounds exhibit only one α region and are designated Eql(1/1) for equilibrium compounds with one state and one region. The electron affinities are determined from the slope. The intercepts for all but dibenz(a,j)anthracene and benz(a)pyrene are the same, indicating that Q_{an} is 1.0. Some of the parameters are given in Table 4.1. The range of electron affinities is 0.25 eV to 0.82 eV. Data from [2, 8, 30].

linear region. The E_a of benz[a]pyrene is about 0.80 eV and that for tetramethylbenzene is 0.1 eV. This is called a global plot because it shows the extrapolation outside of the region of the data to higher temperatures (lower 1,000/*T*). The intercepts for several aromatic hydrocarbons, acetophenone, and benzaldehyde are the same within the experimental error. For these compounds the Q_{an} values are 1.

Some of the Eql(1/1) compounds are predicted to have a second region at lower temperatures. If one or two data points exist in the second region, the slope will be lower and the intercept higher than the average value. In the case of benz[e]pyrene the linear least-squares fit gives an intercept of 15.10 ± 1.60, which is higher than

TABLE 4.1 ECD Parameters for Eql(1/1) Compounds

Species	Q	E_a (eV)
Benzonitrile	1.0	0.25(2)
Phenanthrene	1.0	0.305(7)
Acetophenone	1.0	0.338(2)
Chrysene	0.8	0.42(4)
Benzaldehyde	1.0	0.458(10)
Benz{e}pyrene	[1.0]	0.534(10)
Dibenz[a,j]anthracene	0.2	0.68(10)
Benz[a]pyrene	0.1	0.83(10)

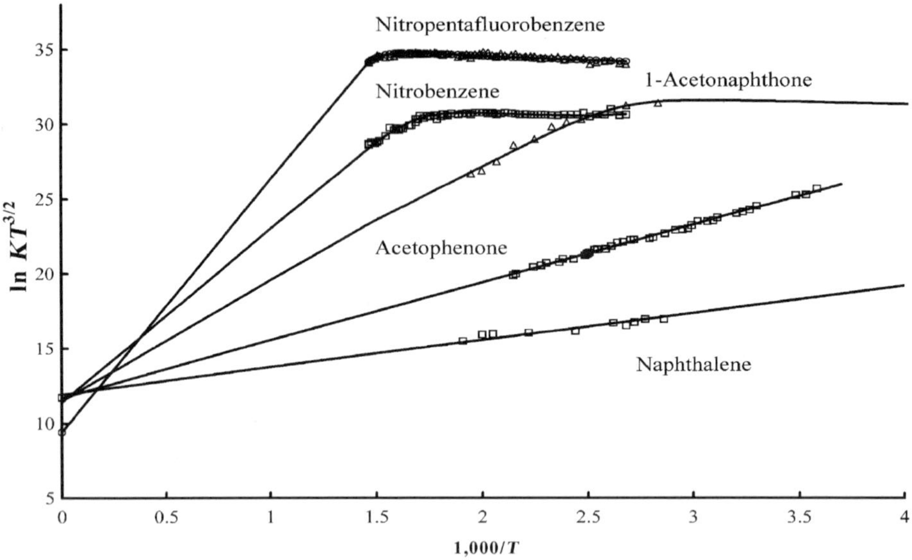

Figure 4.2 Plots of ECD data as ln $KT^{3/2}$ versus $1{,}000/T$. Nitrobenzene, pentafluoronitrobenzene, and 1-acetonaphthone exhibit both an α and a β region. They are designated Eql(1/2) for equilibrium compounds with one state and two regions. The electron affinities are determined from the slope in the α region. The intercepts for all but pentafluoronitrobenzene are the same, indicating that Q_{an} is 1.0. The Q_{an} for the latter is less than 1.0, the parameters are given in Table 4.1. The range of electron affinities is 0.16 eV to 1.50 eV. Data [1, 5, 8].

the average value of 14.7 ± 0.05. Since the higher intercept would imply a Q_{an} greater than unity and the uncertainty encompasses the average value, we fix the intercept and determine the slope from the average intercept with its error. In the case of dibenzanthracene and benz[a]pyrene the intercepts are lower than the average intercept and, therefore, the Q_{an} is lower than unity.

Originally, it was assumed that the Q_{an} should always be unity and that the electron affinity should be obtained from the "fixed" intercept. Other experimental determinations of the electron affinities for CS_2, CH_3NO_2, tetracene, and benz[a]-pyrene demonstrate that Q_{an} can be lower than 1. It is not expected that Q_{an} is greater than 1 so the electron affinity is obtained by assuming it is 1, as in the case of benz[e]pyrene. If the intercept is lower than the average intercept, then the actual slope through the data should be used to obtain the electron affinity along with its associated errors [2–4].

In Figure 4.2 α and β regions are shown for nitrobenzene, pentafluoronitrobenzene, and acetonaphthone. These are designated Eql(1/2). They are compared to the Eql(1/1) molecules, naphthalene and acetophenone. Independent investigators using a "constant current" mode of electron collection obtained the data for naphthalene [5]. The data for the nitrocompounds were obtained in our laboratories

[1]. The maximum value of A_1 for the reactions with thermal electrons is the DeBroglie A_1 value (DeBA). The value of ln(DeBA) is about 36 at 400 K. The A_1 value for $C_6F_5NO_2$ is equal to the DeBA. The electron affinity of $C_6F_5NO_2$ determined by thermal charge transfer experiments was used to obtain the calculated curve, and the data are then utilized to obtain the Q_{an} value, which is less than 1. From Figure 4.2 we see that the A_1 values for $C_6H_5NO_2$ and acetonaphthone are lower than that for $C_6F_5NO_2$. The parameters for these compounds are given in Table 4.1. The β regions are well defined for the nitrocompounds, but is less apparent for acetonaphthone. The E_a range from 0.16 eV to 1.5 eV [1, 4–7].

In the β region $K = k_1 2k_D$ and it is possible to obtain values for A_1 and E_1. We generally use a global plot where ln $KT^{3/2}$ is plotted even though the actual expression has a different T dependence in the β region. The different values of A_1 result from the fact that electron attachment is a two-step process that requires stabilization to the ground state. Thus at sufficiently high pressures or third-body concentrations, A_1 will approach DeBA when the stabilization occurs on every collision. The stabilization term is analogous to the Z factor in traditional collision theories of kinetic rate constants. When the experimental A_1 is less than DeBA, the value of Z is low, most notably for O_2, COS, SO_2, and NO_2 [4].

For the two-state nondissociative case, where $k_2 = 0$, there are eight parameters to be determined, two each for the four rate constants k_1, k_{-1}, k_{1x}, and k_{-1x}. These are designated Eql(2/2) compounds. Each of the values of A_N and A_D is assumed to be equal for the ground and excited states. Two temperature extremes can be considered. At temperatures high enough to populate only the ground state, the region is called the α_g region. At low enough temperatures to populate only the excited state, it is called the α_x region. Here the value of K is

$$\ln KT^{3/2} = \ln(A_N/2A_D) + 12.43 + \ln(Q_{ani}) + E_{ai}/RT \tag{4.13b}$$

where i represents either the ground or excited states. This is the same as equation 4.13, except it refers to either the ground state or excited state.

In going to lower temperatures from the α_g region, there may be an apparent plateau, called the β_g temperature region, followed by a rise with a slope corresponding to the excited-state electron affinity, the α_x temperature region. At still lower temperatures another plateau region, the β_x temperature region, is observed, where $K = (k_1 + k_{1x})/(2A_D)$. In this region the response will be relatively temperature-independent when k_{1x} is larger than k_1. Thus, the four kinetic parameters for the rate constants k_1 and k_{1x} can be determined from the temperature dependence in the two β regions. The electron affinities and Q_{an} values for the two states can be determined from the two α regions.

It was believed that the population of excited states in the ECD was quite rare, but it now appears more common. The first excited states observed in the ECD were for C_6F_6 and CS_2. The data for these two compounds are shown in Figure 4.3. The structure is apparent in the data for CS_2. There may be some fine structure in the data. This set of data for C_6F_6 is not as precise, but the lower-temperature data points are clearly higher than the higher-temperature points

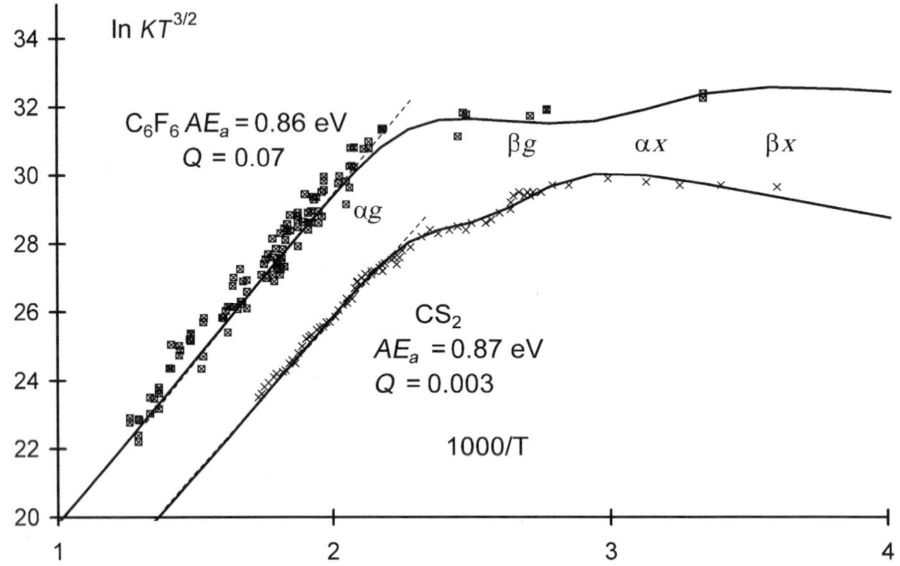

Figure 4.3 Plots of ECD data as ln $KT^{3/2}$ versus $1{,}000/T$. CS_2 and C_6F_6 exhibit two α and two β regions. They are designated Eql(2/2) for equilibrium compounds with two states and two regions. The electron affinities are determined from the slope in the α regions. The ground-state Q_{an} are much less than 1.0, the parameters are given in Table 4.2. Data from [1, 8, 15].

in the excited-state region. The curves were obtained from a least-squares analysis and include parameters determined in multiple ECD experiments and electron affinities determined from TCT and PES experiments. The Q_{an} for these compounds are about 10^{-3} and are among the lowest values observed to date [1, 8, 9].

With the recognition of multiple states and the measurement of Q_{an} less than unity, revisiting the electron affinities of the aromatic hydrocarbons originally reported in the 1960s became possible. This was done in 1998, 1999, and 2002 [3, 10, 11]. The National Institute of Standards and Testing (NIST) list of electron affinities for the aromatic hydrocarbons determined in the gas phase by various techniques has been evaluated [12]. Data for some of the aromatic hydrocarbons with revised E_a are shown in Figure 4.4. Several sets of anthracene data are shown. The data for phenanthrene are given for comparison. Table 4.2 provides the parameters for these molecules. The Q_{an} for the excited state of tetracene is less than 1. The E_a of tetracene was reported to be 0.88 eV for many years based on the slope through the low-temperature data. This required a low value for the intercept. Such was not reported [13]. A higher E_a is definitively measured as 1.08 eV and the lower value is assigned to an excited state. In addition, a third state is postulated by analogy to anthracene and the existence of three distinct types of C—H bonds in both anthracene and tetracene. Average values are given for the multiple sets of anthracene data [6, 7, 11–13].

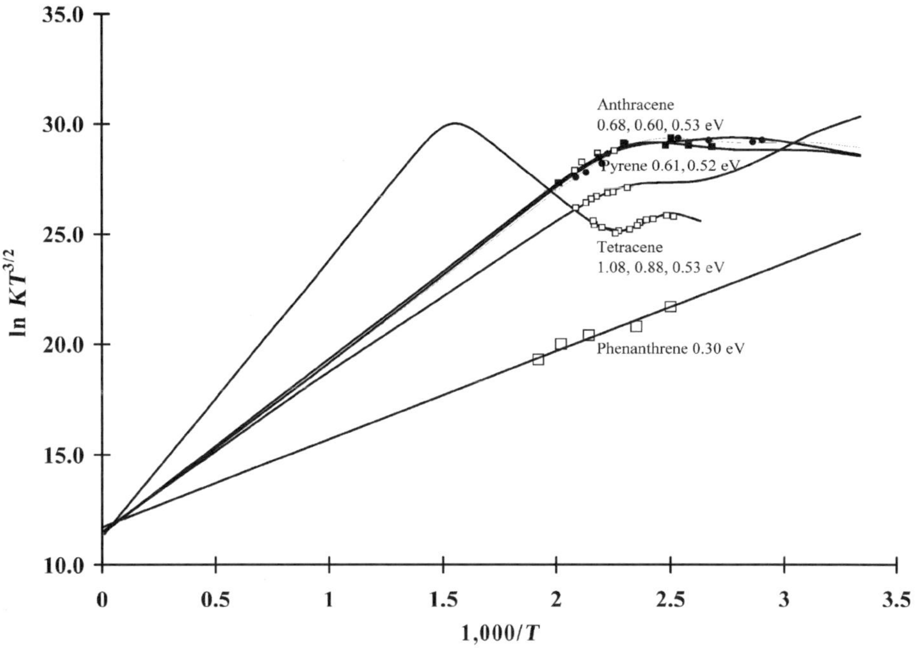

Figure 4.4 Plots of ECD data as ln $KT^{3/2}$ versus $1,000/T$. Anthracene and tetracene exhibit multiple α and β regions. They are designated Eql(n/2) for equilibrium compounds with n states and two regions. The curve for pyrene is shown with two states and two regions. The data for phenanthrene are given for comparison Table 4.2 lists the parameters. Data from [1, 11].

TABLE 4.2 ECD Parameters for Eql(2/n) Compounds

Species	ln(A_1)	E_1 (eV)	Q	E_a (eV)	$E_{-1}(eV)$
Hexafluorobenzene	34.7(1)	0.04(1)	0.07(1)	0.86(2)	0.90(2)
Hexafluorobenzene(ex)	34.9(2)	0.00(1)	1.0(1)	0.61(5)	0.61(5)
Carbon disulfide(bent)	31.2(4)	0.10(1)	0.003(1)	0.87(2)	0.97(2)
Carbon disulfide(linear)	31.9(2)	0.03(2)	0.40(1)	0.61(4)	0.64(4)
Tetracene	37.0(1)	0.95(5)	1.0(1)	1.10(4)	2.05(4)
Tetracene(ex)	36.0(1)	0.80(5)	10^{-4}	0.88(4)	1.66(4)
Tetracene(ex)	36.0(1)	0.65(5)	1.0(1)	0.53(4)	1.18(4)
Anthracene	35.2(1)	0.17(1)	1.0(1)	0.69(2)	0.86(4)
Anthracene(ex)	35.2(1)	0.10(5)	0.9(1)	0.60(2)	0.70(2)
Anthracene(ex)	35.2(1)	0.14(5)	1.0(1)	0.52(5)	0.66(5)
Pyrene	35.1(1)	0.26(1)	0.7(1)	0.61(2)	0.87(4)
Pyrene(ex)	35.1(1)	0.39(1)	1.0(1)	0.50(2)	0.98(5)
Naphthalene	34.2(5)	0.21(1)	1.0(1)	0.16(2)	0.37(4)
Naphthalene(ex)	33.8(5)	0.62(1)	0.8(1)	0.13(2)	0.75(5)

In the case of negative-ion mass spectrometry it is possible to obtain an expression for both the parent negative ion and the products of dissociative capture using a kinetic model similar to that in the ECD. The reactions and rate constants are attachment k_1, detachment k_{-1}, dissociation k_2, recombination of both the electrons k_D, and the negative ions k_N. In some cases excited states are present, but for simplicity we will not include these here.

$$[AB(-)] = \frac{k_1 a[e^-]}{(k_{-1} + k_N + k_2)} \tag{4.15}$$

For nondissociative capture the molar response of the mass spectrometer is

$$[AB(-)] = \frac{k_1 a[e^-]}{(k_{-1} + k_N)} \tag{4.16}$$

The electron affinities of SF_6, nitrobenzene, tri- and tetrachloroethylene, and C_6F_6 have been determined from the kinetic model by measuring the temperature dependence of NIMS responses. The similarity of the global plots of the parent negative ions versus 1,000/T is an indication of the validity of the model. This can be seen in Figure 4.5 for SF_6 and nitrobenzene. The A_1 value for SF_6 is larger

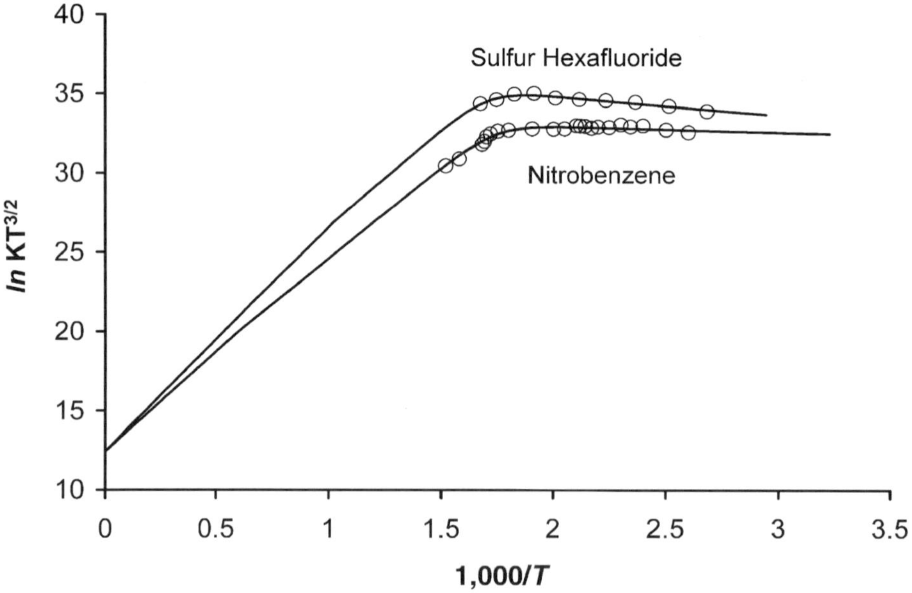

Figure 4.5 Negative-ion mass spectrometry data plotted as ln $KT^{3/2}$ versus 1,000/T for nitrobenzene and sulfur hexafluoride. The data exhibit one α and one β region. The magnitude was scaled to the value of the k_1 for SF_6 at room temperature. The curves are calculated using the measured electron affinities of SF_6 and $C_6H_5NO_2$. The data determine the Q_{an} values to be 1.0. The responses were obtained by injecting a solution with a known amount of the two compounds into the mass spectrometer.

than that for nitrobenzene. The activation energy for thermal electron attachment to SF_6 is small. In NIMS the ion abundance can be scaled to measured values of A_1 to obtain a measure of Q_{an} directly for compounds that exhibit both α and β regions. The values of the Q_{an} are unity for these data since the A_1 for nitrobenzene is less than that for SF_6. The intensity of the departing group in dissociative electron capture can be determined as a function of temperature and used to infer the presence of excited states [14–16].

A similar kinetic model has been developed for the measurement of ion complex formation kinetics and energies involving both dissociative and nondissociative electron capture, for example, the hydration of halide ions and of $O_2(-)$. The ratio of negative ions observed in NIMS can be used to determine energies of complex formation. In this case the sequential formation of the higher complexes must be added to the kinetic model. These studies are important because they demonstrate that the API mass spectrometer can be used to measure thermodynamic quantities. When we use the data for hydrates of $O_2(-)$ as an example, the kinetic expression is given by

$$[O_2 \cdot [H_2O]_I(-)]/[O_2][O_2 \cdot [H_2O]_{i-1}(-)] = k_i/k_{-i}\{k_{-i}/\{k_{-i} + k_N CC\} \tag{4.17}$$

where the k's are forward and reverse rate constants for the formation of the ith complex, CC is $1 + \Sigma R_i$, and R_i is a measured ion ratio.

At low temperatures this reduces to a pseudo-β region in which the temperature dependence is small:

$$[O_2 \cdot [H_2O]_I(-)]/[O_2][O_2 \cdot [H_2O]_{i-1}(-)] = k_i/k_N CC \tag{4.18}$$

At high temperatures the expression becomes

$$[O_2 \cdot [H_2O]_I(-)]/[O_2][O_2 \cdot [H_2O]_{i-1}(-)] = k_i/k_{-i} \tag{4.19}$$

which is the standard equilibrium expression.

These data can be reduced by nonlinear least squares using the relationship

$$\ln\{[O_2 \cdot [H_2O]_I(-)]/[O_2][O_2 \cdot [H_2O]_{i-1}(-)]\} + \ln CC = -\Delta S/R - \Delta H/RT \tag{4.20}$$

to give the enthalpy and entropy of the complex formation [16].

Figure 4.6 shows the ion ratios measured in the determination of the energies of the hydrates of $O_2(-)$. The ratios are normalized to the most intense ion, $n = 1$, $m/z = 50$ ions. At 589 K the ratio of the 50/32 ions is about 1. When the temperature is 480 K, the $m/z = 32$ ion is only about 1% of the $m/z = 50$ ion, while the $m/z = 68$ ion is about the same as the $m/z = 50$ ion.

Figure 4.7 gives the plots of $\ln(K_p)$ versus $1{,}000/T$ for the 0-1 hydrates and 6-7 hydrates. The entropy changes for these complexes are essentially the same, but the enthalpy changes are 0.85 eV and 0.45 eV. The values for the 0-1, 1-2, and

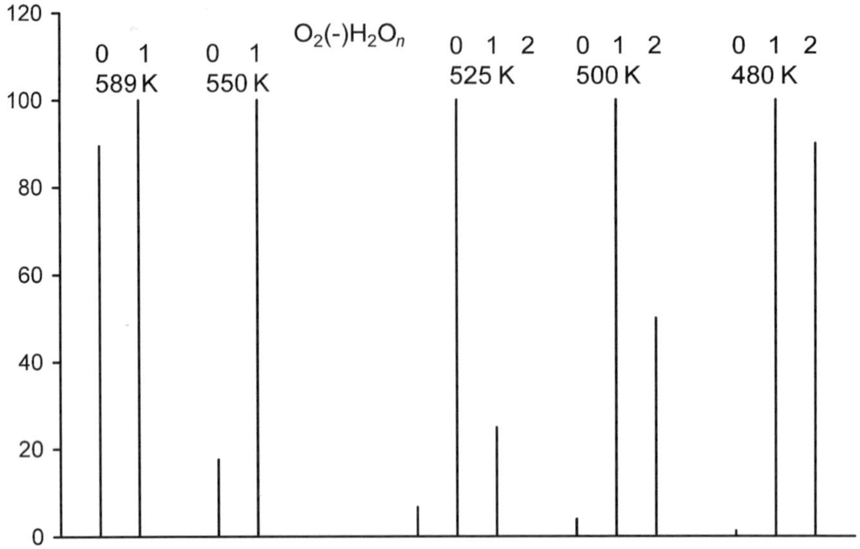

Figure 4.6 Atmospheric pressure ionization ion ratios measured in the determination of the energies of the hydrates of $O_2(-)$. The ratios are given as a function of temperature for the $n = 0$, 1, 2 hydrates. The mass spectrometer is described in detail in Chapter 5.

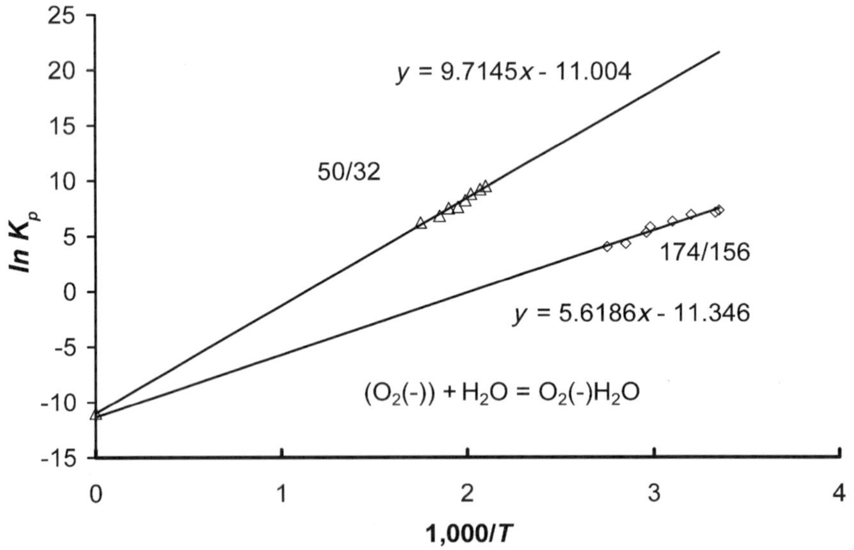

Figure 4.7 Negative-ion mass spectrometry data for $O_2\ (-)\cdot(H_2O)_n$ for equilibrium constants obtained from the ion intensities as shown in Figure 4.6, plotted as ln K_p versus $1{,}000/T$. The intercept multiplied by R is the standard entropy, while the slope multiplied by R is the heat of reaction. The plots for the $n = 0$ to $n = 1$ complex and the $n = 6$ to $n = 7$ complex have approximately the same intercept. Data from [16].

2-3 complexes have been measured by negative-ion mass spectrometry, but the higher hydrates were not studied [19]. This type of research was also carried out for the hydrates and methanolates of the halides. The hydrates are important to atmospheric processes and will be discussed in Chapter 11.

4.4 DISSOCIATIVE ELECTRON ATTACHMENT

Dissociative electron capture is observed with hyperthermal electrons in NIMS electron impact experiments. In order for dissociative electron capture to take place with thermal electrons, there must be a dissociative pathway that is accessible by the thermal activation of the neutral molecule or a low-lying negative-ion state. The quantity $D(R - \text{Le}) - E_a(\text{Le})$ must be less than about 1.0 eV. This limit has been established empirically. Two types of dissociative thermal electron attachment have been observed in NIMS and ECD. The first occurs by unimolecular dissociation in which there is only one temperature region for many compounds. In the original work a low-temperature low-slope region was observed but unexplained. We now believe this could represent the formation of a molecular ion with an electron affinity of about 0.1 eV. The exact nature of this ion is not known, but it could represent stabilization to an excited state. In Figure 4.8 ECD data are plotted for several

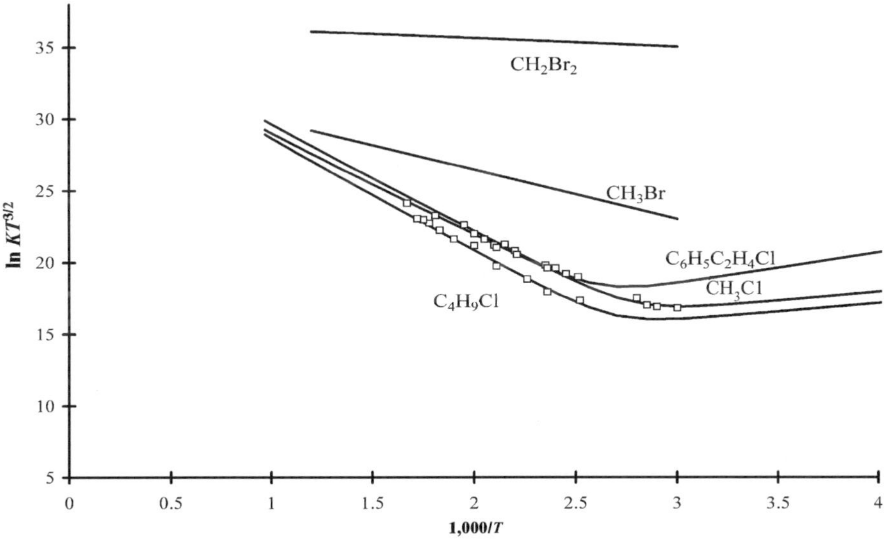

Figure 4.8 ECD data plotted as ln K_p versus $1{,}000/T$. These alkyl halides dissociate via activation of the molecule. They are designated DEC(1) for dissociative electron capture via activation of the molecule. The slope multiplied by R is equal to the activation energy in the high-temperature region. The low-temperature data were originally not explained, but could be an indication of a low molecular electron affinity. The curves were fit using both dissociation and molecular ion formation. Data from [16–19].

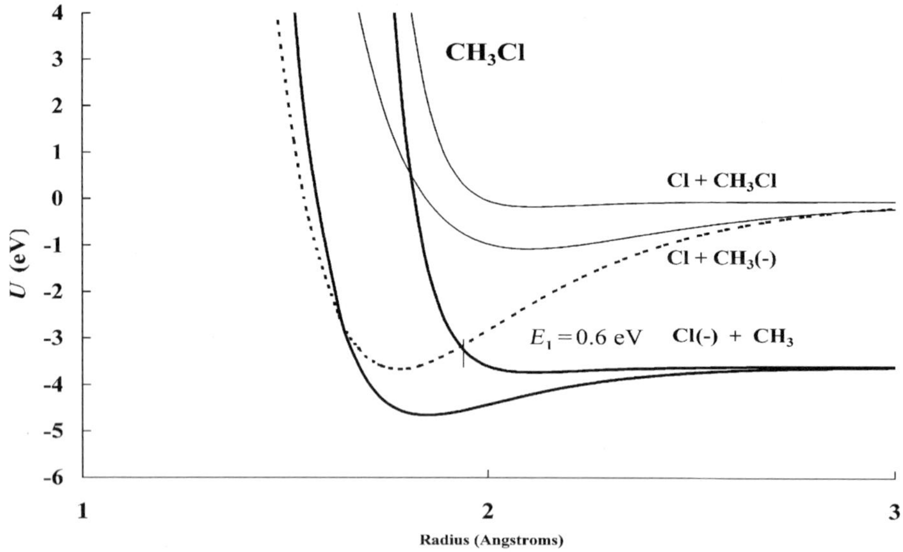

Figure 4.9 Morse potential energy curves for chloromethane and its ions. The curves are calculated using the activation energy determined from data in Figure 4.8. The high-temperature data is for unimolecular dissociation via the curve crossing on the approach side of the molecule. Only the VE_a is negative and dissociation occurs in the Franck Condon transition. The thermal energy dissociation occurs through the thermal activation of the molecule, as is the case for all DEC(1) molecules.

compounds that dissociate unimolecularly. As shown in the graphs, the intercepts range from 33 to 37 and encompass the ln(DEBA) value of 36. The EDEA for many of these reactions are positive so dissociative thermal electron attachment is exothermic. These are designated as DEC(1) molecules [17–20].

In Figure 4.9 pseudo-two-dimensional Morse potentials are shown for CH_3Cl. The quantity $D(R - \mathrm{Cl}) - E_a(\mathrm{Cl})$ is about -0.05 eV so dissociative thermal electron attachment is slightly exothermic. The excited-state Morse potential energy curve that crosses on the right-hand side leads to unimolecular dissociation. However, the excited-state electron affinity is small but positive. If it were possible to stabilize the excited-state ion, then the equilibrium between that anion and the neutral could be responsible for the low-temperature data illustrated in Figure 4.8. Dissociation from the stabilized ground-state negative ion would have an activation energy of dissociation, $E_2 = 1.0$ eV. This is based on the adiabatic electron affinity of CH_3Cl, assumed to be about 1 eV after comparison to the other chloromethanes [21].

In unimolecular dissociation the experimental activation energy is an upper limit to the quantity -EDEA $= D(R - \mathrm{Cl}) - E_a(\mathrm{Cl})$. It was empirically observed that the activation energy was linearly related to EDEA with a slope of unity and approached zero for an exothermicity of about 13 kcal/mole or 0.6 eV. Thus,

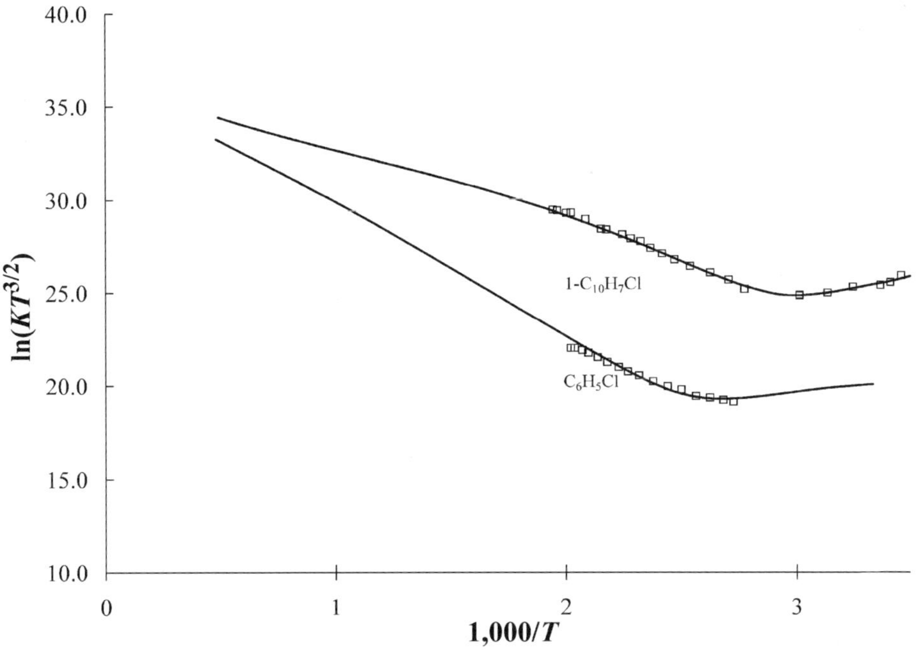

Figure 4.10 ECD data plotted as ln K_p versus $1{,}000/T$. Chlorobenzene and chloronaphthalene dissociate via an intermediate molecular ion. They are designated DEC(2) for dissociative electron capture by a two-step process. The slope in the high-temperature region multiplied by R is equal to the EDEA. Given the electron affinity of the dissociating species, in this case Cl(−), the C—Cl bond dissociation energy can be measured. Data from [17–19].

estimates of bond dissociation energies could be obtained from the ECD data for these compounds [20].

Another type of dissociative thermal electron attachment observed in the ECD is the sequential dissociation of a stabilized negative ion. These molecules are designated as DEC(2) molecules. Chlorobenzene, chloronaphthalene, nitromethane, and acetic anhydride are DEC(2) molecules. At high temperatures the dissociation proceeds via the intermediate molecular ion with an activation energy equal to -EDEA. At low temperatures the data are the same as for the Eql(1/1) molecules. The data for chlorobenzene and 1-chloronaphthalene can be seen in Figure 4.10. The electron affinities of the radicals, the C—Cl bond dissociation energies, and electron impact distributions have been measured for both of these compounds so that one may examine the equality between the activation energy and –EDEA. In addition, the electron impact spectra for the formation of the chloride ion have been measured. The molecular electron affinities are obtained from ECD data and confirmed by reduction potentials. These data can be used to calculate negative-ion Morse potentials. There are two low-lying dissociation limits in the C—Cl dimension, leading to Cl(−) + R and Cl + R(−). A bonding and an antibonding curve

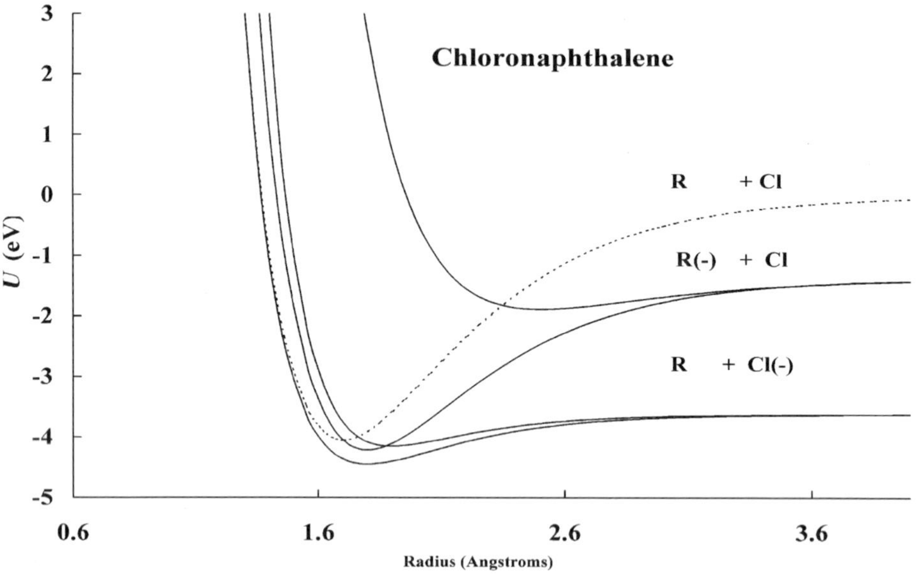

Figure 4.11 Morse potential energy curves for chloronaphthalene and its ions. The curves are calculated using the activation energy and electron affinity determined from the data in Figure 4.10. The high-temperature data are for sequential dissociation. For the two lowest curves the VE_a and EDEA are negative, but molecular ion formation precedes dissociation in the Franck Condon transition.

leading to each limit is drawn. The curves for 1-chloronaphthalene are shown in Figure 4.11. By comparing these curves to those for chloromethane, we see that the major difference is the electron affinity of *R*. In the case of the methyl group it is about zero, whereas in the case of the naphthyl radical it is 1.40(1) eV [22]. In both cases the ground-state curve crosses on the "backside" and thermal electron attachment will take place via an excited-state curve. In the case of chloronaphthalene there are at least two low-lying curves. In addition, there are curves leading to the radical anion and a hydrogen atom, and the hydrogen anion and radical in the C—H dimension. The intermediate negative ions can be stabilized to the ground-state anion, from which dissociation takes place. The higher electron affinity of the naphthyl radical lowers the excited-state anion curve to the region where thermal electron attachment may occur.

Only compounds that undergo sequential dissociative thermal electron attachment can exhibit a γ region, where $(k_2 \gg k_N)$ and $(k_{-1} \gg k_2)$, and $K = [k_2/2k_D][k_1/k_{-1}]$, and the δ region, where $(k_2 \gg k_{-1} + k_N)$ and $K = k_1/2k_D$. In the γ region the slope is approximately equal to $(E_a - E_2)$ since $K = [k_2/2k_D][k_1/k_{-1}]$:

$$\ln KT^{3/2} = \ln(T/2A_D) + \ln(A_2A_1/A_{-1}) + (E_a - E_2)/RT \qquad (4.21)$$

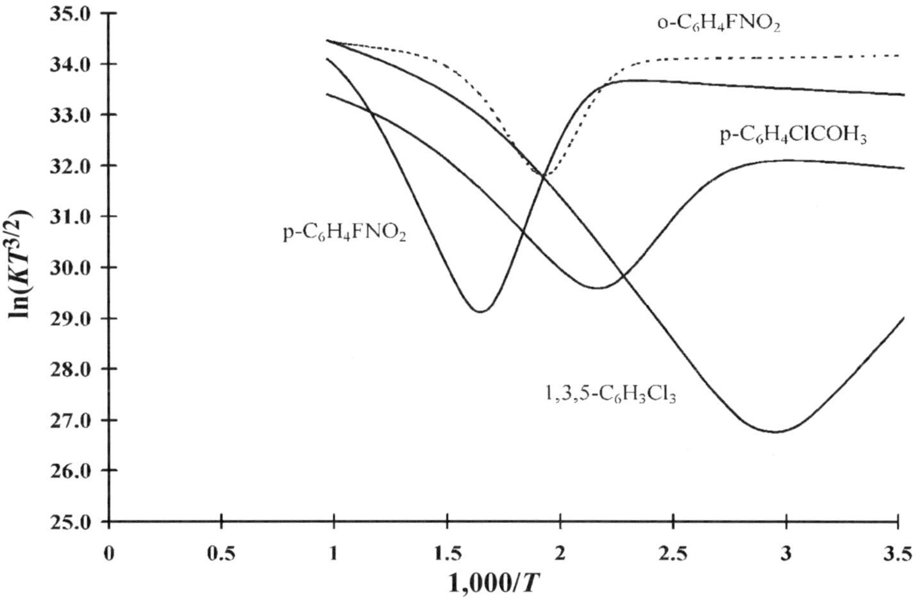

Figure 4.12 ECD data plotted as ln $KT^{3/2}$ versus $1{,}000/T$. The compounds in this graph are DEC(2). The electron affinities of the molecules are higher so that as many as four regions are observed: α, β, γ, and δ for o-F-nitrobenzene. The other molecules exhibit three regions. Data from [23, 24].

Since E_2 is greater than E_a, there will be a change in the sign of the slope in the transition from the α to γ region. The experimental quantity $(E_a(R\text{Le}){-}E_2)$ is $D(R - \text{Le}) - E_a(\text{Le})$. In the β and δ regions $K = k_1/2k_D$, and the ECD data define A_1 and E_1. The specific regions observed will depend on the electron affinity of the dissociating species and the electron affinity of the compound.

Compounds with a large electron affinity and low activation energy can exhibit all four regions. Figure 4.12 is a plot of ln $KT^{3/2}$ versus $1{,}000/T$ for two F-nitrobenzenes, 1,3,5-$C_6H_3Cl_3$ and p-$ClC_6H_4COCH_3$, obtained using the ECD. All six of the kinetic and thermodynamic parameters have been obtained for o-F nitrobenzene. The chloronaphthalene and chlorobenzene data shown in Figure 4.10 only have data in two regions. The curve for 1-chloronaphthalene at high temperatures was drawn with nominal values of A_1 and E_1.

The data for o-FNB is the prototypical data for a compound that exhibits four temperature regions. In the case of o-F-nitrobenzene the A_1, E_1, and E_a values have been measured using other techniques, and the values of A_2 and E_2 have been measured experimentally or calculated [23, 24]. The only parameter that has not been measured by an independent procedure is Q_{an}. All the quantities obtained from the ECD agree with values obtained by other techniques. In other words, the ECD response could have been calculated from properties measured in independent experiments if the value of Q_{an} was assumed to be unity.

4.5 ELECTRON AFFINITIES AND HALF-WAVE REDUCTION POTENTIALS

The first experimental method for obtaining relative electron affinities was the measurement of half-wave reductions potentials in aprotic solvents. The electron affinities are related to the half-wave reduction potentials by a simple relationship

$$E_a(\mathrm{eV}) = E(\mathrm{ref}) - (-\Delta\Delta G) + E_{1/2} \tag{4.23}$$

where $-\Delta\Delta G$ (mdd G) is the solution energy difference for the reaction in the gas phase and in solution, and E(ref) is the energy for the reference electrode (4.71 eV for the SCE electrode). The inclusion of the negative sign in the term $-\Delta\Delta G$ allows us to discuss absolute magnitudes. The gas phase E_a is obtained from half-wave reduction potentials by using the appropriate reference energy and mdd G. The $-\Delta\Delta G$ is a function of the solvent and the specific counter ion, but is constant for a given aprotic solvent.

When the ECD electron affinities were measured, the $-\Delta\Delta G$ for the aromatic hydrocarbons (2.0 eV) and the aromatic aldehydes and ketones (2.3 eV) were observed to be approximately constant within the class of molecules, but different from each other. The fullerenes ($-\Delta\Delta G = 1.76$ eV) have a lower charge density and lower $-\Delta\Delta G$ than the majority of other compounds. With the determination of more gas phase electron affinities, the $-\Delta\Delta G$ values range from 1.7 eV for larger fullerenes to 2.7 eV for small anions [3, 10].

The majority of $-\Delta\Delta G$ obtained from experimental gas phase E_a's fall into three groups that range from 2.0 to 2.4 eV. The average value for all these molecules is 2.2 ± 0.2 eV. Five groups starting at 1.8 eV, with an increment of 0.2 eV for each group, are defined. For the A group, which includes the aromatic hydrocarbons, the charge is delocalized and $-\Delta\Delta G = 2.00$ eV. The B group, containing benzoquinones, is assigned the value $-\Delta\Delta G = 2.20$ eV. The C group, containing chlorinated-nitro aromatics, has higher charge densities with $-\Delta\Delta G = 2.40$ eV. For the D molecules, such as nitrotoluenes, $-\Delta\Delta G = 2.60$. For the fullerenes or F group $-\Delta\Delta G = 1.80$ eV and the charge disperses. The charge is localized in the high-extreme $\mathrm{E_H}$ group, such as acetate ion, with $-\Delta\Delta G = 2.70$ eV, while some of the larger fullerenes are at the lower extreme ($\mathrm{E_L}$) with $-\Delta\Delta G = 1.70$ eV. The groups and are their $-\Delta\Delta G$ $\mathrm{E_L}$, 1.70; F, 1.80; mAF, 1.90; A, 2.00; mAB, 2.1; B, 2.20; mBC, 2.30; C, 2.40; mCD, 2.50; D, 2.60; and $\mathrm{E_H}$, 2.70, respectively. The classifications yield a series of linear equations:

$$\begin{aligned} EA(\mathrm{eV}) &= E(\mathrm{ref}) - 2.20 - n\{\mathrm{dev}(-\Delta\Delta G)\} + E_{1/2} \\ &= E(\mathrm{ref}) - 2.20 - n\{0.10\} + E_{1/2} \end{aligned} \tag{4.24}$$

$$n = -5, -4, -3, -2, -1, 0, 1, 2, 3, 4, 5$$

This may seem to involve merely grouping molecules according to their measured deviations. This would be true if there were no fundamental basis for the grouping.

However, it has been postulated that the different groups have different degrees of charge localization. The charge densities on the atoms in molecules can be calculated quantum mechanically. For the nitrobenzene anion it is about 0.8q on the nitro group and adjacent carbon atom. In the case of the acetophenone anion the charge in the vicinity of the carbonyl group is 0.5q, while that for each carbonyl group in the benzoquinone anion is 0.3q. In the anthracene anion the charges on the C—H bonds range from 0.08 to 0.15q. The maximum charge on two adjacent C atoms in the C_{60} anion is about 0.04q. The change in charge density, 0.04 to 0.8q, mirrors the $-\Delta\Delta G$ change, 1.8 to 2.6 eV. To normalize this to unit charge, the $-\Delta\Delta G$ values extend from 1.70 to 2.70. With more data perhaps a continuous relationship can be found between charge density and $-\Delta\Delta G$. Compounds for which gas phase E_a and reduction potentials have been measured define the set. The new problem is to accurately and objectively assign molecules and determine n [25].

In 1970 the MINDO/3 and $E_{1/2}$ values were correlated with two parameters in the linear equation, $E_{1/2} = m\ (E_a) + b$, to give a reported standard deviation of 0.028 eV. The slope was 0.92. In this data analysis outliers must have been statistically eliminated since some of the deviations are greater than 0.1 eV, three times the reported standard deviation. Using all the MINDO/3 E_a and a unit slope, we obtain an average $-\Delta\Delta G$ value of 1.94 ± 0.08 eV. The greater deviation results from the use of one less parameter and all the data [26].

The aromatic hydrocarbons have been classified into subgroups based on the average and standard deviation, $-\Delta\Delta G = 2.00 \pm 0.05$ eV. For five groups the error in the $-\Delta\Delta G$ is reduced from 0.05 eV to 0.02 eV. For a larger set of 80 compounds the average deviation in the $-\Delta\Delta G$ is reduced from 0.08 eV to 0.015 eV by this classification into eight groups. Clearly, the number of degrees of freedom in this case is more than 70. If the integer range in equation 4.24 is extended from -10 to $+10$, 21 subgroups are defined for the entire range, 1.70 eV to 2.70 eV. The groups can be designated, for example, by $A+$ for an $-\Delta\Delta G$ of 2.05 eV.

In order to use these subgroups, enough accurate data must be available. Since the relative reduction potentials can be measured to within 0.01 volts (V), the errors in the E_a must be reduced. In the case of the aromatic hydrocarbon set there are enough such data. The deviation is defined as $\sigma(E_a) = (E_a(\text{experimental}) - E_a(\text{calculated}))$. The ultimate goal is the reduction of $\sigma(E_a)$. These E_a can then be compared with the theoretical values or other experimental values such as those obtained from electronegativities. The procedure can be iterated to self-consistency. Groups separated by 0.05 eV are postulated for the fullerenes and aromatic hydrocarbons based on experimental data.

Table 4.3 lists $-\Delta\Delta G$ values, the ECD E_a, and the E_a from $E_{1/2}$, of several aromatic hydrocarbons obtained in this manner. The electron affinity for pentacene was determined by TCT, while that of coronene is the value obtained from reduction potentials [6]. The E_a are verified using CURES-EC. The calculated values are given in Table 4.3. Also listed are the weighted average of the values that cluster about the "current evaluated values" from a 1983 compilation [27]. The consistency of the E_a values in this table support the gas phase experiment and the assignments of lower values to excited states.

TABLE 4.3 Electron Affinities (in eV) and Solution Energies of Aromatic Hydrocarbons

Species	ECD/TCT	$E(1/2)$	CURES	1984	$-\Delta\Delta G$
Naphthalene	0.16(2)	0.17(3)	0.15	0.15	2.10
Anthracene	0.68(2)	0.72(2)	0.70	0.68	2.05
Tetracene	1.08(4)	1.09(3)	1.08	1.08	2.05
Pentacene	1.39(5)	1.37(3)	1.34	1.35	1.95
Phenanthrene	0.30(2)	0.31(3)	0.32	0.31	1.95
Benz[a]anthracene	0.72(2)	0.72(3)	0.74	0.69	1.90
Triphenylene	0.29(2)	0.29(3)	0.27	0.29	1.95
Dibenz[a,h]anthracene	0.69(3)	0.65(3)	0.66	0.69	1.95
Chrysene	0.42(4)	0.42(3)	0.43	0.42	2.00
Benz[c]phenanthrene	0.58(1)	0.58(3)	0.60	0.55	1.90
Picene	0.54(3)	0.49(3)	0.50	0.52	1.90
Pyrene	0.61(2)	0.63(3)	0.62	0.60	2.00
Perylene	0.98(1)	1.01(3)	1.00	0.96	1.95
Benz[a]pyrene	0.82(4)	0.79(3)	0.80	0.75	1.90
Fluoranthene	0.82(4)	0.83(3)	0.81	0.76	2.00
Biphenyl	0.13(2)	0.10(3)	0.10	0.15	2.00
Styrene	0.10(5)	0.12(3)	0.10	0.05	2.10
Azulene	0.84(5)	0.83(5)	0.78	0.75	2.25
Coronene	—	0.79(5)	0.80	0.82	1.80

4.6 ELECTRON AFFINITIES AND IONIZATION POTENTIALS OF AROMATIC HYDROCARBONS

The electronegativity of hydrocarbons is half the sum of the ionization potential plus the electron affinity, the Mulliken definition of the electronegativity of atoms. In the simplest approximation the electronegativity is constant and equal to the work function of graphite. This gives the relationship:

$$2\text{EN} = \text{IP} + E_a \tag{4.28}$$

If two of these quantities are measured, then the third can be obtained.

As more experimental values of both electron affinities and ionization potentials were measured, this relationship was tested. For the alternate aromatic hydrocarbons the EN is approximately 4.02 eV, as opposed to the work function of graphite that is 4.39 eV. The EN for the smaller aromatic hydrocarbons is 4.1 eV. The EN for hydrocarbons with five-membered rings, 4.4 eV, and C_{60}, 4.5 eV, is closer to the work function of graphite. Table 4.4 gives the E_a, IP, and EN values for several hydrocarbons. From a larger set of data the EN is not constant. If the values for styrene, fluoranthene, naphthalene, styrene, and azulene are not included, then $\text{EN} = 4.02 \pm 0.02$ eV can be used to calculate either the E_a or IP. The calculated E_a are compared to the ECD values in Table 4.4 [10].

TABLE 4.4 Electron Affinities, Ionization Potentials, and Electronegativity of Aromatic Hydrocarbons (in eV)

Species	ECD/TCT	2EN-IP	IP	EN
Anthracene	0.68(2)	0.68	7.40	4.04
Tetracene	1.08(2)	1.11	6.97	4.04
Pentacene	1.39(5)	1.47	6.61	4.04
Phenanthrene	0.30(2)	0.22	7.86	4.04
Benz[a]anthracene	0.72(2)	0.65	7.43	4.04
Triphenylene	0.29(2)	0.24	7.84	4.04
Dibenz[a,h]anthracene	0.69(3)	0.78	7.30	4.04
Chrysene	0.42(4)	0.57	7.59	4.04
Benz[c]phenanthrene	0.58(1)	0.48	7.60	4.04
Picene	0.54(3)	0.54	[7.5]	4.04
Pyrene	0.61(2)	0.67	7.41	4.04
Perylene	0.98(1)	1.16	6.92	4.04
Benz[a]pyrene	0.82(4)	0.96	7.12	4.04
Fluoranthene	0.82(4)	—	7.95	4.38
Biphenyl	0.13(2)	0.13	7.95	4.04
Styrene	0.10(5)	—	8.43	4.27
Azulene	0.84(5)	—	7.41	4.13
Coronene	—	0.79	7.29	4.04

It was possible to obtain better resolution for these compounds because there are three independent experimental methods for estimating the electron affinities that could be iterated to consistency. These were the ECD measurements, half-wave reduction potential measurements, and electronegativity values. In addition, these electron affinities had been calculated with the MINDO/3 procedure. We also calculated the values using the CURES-EC procedure and obtained estimates of the charge densities. These procedures were extended to a larger set of 80 compounds, some without gas phase E_a that will be discussed in Chapter 10.

4.7 ELECTRON AFFINITIES AND CHARGE TRANSFER COMPLEX ENERGIES

The theory of charge transfer complexes relates the maximum in the absorption spectrum, the charge transfer energies E_{CT}, and energies for complex formation ΔG_{CT} to the vertical ionization potential of the donor and the vertical electron affinities of the acceptor. The relationship uses constants related to the geometry of the complexes. Mulliken described the theory of charge transfer as follows:

> The basic idea is the qualitatively familiar one that any two bits of matter, whether they are atoms, or positive or negative ions, or molecules or even solids have a

tendency to exchange electrons. Quantum mechanically the interaction may be formulated by saying that when any two such entities get together their joint or combined wave function may be expressed as:

$$\psi = a\psi(D, A + b\psi(D^+, A^-) \tag{4.25}$$

Sometimes, the donor and acceptor are nearly equal in strength and a $\approx$ b. For the strengths, one must **not** use the adiabatic ionization potential and (adiabatic) electron affinity (AE_a) corresponding to the passage of D with **its** natural skeleton to D with its natural skeleton or from A with its natural shape to A(−) in its natural shape; instead, one must take the so-called vertical values of IP (VIP) and E_a (VE_a) corresponding to **no change** in skeleton. Moreover, one must take VIP and VE_a for such deformed skeletons for both D and A as in the final compromise skeleton of the complex.... Values of E_a and especially VE_a are less available. [28]

The energy for the formation of a charge transfer complex is related to the VIP and VE_a of the donor. To examine this relationship for the aromatic hydrocarbons with the measured electron affinities, the energies of complexes of methylbenzenes as the donors and aromatic hydrocarbons as the acceptors were determined. The charge transfer bands for these complexes were not observed [29]. Therefore, only the relation between the energy for complex formation and the VIP and VE_a could be examined. The equation is

$$-RT \ln K_{CT} = C1 + 1/(\text{VIP} - VE_a - C_2) \tag{4.26}$$

Figure 4.13 is a plot of the experimental data for a series of methylbenzenes with $n = 1$ to 6 with the aromatic hydrocarbons anthracene, A; pyrene, Py; phenanthrene, P; chrysene, C; and triphenylene, T; versus $0.1/(\text{IP} - E_a - C_2)$ with the least-squares value of $C_2 = 2.90 \pm 0.10$ eV. If we consider the narrow range of values for these complexes, the correlation is quite good.

The maximum energy of the charge transfer transition is given by

$$E_{CT} = \text{VIP} - VE_a - C_2 + C_1/(\text{VIP} - VE_a - C_2) \tag{4.27}$$

For a series of acceptors with a given donor, $-C2 - VE_a$ is constant so that a plot of E_{CT} versus the VIP will be represented by the above equation. In some cases this is approximately linear, so that given an experimental value for E_{CT}, the IP can be determined from the linear plot. This implies that the C_1 value is small. Plots for a series of donors with different acceptors are given in Figure 4.14. The horizontal displacement of the curves is due to the differences in the VE_a. Similarly, a plot of E_{CT} for a given donor with various acceptors can be used to estimate the electron affinity. Thus, given the VIP, E_{CT}, and C_2, the E_a can be calculated. If the data are scaled to the AE_a so that the rearrangement energies are included in the C_2 term, adiabatic electron affinities can be estimated. Table 4.5 gives the values of C_1 and

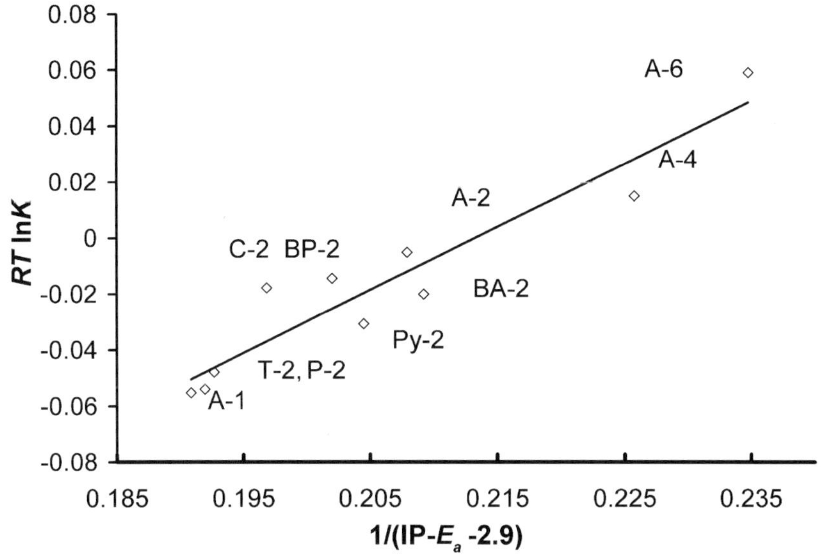

Figure 4.13 Free energy for complex formation by methyl benzene and aromatic hydrocarbons is plotted versus $1/(IP{-}E_a - 2.90)$. The aromatic hydrocarbons are A, anthracene; BA, benzanthracene; C, chrysene; Py, pyrene; T, triphenylene; and P, phenanthrene. The numbers refer to the number of methyl groups on benzene. Toluene is 1, while s-tetramethylbenzene is 4. Original data from [29].

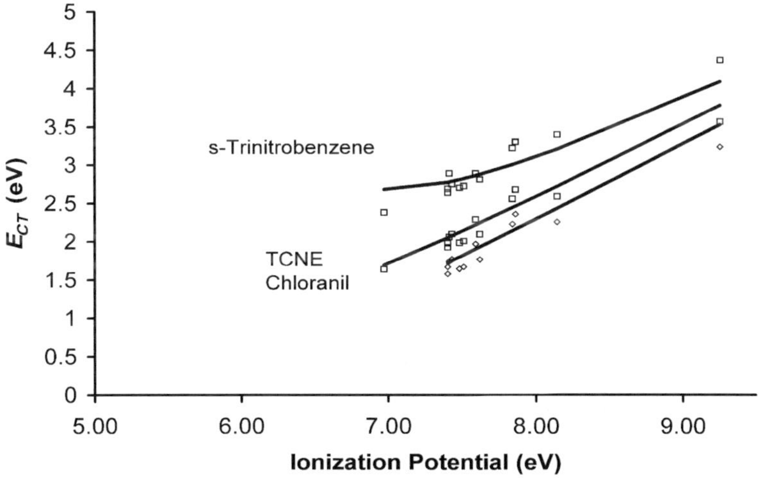

Figure 4.14 Energy for the maximum absorbance for charge transfer complexes of s-trinitrobenzene, tetracyanoethylene, and chloranil with various donors plotted against the adiabatic ionization potential of the donor. Recent ionization potentials from the NIST tables were used. The vertical displacement results from the differences in the E_a of the molecules. The calculated curves were obtained by using a two-parameter nonlinear least squares. The values of the constants are given in Table 4.5, where they are compared with published values. Data from [8, 30, 32].

TABLE 4.5 Charge Transfer Complex Constants C_1 and C_2 for equation 4.27 (in eV)

	C_2	C_1	$\sigma C_1 C_2$	C_2	C_1
Species	This Work			[32]	
Tetracyanoethylene	5.86(11)	0.31(23)	−0.026	6.10	0.54
Tetracyanoquinodimethane	5.81(15)	0.06(27)	−0.037	—	—
Chloranil	5.71(8)	0.63(15)	−0.010	5.70	0.44
s-Trinitrobenzene	5.36(7)	1.55(21)	−0.015	5.00	0.70
Trinitrofluorenone	5.34(8)	0.60(.20)	−0.015	—	—
Benzoquinone	4.91(30)	0.0(15)	−0.26	—	—
Iodine	4.61(18)	0.26(18)	−0.11	5.20	1.50
Tetrachlorophthalic anhydride	4.60(4)	0.15(20)	−0.01	4.9	—
Maleic anhydride	4.43(3)	0.55(13)	−0.04	4.4	—

C_2 determined for acceptors using a nonlinear least-squares fit to equation 4.27 and data compiled by Briegleb [32]. These constants are compared to similar values given by Briegleb.

In 1975 when more reliable electron affinities for the reference compounds were available, a large number of calculated E_a were published. Since organic acceptors

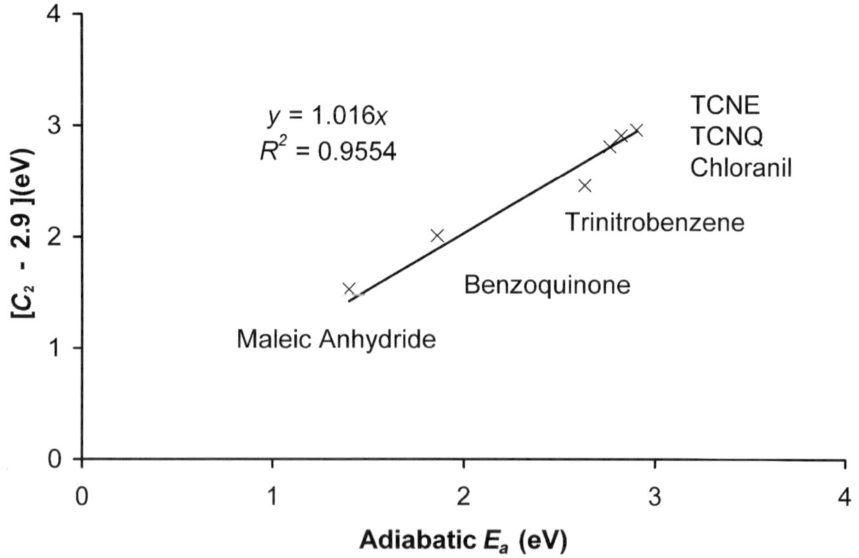

Figure 4.15 Electron affinities of charge transfer complex acceptors calculated from $C_2 = 2.9$ versus the current "best" adiabatic electron affinities. This is a precision and accuracy plot. The zero intercept slope indicates that the same quantities are measured. The compounds are maleic anhydride, tetrachlorophthalic anhydride, benzoquinone, trinitro-flourenone, s-trinitrobenzene, chloranil, tetracyanoquinodimethane, and tetracyanoethylene in order of their electron affinities.

typically contain electronegative elements such as N, O, and the halogen atoms, this offers an additional method for estimating the substitution and replacement effects. In Figure 4.15 the E_a calculated from $AE_a = C_2 - 2.9$ eV (using the C_2 given in Table 4.5) are plotted against the current best adiabatic values. The deviations are within the experimental errors and the slope is approximately 1. The E_a are adiabatic values since the data have been scaled to adiabatic E_a [29–31]. Based on the values of C_2, the AE_a of trinitrofluorenone is 2.4 ± 0.1 eV, while that of tetrachlorophthalic anhydride is 1.7 ± 0.1 eV. These values have not been measured in the gas phase. The electron affinity of phthalic anhydride is 1.25 eV and the addition of four Cl atoms could easily raise the E_a to 1.7 eV. The electron affinity of fluorenone has not been measured.

4.8 SUMMARY

Molecules studied in the ECD can be grouped into three classes. If the EDEA is less than about −35 kcal/mole (−1.5 eV), the compounds fall under group Eql(1/1) (equilibrium 1 with one temperature region). Molecules with two states and two regions are designated Eql(2/2). For the Eql(1/1) group the VE_a and AE_a can be positive. For molecules in group DEC(1) (dissociative electron capture, unimolecular) the EDEA is positive, and for molecules in group DEC(2) (dissociative electron capture, unimolecular) it is an intermediate value.

Compounds such as tetramethylbenzene, phenanthrene, chrysene, acetophenone, and benzaldehyde belong to group Eql(1/1) since only the α region is observed in the temperature region. Compounds that exhibit both α and β regions, such as nitrobenzene, pentafluoronitrobenzene, and acetonaphthone, belong to group Eql(1/2). Since thermal electron reactions with C_6Cl_6, C_6F_6, and CS_2 involve two low-lying bound negative-ion states, they belong to group Eql(2/2). The chloromethanes are classified under group DEC(1) since thermal electron attachment is exothermic. The $C_6H_nCl_{6-n}$ for $n = 1$ to 4 are in group DEC(2) since the reaction of thermal electrons leads to dissociation via an intermediate molecular ion and all four temperature regions may be observed.

Examples of the temperature dependence for different classes of molecules are given as global plots of ln $KT^{3/2}$ versus $1{,}000/T$. The curves that are drawn used the equations for the complete model. Excited-state E_a have been measured with the ECD. The clearest indication of an excited state is structure in the data, as illustrated for carbon disulfide and C_6F_6. The temperature dependence of the ions formed in NIMS of the chloroethylenes indicate multiple states. NIMS also supports AE_a, as in the case of SF_6 and nitrobenzene. The quantity $D - E_a$ can be obtained from ECD data for DEC(2) dissociative thermal electron attachment. If one is measured, then the other can be determined. In the case of the chlorinated benzenes this quantity gives the C—Cl bond dissociation energy. The highest activation energy of 2.0 eV has been observed for the dissociation of the anion of o-fluoronitrobenzene.

For compounds with positive slopes in a plot of ln $KT^{3/2}$ versus $1{,}000/T$ with an intercept in the region of 11.7 ± 3, a nondissociative mechanism is indicated. The slope is E_a/R. At lower temperatures, Eql(1/1), the data could level off with an intercept in the region of 34 ± 3. The slope will be E_1/R and the intercept ln $[A_1/2A_D]$, Eql(1/2). The maximum value of A_1 is determined by the DeBroglie wavelength of the electron and the stabilization term Z. The value of A_1 when $Z = 1$ is 5×10^{15} $K^{-1/2}$-l-mole^{-1}, which gives ln $(A_1) = 36$. Excited states could exist at lower temperatures, Eql(2/2).

For compounds that dissociate unimolecularly, DEC(1), the slope will be negative and the intercept will fall in the region 34 ± 3. The slope is E_1/R and the intercept ln$[A_1/2A_D]$. The magnitude of E_1 is a function of the $D - E_a$ for the molecule. In the case of CCl_4 and CH_3I the E_1 are approximately zero. For CH_3Cl, the activation energy is greater than 0.5 eV and there is a large increase in K with increasing temperature. For some DEC(1) molecules an excited-state molecular ion is formed with a low electron affinity.

For compounds that dissociate via an intermediate negative ion, DEC(2), a change in slope from positive to negative will occur at higher temperatures. This indicates the opening up of the dissociative pathway for the loss of ions in competition with recombination but lower than detachment. The lower-temperature portion will be the same as for the formation of a stable negative ion. At a higher temperature the negative slope will be equal to $D - E_a$. In this case there are six parameters, one pre-exponential, and one exponential (energy) parameter for each of k_1, k_{-1}, and k_2. In order to observe this type of behavior, the A_2 must be large. Thus, the intercepts in the high-temperature region will be ln$\{A_1A_2/2A_{-1}A_N\}$ and can rise as high as 45. At still higher temperatures the dissociation overwhelms detachment and K will be $k_1/2k_D$, as in the low-temperature case.

The kinetic model for the determination of the energies of complex formation was described. Examples of negative-ion mass spectrometry data for the mono- and di-hydrates of $O_2(-)$ were given, and typical plots of the equilibrium constants for the 0-1 and 6-7 complexes were presented. Once the equilibrium constants are determined, the equations used to obtain the entropy and energy for the consecutive reactions become the standard.

The basic equations for the calculation of electron affinities from half-wave reduction potentials in aprotic solvents were presented. More accurate values of the E_a of organic molecules can be obtained by using variable-solution energy differences. The electronegativity concepts and CURES-EC calculations support the use of these modified equations. Equations relating the free energy of charge transfer complex formation to the electron affinity of the acceptor and ionization potential of the donor were summarized. Complexes with aromatic hydrocarbons as acceptors and methylbenzenes as donors were used as an example. The good correlation of the complex formation energy supports the E_a of the aromatic hydrocarbon acceptors. The maximum energy in the new absorption band for charge transfer complexes for trinitrobenzene, tetracyanoethylene, and other acceptors was related to the ionization potentials of the donor and the electron affinities of the acceptors. By using adiabatic, rather than vertical E_a, the rearrangement energy is automatically included in the correlation terms. Given additional E_{CT} data, the adiabatic E_a can be obtained. The E_a of trinitrofluorenone (2.5 ± 0.1 eV) and tetrachlorophthalic anhydride (1.7 ± 0.1 eV) were determined in this manner.

REFERENCES

1. Chen, E. C. M.; Carr, S. D.; Wentworth, W. E.; and Chen, E. S. D. *J. Chromatogr. A* **1998**, 827, 91.
2. Chen, E. C. M.; Wentworth, W. E.; and Lovelock, J. E. *J. Phys. Chem.* **1966**, 70, 445.
3. Ruoff, R. S.; Kadish, K. M.; and Chen, E. C. M. *J. Phys. Chem.* **1995**, 99, 8843.
4. Chen, E. C. M. and Wentworth, W. E. *J. Phys. Chem.* **1983**, 87, 45.
5. Wojnarvits, L. and Foldiak, G. *J. Chromatogr.* **1981**, 206, 511.
6. Crocker, L.; Wang, T. B.; and Kebarle, P. *J. Amer. Chem. Soc.* **1993**, 115, 7818.
7. Scheidt, J. and Weinkauf, R. *Chem. Phys. Lett.* **1997**, 266, 201.
8. Chen, E. C. M. Ph.D. dissertation, Universtiy of Houston, **1967**.
9. Chen, E. C. M. and Wentworth, W. E. *Mol. Cryst. Liq. Cryst.* **1989**, 171, 271.
10. Chen, E. S. D.; Chen, E. C. M.; Sane, N.; Talley, L.; Kozanecki, N.; and Shultze, S. *J. Chem. Phys.* **1999**, 110, 9319.
11. Chen, E. C. M. and Chen, E. S. D. *J. Chromatogr. A* **2002**, 952, 173.
12. National Institute of Standards and Technology (NIST). *Chemistry WebBook*, **2003**. Available at http://webbook.nist.gov.
13. Lyons, L. E.; Morris, G. C.; and Warren, L. J. *J. Phys. Chem.* **1968**, 72, 3677.
14. Wiley, J. R.; Chen, E. C. M.; and Wentworth, W. E. *J. Phys. Chem.* **1993**, 97, 1256.
15. Wentworth, W. E.; Limero, T.; Batten, C. F.; and D'sa, E. D. *J. Chem. Phys.* **1988**, 88, 4711.
16. D'sa, E. D. Ph.D. dissertation, University of Houston, **1987**.
17. Wentworth, W. E.; Becker, R. S.; and Tung, R. *J. Phys. Chem.* **1967**, 71, 1652.
18. Wentworth, W. E.; George, R.; and Keith, H. *J. Chem. Phys.* **1969**, 51, 1791.
19. Steelhammer, J. C. and Wentworth, W. E. *J. Chem. Phys.* **1969**, 51, 1802.
20. Chen, E. C. M.; Albyns, K.; Dussak, L.; and Wentworth, W. E. *J. Phys. Chem.* **1989**, 93, 6827.
21. Page, F. M. and Goode, G. C. *Negative Ions and the Magnetron.* New York: Wiley-Interscience **1969**.
22. Ervin, K. M.; Ramond, T. M.; Davico, G. E.; Schwartz, R. L.; Casey, S.M.; and Lineberger, W. C. *J. Phys. Chem. A* **2001**, 105, 10822.
23. Chen, E. S. D.; Chen, E. C. M.; and Wentworth, W. E. *J. Phys. Chem.* **1991**, 95, 520.
24. Han, C. C. and Brauman, J. I. *J. Phys. Chem.* **1990**, 94, 3415.
25. Chen, E. S. D.; Chen, E. C. M. Sane, N.; and Shulze, S. *Bioelectrochem. Bioenerget.* **1999**, 48, 69.
26. Dewar, M. J. S.; Hashmall, J. A.; and Trinajstic, N. *J. Amer. Chem. Soc.* **1970**, 92, 5555.

27. Christodoulides, A. A.; McCorkle, D. L.; and Christophorou, L. G. "Electron Affinities of Atoms, Molecules and Radicals" in *Electron-Molecule Interactions and Their Applications*. New York: Academic Press, **1984**.
28. Mulliken, R. S. and Person, W. B. *Molecular Complexes, a Lecture and Reprint Volume*. New York: Wiley, **1969**.
29. Chen, E. C. M. and Wentworth, W. E. *J. Phys. Chem.* **1963**, 67, 2201.
30. Chen, E. C. M. and Becker, R. S. *J. Chem. Phys.* **1966**, 45, 2403.
31. Briegleb, G. *Angew. Chem. Internat. Edit.* **1964**, 3, 617.
32. Briegleb, G. *Elektronen Donator Acceptor Komplexe.* Berlin: Springer-Verlag, **1961**.

CHAPTER 5

Experimental Procedures and Data Reduction

5.1 INTRODUCTION

> All electron affinities are of course relatively small (perhaps mostly 1–3 eV). From chemical behavior and stability of molecular complexes, we may reach some qualitatively conclusions about relative electron affinities. . . . Quantitative methods for determining molecular electron affinities are not very well developed but there seem to be possibilities for the future.
>
> —Robert Sanderson Mulliken
> Molecular Complexes, a Lecture and Reprint Volume

In this chapter the experimental ECD and NIMS procedures for studying the reactions of thermal electrons with molecules and negative ions are described. Gas phase electron affinities and rate constants for thermal electron attachment, electron detachment, anion dissociation, and bond dissociation energies are obtained from ECD and NIMS data. Techniques to test the validity of specific equipment and to identify problems are included. Examples of the data reduction procedure and a method to include other estimates of quantities and their uncertainties in a nonlinear least-squares analysis will be given. The nonlinear least-squares procedure for a simple two-parameter two-variable case is presented in the appendix.

The procedures for the measurement of reduction potentials are described in original articles and will not be repeated here. Carefully dried aprotic solvents must be used and measurements are referenced to the same electrode. Once it is established that the process is reversible, the data can be used to obtain E_a [1].

Both the GC/ECD and GC/NICI instruments commonly analyze compounds that form negative ions such as pesticides and polychlorinated biphenyls, making devices to carry out the study of thermal electron attachment reactions commercially available. These include the radioactive and nonradioactive ECD,

The Electron Capture Detector and the Study of Reactions with Thermal Electrons
by E. C. M. Chen and E. S. D. Chen
ISBN 0-471-32622-4

atmospheric pressure ionization sources, methane thermal electron chemical ionization sources, an electron monochrometer source for GC/NIMS, and the ion mobility spectrometers. The GC/NIMS and ion mobility spectrometers have not been used extensively to measure fundamental properties. However, such information can be obtained by measuring responses as a function of temperature [2].

The ECD used to establish the kinetic model was custom-built. Affordable commercial ECD and negative-ion gas chromatograph chemical ionization mass spectrometers are now available. Thus in order for an investigator to use these techniques, it is simply a matter of calibrating the instrument by reproducing the experimental results to verify that no artifacts result from the equipment. The major work described in this book was conducted with the radioactive ECD. The mechanisms for the pulsed discharge electron capture detector are the same as with the radioactive ECD that is now commercially available [4]. We have used commercial detectors and a quadrupole mass spectrometer with a home-made data collection system to determine electron affinities and to study the complexes of negative ions [5–9].

In many cases the mere observation of a parent negative ion in mass spectrometry or ion mobility spectrometry is evidence of the positive electron affinity of a molecule. The ECD kinetic model is applicable to the ions observed in NICI experiments so the same quantities measured in the ECD can also be measured with this technique. There is a large body of NIMS data taken at two temperatures for compounds significant to those used in environmental chemistry that can be analyzed to obtain approximate electron affinities and activation energies [10].

5.2 EXPERIMENTAL ECD AND NICI PROCEDURES

The objective of the ECD and NIMS experiments is to measure the molar response of different compounds as a function of temperature. From these data the fundamental kinetic and thermodynamic properties of the reaction of thermal electrons with molecules and negative ions can be determined. The measurement is carried out in the same manner as the calibration of any detector. Known amounts of a compound are injected into the chromatograph and purified on a column, they then enter the detector. The response of the detector is normalized to the number of moles injected. When obtaining physical parameters, the detector temperature is changed and the procedure repeated. Since the molar response can vary by three to four orders of magnitude, the concentrations of the test molecule and the conditions in the detector at different temperatures must be taken into account.

Once an ECD/GC or NIMS has been obtained, its suitability for the determination of fundamental properties can be examined by measuring the temperature dependence of the response for standard compounds. Although strictly speaking, it would only be necessary to measure the temperature dependence of three compounds—CCl_4, acetophenone, and CH_2Cl_2—to obtain A_D and A_N, other compounds should also be studied. In addition, is valuable to determine the temperature dependence of different internal standards. We have used $C_6F_5NO_2$, C_7F_{14}, CCl_4, and $C_6H_4Br_2$ because they have little temperature dependence. Acetophenone has been used because it has a large α region that establishes Q_{an} and E_a. Others have used nitrobenzene and naphthalene. The use of an internal standard will automatically cancel out instrumental variations. In NIMS this is especially important since it is difficult to know the exact amount of sample introduced into an ion source but relative concentrations can be easily prepared.

Because of the high sensitivity of the ECD, it is necessary to make sure that glassware, syringes, and solvents used to prepare samples are clean. In preparing standard solutions of CCl_4, $CHCl_3$, and CH_2Cl_2, pesticide-grade solvents should be used. The solvents should be placed into glassware tested with the ECD. The syringes used to prepare the dilutions must be cleaned until no peaks are observed in the ECD. The response of CH_2Cl_2 is a million times lower than that for CCl_4, and often, enough CCl_4 is present in the "pure" material to give a measurable peak when CH_2Cl_2 is diluted.

The basic procedure for preparing dilutions of CCl_4 will be given in detail and should be used for other samples. A 10-μL sample of pure CCl_4 should be diluted in 10 mL of solvent using a syringe and a 10-mL volumetric flask (10^{-3}). After making sure that the syringe is clean, you should further dilute 10-μL of that sample in 100 mL of solvent. This gives a total dilution of 10^{-7} mL CCl_4/mL solvent. The density of CCl_4 should be used to find the molar concentration of CCl_4 in the solutions. It will be 1.8×10^{-7}/mL in the final solution. Larger amounts of the final concentration can be utilized to prepare other samples with a fixed amount of CCl_4 that can be used as an internal standard. For example, to prepare a solution 10^{-5} in $CHCl_3$, 100 μL of a 10^{-3} solution of $CHCl_3$ is diluted into 10 mL of the 10^{-7} solution of CCl_4.

Figure 5.1 shows a series of simulated chromatograms where the peaks represent a concentration change of two orders of magnitudes. The standing current is equal to I_b and is 100 units. The highest direct ECD peak is 47 units, while the converted signal is 0.887 since $I_b - I_e = 47$ and $I_e = 53$ units. When the number of moles injected is reduced by a factor of 8, the peak height for the corrected response decreases to 0.11, while the direct ECD peak is 9.5 units. This only represents a decrease of a factor of 5. For the next decrease of a factor of 8 in concentration, both the direct and converted signal are decreased by a factor of 8. Thus, the ECD response factors should be measured using the converted response or in the linear range of the detector. This linearity test is the first step in characterizing a detector. Modern detectors automatically convert the signal to a linear one so that the molar response need only be determined in the linear range.

If peaks are symmetrical, as is frequently the case with modern capillary columns, only the peak height must be converted. When recording the chromatograms on a strip chart recorder, the chart speed on the recorder should be increased to obtain an accurate measurement of the peak width. The area can then be obtained by triangulation. Other required data are the flow rate through the detector and the moles of sample in the detector. For samples that are split the measurement of the split ratio is also required to obtain the number of moles injected. The

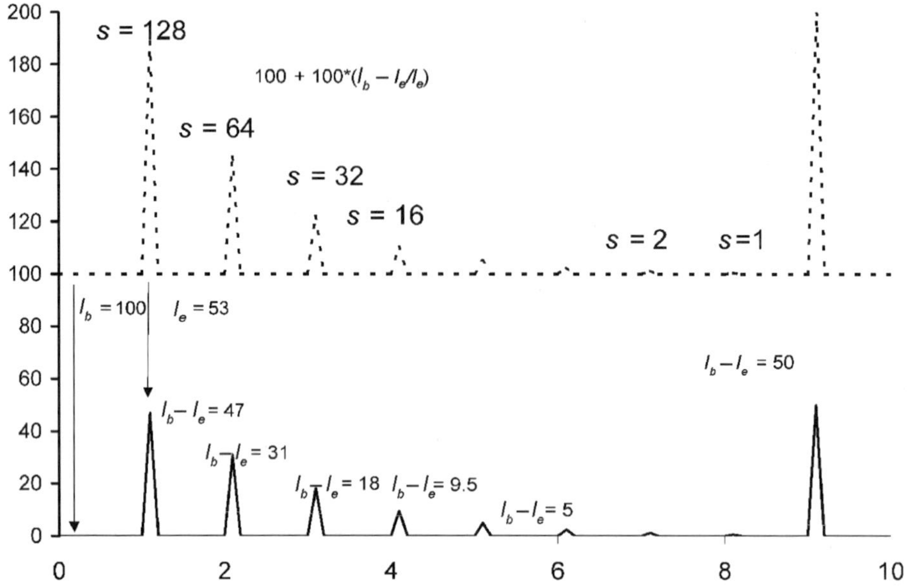

Figure 5.1 Simulated chromatograms showing the difference between the concentration dependence of the absolute signals and the corrected signals.

calculation for K, the molar response, is

$$K_{EC} = \{[I_b - I_e]/I_c\}_{\max}\{t_{1/2}\}V/n \tag{5.1a}$$

The $\{[I_b - I_e]/I_e\}_{\max}$ is the corrected height at the peak maximum and is dimensionless. The measured peak width at corrected half-height is converted to half-time by dividing by chart speed. The volumetric flow rate or total flow rate through the detector is measured with a bubble flow meter and is typically 50 to 150 mL/min.

The moles injected are determined from the sample size, concentration, and split ratio and are typically in the range of 1.0 femtomole for a high-capturing compound such as CCl_4 with a K of 10^{11} L/mole. The molar response of a negative-ion mass spectrometer can be determined in the same manner, but the moles into the mass spectrometer must be estimated.

Once the molar response is measured at a given temperature, the temperature of the detector can be changed and the measurement repeated. The temperature must be that in the actual ion source of the mass spectrometer or in the ECD detector. Changing the temperature slowly or waiting for equilibration can accomplish this.

Table 5.1 presents data taken to characterize a high-temperature Ni-63 detector using C_6F_6. A series of solutions were prepared at concentrations of 5×10^{-5}, 5×10^{-6}, and 5×10^{-7} moles of C_6F_6/mL heptane. The samples were injected into the chromatograph with a 10-μL syringe without splitting. The total flow rate was 150 mL/min at 300 K. The data were recorded on a strip chart recorder and the peak width converted to time using the chart speed of the recorder.

TABLE 5.1 Calculation of ECD Responses from Data for Hexafluorobenzene

b	*e*	*w*	*n*	*T*	*F*	*K*
19.1	13.9	0.650	2.15E-10	515	0.15	3.08E + 08
19.9	17.0	0.650	2.15E-10	529	0.15	1.38E + 08
18.6	15.6	0.630	3.22E-10	540	0.15	1.10E + 08
18.8	17.1	0.630	3.22E-10	556	0.15	5.81E + 07
18.8	17.2	0.625	3.22E-10	571	0.15	5.54E + 07
18.5	17.9	0.625	3.22E-10	588	0.15	2.09E + 07
21.4	20.2	0.625	6.44E-10	606	0.15	1.65E + 07
21.2	20.5	0.625	6.44E-10	625	0.15	9.87E + 06
21.1	18.3	0.615	6.44E-09	654	0.15	4.57E + 06
20.5	19.1	0.615	6.44E-09	676	0.15	2.33E + 06
20.8	17.5	0.600	3.22E-08	709	0.15	1.21E + 06
21.2	19.5	0.600	3.22E-08	730	0.15	5.65E + 05
21.2	19.2	0.600	4.30E-08	746	0.15	5.17E + 05
19.4	17.6	0.600	6.44E-08	769	0.15	3.81E + 05

The values of I_b and $I_b - I_e$ (the peak height) were measured with a ruler and the temperature read from a thermometer next to the detector in a metal heating block. The column and injector temperatures were kept constant. The response was corrected for the small change in the standing current by multiplying by $I_{\max}/I_b = 21.2/I_b$. The volumetric flow rate was corrected by multiplying by $T/300$ K.

The final equation for K from the raw data is

$$K = \{[I_b - I_e]/I_e\}_{\max}\{t_{1/2}\}(V/n)(T/300)(I_{\max}/I_b) \tag{5.1b}$$

where $n =$ (sample size) (concentration) and $t_{1/2} = w$/chart speed.

The calculations shown in Table 5.1 are typical for the determination of absolute molar responses. Note that the determinations at lower temperatures required the injection of a lower concentration. The values of n and w are rounded for this example. In the event that the integrated response is from a commercial detector, the data must be in the "linear range." A scale factor will give the values of the absolute response factors. An example for this type of calculation will be provided later. Figure 5.2 is a plot of these data overlaid by a calculated curve for C_6F_6 [11]. The E_a determined from a plot of ln $KT^{3/2}$ versus $1,000/T$ is the gas constant times the slope. The electron affinity from this limited data is determined to be 0.86 ± 0.10 eV. The intercept gives a Q_{an} of 0.07 ± 0.05. This is within the uncertainty for the weighted averages determined with more data.

Modern chromatographs report linearized areas for the ECD. In these cases a single solution can be used to determine the temperature dependence of several compounds. The relative responses can be standardized to a value for an internal standard such as CCl_4. The use of relative responses cancel the flow rate, sample

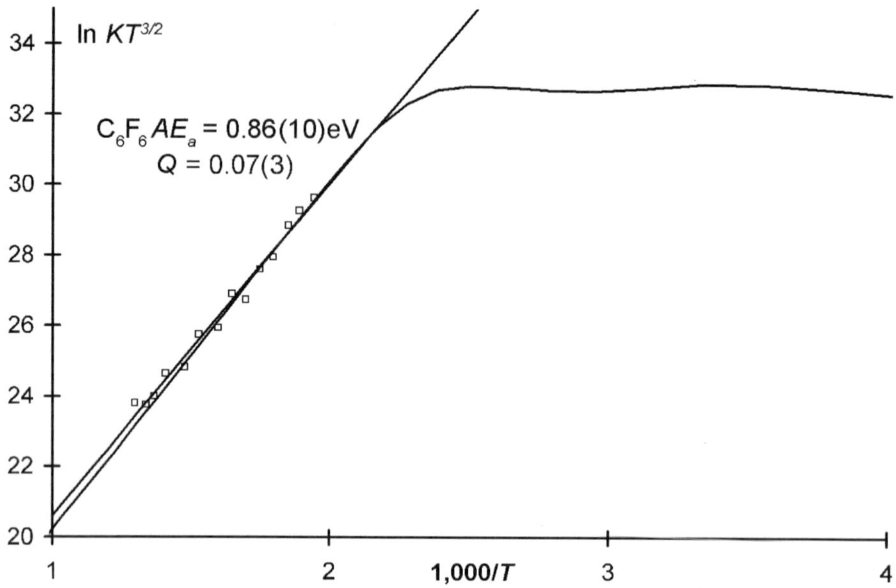

Figure 5.2 Illustration of data for C_6F_6 in the linear region obtained from the raw data shown in Table 5.1. The slope in this region gives an E_a of 0.86(10) and a Q_{an} of 0.070(5) by a simple least squares through the data. The calculated curve was obtained from data in four regions.

size, temperature, and standing current corrections since the same injection is utilized. More precise relative values can be obtained than if individual injections are made. The only other experimental quantities that must be known are the relative values of the concentrations. Table 5.2 provides an example of such a calculation for CH_2Cl_2, $CHCl_3$, and CCl_4. The areas are in the linear range of the detector and are reported on an integrator. With modern computer data systems the actual signals can also be recorded so that an absolute value of K may be obtained for CCl_4, it is 7.2×10^{10} L/mole at 328 K. This is equal to $k_1/2k_D$. The value for k_1 is well known to be 2.1×10^{14} L/mole-s. Consequently, the value for $A_D = k_D = 1.1 \times 10^{14}/7.2 \times 10^{10} = 1,500\,\text{s}^{-1}$. By knowing the concentration ratios for the other compounds, the relative responses can be converted to absolute responses. The maximum value for k_N for the system can be determined from the intercept for the acetophenone data. The acetophenone data were taken separately from the chloromethanes. The value of Q_{an} is 1.00 for acetophenone. The experimental intercept is 12.5. This is equal to $(11.73) + \ln(A_N/A_D)$, which gives a value of $A_N = A_D \exp(12.5 - 11.73) = 2 \times 1,500\,\text{s}^{-1}$. These values are typical for the original detectors. However, modern detectors have smaller values of $A_D(500\text{–}1,000\,\text{s}^{-1})$ and smaller ratios of $A_N/A_D = 1$ operated in the constant current mode because of the "cleaner" carrier gases and systems. These data are plotted

TABLE 5.2 Calculation of *K* for Chloromethanes Using Carbon Tetrachloride as an Internal Standard

Area 1 (CH_2Cl_2)	Area *R* (CCl_4)	Area 2 ($CHCl_3$)	$T(K)$	ln $KT^{3/2}$ (CCl_4)	ln $KT^{3/2}$ ($CHCl_3$)	ln $KT^{3/2}$ (CH_2Cl_2)
3.7182	0.4357	1.8739	326	34.59	31.13	25.68
3.9381	0.4399	1.8612	326	34.59	31.15	25.74
3.9593	0.4442	1.8654	326	34.59	31.16	25.74
3.2106	0.4040	1.9035	281	34.48	30.93	25.40
3.1894	0.3989	1.8824	281	34.48	30.92	25.40
1.9500	0.3316	1.9331	254	34.40	30.64	24.81
1.9247	0.3244	1.8993	254	34.40	30.63	24.81
1.9966	0.3350	1.9712	254	34.40	30.63	24.81
1.0321	0.2826	2.1700	204	34.25	30.21	23.91
0.9983	0.2750	2.1235	204	34.25	30.21	23.90
0.4090	0.2377	2.5930	149	34.07	29.68	22.62
0.4437	0.2517	2.7326	149	34.07	29.68	22.65
0.1730	0.2356	3.4982	100	33.88	29.18	21.27
0.1700	0.2314	3.5701	100	33.88	29.15	21.24
0.0711	0.2348	5.1902	55	33.69	28.59	19.80
0.0656	0.2466	5.4229	55	33.69	28.60	19.67
0.0651	0.2335	5.2410	55	33.69	28.58	19.70

The concentration ratio of $CHCl_3$ to CCl_4 is 7.5 moles of $CHCl_3$/mole CCl_4.
The concentration ratio of moles of CH_2Cl_2 to CCl_4 is 60,000 moles CH_2Cl_2/mole CCl_4.

in Figure 5.3. The electron affinity for acetophenone determined from a simple linear least squares is 0.333 ± 0.02 eV and the intercept 12.6 ± 0.2. The activation energy for the electron attachment to CCl_4 was zero, and the values for CH_2Cl_2 and $CHCl_3$ were 0.33 ± 0.02 eV and 0.14 ± 0.02 eV, respectively. These are in agreement with previous determinations and indicate that the electron energy distribution is thermal [12–14].

The combination of a gas chromatograph with the ECD or negative-ion mass spectrometer is very important. This ensures that only a small amount of highly purified sample is responsible for the response. However, in the case of compounds that are not highly reactive with thermal electrons, it must be certain that the peak observed in the ECD or NIMS is due to the major component in the sample. This can be accomplished by using a nonselective detector in parallel or series. A good detector for this purpose is the flame ionization detector or a pulsed discharge ionization detector. In the case of the mass spectrometer the positive-ion mode can be used to locate the major component. It is also important to verify that the specific detector gives the same type of results as have been obtained previously. Good DEC(1) compounds for this purpose are CCl_4, $CHCl_3$, and CH_2Cl_2. Acetophenone, nitrobenzene, and $C_6F_5NO_2$ are good Eql(1/1) and Eql(1/2) compounds, whereas C_6F_6, C_6F_5Cl, and CS_2 are good Eql(2/2) compounds [11]. These compounds have been chosen to give ranges of values for the ECD parameters.

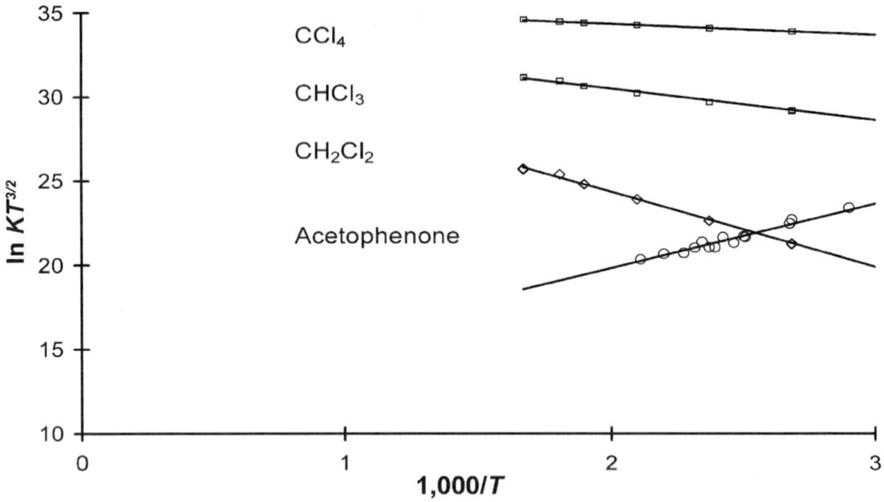

Figure 5.3 Plots of ECD data as ln $KT^{3/2}$ versus 1,000/T. The data are typical for the chloromethanes. An ECD should be able to reproduce these data. They were obtained using a single injection at each temperature. The calculations of the data are illustrated in Table 5.2 using carbon tetrachloride as an internal standard. The data for acetophenone were obtained in a separate experiment.

The first step in verifying the validity of data from a specific detector is to determine the relative responses of the above compounds at a fixed detector temperature. This can be accomplished easily for a solution designated A (CCl_4, $CHCl_3$, CH_2Cl_2, C_6F_6, and CS_2) at one column temperature and another designated B (naphthalene, acetophenone, nitrobenzene, C_6F_5Cl, and $C_6F_5NO_2$) at a higher column temperature. The detector temperature should be stabilized at about 423 K(150°C). Then appropriately prepared solutions should be injected and the relative molar responses calculated. It is best to prepare solutions of only one component for this constant-temperature determination. In this manner the concentration of CCl_4 in the $CHCl_3$ or CH_2Cl_2 solutions can be determined using the ECD data. The molar response for maximum capturing compounds such as CCl_4 and $C_6F_5NO_2$ will be relatively temperature-independent. From values of $K = 2.1 \times 10^{11}$ L/mole and $k_1 = 4.2 \times 10^{14}$ L/mole-s for CCl_4, the value of $k_D = k_1/2K = 1{,}000\,\mathrm{s}^{-1}$ can be calculated. This is a nominally "clean" ECD. At this temperature the relative responses should be 8,000:800:1 for CCl_4: $CHCl_3$: CH_2Cl_2. The responses of C_6F_6 and CS_2 should be about the same as that for $CHCl_3$. Compounds with higher molecular weights in solution B will be chromatographed at a higher column temperature. The molar response of $C_6F_5NO_2$ should be roughly the same as that for CCl_4, whereas C_6F_5Cl and $C_6H_5NO_2$ should be an order of magnitude lower. For acetophenone the response should be three orders of magnitude lower, for naphthalene six orders of magnitude lower. Once it is established that the detector gives the proper relative response and is relatively "clean," the temperature dependence of the above compounds should be measured. If the actual electron currents are not measured, then the linearized response of the detector can be used to obtain fundamental properties. If the response for a given compound is too large, then the concentration must be reduced or the size of the injected sample decreased. A convenient technique to adjust for variations in the standing current, or I_b, is to measure the response of a compound relative to an internal standard, for example, CCl_4 or $C_6F_5NO_2$, and to adjust for the concentration difference. The reported response for the reference compound is used to obtain absolute values, as illustrated in Table 5.2.

The general procedure for collecting data is as follows for a mixture of CCl_4, $CHCl_3$, and CH_2Cl_2. Prepare 25 mL of a standard solution of CCl_4 in a nanograde solvent. Inject about 50 femtograms of CCl_4 into the detector at 423 K. Determine the area per mole response of the detector. Determine a response factor for CCl_4 such that area/mole $= RF \times 10^{11}$ L/mole. This response factor should be about 1.0 if the area is calculated using $\{[I_b - I_e]/I_e\}_{\max}\{t_{1/2}\}V$. Record the values of I_b or f_0, the base frequency. The frequency modulated mode is commonly used in commercial detectors. Prepare a 25-mL solution of $CHCl_3$ that is about 10 times that of CCl_4 and determine the molar response. Be sure to note any CCl_4 that appears in this solution and determine its concentration using the molar response determined above. Finally, prepare a solution of CH_2Cl_2 10,000 times that of CCl_4 and analyze that mixture for $CHCl_3$ and CCl_4 using the measured response factors. That solution can often be used to determine relative responses from 150°C to room temperature. If there is insufficient CCl_4 or $CHCl_3$ in this solution, then mix known amounts of the standard solutions to prepare a known concentration of each from the stock solutions. Raise the temperature of the detector to its highest temperature and condition the detector. Measure the temperature of the detector and determine the relative response factor of the compounds. Gradually lower the temperature, record the temperature, and determine the relative responses. In the past this has been done with a strip chart recorder, but with modern computers and data storage, the information can be merely stored for later data reduction. In the original work the temperature of the detector was recorded with a thermometer inside the detector. However, temperatures from a calibrated thermocouple can now be recorded continuously as the sequential samples are injected. As the temperature of the detector is lowered, the relative response of CH_2Cl_2 will decrease so that different solutions will be required to remain in the linear detectable region. Compare the results with those in Table 5.2. If the temperature dependence is about the same, the detector is suitable for fundamental studies.

The data collection procedure for negative-ion mass spectrometers is very similar to that for the ECD. The ion signals are directly proportional to the concentration. The molar response will be the integrated area divided by the number of moles in the gas phase. The number of moles will cause a problem because the exact concentration in the gas phase is not easily measured in a GC/MS. Consequently, the response to a given compound with a known molar response is used to calibrate the detector. The procedures will be the same as for the ECD described earlier. Table 5.3 provides data for SF_6, nitrobenzene, and m-dinitrobenzene determined

TABLE 5.3 Calculation of K for SF_6, $C_6H_5NO_2$, and $C_6H_4(NO_2)_2$ Using $C_6H_4(NO_2)_2$ as a Calibration Point

T	Area R	Area 1	Area 2	$\ln KT^{3/2}$	$\ln KT^{3/2}$	$\ln KT^{3/2}$
373	310	145	371	34.62	33.86	32.50
398	192	185	294	34.24	34.20	32.36
423	176	200	296	34.24	34.37	32.46
448	147	210	286	34.15	34.50	32.51
473	122	200	303	34.04	34.54	32.65
498	108	210	280	34.00	34.66	32.65
523	91	250	295	33.90	34.91	32.77
548	78	190	182	33.82	34.71	32.36
573	72	146	94	33.80	34.51	31.77
598	45	90	58	33.40	34.09	31.35

The concentration of nitrobenzene is a factor of 10 lower than that of m-dinitrobenzene, whereas that of SF_6 is about the same.

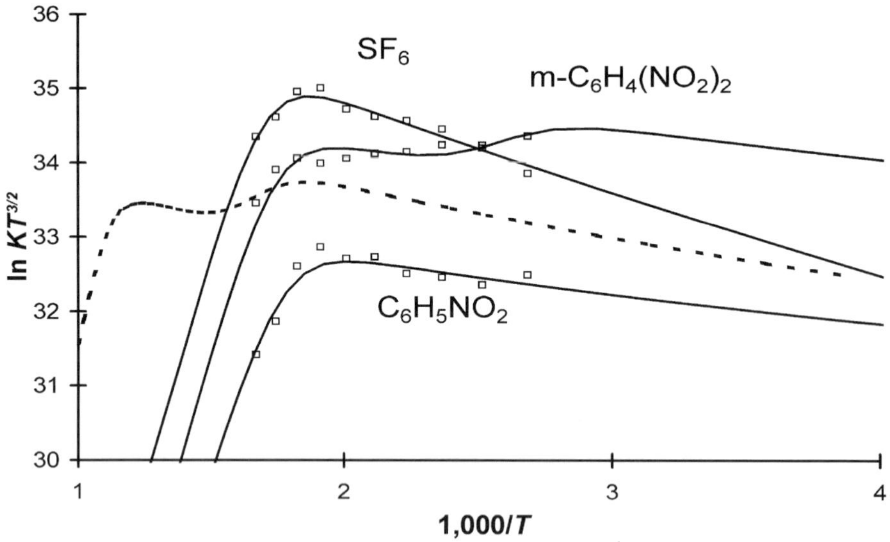

Figure 5.4 Negative-ion mass spectrometry data plotted as $\ln KT^{3/2}$ versus $1{,}000/T$ for nitrobenzene m-dinitrobenzene and sulfur hexafluoride. The magnitude was scaled to the value of the k_1 for SF_6 at room temperature. The data for m-dinitrobenzene exhibits two states with α and β regions. Neither of these is the ground state. The curves are calculated using the measured electron affinities of SF_6 and $C_6H_5NO_2$ and m-$C_6H_4(NO_2)_2$. The Q_{an} values are determined by the data to be 1.0. The responses were obtained by injecting a solution with a known amount of the three compounds into the mass spectrometer. These data were collected at higher temperatures than those in Figure 4.5.

using an API source [7]. The E_a of SF_6 and nitrobenzene can be determined from the NIMS data. A solution containing equal moles of SF_6 and m-dinitrobenzene and a factor of 10 times as many moles of nitrobenzene was injected into an atmospheric pressure source with Ar/10% methane as the carrier gas. The parent ion intensities were measured at a series of temperatures and are recorded in Table 5.3. The scale factor was determined to yield a value of $\ln(AT^{3/2})$ of about 34 for m-dinitrobenzene at the lowest temperature. The same scale factor was then used for the other data. Because the relative areas are multiplied by the same scale factor, the actual concentrations in the gas phase need not be known. The relative responses will only depend on the concentrations. The data are plotted in Figure 5.4. The E_a for m-dinitrobenzene measured utilizing TCT experiments is 1.65 eV and was used in standard two-state equations to calculate the dotted curve at higher temperatures. Excited-state E_a of 1.0 ± 0.05 eV and 0.67 ± 0.05 eV fit the data at lower temperatures. The E_a of nitrobenzene is determined to be 1.00 ± 0.06 eV, whereas that of SF_6 is determined to be 1.07 ± 0.07 eV. The activation energies for attachment to the ground state are 0.06 ± 0.02 eV for SF_6 and 0.02 ± 0.02 eV for nitrobenzene. All the Q_{an} are 1.0 ± 0.1 and the $\ln A_1$ are 35.8 ± 0.2 for SF_6, 34.8 ± 0.2 for m-dinitrobenzene, and 33.5 ± 0.2 for nitrobenzene. In a separate NIMS determination the E_a for nitrobenzene was determined to be 1.0 ± 0.07 eV, the $\ln A_1$ 33.0 ± 0.5, and the Q_{an} 0.9 ± 0.2.

5.3 REDUCTION OF ECD DATA TO FUNDAMENTAL PROPERTIES

5.3.1 Introduction

The data reduction procedure for ECD data is simple if there is only one temperature region. In the case of compounds that only have one α region, a simple least-squares procedure can determine the slope and intercept. As shown in Figure 4.1, from a plot of $\ln KT^{3/2}$ versus $1{,}000/T$, the electron affinity is the slope times R and the quantity $Q_{an}(A_N/2A_D)$ is the intercept. Recall that Q_{an} is the partition function ratio, whereas A_N/A_D is the ratio of the recombination rate constants. The problem is to establish that only one α region is present. This is normally done by comparing the intercept with an average value or the value for a compound with a Q_{an} of 1. Alternatively, if the temperature dependence for acetophenone is obtained for an untested detector, the established electron affinity of acetophenone, 0.338 ± 0.002 eV, can be used to calibrate the intercept. The intercept will be an upper limit for unknown compounds. Lower values can be attributed to a lower partition function ratio. For compounds that have a higher intercept the value for acetophenone should be used to determine the E_a. If this quantity has not been estimated, then the nominal intercept of 12.43 ± 1.0 can be used to obtain an E_a. This assumes a $Q_{an} = 1$ and $(A_N/A_D) = 2.0$. Likewise for DEC(1) compounds with only a negative slope, such as CCl_4, the instrument can be calibrated for a value of A_D since E_1 is zero, A_1 is known, and $K = A_1/(2T^{1/2}A_D)\exp(-E_1/RT)$. The graphs in Figure 5.3 illustrate these determinations.

The temperature dependence of the combined mechanisms is more complicated, with as many as eight potential parameters that can be obtained from the temperature dependence. These parameters are the activation energies and pre-exponential terms of the respective rate constants. In the case of compounds with one α and one β region the four parameters are A_1, E_1, Q, and E_a. For compounds with one excited state there are eight parameters, four for each state. Sufficient structure must exist in the ECD data to obtain parameters from it, or independent estimates of some of the parameters must occur.

For C_6F_6, CS_2, and anthracene the electron affinity of the excited state was included in the data reduction procedure with its appropriate errors to define the other parameters. Estimates of the parameters can be obtained from the global plots of the ECD data as ln $KT^{3/2}$ versus $1{,}000/T$. Then the data can be reduced using a rigorous nonlinear least-squares procedure including data from other techniques and their respective errors. This type of data reduction can be completed with standard statistical packages, but has also been implemented in Visual BASIC using EXCEL. A complete description of the general least-squares procedures is given in the appendix, although we will examine examples of adjustments in this section [11, 15–18].

5.3.2 Acetophenone and Benzaldehyde

In Figure 5.5 we demonstrate the use of two parameters to determine the electron affinity of acetophenone, the most precisely determined ECD electron affinity. In 1964 we also measured the ECD data as a function of reaction time, but did not know how the slope would change. Now we know that the slope gives the electron affinity, independent of reaction time. Figure 5.5, shows the plots of ln $KT^{3/2}$ versus $1{,}000/T$ for data taken at 1,000, 1,000, 100, 75, and 50 μs reaction times fit with two parameters. Table 5.4 gives the slopes, intercepts, their uncertainties, and the covariance σ_{ab} terms (see the appendix for the significance of the covariance term). The simple average is 0.348(27). The weighted average is 0.338(2) eV. If this value is used in the least-squares adjustment, the intercepts can be determined more precisely. Figure 5.6 is a plot of the data obtained using the average value of the E_a. In the appendix a numerical example of the calculations for one set of data and the extension of those data by including the average value of the E_a to determine the intercept are given. This shows the consolidation of data from different experiments, which could also be accomplished by simply solving equation 4.13 with a fixed electron affinity at each temperature and taking the weighted average of the values of $\ln(A_N A_1/(2A_D A_{-1}))$. However, the uncertainties in the various parameters will not be obtained and the weighting would be complicated. The simple average of the values obtained for the 50-μs data for acetophenone with an E_a of 0.338 eV is 11.51(24) versus the rigorous weighted value of 11.71(11):

$$\ln KT^{3/2} = \ln(A_N A_1/(2A_D A_{-1})) + E_a/RT \tag{4.13}$$

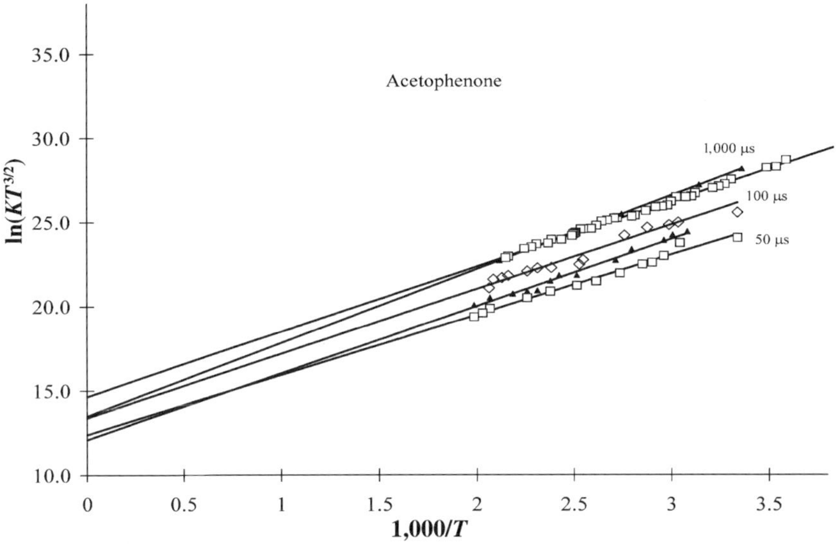

Figure 5.5 Plots of ECD data as ln $KT^{3/2}$ versus 1,000/T for acetophenone taken at different times: 1,000, 1,000, 200, 100, and 50 μs. The lines were calculated by linear least squares through the data. The two intercepts at 1,000 μs are different because of the different amounts of data. The parameters are given in Table 5.4.

The majority of the earlier ECD data were collected at 1,000 μs and established that the average value of the intercept of 14.7(1) corresponds to a ratio of $A_N/A_D = 20$. The intercepts for acetophenone and benzaldehyde were approximately this value, which is $Q_{an} = 1.0$. At the lower reaction times this intercept is clearly lower, as indicated in Figures 5.5 and 5.6. Such implies that the ratio of $A_N/A_D = 8$ at 100 μs and approaches 1 at 50 μs. The intercept can be used to obtain a value of A_N/A_D by determining the intercept from the temperature dependence of the ECD response for acetophenone [11, 18].

5.3.3 Benzanthracene, Benz[a]pyrene, and 1-Naphthaldehyde

If there are sufficient data in either the α or β region, or both, then linear lines can be drawn to determine the slopes and intercepts as well as the parameters A_1, E_1, Q,

TABLE 5.4 Least-square Parameters for Acetophenone as a Function of Reaction Time

Time (μs)	Points	I	E_a (eV)	S_{ab}	$I(E_a$ fixed)	$S_{ab}(E_a$ fixed)
50	12	12.36(32)	0.310(7)	−40	11.71(11)	−3.8
75	12	12.05(40)	0.346(10)	−66	12.35(10)	−3.5
100	14	12.11(50)	0.366(13)	−107	12.88(20)	−14
1,000	4	13.30(55)	0.379(12)	−108	14.85(30)	−21
1,000	36	14.66(12)	0.337(3)	−5.7	14.69(5)	−0.67

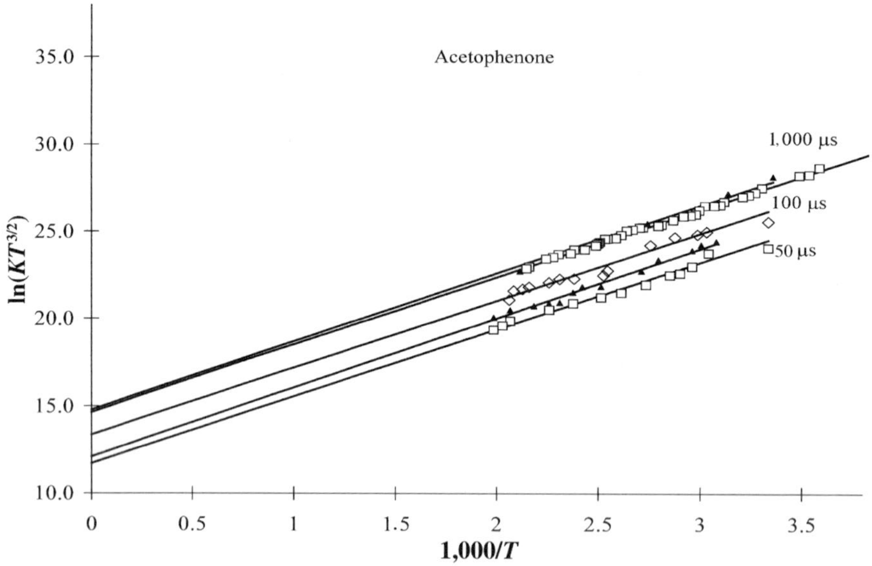

Figure 5.6 Plots of ECD data as ln $KT^{3/2}$ versus $1{,}000/T$ for acetophenone taken at different times: 1,000, 1,000, 100, 75, and 50 μs. The lines were calculated by fixing E_a to the weighted average value. The two intercepts at 1,000 μs are now the same. The value of the intercept at 50μs is that calculated from fundamental constants, 11.73, indicating that the recombination rates of the ions and electrons are equal.

and E_a. Figure 5.7 is a plot of the data for benzanthracene, benz[a]pyrene, and 1-naphthaldehyde at 1,000 μs and 100 μs. For the hydrocarbons there is one point that could be in the β region, as indicated by the equality of the last two data points. The slope through the higher-temperature points for benzanthracene gives an E_a of 0.71 ± 0.04 eV, while the slope through all four points gives an E_a of 0.61 ± 0.06 eV and an intercept higher than the average value. By fitting the data to all points using the four-parameter equation, the quantities are $Q_{an} = 0.25 \pm 0.10$; $E_a = 0.71 \pm 0.04$ eV; $E_1 = 0.03 \pm 0.02$ eV, and $\ln A_1 = 34.0 \pm 0.15$. For benz[a]pyrene the E_a using a linear plot through all points is 0.58 ± 0.07 eV with an intercept of 18, which is higher than the average value of 14.7. By dropping one data point, the E_a increases to 0.81 ± 0.07 eV and the intercept drops to 13.5, indicating a lower $Q_{an} = 0.33(3)$. With these values the $E_1 = 0.01 \pm 0.02$ eV and $\ln A_1 = 35.2 \pm 0.10$. Both the higher and lower fixed intercept values were originally reported. We now know that when the slope through the data gives an intercept significantly lower than the average intercept, the higher E_a is the accurate value. The data taken at two reaction times for 1-naphthaldehyde can be reconciled using the larger $E_a = 0.72 \pm 0.07$ eV. By fitting both sets of data, the parameters are $E_a = 0.72 \pm 0.05$ eV and $Q_{an} = 0.25(8)$. If we use these values, the $E_1 = 0.06 \pm 0.02$ eV and $\ln(A_1) = 32.64 \pm 0.1$. The calculated curves are shown in Figure 5.7. Note that the intercepts for the aromatic

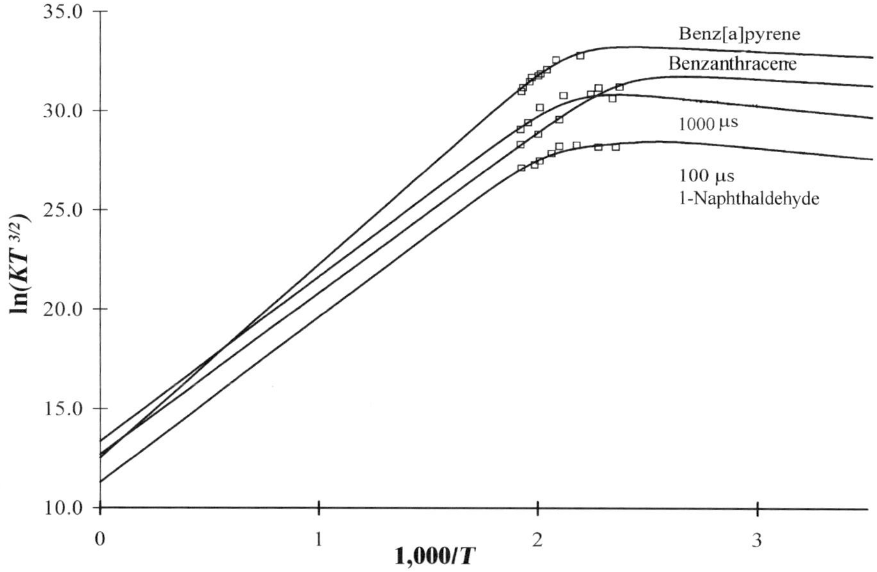

Figure 5.7 Plots of ECD data as ln $KT^{3/2}$ versus $1{,}000/T$ for benz[a]pyrene, benzanthracene, and 1-naphthaldehyde. The latter was taken at 1,000 and 100 μs. The limiting slope is used to determine the E_a. The intercept for benz[a]pyrene gives a Q_{an} less than 1.

hydrocarbons at 1,000 μs fall between the two intercepts for 1-naphthaldehyde separated by two units or a factor of 10 [11, 18].

5.3.4 Carbon Disulfide

In Figure 4.3 the ECD data for C_6F_6 and CS_2 were shown and the parameters given for the high-temperature region [11, 15]. Unlike the case for the m-dinitrobenzene data shown in Figure 5.4, there are other estimates for the excited state E_a for these two compounds. Although the data clearly indicate two states, we cannot discern the parameters for the excited state. However, by using the estimate of the excited state E_a from other methods, the parameters can be obtained. In Figure 5.8 the ECD data for CS_2 are shown with the two resolved states. The initial estimates for the low-temperature data were obtained by fitting the data in the β region for the excited state with two linear plots, as shown in Figure 5.8. Using the same A_1 value for the ground state, we were able to approximate the E_1 by a linear plot. Finally, a line through the high-temperature data was drawn and the slope and intercept determined. Table 5.5 provides the first approximations determined from the linear plots, the final values obtained from the general eight-parameter least squares, and the values reported earlier for a different set of data. All values agree within the experimental uncertainties. This is a good example of two states

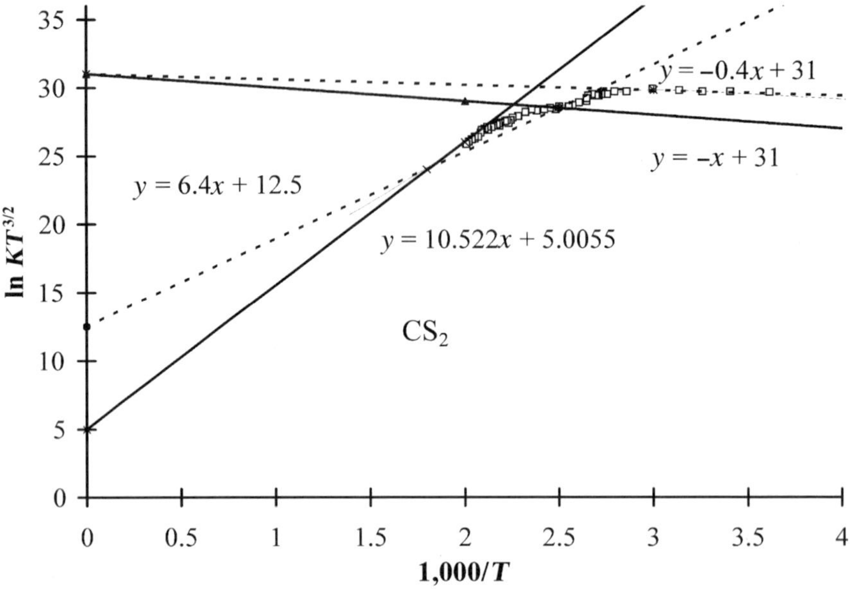

Figure 5.8 Plots of ECD data as ln $KT^{3/2}$ versus $1{,}000/T$. Approximate lines in four regions for CS_2 give the E_a, E_1, A_1, and Q_{an} for two states. These are used as first approximations in a nonlinear least-squares analysis. By including measured values of the E_a, more precise values of the other parameters are obtained. The final results are given in Table 5.5.

observed in a single ECD experiment. Other examples are C_6F_6, anthracene, tetracene, and C_6F_5Cl [11].

5.3.5 Nitromethane

A similar analysis was carried out for two sets of nitromethane data compiled by two different students years apart. Here the two temperature regions involve

TABLE 5.5 ECD Parameters for Carbon Disulfide

Form	ln A_1	E_1 (eV)	E_a (eV)	Q_{an}	Intercept	Source
Linear	—	0.04	—	—	32	lines
Linear	—	—	0.62	0.40	11.5	lines
Bent	—	0.10	—	—	32	lines
Bent	—	—	0.89	1×10^{-3}	5.3	lines
Linear	30.9(3)	0.10(1)	0.88(3)	1.6×10^{-3}	5.3(7)	least squares
Bent	32.0(3)	0.04(1)	0.62(2)	0.3	10.4(7)	least squares
Linear	31.2(4)	0.10(1)	0.87(3)	1.0×10^{-3}	6.1(7)	published
Bent	31.9(2)	0.03(2)	0.61(4)	0.4	11.5(7)	published

TABLE 5.6 ECD Parameters for Nitromethane

Parameter	Data Set 1	Data Set 2	Fixed E^*	Estimated
ln A_1	36.31 (fixed)	36.31(fixed)	35.55(92)	36
E_1 (eV)	0.20 (fixed)	0.20 (fixed)	0.16(1)	0.2
E_a (eV)	0.49(2)	0.50(2)	0.49(1)	0.5
Q_{an}	6×10^{-3}	6×10^{-3}	6.5×10^{-3}	10^{-2}
ln A_2	29.33(41)	28.35(160)	28.70(24)	29
E_2 (eV)	0.99(5)	0.99(10)	1.00(2)	1.00
D(CN) − $E_a(NO_2)$ (eV)	0.30(5)	0.29(10)	0.35(1)	0.25

molecular ion formation and dissociative electron attachment via the intermediate molecular negative ion. Assumed values for A_1 and E_1 were fixed to fit the data to the equation with six parameters. The other four parameters were obtained from the ECD data. The results of this analysis have been published and the activation energy for dissociative electron attachment determined from E_2. The electron affinity of the NO_2 radical and the bond dissociation energy were both measured so that the EDEA could also be calculated. By using the value of E^* and its uncertainty, the other five parameters can be estimated and their uncertainties determined. In Table 5.6 the parameters obtained from the two sets of data are compared with a new fit using the experimental value of $E^* = D(\text{C—N}) - E_a(NO_2) = 2.65 - 2.27 = 0.38 \pm 0.05$ eV. The values are clearly the same within the experimental uncertainties, but now estimates of the E_a are more precise. In Figure 5.9 the data and curves obtained for these data are plotted for one set of nitromethane data. Also shown in Figure 5.9 are linear plots through the different regions, which can be used to estimate parameters. Linear plots give an E_a of 0.5 eV and activation energy of 0.35 eV. The final least-squares values are 0.50(2) eV for the E_a and 0.35(1) eV for the activation energy. The simple graphical method yields reasonable results [19].

5.3.6 Consolidation of Electron Affinities for Molecular Oxygen

The superoxide anion was the first homonuclear diatomic anion observed. According to a 1953 review, the AE_a of O_2 was 0.9 eV and low-energy electron attachment leads to an excited state. Experimental values now range from 0.15 to 1.07 eV: photodetachment, with E_a of 0.1_5, 1958; CTC, 0.7_5, 1961 to 1971; electron swarm, 0.4_5, 1961; EB, 0.6, 1_{-1}, 1961 to 1971; AMB, 0.2, 0.3, 0.5, 0.8, 1_{-3}, 1970 to 1977; IB, 1.0_7, 1970; PES, 0.43, 0.45, 1971 to 1995, and ECD, 0.4_5, 0.5, 0.9, 1970 to 2002. This large range of values indicates significant differences beyond random uncertainty [20–33]. The complete temperature dependence for a given state in the ECD is determined by four parameters: the E_a, the partition function ratio Q_{an} for the anion and neutral, A_1, and E_1. The four parameters for the excited state with an E_a of 0.430, 0.450 ± 0.002 eV have been determined by the swarm and PES methods. In the PES method two spin orbital coupling states

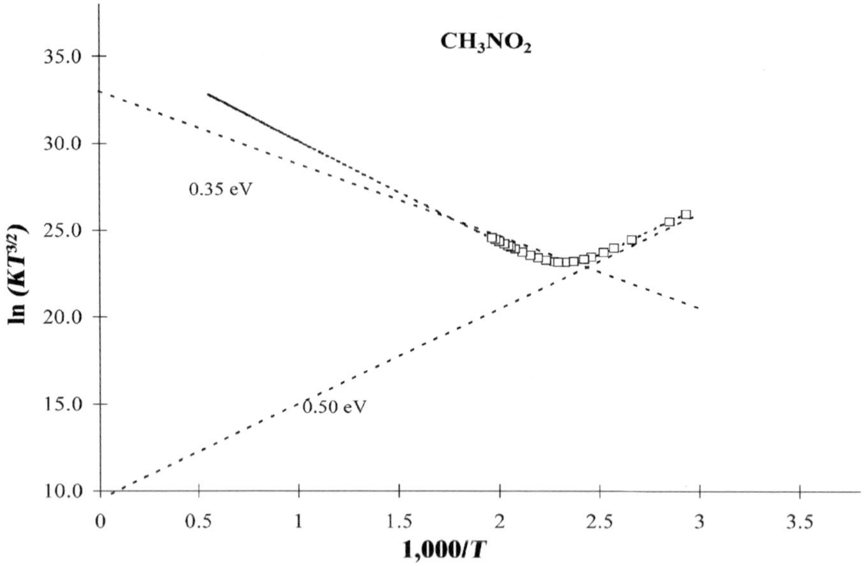

Figure 5.9 Plots of ECD data as ln $KT^{3/2}$ versus $1{,}000/T$ for nitromethane. The limiting slope at low temperatures is used to determine the E_a. The intercept gives a Q_{an} less than 1. The high-temperature slope gives the quantity D(C—N) – $E_a(NO_2)$. By using the measured value of this quantity in the least-squares analysis, more precise values of the other parameters are obtained. These are given in Table 5.6.

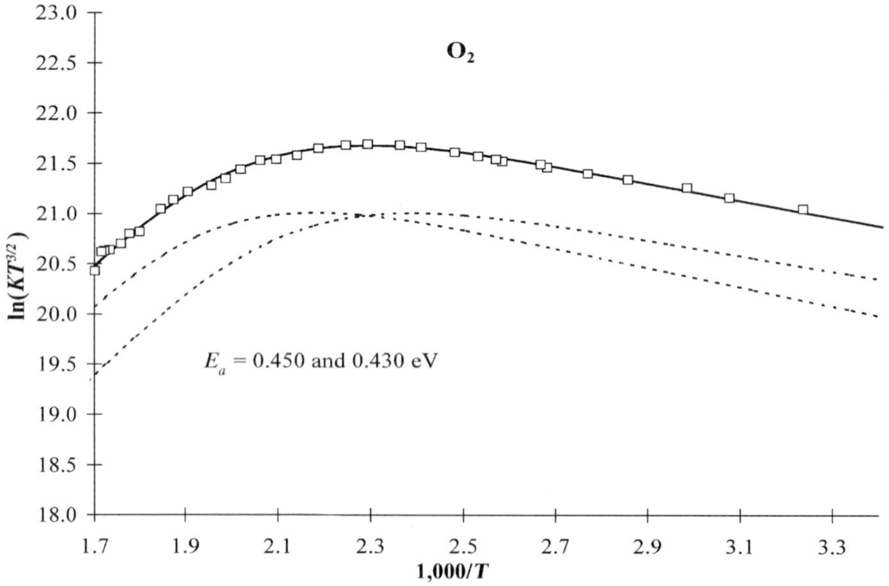

Figure 5.10 Plots of ECD data as ln $KT^{3/2}$ versus $1{,}000/T$ for molecular oxygen. The E_a of two spin orbital coupling states are resolved by fixing the E_a for the two states.

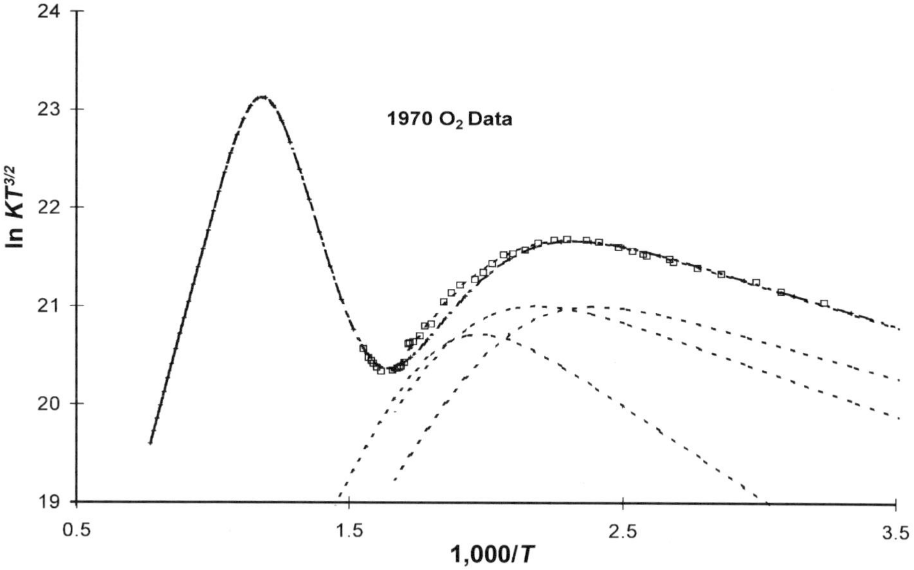

Figure 5.11 Plots of ECD data as ln $KT^{3/2}$ versus $1{,}000/T$ for molecular oxygen. The E_a of the low-temperature data are used to estimate the E_a of two additional excited states based on higher E_a measured in other experiments.

were resolved [22, 29]. The low-temperature ECD data compiled in our laboratory agree with these values, and the sum of the squares of the deviations is reduced when two spin orbital coupling states are assumed. However, in the ECD data there is an increase in the response at higher temperatures, indicating excited states [22].

Figure 5.10 is a plot of the ECD data with the curves calculated using the experimental parameters measured in other experiments, including the high-resolution PES data that resolve the two spin orbital coupling states. By using these values and their associated uncertainties, it is possible to identify excited states at higher temperatures. The excited-state parameters that have not been measured in other experiments are obtained by sequentially fitting the data with the values that have been measured. Two additional postulated curves are shown in Figure 5.11. The quantities that define these states will be discussed further in Chapter 9.

5.4 REDUCTION OF NEGATIVE-ION MASS SPECTRAL DATA

We have utilized two different negative-ion mass spectrometers. The first was an Extranuclear Spectra EL quadrupole mass spectrometer equipped with a custom atmospheric pressure ionization source. The data were obtained with a custom data collection system or on an X-Y recorder. The temperature was measured using a thermocouple imbedded in the ion source. The carrier gas was a mixture of argon and methane. Other details of this apparatus can be found in various publications and theses [5–10].

In Figure 5.4 NIMS data for SF_6, m-$C_6H_4(NO_2)_2$, and $C_6H_5NO_2$ are plotted. These data were collected in the 1980s and have not been published [7]. The majority of the data are in the β region for all three compounds. The data for SF_6 and $C_6H_5NO_2$ have small slopes at low temperatures. However, structure exists in the low-temperature data for m-$C_6H_4(NO_2)_2$. In addition, the E_a indicated by the slope in the high-temperature region is lower than the E_a for m-$C_6H_4(NO_2)_2$ measured by TCT. From these data it was suspected that there are two excited states. The E_1 and A_1 for SF_6 and $C_6H_5NO_2$ were determined from the low-temperature data and the E_a fit with a value of $Q_{an} = 1$. This essentially fixes the intercept. Each high-temperature data point defines a value of the E_a. The actual reported values were fit to all the data.

The second system used was a Hewlett Packard 5988A GC/MS equipped for chemical ionization negative-ion mass spectra. We carried out our studies in the early 1990s. E. A. Stemmler and R. A. Hites used a similar system to collect mass spectra for over 300 compounds at 373 K and 523 K [10]. In our experiments CO_2 was used as the cooling gas, whereas methane was used in the other studies in the literature. The pressures in the ion source were about 1 torr in each case. The electron energies were thermal, as indicated by the absence of dissociation of molecules with large EDEA. In the case of our studies this was also verified by the ratio of SF_5(-)/SF_6(-) measured as a function of temperature. A small change in this ratio at temperatures below 473 K indicates thermal electrons. The temperature was measured by a thermocouple in the block. The data were collected using the commercial data system. The details of the instruments' calibration can be found in the original publications.

The 1-torr CO_2 chemical ionization mass spectra were taken after the instrument was tuned using peaks for perfluorotributylamine, (PFTBA). The tuning was optimized at 473 K and kept the same throughout the temperature variations. Tuning profiles were checked to verify the results at different temperatures. The spectra were taken from a single mixture containing SF_6, C_2Cl_4, and C_2HCl_3 in concentrations to give peaks in the linear range at low temperatures. Only one solution was used so that at higher temperatures the parent negative ion of the C_2HCl_3 was not observed. The ion source temperature was allowed to equilibrate for approximately 45 min. Integrating the ion intensities for each of the isotopes over the chromatographic peak and calibrating to the SF_6 peak at low temperatures give the molar response. This assumes that the instantaneous negative-ion concentration is proportional to the concentration of the neutral, and that the ionic reactions are fast relative to the elution time of the chromatographic peak [9].

Figure 5.12 illustrates the negative-ion abundances for SF_6, C_2Cl_4, and C_2HCl_3 measured as a function of temperature using this system. In the original analysis only one state was assumed and the slope and intercepts determined using four

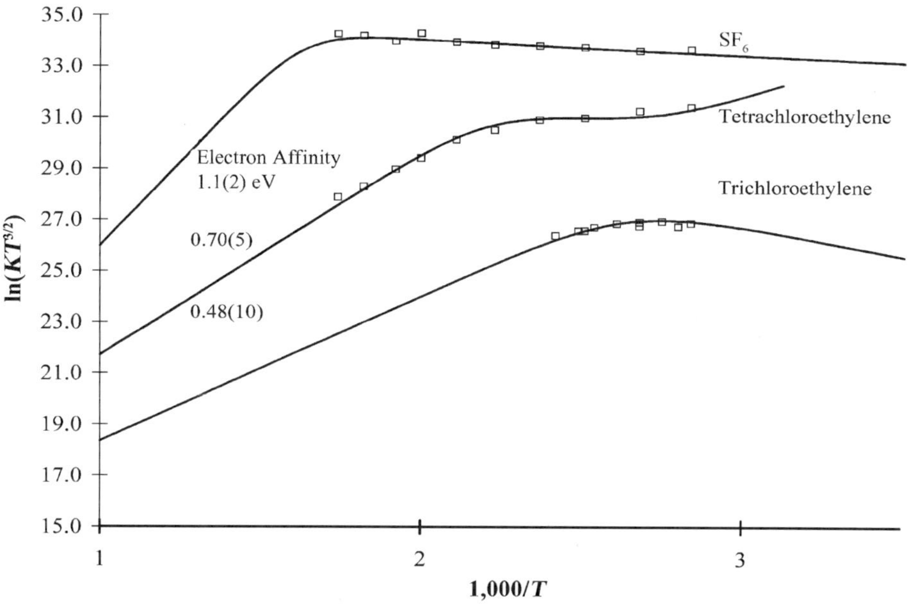

Figure 5.12 Negative-ion mass spectrometry data plotted as ln $KT^{3/2}$ versus $1{,}000/T$ for tetrachloroethylene, trichloroethylene, and sulfur hexafluoride. The magnitude was scaled to the value of k_1 for SF_6 at room temperature. The data for the chloroethylenes exhibit two states with α and β regions. The Q_{an} values are determined by the data to be 1.0. The responses were obtained by injecting a solution with a known amount of the three compounds into the mass spectrometer. These data were collected with a chemical ionization mass spectrometer.

parameters. The intercepts for C_2Cl_4 and C_2HCl_3 were higher than for SF_6 when the same scale factor was used. Now it is believed two bound states contribute to the ECD and NIMS data and unimolecular dissociation takes place. Thus, three low-lying curves are involved. If we use two states and fix the Q and A_1 values, two electron affinities for C_2Cl_4 and C_2HCl_3 are obtained. The ground-state E_a are 0.70(5) eV and 0.48(10) eV, while the first excited-state values are 0.52(10) eV and 0.35(10) eV. The activation energies for C_2HCl_3 are 0.18(5) eV for both states. The E_1 for the ground state for C_2Cl_4 is 0.10(5) eV and for the excited state is 0.01(5) eV [9].

These results suggest a method for converting the NIMS data in the literature to limits for the electron affinities. For this particular system the parent negative ion is observed at the higher temperature of 523 K and the lower temperature of 373 K when the electron affinity is greater than 0.7 eV. If the parent negative ion is not observed at the higher temperature but is observed at the lower temperature, 373 K, the electron affinity should be lower than 0.5 eV. If the parent negative ion is not observed at either temperature, the electron affinity of the molecule cannot be estimated. The parent negative ions of benz[a]pyrene, fluoranthene, and benz[g,h,i]perylene are observed at both the high and low temperatures. The electron affinities of these molecules are 0.80(3) eV, 0.82(4) eV, and 0.90(10) eV, all in agreement with the value of greater than 0.70 eV. The parent negative ions of chlorobenzene and chloronaphthalene are not observed at either temperature. Their electron affinities are both less than 0.5 eV. For other molecules such as dimethyl phthalate, the parent negative ion is observed at 373 K but not at 523 K. The electron affinity is 0.60(10) eV. Using these examples, we determine that the data will provide an approximate value of the E_a. Substitution effects or CURES-EC calculations can refine these estimates. For example, the electron affinities of the nitrophenols have not been measured. The parent negative ion is the base peak at both temperatures, implying an electron affinity greater than 0.7 eV. The fact that they are hydroxy-substituted nitrobenzenes gives an electron affinity greater than 1.0 eV. The addition of a hydroxyl group to an aromatic molecule increases the E_a by about 0.2 to 0.4 eV. This makes the electron affinities of the nitrophenols between 1.2 and 1.4 eV. The CURES-EC calculated values for the meta, ortho, and para nitrophenols are 1.25, 1.35, and 1.45 eV, with an uncertainty of 0.1 eV.

Some compounds undergo dissociative electron attachment and cannot be analyzed using this procedure. An HF molecule is eliminated in pentafluoroaniline with an electron affinity of about 0.8 eV and the parent negative ion is not observed at either temperature. For C_6F_5I with E_a greater than 1.5 eV, dissociative electron attachment predominates over molecular ion formation and the largest peak in NIMS is the C_6F_5 radical anion. The acetyl derivative of m-nitrophenol with an E_a greater than 1.0 eV forms an acetate anion radical and the parent negative ion is not observed. The acetyl derivative of 2,4, dinitrophenol loses an acetate radical and forms the dinitrophenyl radical anion and the parent negative ion is not observed. Another example is the dissociative electron attachment to tetrabromoethylene. The parent negative ion of this compound is not observed at the high or low temperature. The base peak is the 158, 160, 162 triplet for $Br_2(-)$, while the (M—Br)(−) and Br(−) ions are of much lower intensity. The elimination of $Br_2(-)$ from the parent negative ion is preferred over the simple dissociation to give the Br(−) ion. The absence of a parent negative ion is not just evidence of low electron affinity. A parent negative ion can undergo dissociative electron attachment, involve ion molecule reactions, as well as detach an electron due to its low electron affinity.

5.5 PRECISION AND ACCURACY

The concepts of precision and accuracy, "P and A plots," and timelines are applied to the evaluation of the accuracy and precision of the electron affinities of selected atoms and molecules. The adiabatic electron affinities of the elements have been measured with a variety of techniques. Thus, the most accurate and precise values will be the weighted averages, which is also the least-squares solution.

The measurement of any property or measurand is the result of a series of steps that result in an end value. The required accuracy of the measurand is dependent on its use, but the actual uncertainty, that is, its deviation from the truth, is unknowable. However, the reported value of a measurand cannot be used effectively without some estimate of this uncertainty. Thus, limits to the error must be inferred from the repeatability or precision of the method of measurement and from reasonable estimates of the bias or systematic error of the measurement process, their accuracy. Both precision and accuracy are characteristics of the measurement procedure, not the value. The random errors can be established by multiple determinations if we assume there are no systematic errors. Then if a more precise procedure is developed, the differences in the values can be used as a measure of the systematic error. If these two procedures are used to measure other quantities of the same type and the differences persist, then a systematic error between the two methods of measurement can be inferred. Ideally, the two procedures give the same value within the mutual uncertainties, which would imply that there are no systematic uncertainties and the "best" value according to the least-squares principle is the weighted average.

There are four possible scenarios given the uncertainty required for a specific application of the measurand. In the worst case the random and systematic errors are larger than the required error and both the systematic and random errors must be reported. In the best case the random and systematic errors are smaller than the required error, in which case the number can be specified with the proper number of significant figures based on the combined errors. One method of specifying this quantity is to indicate the uncertainty in the final quoted figure in parentheses after the value. Another technique is to subscript the first insignificant figure. In the second best case the systematic errors are negligible and the standard deviations can be given with the explanation that these are random uncertainties. If the random errors are negligible and the systematic errors are not, then the systematic uncertainties must be estimated and given along with the random errors. In general, uncertainties should not be ascribed to more than two figures.

5.6 EVALUATION OF EXPERIMENTAL RESULTS

The objective of any review of experimental values is to evaluate the accuracy and precision of the results. The description of a procedure for the selection of the evaluated values (EvV) of electron affinities is one of the objectives of this book. The most recent precise values are taken as the EvV. However, this is not always valid. It is better to obtain estimates of the bias and random errors in the values and to compare their accuracy and precision. The reported values of a property are collected and examined in terms of the random errors. If the values agree within the error, the weighted average value is the "most appropriate value." If the values do not agree within the random errors, then systematic errors must be investigated. In order to evaluate bias errors, at least two different procedures for measuring the same quantity must be available.

A very useful tool in establishing the accuracy and precision of a measurement is a timeline of all the measurements. Very often, a technique will improve over time

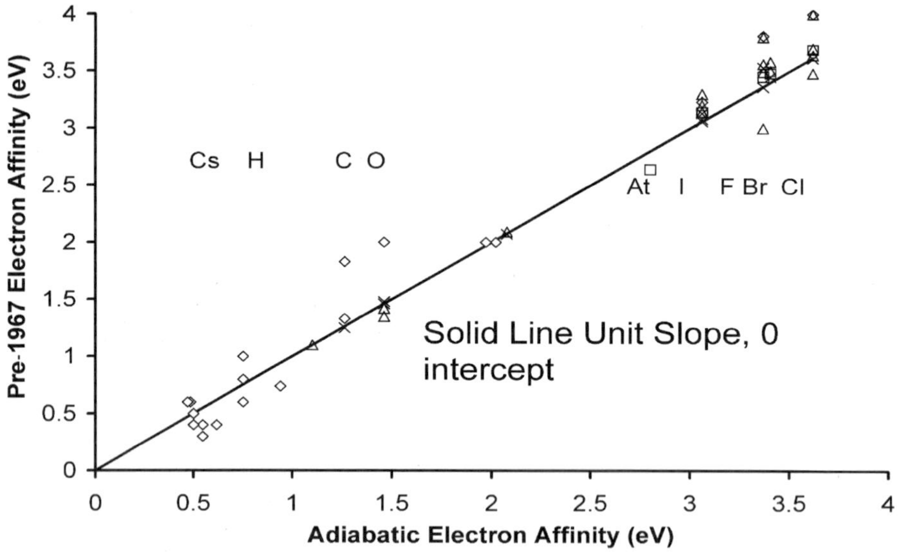

Figure 5.13 A precision and accuracy plot of the atomic electron affinities determined before 1967 versus the current "best" values. The deviations from the unit slope and zero intercept line result from random and systematic errors.

so that its precision improves as a result. At some point this will not be true, and an estimate of the ultimate accuracy and precision of the value can then be established. The clustering of the values about a mean with a random distribution of the errors will signal this. Such is the current situation with the electron affinities of the main group atoms. The atomic electron affinities have been reported in numerous handbooks, the widely used NIST tables, and historical reviews.

In a 1953 review of electron affinities the values for hydrogen, the halogens, carbon, and oxygen using flame, surface ionization, magnetron equilibrium techniques, lattice energy calculations, and electron impact determinations were evaluated. By 1960 the first accurate and precise electron affinity of S, C, and O were reported using photodetachment studies [23]. Before 1970 other photon methods were applied to the halogens and other main groups of elements.

The values of the atomic electron affinities determined before 1967 are plotted against the current EvV in Figure 5.13. This is a precision and accuracy (or P and A) plot. The plot of the electron affinities of the charge transfer acceptors (Figure 4.15) was also a P and A plot. It is used to concisely illustrate the quality of the experimental data. By comparing the data to a line with unit slope and zero intercept, an immediate picture of the precision (random) errors and accuracy (bias) errors can be visualized and outliers identified. By inspection the electron affinities of O (2 eV), Cl (4 eV), and F (4 eV) are suspected outliers. The higher values deviate from the most precise confirmed values by more than twice the standard deviation of the averages. They are outliers and not included in the averages. If we use the "best" values of the E_a at the time, the slope of the linear regression line is 1.03 ± 0.05, while the intercept is -0.08 ± 0.10 eV, which includes the unit slope and zero intercept.

In the past decade many of the atomic E_a have been determined by photodetachment threshold techniques with a precision of parts per million. The earlier values are used to verify the magnitude of these values since they overlap the more precise values within the uncertainty. The electron affinities of Br are as follows 3.76(50) eV, 1927, flame method; 3.49(2), 1944, magnetron; 3.45(20), 1960, lattice energy; 3.53(12), 1960, photoionization; 3.363(3), 1963, photodetachment; 3.363588(2), 1989, photodetachment. The simple average is $A = (3.76 + 3.49 + 3.45 + 3.53 + 3.363 + 3.363588)/6 = 3.49(15)$. This is less accurate and less precise than the 1989 photodetachment value. However, if the weighted average is taken using equations 5.2, this is no longer true:

$$y(\text{avg}) = \sum(w_i y_i)/\left[\sum(w_i)\right]; w_i = 1/s_i^2 \tag{5.2}$$

$$s_y^2 = 1/\left[\sum(w_i)\right] \tag{5.2b}$$

The weighted average is $A = [3.76/(0.50)^2 + 3.49/(0.02)^2 + 3.45/(0.2)^2 + 3.53/(0.12)^2 + 3.363/(0.003)^2 + 3.363588/(2 \times 10^{-7})^2]/(N)$, where $N = 1/(0.5)^2 + 1/(0.02)^2 + 1/(0.2)^2 + 1/(0.12)^2 + 1/(0.003)^2 + 1/(2 \times 10^{-7})^2$. $A = 3.363588 \pm 0.0000002$ or 3.363588(2), the same as the last value. Before the photon values the weighted average was 3.49 ± 0.02 eV. With the 1963 photoionization value it became 3.366 ± 0.003 eV, which is equal to the current EvV. The P and A plot in Figure 5.13 shows the systematic uncertainties of 0.15 eV in the earlier values for all the halogens.

The precision of the experimental methods can be examined for different species by using a timeline. Plots of the value reported versus the date of the determination for nitrobenzene and SF_6 are shown in Figures 5.14 and 5.15. In the NIST compilation six electron affinities for SF_6 cluster around 1.07 eV: 1.07(7) (NIMS, 1994); 1.05(10) (TCT, 1985); 1.0(5) (TCT, 1971); 0.9(2) (collisional ionization, 1978), 1.1(2) (electron swarm, 1966), and 1.49(22) (magnetron, 1964). The simple average is 1.11(17) eV, while 1.08 ± 0.05 eV is the weighted average of these values. In the determination of the magnetron values one low value was discarded. If it is included in the average, the magnetron value is 1.4(4) eV and the weighted average is 1.07(5). This is the "best" value and its random uncertainty. Other NIST values range from 0.3 to 0.8 eV up to 3.16 eV. The lower values are different from the largest value by more than the uncertainty and can be assigned to excited states, whereas the photodetachment value is a different quantity, the vertical detachment energy.

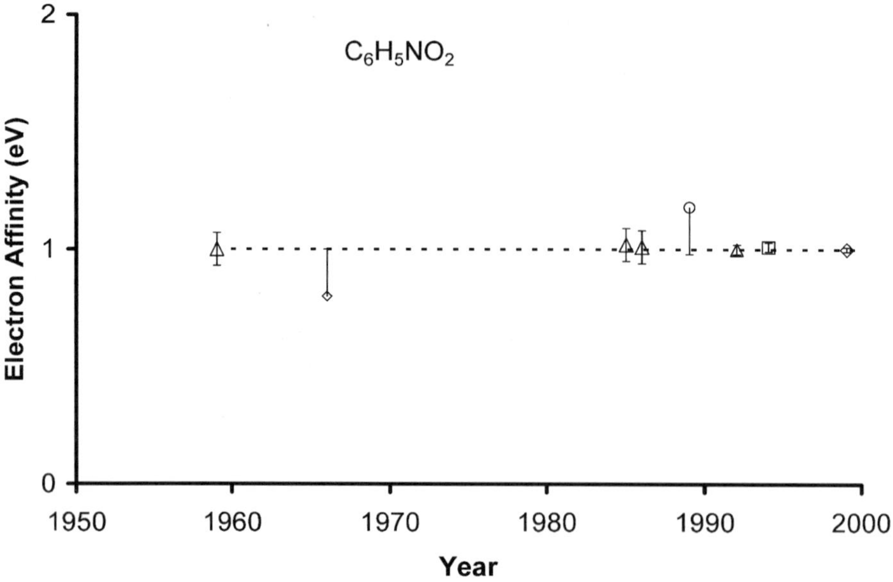

Figure 5.14 Timeline for the electron affinities of nitrobenzene. The values that deviate from the lines are limits, not precise values.

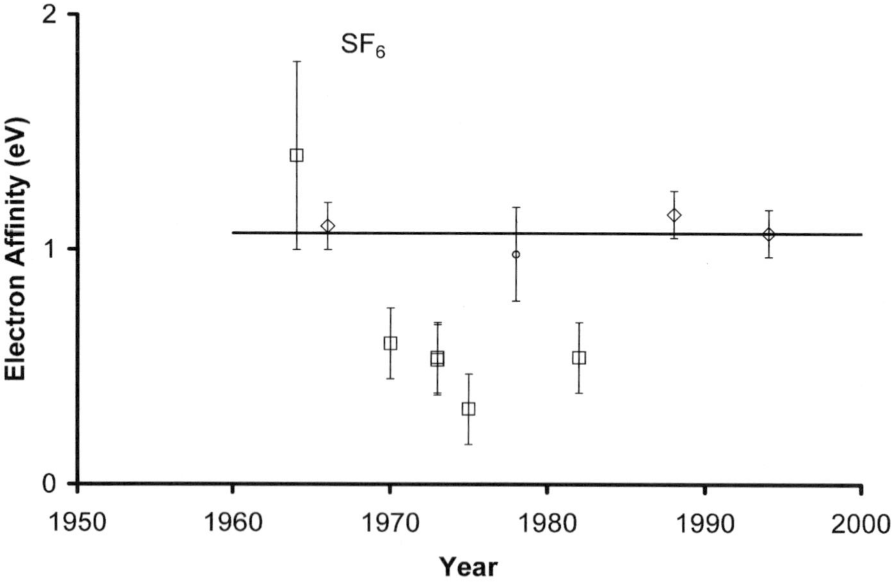

Figure 5.15 Timeline for the electron affinities of sulfur hexafluoride. The lower values can be assigned to excited states. The one higher value is known to have a larger random uncertainty.

5.7 SUMMARY

The experimental procedures for obtaining ECD and NIMS data have been described. Examples of the calculations are given for the various classes of molecules. For each group specific test molecules are provided. The aromatic hydrocarbons and aldehydes are Eql(1/1or 1/2) molecules, CS_2 is a Eql(2/2) molecule, haloalkanes are DEC(1) molecules, and the halobenzenes and nitromethane are DEC(2) molecules that dissociate via a molecular ion. A graphical procedure for obtaining parameters from ECD data and the calibration of NIMS data using SF_6 and nitrobenzene is presented. The use of multiple electron affinities of O_2 to define negative-ion states from ECD data is illustrated. A method for the analysis of published NIMS spectra measured at two temperatures reveals the electron affinities of molecules when combined with substitution effects. We then explored the use of precision and accuracy plots and timelines for the evaluation of electron affinities.

REFERENCES

1. Wiley, J. R.; Robinson, J. M.; Ehdaie, S.; Chen, E. S. D.; Chen, E. C. M.; and Wentworth, W. E. *Biochem. Biophys. Res. Comm.* **1991**, 180, 841.
2. Laramee, J. A.; Mazurkiewicz, P.; Berkout, V., and Deinzer, M. L. "Discrete Electron Capture Negative Ion Mass Spectrometry," in *Encyclopedia of Analytical Chemistry*, New York: Wiley, **2000**.
3. Wentworth, W. E.; Chen, E. C. M; and Lovelock, J. E.; *J. Phys. Chem.* **1967**, 70, 445.
4. Huang, J.; Sun, K.; Zhang, Y.; Rao, H.; Cai, H.; and Stearns, S. D. *J. Chromatogr. A* **1999**, 842, 229.
5. Lin S. N. Ph.D. dissertation, University of Houston, **1969**.
6. D'sa, E. D. Ph.D. dissertation, University of Houston, **1978**.
7. Shuie, L. R. Ph.D. dissertation, University of Houston, **1984**.
8. Chen, E. C. M.; Shuie, L. R.; D'Sa, E. D.; Batten, C. F.; and Wentworth, W. E. *J. Chem. Phys.* **1988**, 88, 4711.
9. Chen, E. C. M.; Wiley, J. R.; Batten, C. F.; and Wentworth, W. E. *J. Phys. Chem.* **1994**, 98, 88.
10. Stemmler, E. A. and Hites, R. A. *Electron Capture Negative Ion Mass Spectra*. New York: New York, VCH, **1988**.
11. Chen, E. C. M.; Carr, S. D.; Wentworth, W. E., and Chen, E. S. D. *J. Chromatogr. A* **1998**, 827, 91.
12. Wentworth, W. E.; Becker, R. S.; and Tung, R. *J. Phys. Chem.* **1967**, 71, 1652.
13. Wentworth, W. E. and Chen, E. C. M. *J. Gas Chromatogr.* **1967**, 5, 170.
14. Wentworth, W. E; George, R.; and Keith, H. *J. Chem. Phys.* **1969**, 51, 1791.
15. Chen, E. C. M.; George, R.; Carr, S. D.; Wentworth, W. E.; and Chen, E. S. D. *J. Chromatogr. A* **1998**, 811, 250.
16. Chen, E. C. M. and Chen, E. S. D. *J. Chromatogr. A* **2002**, 952, 173.
17. Wentworth, W. E. *Mol. Cryst. Liq. Cryst.* **1989**, 171, 271.

18. Chen, E. C. M. Ph.D. dissertation, University of Houston, **1966**.
19. Chen, E. C. M.; Welk, N.; Chen, E. S.; and Wentworth, W. E. *J. Phys. Chem. A* **1999**, 103, 9072.
20. Massey, H. S. W. *Negative Ions*. New York: Cambridge University Press, **1976**.
21. Pritchard, H. O. *Chem. Rev.* **1953**, 52, 529.
22. Chen, E. S.; Wentworth, W. E.; and Chen, E. C. M. *J. Mol. Struc.* **2002**, 606, 1.
23. Burch, D. S.; Smith, S. J.; and Branscomb L. M. *Phys. Rev.* **1958**, 112, 171.
24. Pack J. L. and Phelps, V. *Phys. Rev. Lett.* **1961**, 6, 111.
25. Vogt, D.; Hauffe, B.; and Neuert, H. *Z. Physica.* **1970**, 232, 439.
26. Bailey, T. L. and Mahadevan, P. *J. Chem. Phys.* **1970**, 52, 179.
27. Stockdale, J. A. D.; Compton, R. N.; Hurst, G. S.; and Reinhardt, P. W. *J. Chem. Phys.* **1969**, 50, 2176.
28. Celotta, R. J.; Bennett, R. A.; Hall, J. L.; Siegel, M. W.; and Levine, J. *Phys. Rev. At. Mol. Opt. Phys.* **1972**, 6, 631.
29. Schiedt, J. and Weinkauf, R. *Z. Natforsch.* **1995**, 50a, 1041.
30. Baeda, A. P. M. *Physica* **1972**, 59, 541.
31. Nalley, S. J. and Compton, R. N. *Chem. Phys. Lett.* **1971**, 9, 529.
32. Tiernan, T. O.; Hughes, B. M.; and Lifschitz, C. *J. Chem. Phys.* **1971**, 55, 5692.
33. Lacmann, K. and Herschbach, D. R. *Chem. Phys. Lett.* **1970**, 6, 106.
34. Wentworth, W. E. *J. Chem. Ed.* **1965**, 42, 96.
35. Wentworth, W. E. *J. Chem. Ed.* **1965**, 42, 162.
36. Deming, W. E. *The Statistical Adjustment of Experimental Data*. New York: Dover, **1964**.
37. Wentworth, W. E.; Hirsch, W.; and Chen E. C. M. *J. Phys Chem.* **1967**, 71, 218.

CHAPTER 6

Complementary Experimental and Theoretical Procedures

6.1 INTRODUCTION

> Any physical measurement is the result of applying a given procedure. . . . A preferred procedure is distinguished by the fact that it gives results nearest to what are needed for a particular end; and also by the fact that it is more expensive or more time consuming, or even impossible to carry out. . . . In 1939, Shewhart showed that every published determination of the speed of light showed a lower value than was ever determined before. The results depend on the method used. The value of 3×10^{10} cm/sec is still good enough for most purposes. . . .
>
> —W. E. Deming
> *Sample Design in Business Research*

> A pure scientist makes a series of measurements and upon the basis of these makes the best estimates of accuracy and precision, regardless of the number of measurements. It is readily admitted that future studies may prove such estimates to be in error. All that is claimed that they are as good as any reasonable scientist could make upon the basis of the data available at the time.
>
> —W. A Shewhart
> *Statistical Method from the Viewpoint of Quality Control*

The purpose of this chapter is to describe in detail the general methods of obtaining complementary experimental and theoretical estimates of negative-ion properties obtained from the ECD and NIMS techniques. The nominal precision and accuracy of the methods are established from random and systematic uncertainties observed in selected results. A listing of the molecular E_a determined by each method will be presented in Chapter 10 and the appendices.

The equilibrium, beam, and photon methods have been used to measure E_a. The electron capture detector (ECD), magnetron (MGN), and swarm equilibrium

The Electron Capture Detector and the Study of Reactions with Thermal Electrons
by E. C. M. Chen and E. S. D. Chen
ISBN 0-471-32622-4 © 2004 John Wiley & Sons, Inc.

methods provide E_a from measurements of the equilibrium constant for thermal electron reactions at different temperatures. The molecular electron affinities obtained from the ECD and NIMS and the "direct capture" procedures with an MGN mechanism are absolute values. The term "absolute" means that the value is obtained directly from experimental data and fundamental constants. The equilibrium thermal charge transfer (TCT) method gives relative E_a based on the kinetics and/or thermodynamics of electron transfer reactions between molecules and anions. The thresholds for reactions with electron beam (EB) or alkali metal beams (AMB) are combined with bond dissociation energies or ionization potentials to obtain E_a. Photodetachment (PD), photoelectron spectroscopy (PES), and photoabsorption (PA) methods give E_a from measurements and the photon energy. The reduction potential and solution charge transfer methods cover the E_a range from 0.1 eV to 3.5 eV and confirm many of the values obtained in the gas phase. These are especially important when it is suspected that excited states are involved in gas phase measurements.

The ECD and NIMS procedures, last reviewed in 1989, were updated in Chapter 5 [1]. Reviews covering the MGN[2], TCT[3], AMB[4], EB[5], and ET[6] studies have been published. The PD and PES E_a for approximately 1,000 species have been summarized and compared with theoretical calculations. Few of these are large organic molecules [7]. The accuracy and precision of the methods are determined using precision and accuracy (P and A) plots. The sources of the uncertainties are identified. The EB and electron transmission (ET) techniques yield the vertical electron affinities and cross-sections for ion formation.

Only 10 to 15 molecular E_a have been measured by more than two techniques. The electron affinities of the diatomic halogens, oxygen, and nitrobenzene have been measured by four or more methods. The experimental values that are significantly lower than the evaluated value (EvV) are assigned to excited states based on precision and accuracy plots. The MGN values for SF_6 and C_6F_6 are higher than the EvV but within two standard deviations. The EnCT values for the diatomic halogens are equal to the EvV. However, the values for some fluorocarbons are unexplainably higher than the EvV. With these stipulations the molecular electron affinities reported in the literature agree within twice the nominal uncertainties described in this chapter.

The electron affinities of the aromatic hydrocarbons have been calculated using Huckel theory and MINDO/3 procedures. The electron affinities of benzene, naphthalene, anthracene, and tetracene have been calculated by density functional and ab initio procedures [8]. The relationship between the experimental and calculated values is examined. The electron affinities of other organic compounds have been calculated using MNDO, density functional, and ab initio procedures [9]. A more thorough discussion of these experimental and theoretical methods can be found in Electron and Molecule Interactions and Their Applications, Volume 2, Chapter 6. The experimental and theoretical electron affinities of atoms, molecules, and radicals up to 1984 are listed but not evaluated [10]. The NIST site briefly discusses the various methods for determining electron affinities and gives an unevaluated list of electron affinities that can be searched by combinations of elements [11].

The complementary techniques for determining rate constants for thermal electron attachment, detachment, and dissociation are the flowing afterglow, the microwave technique, the ion cyclotron resonance procedures, the swarm and beam procedures, the shock tube techniques, the detailed balancing procedures, the measurement of ion formation and decay, and the high-pressure mass spectrometer procedures. In all cases the measurement of an ion or electron concentration is made as a function of time so that kinetic information is obtained. In the determination of lifetimes for ions, a limiting value of the ion decay rate or k_{-1} is obtained.

The electron attachment reactions for inorganic molecules were reviewed in 1974. Those for organic molecules were summarized in 1984 [12, 13]. In many cases activation energies were not measured. If a nominal value for A_1 and A_{-1} is assumed, the activation energies can be estimated. Recently, the flowing afterglow procedure has been extended to include electron and gas heating so that the dependence of the rate constants on thermal electron attachment can be examined for bulk temperature and electron temperature [14].

6.2 EQUILIBRIUM METHODS FOR DETERMINING ELECTRON AFFINITIES

In the equilibrium methods the electron is treated the same as any other chemical reactant. The measurement of the electron affinity of a molecule involves the measurement of the equilibrium constant for the reaction of thermal electrons with a molecule $(AB + e(-) = AB(-))$ at some specific temperature or series of temperatures. The equilibrium technique requires (1) a source of thermal electrons, (2) a source of the test species, (3) a method of measuring the equilibrium concentrations of the neutral species, negative ions, and free thermal electrons, and (4) a temperature measurement. The equilibrium constant is directly related to the E_a by this equation:

$$\ln K_{eq}T^{3/2} = \ln A + \ln (Q_{an}) + E_a/RT \tag{6.1}$$

where ln A is 12.43 when evaluated from fundamental constants. If the partition function ratio Q_{an} is independent of temperature, then the slope of a plot of $\ln KT^{3/2}$ versus 1,000/T will give the electron affinity when multiplied by R, the gas constant. If the measurement is made at a single temperature, heat capacity and entropy data must be available for the reactants and products, or assumptions must be made concerning Q_{an} to determine the energy at absolute zero.

The three general equilibrium methods are ECD/NIMS, surface ionization, including the magnetron direct capture techniques, and the kinetic or detailed balancing measurement of the rate constants. The ECD method has been applied to about 150 molecules. The NIMS method involves the application of the ECD

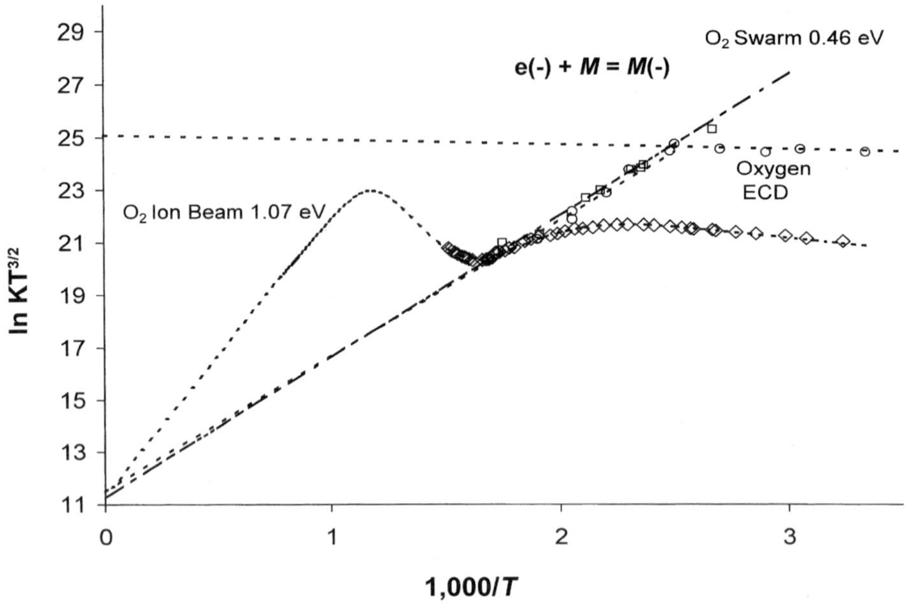

Figure 6.1 Plots of ECD and electron swarm data as ln $KT^{3/2}$ versus 1,000/T for molecular oxygen. The ECD data with the higher β region were published in [18], while the other data appeared in [17]. The electron swarm data derive from [16]. This shows the equivalence of the ECD and electron swarm data. The calculated curve through the ECD data gives an AE_a of 1.07 eV as determined by three techniques [111–113].

model to atmospheric pressure ionization and chemical ionization mass spectrometry. It has only been applied to a few molecules, but can be extended using additional published data [15]. The magnetron direct capture technique has been applied to approximately 20 molecules. The kinetic detailed balancing techniques have only been applied to a few molecules.

The first accurate excited-state electron affinity of molecular oxygen was determined using the equilibrium swarm method [16]. Figure 6.1 is a global plot of the swarm data for oxygen as ln $KT^{3/2}$ versus 1,000/T. This is compared to two sets of ECD data used to determine the excited-state electron affinity of oxygen. One set has a larger α region and a higher value of K in the β region. Independent investigators used a highly purified carrier gas to reduce the recombination rate constant to obtain these data [17, 18]. Lines drawn through the two regions illustrate how first approximations to the parameters can be obtained for least-squares analysis. Also shown in Figure 6.1 are the calculated curves and high-temperature ECD data obtained in this laboratory and used to support the higher AE_a for O_2 [10, 11]. The ECD results were supported by atmospheric pressure ionization experiments in which the molecular negative ion was observed at 550 K where the change in slope is observed. The low response in the β region in the ECD data results from the low k_1 value that was also measured in the swarm data. The swarm data gives a

Q_{an} ratio of unity. This was one of the first determinations of a Q_{an} ratio for electron molecule reactions. The ECD values of the excited-state electron affinity, Q_{an}, and A_1 and E_1 agree with the swarm values. The low-temperature response of the ECD can be calculated from parameters measured in swarm experiments.

In surface ionization experiments the electron source is a heated filament where reactions take place. The reactions are studied at high temperatures and relatively low pressures. The filament has been heated by a flame and resistance heating. The experimental data are the electron and ion currents. These have been measured using an electric field, a magnetic field, or combinations of both [19–22]. The magnetron method is a surface ionization technique in which a magnetic field is used to measure the ion and electron currents. Both ions and electrons carry the current from the filament to a concentrically mounted electrode in the absence of an intense magnetic field. In the presence of a sufficiently intense magnetic field, the electrons are driven into concentric paths between the filament and the anode so that only the ions are collected. The ratio of the electrons and ions can be measured by applying a magnetic field. In the 1960s the magnetron method was used to determine the electron affinity of several organic molecules. This direct capture process is only one of four possible mechanisms observed in the magnetron. Another reaction involves gas phase dissociation prior to electron capture. This yields electron affinities of radicals. Many of these values are equal to the current E_a within the experimental error. Two other mechanisms involve the adsorption of a radical on the filament. These give values that are higher than the current evaluated values for the phenyl and alkyl radicals. These high values cannot be currently explained. When the equilibrium mechanism can be established, the magnetron E_a is the AE_a or an excited state E_a.

In Figure 6.2 global plots of the magnetron data as ln $KT^{3/2}$ versus 1,000/T for tetracyanoethylene, tetracyanobenzene, and dicyanoethylene studied in the direct capture method are presented. These data were taken from the review on the magnetron method [2]. The similarity of these data to ECD and NIMS data is clear. The magnetron method has also been used to determine Q_{an} values. For tetracyanoethylene there is a change in slope at the lower temperatures so that other ion losses compete with detachment. Shown in Figure 6.3 are the electron swarm data for O_2, the ECD and magnetron data for NO and C_6F_6, and NIMS and magnetron data for SF_6. The calculated equilibrium method data for an E_a of 0.02 eV for NO are also shown. These illustrate different temperature ranges and the similarity of data from the equilibrium methods. An ideal equilibrium method would extend the temperature range from the ECD/NIMS range $(1{,}000/T = 1.5\text{–}3.0)$ to the magnetron range $(1{,}000/T < 1.0)$. The magnetron values for SF_6 and C_6F_6 are higher than for the evaluated values by about 0.3 eV. Two states are observed in the ECD data for C_6F_6, as indicated by the structure at lower temperatures [3, 23].

The electron affinities of molecules determined from the direct capture magnetron method are plotted in a P and A graph in Figure 6.4. The zero intercept slope is 1.01(5). The standard deviation is ± 0.13 eV. This gives a nominal precision of ± 0.15 eV, including the low value for $C_4H_2N_2$ and high values for SF_6 and C_6F_6. The magnetron values for anthraquinone, benzoquinone, fluoranil, and chloranil are

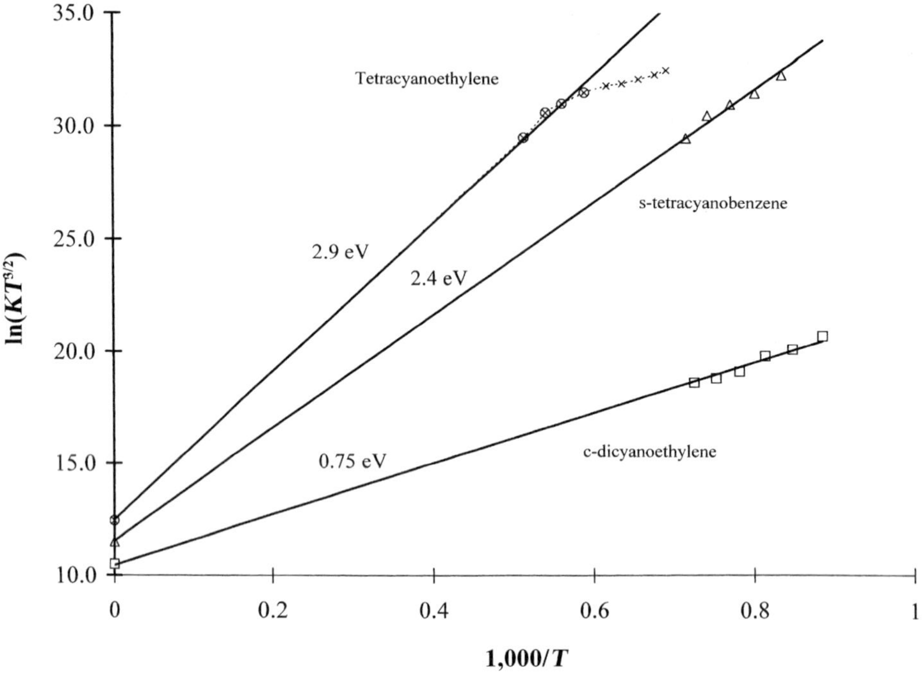

Figure 6.2 Plots of magnetron data as ln $KT^{3/2}$ versus 1,000/T. The similarity of these data compared to the ECD data should be noted. The data were taken from [27] and original references cited therein. This shows the range of the electron affinities that can be obtained using the magnetron which is 0.75 to 3.3 eV for hexacyanobutadiene (not shown).

lower (0.42 eV) than the EvV. Since this is more than twice the nominal uncertainty, these are assigned to excited states.

The equilibrium methods yield electron affinities directly from experimental data and fundamental constants. The ECD is the most extensively applied equilibrium method. It is limited to molecules with electron affinities less than about 1.5 eV ($C_6F_5NO_2$) by the upper temperature limit. Figure 6.5 is a plot of data that illustrates the range of electron affinities which have been measured in the ECD. If the temperature range can be extended, higher electron affinities can be measured. The NIMS method has the potential of being applied to molecules with electron affinities higher than 1.5 eV. For molecules, such as acetophenone, with a large enough linear temperature range and only one negative-ion state, the ECD or NIMS has attained a precision of ± 0.01 eV for E_a. By repeating the determinations, this precision can be improved. The swarm technique used for the determination of an electron affinity of oxygen and the detailed balancing procedures have the same potential precision as the ECD and NIMS procedures. The possibility of anionic excited states must be considered in all cases. The magnetron method

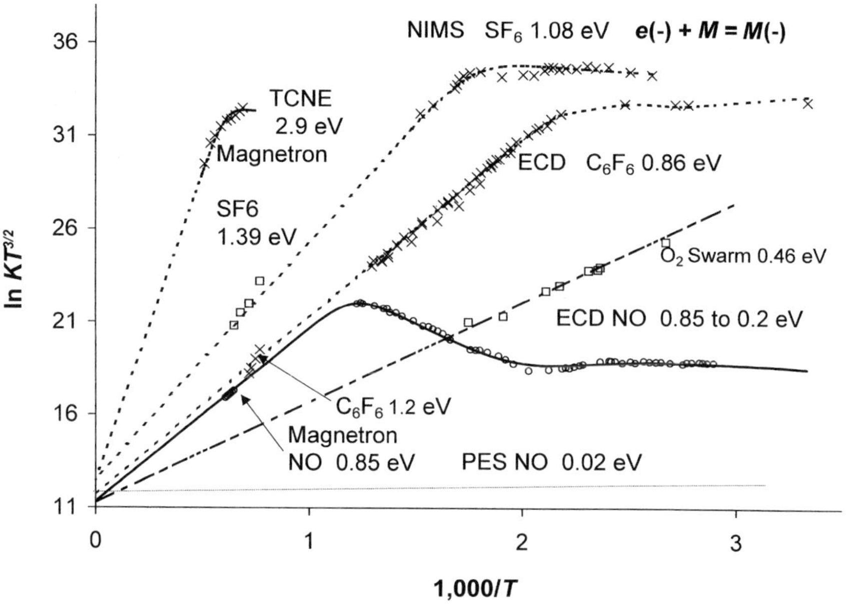

Figure 6.3 Plots of ECD, NIMS, and magnetron data as ln $KT^{3/2}$ versus 1,000/T. The similarity between the types of data should be noted. The higher E_a for the magnetron values are apparent. The magnetron data derive from [2] and original references cited therein. The ECD and NIMS data were taken from [17, 72, 114]. The low value for NO has been obtained by many methods (see [11]).

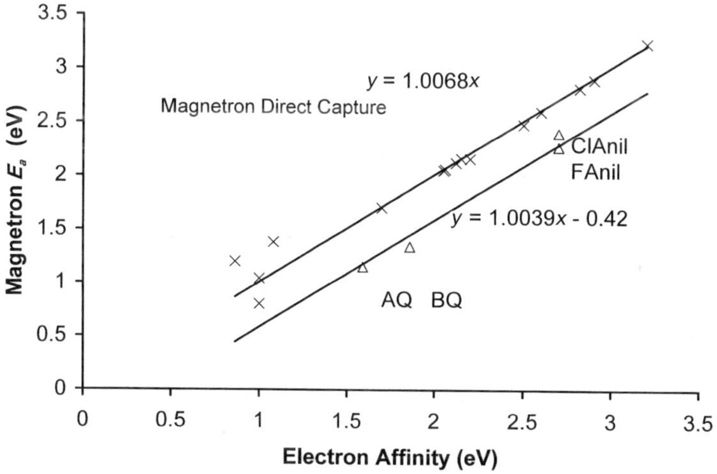

Figure 6.4 Precision and accuracy plot for direct capture magnetron data [2]. The values for the quinones are systematically lower than the others by 0.42 eV. The random uncertainty in the other values is ± 0.15 eV.

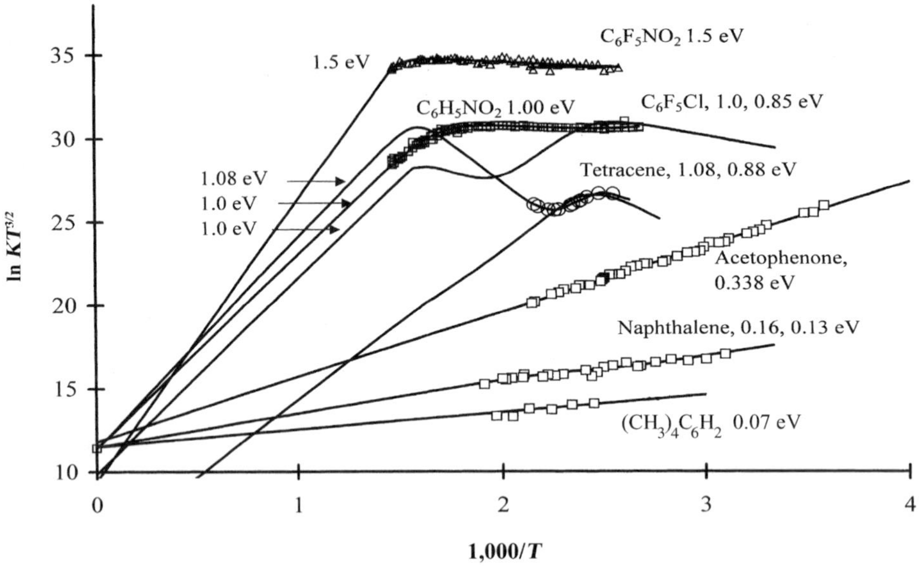

Figure 6.5 ECD data plotted as ln $KT^{3/2}$ versus 1,000/T illustrating the range of electron affinities measured in the ECD. This is 0.05 eV to 1.5 eV. The data have been published, but not in this specific combination [31, 34, 40].

can be used to obtain accurate electron affinities of molecules with a demonstrated precision of ± 0.15 eV. The major problems with the magnetron method are the establishment of the mechanism, the identification of the state of the anion, and the correction of the results from the high temperature of the measurement to absolute zero. The magnetron procedure can be improved by obtaining a more accurate measurement of the temperature and concentration of the reactant. The magnetron electron affinities have been generally disregarded because of the lack of mass analysis. The major ion formed for TCNQ and TCNE at a heated filament is the parent negative ion. The use of mass spectrometry will not eliminate the problem of identifying the state of the anion.

6.3 PHOTON TECHNIQUES

The reaction $AB(-) + h\nu \rightarrow AB + e^-$ is the basis of photoelectron spectroscopy and photodetachment methods. Many precise and accurate ionization potentials of molecules have been obtained by studying the photoionization of neutral molecules. The same principles apply to the photon methods for determining electron affinities, except that negative ions are studied. The electron affinities of over 1,000 atoms, radicals, clusters, and small molecules have been determined using

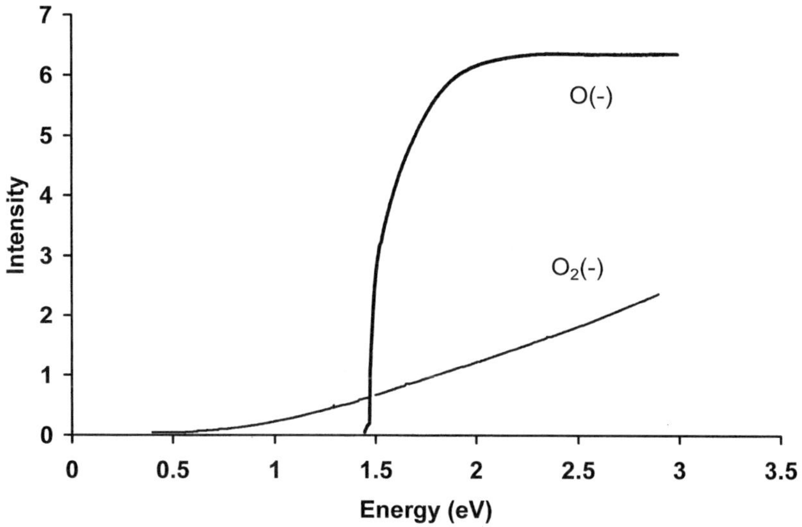

Figure 6.6 Plots of ion intensities versus photon energy for O(−) and O_2(−) from [26]. The atomic threshold is very sharp but the molecular threshold is gradual.

these techniques. Only about a dozen electron affinities of large organic molecules have been determined by photon methods. The combination of photodetachment with an ECD has been used to estimate the electron affinities of molecules [24]. These experiments require a monochromatic light source, source of negative ions, and way to measure the intensity and energy of the electrons. In photodetachment experiments the threshold for electron formation is related to adiabatic electron affinity. In photoelectron spectroscopy the intensity and energy of the electrons formed when photons of a fixed energy interact with the anions are measured. The absorption and emission spectra of atoms or their anions have also been used to obtain the electron affinities of atoms [7]. Bound excited states of anions have been studied using two-photon photodetachment spectroscopy [25].

The first accurate and precise determinations of the electron affinity of atoms other than the halogens were carried out using the photodetachment procedure in the late 1950s. With a high-intensity carbon arc lamp and interference filters with a glow discharge source, the photodetachment spectrum of O(−) was measured and the electron affinity determined from the threshold [26]. Figure 6.6 is a graph of the data for O(−) and a graph for O_2(−) [27]. The difference in the shape of the threshold for the molecule and atom is important. The onset for the photodetachment of the atomic anion is very sharp and gives an accurate and precise estimate of electron affinity. However, the onset for the molecule is gradual and extrapolates to well below adiabatic electron affinity. This was believed to result from the population of excited states of the anions [28]. In the 1960s photoabsorption and photoemission

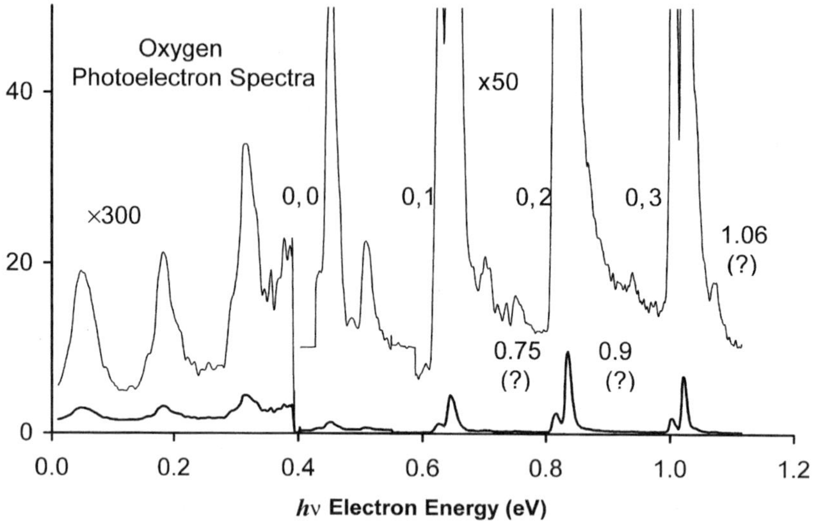

Figure 6.7 High-resolution photoelectron spectra of $O_2(-)$. The large peaks form a progression from one negative-ion state. The small peaks were unexplained in the original article, but coincide with electron affinities measured by other techniques. The PES spectrum was taken from [33].

spectra of the halogen atoms or anions were used to obtain electron affinities. Some of these values remain the most accurate and precise E_a [7, 29].

With the development of laser light sources the PES and PD methods have become the preferred method of measuring atomic electron affinities. The electron affinities of most of the atoms have been determined to precisions of 0.1% using laser PES. The use of laser photodetachment spectroscopy has improved the precision to parts per million. The evaluated values of the electron affinities of the atoms and their relationship to the Periodic Table will be discussed in Chapter 9. The problems in establishing the threshold and in assigning the onset to a given state make the determination of molecular electron affinities by photon techniques more difficult. The ground-state negative ions of many homonuclear diatomic molecules have been characterized by PES [7]. In the photoelectron spectra of diatomic molecules, hot bands due to excited vibrational levels of the negative ion have been observed. The highest resolution, 0.02 eV, photoelectron spectra for oxygen, is shown in Figure 6.7 [30]. The internuclear distance and fundamental frequency for the excited-state anion are obtained from a detailed analysis of this spectrum. The most intense progression is for the excited state, with an electron affinity of 0.430(2) eV and 0.450(2) eV. The doublets result from spin orbital coupling. However, other peaks in the spectrum were unexplained. We have interpreted these peaks as arising from the higher-energy, lower-population anion states observed in other experiments [31].

The photoelectron spectroscopy data for CS_2 led to the first accurate and precise value for the electron affinity for the ground-state bent anion, 0.895 ± 0.02 eV [32].

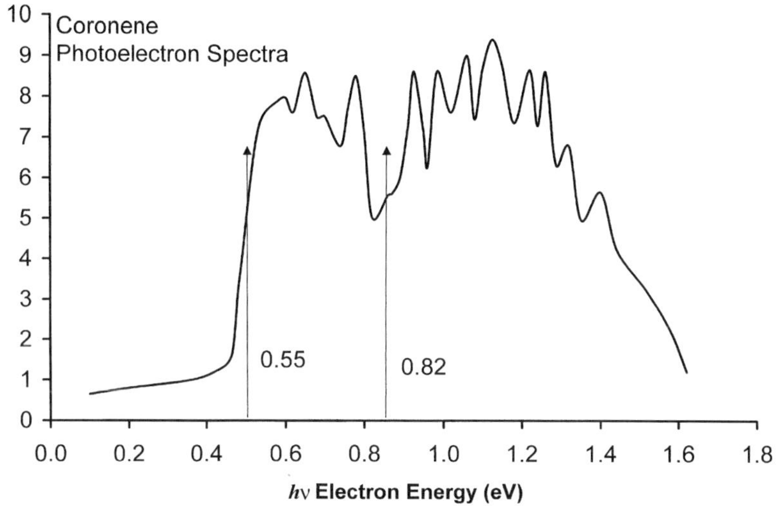

Figure 6.8 Photoelectron spectra of coronene(−), from [38]. The electron affinity was determined from the first peak in the first series of peaks. If the second series of peaks is from the ground state of the anion, the electron affinity is 0.8 eV, in agreement with the reduction potential and CURES-EC value.

A second higher-resolution PES study gave an upper limit of 0.8 eV for the E_a. The high-resolution PES spectrum is shown in Chapter 9 [33]. In neither case was an electron affinity for the excited state reported, although both spectra have peaks at low energy attributed to "hot" bands. If these bands are assigned to the excited linear state, the excited-state E_a can be estimated from the initial onset, while the ground-state value can be obtained from the identification of the 0-0 band of the second series [31–33]. The examples of O_2 and CS_2 illustrate the problems encountered with excited states in PES. Both the ground-state electron affinity and linear excited-state electron affinity have been determined using the ECD [34] (see Figure 4.3). The excited-state electron affinity has been measured in TCT and AMB studies, whereas the higher E_a has been measured in a separate AMB and TCT study [3, 35–37].

The data obtained from the ECD and reduction potentials can be used to interpret PES data. Three examples for molecules are the "lower" values for the electron affinities of nitromethane, anthracene, and coronene. Based on the observation of excited states in anthracene and tetracene in the ECD data, it is reasonable to assume that the lower value for coronene derives from the population of an excited state. In Figure 6.8 the PES of coronene is shown with two sets of peaks. If the ground-state E_a for coronene is taken from the initial onset, it is much lower than the value obtained from reduction potential or electronegativity data. In addition, the second onset must be explained [38–42].

The electron affinity of nitromethane has been determined by ECD, PES, AMB, and TCT. The ground-state electron affinity for nitromethane has been measured by

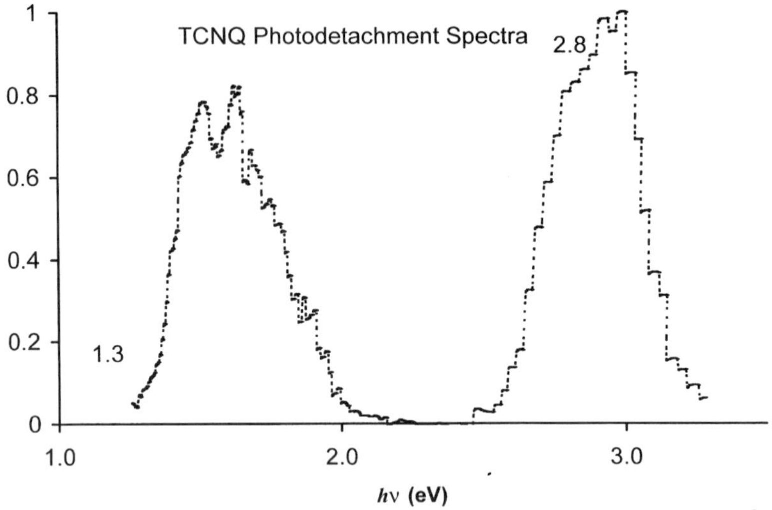

Figure 6.9 Photodetachment spectrum of tetracyanoquinodimethane(−), from [25]. The two peaks represent detachment from an excited state by way of a two-photon process.

ECD, TCT, and AMB studies. In the case of the PES data there are low-energy bands that can be attributed to excited valence states, one of which is a dipole bound state. The new question becomes, "Why are the excited states not stabilized to the ground state?" The answer could be that the pressure is not sufficiently high [43–46].

Photon experiments have produced convincing evidence that bound excited anionic states of atoms, small molecules, and large organic molecules such as tetracyanoethylene (TCNE) and tetracyanoquinodimethane (TCNQ) exist. One of the most convincing studies made use of electron photodetachment and one-photon and two-photon intensity dependence studies to establish at least one excited state of TCNQ(−) that is lower in energy than electron detachment. The most direct evidence lies in the photodetachment spectra shown in Figure 6.9 [25]. The similarity between this spectrum and that in Figure 6.8 is striking. Here the low-energy process is attributed to a two-photon absorption in which the anion is excited and then the electron is photodetached with a second photon. The existence of a bound excited state was also used to explain the difference between the photodetachment value for the E_a of TCNE (1.7–2.3 eV) and the TCT and magnetron values of 2.9 eV [47–49]. It was further concluded that this appears to be a general phenomenon which should apply to a whole class of radical anions [25].

In the case of the photoelectron spectra of cyclooctatetraene (COT) the ECD and TCT values for the ground-state electron affinity can be used to assign the onset to the formation of a transition state. Thus, the onset occurs at an energy higher than the adiabatic electron affinity. In the case of the photodetachment data for COT the onset occurs between the adiabatic electron affinity and the onset for photoelectron spectroscopy [50]. These will be considered more extensively

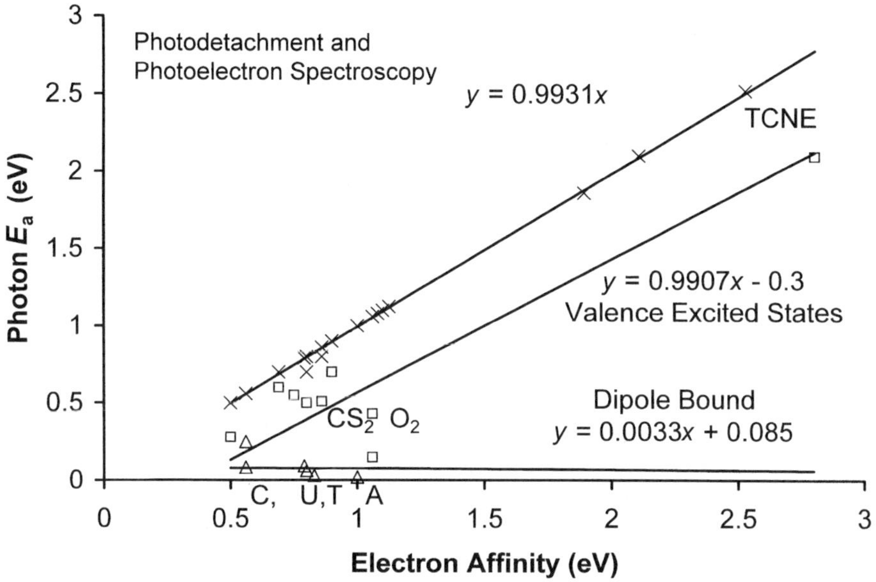

Figure 6.10 Precision and accuracy plot for photon data. The nitromethane, CS_2, and anthracene values are lower than the current "best" value by 0.3 eV. The dipole bound E_a are all less than 0.25 eV. The random uncertainty in the other values is ± 0.01 eV.

in Chapter 10. In another important case the PES data give the dipole bound E_a of the purines and pyrimidines. Only the dipole bound state is observed in rare gases, but in the presence of water, the ground state can be accessed. In addition, two-photon absorption of the hydrates leads to the dipole bound anion. These will be described in Chapter 12 [51].

Figure 6.10 is a precision and accuracy plot for the E_a of several molecules determined using photon methods. The excited-valence-state values (shown in the squares) are clearly below the unit slope zero intercept line. The dipole bound states (shown in the triangles) are all less than 0.25 eV. The ground-state values are within ± 0.01 eV of the unit slope zero intercept line. Thus, the photon methods are capable of measuring ground-state values within this uncertainty. When a value is much lower than the largest accurate value, excited states must be considered. In Chapters 9 to 12 the use of this concept to interpret photon data for other molecules in the same manner as for oxygen and coronene will be discussed.

The absorption, emission, photodetachment, and photoelectron spectroscopy experiments are capable of providing accurate and precise values for the electron affinities of atoms. The best precision is about 1 part per million, more than precise enough for chemical purposes. The state of the ion must be identified and some excited-state electron affinities of atoms have been reported. The photoelectron spectroscopy and photodetachment procedure can give the accurate and precise electron affinities of molecules and radicals when the state of the anion is assigned.

In the case of tetracyanoquinodimethane, carbon disulfide, nitromethane, and the purines and pyrimidines, two or more negative-ion states have been observed. In some cases the photoelectron spectrum can be assigned to an excited state and reveal an electron affinity lower than adiabatic electron affinity. In the case of cyclooctatetraene the onset in the PES spectrum is higher than adiabatic electron affinity because of the significant change in the geometry of the anion.

6.4 THERMAL CHARGE TRANSFER METHODS

In the thermal charge transfer methods the electron affinity of a molecule is determined by bracketing the electron affinity of a test species between that of two species with known electron affinities. When the reaction studied is $AB(-) + CD \Leftrightarrow AB + CD(-)$ direct charge transfer, the relative electron affinities are obtained. A variation on this method is the measurement of the intensity of ions formed by the collisional ionization of an electron bound dimer. This has been applied to a number of aromatic and halogen-substituted aromatic hydrocarbons, but can give E_a for excited states. When the reaction studied is proton transfer, $A(-) + BH \rightarrow AH + B(-)$, the relative gas phase acidities are obtained. With experimental values of the AH and BH bond energies the relative electron affinities of the radicals may be determined [52]. In order to carry out these experiments, there must be a source of the test material and a source of the reactant anion. There must also be a method for measuring the concentration of the test anion and the product anion. The electron affinities or gas phase acidities of the reference compounds and the accuracy and precision must be known. The general procedure has been to develop a ladder of bracketing reactions and to reference that ladder to an accurate and precise acidity or electron affinity. These reactions can be studied by observing the direction of charge transfer from kinetic experiments or by measuring the equilibrium concentrations of the anions and neutrals. In either case the relative values of the electron affinity or gas phase acidity are obtained. The error in the measured electron affinity is no smaller than the errors in the electron affinity of the bracketing species. The determination of the equilibrium concentrations can be completed at one temperature or multiple temperatures. When the reaction is carried out at a single temperature, it is necessary to assume that the entropy changes for the two reactions are equal so as to eliminate the need for temperature corrections. If the entropy and energy changes are determined, the gas phase acidity or electron affinity must be corrected to absolute zero.

The first thermal electron transfer measurements to give molecular electron affinities were observations of the direction of charge transfer using negative-ion mass spectrometry. The electron affinity of nitrobenzene was bracketed between that of SO_2 and NH_2 in 1959 [53]. In 1961 the electron affinities of CS_2 and SO_2 were bracketed between that of the O atom and NH_2 [54]. In the 1970s the use of negative-ion mass spectrometry to study equilibrium reactions began to flourish. This coincided with the development of the flowing afterglow [55], ion cyclotron resonance (ICR) mass spectrometer, and high-pressure mass spectrometer (HPMS)

[3, 56]. Among the reactions studied extensively were electron transfer to small molecules and radicals and proton transfer for inorganic and organic acids. The difficulty in generating molecular ions of organic compounds precluded the measurement of their electron affinities. At the low pressures of typical mass spectrometer sources, the molecular anions that are initially formed can undergo autodetachment before entering the mass analyzer.

In 1978 the following was reported:

> Recently we have also observed rapid electron transfer reactions between a wide variety of stabilized negative ions and neutral molecules. For example $CH_3O(-)$ generated by dissociative electron attachment to CH_3ONO rapidly transfers an electron to benzophenone to produce exclusively the negative ion radical of benzophenone.... This result indicates that the electron affinity of benzophenone is greater than 36.7 kcal/mole, the measured electron affinity of CH_3O. A scale of relative electron affinities constructed in this way is shown below. CH_3O < benzophenone < C_6F_6 < nitrobenzene < p-F-nitrobenzene < phthalic anhydride < m-Cl-nitrobenzene < maleic maleic anhydride < naphthoquinone < benzoquinone.... By studying many equilibrium transfer reactions of this type, it will be possible to develop a scale of electron affinities for a wide variety of molecules. [56]

This was the first description of the general TCT method for the determination of relative molecular electron affinities. The problem with the scale was the reference point. The lower limit of the scale was established by the experimental value of the methoxy radical, 1.59(5) eV, determined by photodetachment experiments. Because the methoxy radical was generated by dissociative electron attachment, it could not exist in equilibrium with molecular species created by electron transfer. The relative order was correct, but the scale was not. The results were precise, but not accurate. At the time of the experiments the electron affinity of nitrobenzene, hexafluorobenzene, maleic anhydride, and 1.4 benzoquinone had been reported. The values for benzoquinone and maleic anhydride agree with the current EvV. By scaling to these E_a, all values agree with the current values to ± 0.1 eV. These are (0.8 eV, benzophenone) < (0.9 eV, C_6F_6) < (1.0 eV, nitrobenzene) < (1.1 eV, p-F-nitrobenzene) < (1.2 eV, phthalic anhydride) < (1.3 eV, m-Cl-nitrobenzene) < (1.4 eV, maleic anhydride) < (1.82 eV, naphthoquinone) < (1.85 eV, benzoquinone). These are independent ICR determinations of the electron affinities for the molecules. The electron affinities were obtained by incrementing the reference values by 0.1 eV and assuming an uncertainty of 0.1 eV. The electron affinity of chloronitrobenzene is the maleic anhydride value, 1.44 eV, decreased by 0.1 eV to yield 1.34(10) eV.

From the known concentrations of the molecules and measured ion abundances, the equilibrium constant for the electron transfer reaction can be obtained. Figure 6.11 presents the kinetic data for thermal charge transfer reactions in the ICR determination. The concentrations of the two molecules have a ratio of BQ/NQ = 1.33. The rate constant for the electron transfer can be obtained from the initial rise of the BQ intensity and the decline of the NQ intensity. This is reported as approximately 6×10^{-10} cc/molecule-s. At about 0.15 s, the ratio becomes

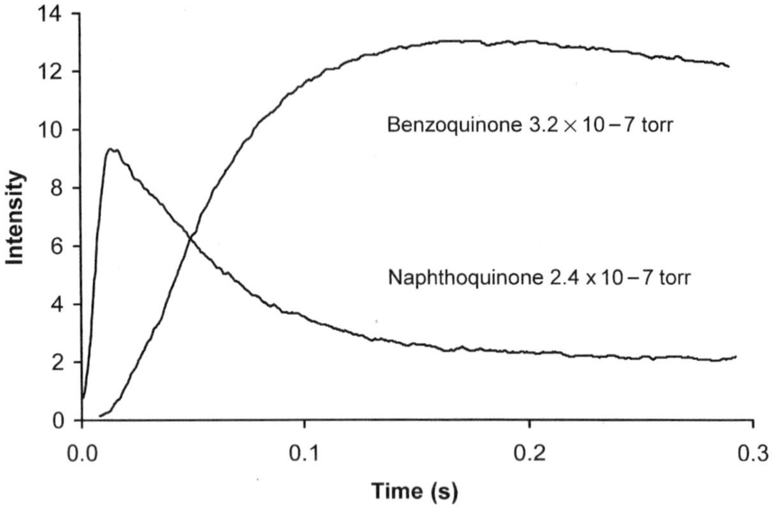

Figure 6.11 Ion intensities in an ICR cell containing benzoquinone and naphthoquinone at the indicated pressures, from [56]. The equilibrium constant for electron transfer gives relative electron affinities.

relatively constant and the equilibrium constant can be calculated. It is determined from the concentration ratio and ion ratio as $K = [NQ]/[BQ](IBQ/INQ) = 4.4/1.33 = 3.2$. The original article on the subject stated, "For the electron transfer reaction between naphthoquinone and benzoquinone, we find $K = 3.2$ and $\Delta G^0 = -RT \ln K = -0.7$ kcal/mole. Since entropy effects are negligible, the electron affinity of benzoquinone is 0.7 kcal/mole greater than the electron affinity of naphthoquinone." The above E_a reflect these data [56].

In 1983 the HPMS TCT method for determining molecular electron affinities was introduced. It is based on the same fundamental concepts as the ICR method, but the absolute value was anchored to the electron affinity of benzophenone obtained from the ECD. Later, the scale was anchored to the electron affinity of SO_2. At present about 300 electron affinities of organic molecules have been determined by the TCT and/or ECD methods [1, 3, 55–59]. The TCT method has been used to measure E_a between 0.50 eV for nitromethane and 3.2 eV for tetracyanoethylene, as shown in Chapter 10.

Many TCT values were determined by both the ion cyclotron resonance (ICR) and high-pressure mass spectrometry (HPMS) variants [3, 55–59]. In the ICR method ΔG values are calculated from measured ion ratios and known concentrations at a single temperature and ΔH obtained by making assumptions about ΔS. For example, in the above study of naphthoquinone and benzoquinone, the entropy changes are assumed to be the same [56]. In the HPMS method measurements are made as a function of temperature so that ΔH and ΔS can be obtained ($\Delta H = \Delta G + T\,\Delta S$). Good agreement between the independently obtained values

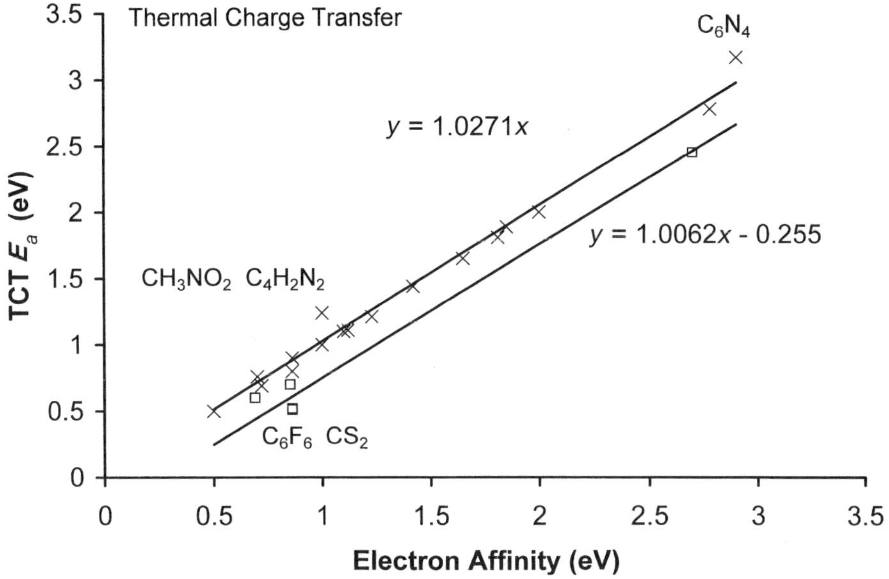

Figure 6.12 Precision and accuracy plot for TCT data. One value for CS_2 and one for C_6F_6 is lower than the current "best" value by 0.3 eV. The value for fluoranil is also lower than the current best value. The random uncertainty in the other values is ±0.10 eV.

has been noted. This is especially important because the ICR pressures are about 10^{-6} torr, whereas the HPMS pressures are about 5 torr. The opportunities for the collisonal deactivation of the ionic reactant are better in HPMS.

The combined random and systematic uncertainties in the absolute values due to the direct measurements and uncertainties in the reference compounds are quoted at ±0.1 eV for the HPMS method and ±0.05 eV from the ICR method. The uncertainties in specific values can be lower or higher. At the extremes the uncertainties in the values can be as large as ±0.2 eV. Some of the molecular E_a determined using the ICR and TCT methods are shown in the precision and accuracy plot in Figure 6.12. The ICR value for C_6F_6 is consistent with the ground-state value of 0.86 ± 0.02 eV obtained from the ECD, NIMS, and PES studies. The data in Figure 6.12 illustrate the precision of the values. The slope of the zero intercept is 1.027(30) and the standard deviation of the values calculated using this equation is ±0.1 eV. The largest deviations occur at the extremes. If these are excluded, the standard deviation is reduced to ±0.07 eV. The average deviation of the excited-state values is 0.245 eV, confirming the assignments to excited states. The ICR value for fluoroanil (2.45(5) eV), and the HPMS values for C_6F_6 (0.52(10) eV), CS_2 (0.62(10) eV), azulene (0.70(10) eV), and anthracene (0.60(10) eV) are assigned to excited states. The latter two are confirmed by ECD and PES data.

The kinetic method for determining the direction of charge transfer by collisional ionization is based on the competitive dissociation of mass-selected electron bound ions, $M_1 \cdot M_2(-)$. If we assume that the activation energy for electron transfer is less than that for dissociation, then the relative abundances of the ions are related to the differences in the E_a and the effective temperature of the dimer by $\Delta E_a = RT_{\mathrm{eff}} \ln\{[M_1(-)]/[M_2(-)]\}$. The effective temperature of the dimer is reported to be over 1200 K by calibrating it to ratios for molecules with measured E_a [60–63].

This method was first applied to relative electron affinities of substituted nitrobenzenes. All but one of these has been measured by HPMS TCT studies. However, the E_a of s-butyl nitrobenzene has only been determined by collisional ionization and is still listed in the NIST tables as 2.17(20) eV. This value is referenced to a "high" value for nitrobenzene and should be about 1 eV lower [60]. The electron affinities of aromatic hydrocarbons have been reported using the collisional ionization method. The value for biphenylene is larger than that obtained from half-wave reduction potentials. The values for pyrene, anthracene, and c-C_8H_8 are consistent with other reported values, but the values for benzanthracene, coronene, and benzo[ghi]perylene are significantly lower than the largest precise value and are attributed to excited states.

In electron bound complexes of C60 with aromatic hydrocarbons, the intensity of the aromatic hydrocarbon anion from collisional ionization is larger than that for the aromatic hydrocarbon. However, it is known that the adiabatic electron affinities of the aromatic hydrocarbons are significantly lower than that of C60. In these cases the relative intensities were attributed to a localized anion. When the benzanthracene and phenanthrene electron bound dimer is dissociated, the intensity of the phenanthrene anion is only about one-half that of the benzanthracene anion, while the electron affinity of benzanthracene is 0.72 eV and that for phenanthrene is 0.30 eV. The experimental data cannot be disputed. The ion ratio gives a ΔE_a of 0.08 eV or the E_a of benzanthracene $= 0.30 + 0.08 = 0.38(10)$ eV. Therefore, some other explanation must be presented. In the case of C60 excited anion states are observed and three degenerate LUMO + 1 are calculated with positive electron affinities. Benzanthracene has an excited-state MO at -0.38 eV that gives an E_a of 0.38 eV. Just as C60 has multiple LUMOs, so also does benzanthracene. Because the collisional ionization values are the most recent, they are returned by a search of the NIST tables for CH molecules. These values are listed in Chapter 10 [61–63].

The TCT method of obtaining relative molecular electron affinities and gas phase acidities has a demonstrated precision of ± 0.05 to 0.10 eV in the midrange of values from 0.5 eV to 3.0 eV. At the extremes the precision is less, ± 0.2 eV. Most of the TCT E_a are ground-state electron affinities. The exceptions are the HPMS electron affinities determined for azulene, anthracene, C_6F_6, and CS_2, and the ICR value for fluoroanil. The TCT method has been applied to more than 200 molecules. About 30 have been determined by the HPMS and ICR methods and many have been confirmed by the ECD method. Many have also been confirmed by the half-wave reduction potential method and/or solution charge transfer complex spectra. These will be discussed in Chapter 10. The collisional ionization method of measuring relative electron affinities can produce inverted orders of intensities and give excited-state E_a rather than ground-state E_a.

6.5 ELECTRON AND PARTICLE BEAM TECHNIQUES

The threshold for the reaction of an electron, anion, or alkali metal beam with a molecule to form an anion has been used to measure the electron affinities of molecules and radicals. In this experiment there must be a source for the reactant beam, a source for the test molecule, and a way to measure the onset. The oldest such procedure is the electron impact method in a mass spectrometer source. This gives the vertical electron affinity of the molecule from the peak in ion intensity. The quantity $D - E_a(X)$ is the onset for ion formation. The distribution of the negative ions reveals information about the negative-ion potential energy curve in the Franck Condon region. The measurement of the transmitted electron current as a function of electron energy also gives the vertical electron affinity. The alkali metal beam experiments determine molecular electron affinities from the ionization potential of the metal and the onset of ion pair formation. These also provide the electron affinities of radicals when the threshold is combined with the ionization potential and bond dissociation energy. The difference in the onset for ion pair formation and radical ion formation is the bond dissociation energy [4–6]. The threshold for endothermic electron transfer reactions gives relative electron affinities. Precision is governed by the ability to establish the threshold, whereas accuracy depends on the assignment of the process. The errors in the AMB values are generally postulated to be ± 0.2 eV, while those of the EnCT method are ± 0.3 eV. Most of the values obtained using these methods agree with the EvV. For example, the E_a for the diatomic halogen molecules easily agree with the evaluated values within this error [64–66].

The electron impact studies provide the electron affinities of radicals and vertical electron affinities. These have been improved by the use of an electron monochrometer. The observation of a parent negative ion at zero energy is indicative of a positive electron affinity [67]. The identification of an onset for dissociative electron attachment at higher energies gives a value of $D - E_a(X)$, from which the electron affinity can be determined if D is known. Or alternatively, if the $E_a(X)$ is known, D can be determined. In addition, the cross-section for anion formation can be determined using electron impact experiments. These data define the vertical electron affinity and slope of a pseudo-two-dimensional curve for the vertical transition. The vertical electron affinity can also be determined in electron transmission spectra. A combination electron transmission and electron impact mass spectrometer now exists that allows the simultaneous determination of both types of data. These data are vital in constructing Morse potential energy curves for dissociative thermal electron attachment.

Experimental data obtained from the electron impact studies of SF_6, C_6H_5Cl, and C_6F_6 are shown in Figures 6.13 to 6.15 [68–70]. The data for SF_6 were obtained in 1979 using a traditional experimental apparatus, whereas the C_6H_5Cl data are an example of data collected using a combined electron transmission and electron impact source. The observation of a negative-ion peak at zero energy is an indication of a positive adiabatic and vertical ground-state electron affinity. The peak in the cross-section for $SF_5(-)$ formation is a measure of an excited-state vertical

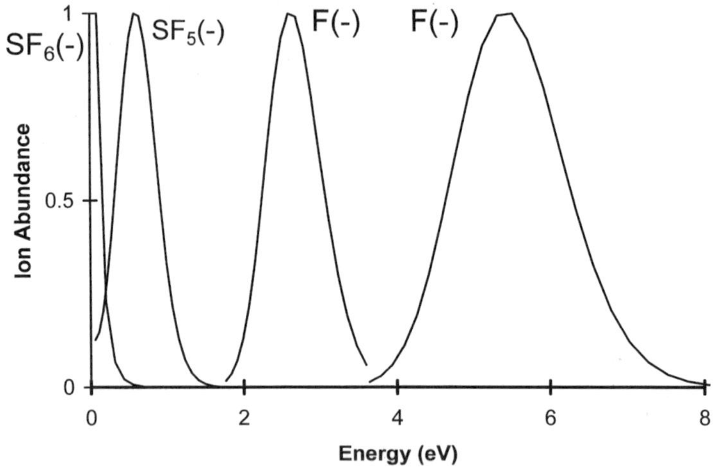

Figure 6.13 Electron impact ion intensities for ions from SF_6 versus the electron energy, replotted from [68].

electron affinity. The data for C_6H_5Cl show a small peak at zero energy. The peak for anion formation at about 0.5 eV is equal to the negative of energy for dissociative electron attachment, $-\text{EDEA} = D - E_a(\text{Cl})$. It gives a value of the C—Cl bond dissociation energy of 4.13 eV since the electron affinity of the Cl atom is well

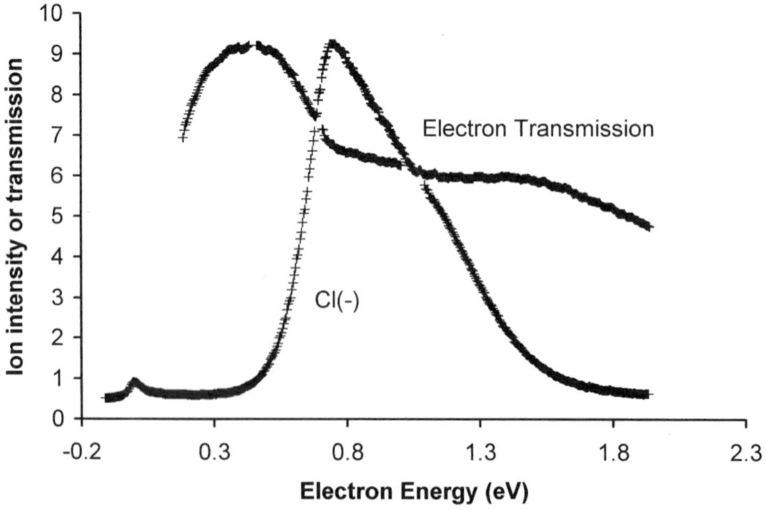

Figure 6.14 Electron impact spectra for Cl(−) from C_6H_5Cl and simultaneous electron transmission spectra versus the electron energy, replotted from [69].

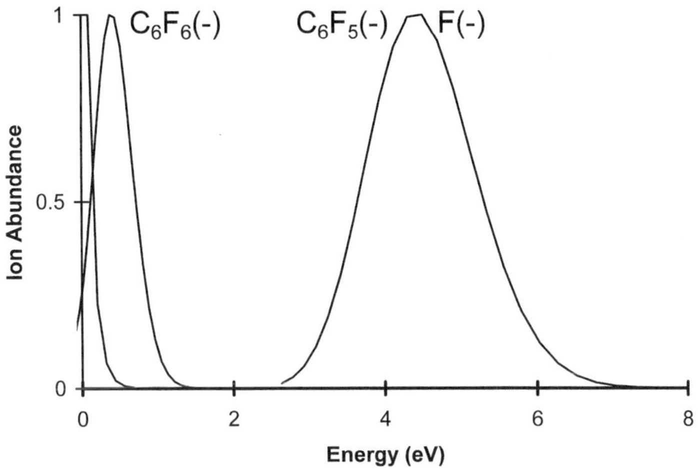

Figure 6.15 Electron impact ion intensities for ions from C_6F_6 versus the electron energy, replotted from [70].

known. The electron transmission data are reported as a direct signal, not the usual derivative signal. The decline at 0.5 eV is a measure of the vertical electron affinity of an excited state. The electron transmisson experiments can only measure negative vertical electron affinities.

The data for C_6F_6 indicate a positive vertical and adiabatic electron affinity for the ground state. The presence of a low-lying excited state was observed in other experiments with a vertical electron affinity of −0.4 eV [71]. The higher-energy resonances also result from excited states. These data do not provide direct estimates of the adiabatic electron affinity of a molecule, but have been used to calculate pseudo-two-dimensional Morse potential curves for the molecular anions using data from various sources including the ECD [72].

In 1974 the electron affinity of tetracyanoquinodimethane (TCNQ) was reported using the alkali metal beam technique [73]. This confirmed the magnetron determination of 2.8 ± 0.2 eV [2]. A precision and accuracy plot for electron affinities determined using the AMB and EnCT method is shown in Figure 6.16. There are clearly values lower than the evaluated values, specifically SF_6, CS_2, and NO. The lower values have been assigned to excited states. The value for fluoranil, 2.92(20) eV, is significantly higher than two TCT values. The EnCT value for C_6F_6 is 1.8(3) eV, significantly higher than other reported values. We do not have an explanation for the high value. The value for $C_6F_5CF_3$ is 1.7(3). For other fluorinated hydrocarbons the endothermic charge transfer values are the only ones available, but parent negative ions have been observed in swarm beam experiments [23, 64, 74–76].

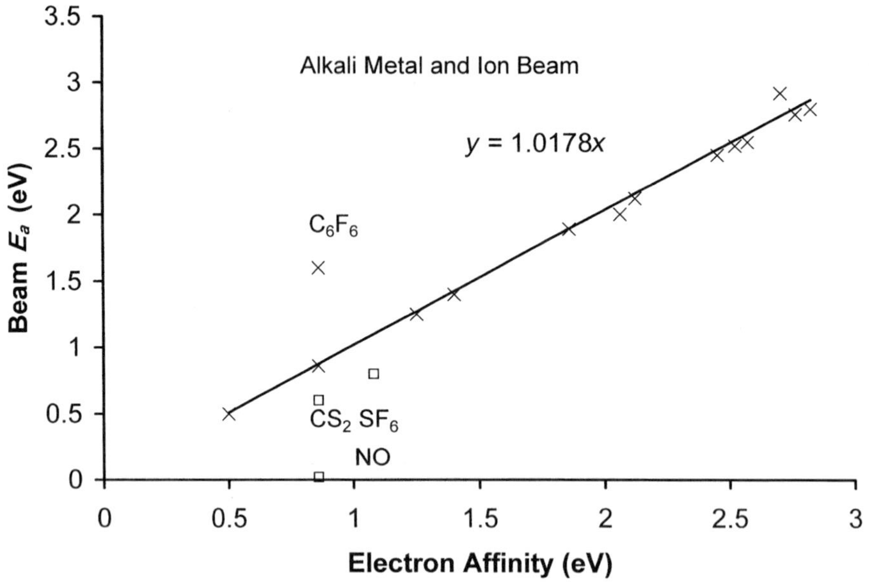

Figure 6.16 Precision and accuracy plot for AMB and EnCT data. The EnCT value for C_6F_6 is an outlier and for fluoranil is higher than the current "best" value. The values for CS_2, SF_6, and NO are significantly lower than the current best value. The random uncertainty in the other values is ±0.10 eV.

6.6 CONDENSED PHASE MEASUREMENTS OF ELECTRON AFFINITIES

The E_a from half-wave reduction potential measurements and energies of charge transfer complex absorption in solution support the gas phase measurements from all the cited techniques. The major difference between the gas phase measurements and the solution or solid phase measurements is the interaction between the solvents or solid phase and the anions. These measurements provide a transition between the low values of valence-state electron affinities measured in the gas phase and the negative valence-state electron affinities. Thus, the valence electron affinities for naphthalene and pyridine are 0.17 eV and 0.0 ± 0.2 eV by solution phase techniques [39, 77].

In Figure 6.17 the electron affinities of several substituted quinones determined from $E_{1/2}$ measurements, and/or charge transfer spectra, are plotted versus the current evaluated gas phase values. They are chosen to give a comparison of the values obtained by the two methods and to note the deviations from the unit slope line. The displacement of the unit slope lines by a constant amount is a systematic uncertainty. This indicates that the solution energy differences and constants used to correlate the charge transfer complex data are both a function of the electron affinity and/or the type of molecules [78]. Thus, it will be possible to reduce deviations by classifying the molecules and identifying the functional relationship.

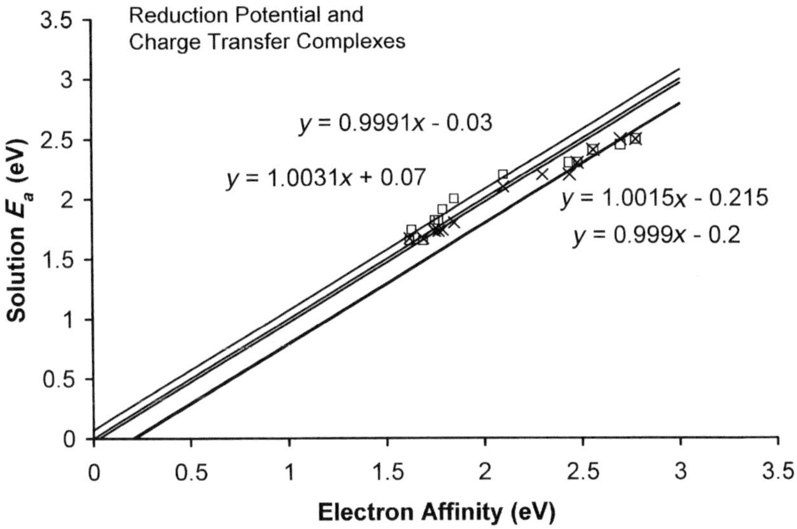

Figure 6.17 Precision and accuracy plot for reduction potential and charge transfer complex E_a for various quinones; data published in 1975 [9, 78]. Systematic uncertainties can be identified as shown by parallel lines.

There are many other types of solution data that support the half-wave reduction potential and charge transfer complex data. These include the measurement of cell potentials or equilibrium constants for electron transfer reactions. Another important condensed phase measurement involving a negative ion is the determination of electron spin resonance spectra. In these studies the existence of a stable molecular anion is established and the spin densities can be measured [79]. The condensed phase measurements support the electron affinities in the gas phase and extend the measurements to lower valence-state electron affinities.

6.7 COMPLEMENTARY THEORETICAL CALCULATIONS

> The underlying physical laws necessary for the mathematical theory of a large part of physics and the whole of chemistry are completely known and the difficulty is only that the application of these laws leads to equations much too complicated to be soluable.
>
> —P. A. M. Dirac
> *Proceedings of the Royal Society (London)*

For the calculation of electron affinities the laws are all known, but it is not possible to rigorously calculate the integrals with exact Hamiltonians and wave functions. It is possible to calculate the electron affinities of some atoms and molecules, but the problem of obtaining agreement with experiment to chemical accuracy remains (0.04 eV). The possibility of predicting the values of electron affinities by exact quantum mechanical principles continues to be a distant one. Thus, many semi-empirical procedures are used to simplify the calculations. In this section the complementary procedures for the theoretical calculation of the electron affinities of atoms, homonuclear diatomic molecules, small molecules, and large organic molecules are briefly discussed. A complete review of theoretical calculations is beyond the scope of this book.

Many of the electron affinities for the elements of the lower main group have been calculated accurately using quantum mechanical procedures. The electron affinity of the hydrogen atom was the first calculated value more accurate than experiment. The electron affinities of elements with large numbers of electrons are not as accurate as experiment. The primary reason for this is electron correlation. This section summarizes the calculation of theoretical atomic electron affinities and the values based on the isoelectronic principle. The simple molecular orbital theory of the homonuclear diatomic molecules is considered. Next, the theoretical effect of geometry will be investigated for triatomic molecules, clusters, and small molecules. Then the theoretical treatment of larger aromatic molecules will be discussed, beginning with the Huckel theory, semi-empirical SCF methods, and closing with a comparison of the ab initio and density functional results for the larger molecules with the CURES-EC procedure.

6.7.1 Atomic Electron Affinities

The problem of calculating the electron affinity of an atom consists of evaluating the total energy of the negative ion and the total energy of the neutral atom:

$$AE_a = E(N+1) - E(N) \tag{6.2}$$

where $E(N+1)$ is the total energy of the ion and $E(N)$ the total energy of the atom. For atoms the geometry is the same in the neutral and valence-state anions. For polarization anions, such as those with filled or half-filled orbitals, the distance between the additional electron and the neutral atom can be greater than the average size of the neutral atom. This separation will be the distance at which the polarization attraction is balanced by the repulsion between the two species. The AE_a of these atoms will be positive but small. The calculation of valence-state AE_a for the remaining elements is more difficult because two ab initio calculations must be carried out and a small difference between large numbers obtained. In addition, unlike the case for the neutral or positive ion, the addition of an electron to the valence shell requires an estimate of the modified correlation energy.

Two general quantum mechanical procedures were applied to the calculation of atomic electron affinities prior to 1960. These were the variational method and Hartree Fock semi-empirical method. Prior to 1960 there were no accurate calculations for elements other than hydrogen. Indeed, only the electron affinities of C and

F were positive without correlation and relativistic effects. However, a simple statement of the variational principle was used to suggest that

> The energy of the wave function for N electrons, over the normal nonrelativistic Hamiltonian describing their motion in the field of a finite number of nuclei fixed within a bounded region, is an upper bound to the ground Hartree Fock energy of an $N + 1$ electron system in the identical nuclear field.... A generalization of this theorem holds, implying in particular that the electron affinities of one-center systems are always non-negative. [81]

Thus, the electron affinities of the rare gases, nitrogen, and group II elements are positive but small, and the AE_a of the remaining elements should be larger than these.

The electron affinity of the hydrogen atom was calculated to chemical accuracy in 1930 using the variational method. A value of 0.74(4) eV is compared with the EvV of 0.75419(2) eV. This was obtained by applying the variational principle to approximate wave functions for the neutral and anion. In 1962 C. L. Pekeris used 444 parameters and obtained a value of 0.75421 eV. Until 1991 this was the most accurate and precise value for the electron affinity of the H atom [82–85]. The calculation of electron affinities of atoms beyond hydrogen were challenges to theoretical chemists until recently. The earliest calculations gave negative electron affinities for the first row elements, except for F and C. The problem was that the Hartree Fock method only considered the correlation of electrons with parallel spins [85].

As more experimental values of atomic electron affinities were obtained, the theoretical calculations agreed with experiment values for many of the elements. For example, the addition of electron correlation to the self-consistent field calculations were used to verify the electron affinities of Li, B, C, N, O, F, Na, Al, Si, S, Cl, K, and some of the transition metals in the 1960s. The standard deviation of these values from the EvV is about 0.15 eV, setting the negative values to zero. However, some of the predictions are higher than the current best experimental values by as much as 0.4 eV. By the 1980s the calculations were extended to higher atomic weights with about the same standard deviation. The maximum deviations were as large as 0.6 eV for some of the higher transition metals [85]. More recently, ab initio calculations of the E_a of the main group elements up to Cl have been improved so that the standard deviation of the values from experiment has been reduced to 0.03 eV [8].

The electron affinities of the atoms can also be estimated by semi-empirical procedures. For the main group elements the values for a given family should be relatively constant since they have the same outer electronic configuration. Except for nitrogen and oxygen, this is true. The values of the group IIA and IIB elements and the rare gases are expected to be close to zero. The experimental values for the halogens range from 3.1 eV to 3.6 eV and average 3.4 ± 0.15 eV. Likewise, the values for H and the alkali metals range from 0.75 eV to 0.5 eV and average 0.55 ± 0.05 eV. The Group IB elements—Cu, Ag, and Au—are another exception to this constant value. The values for Cu and Ag are the same, but that for Au is 1 eV higher. This is larger than the change from N to P, 0.75 eV, or O to S, 0.5 eV. The change from Pd to Pt represents the largest change from one row to another at 1.5 eV. The horizontal analysis can also be used to estimate electron affinities. Except for the first row the periodic behavior of the main group elements is about the same. In the first row the values for N and O are low, but fall in the expected order. The partial filling of the p shell at N is clearly responsible for the near-zero electron affinity.

6.7.2 Polyatomic Molecules

The extension of the quantum mechanical calculations to polyatomic species introduces a further complication, the change in the geometry of the anion relative to that of the neutral. The valence-state electron affinities of the homonuclear diatomic molecules can be calculated from simple molecular orbital theory. For the H_2 molecule there are two bonding electrons. In the molecular anion the additional electron goes into an antibonding orbital so that there is only one net bonding electron. Thus, the bond energy of the anion should be about one-half that of H_2. Since the electron affinity of the H atom is less than one-half the bond energy of H_2, the valence-state electron affinity will be negative. The equation is $E_a(\mathrm{H_2}) - E_a(\mathrm{H}) = D(\mathrm{H_2}(-)) - D(\mathrm{H_2}) = -\frac{1}{2}D(\mathrm{H_2})$, which gives $EA(\mathrm{H_2}) = -2.75 \pm 0.75$ eV or -2 eV. However, just as in the case of atoms, the most stable form of the anion will be an electron at some distance r and a hydrogen molecule. Quantum mechanical calculations give a molecule and "free" electron. This will result in a positive but small AE_a. A similar situation occurs for the N_2 molecule. In the case of N_2 the number of bonding electrons is six, indicating a triple bond. The addition of another electron to an antibonding orbital reduces this to five. Thus, the bond dissociation energy of the anion is $5/6D(\mathrm{N_2})$. When this is combined with the negligible electron affinity of the nitrogen atom, the valence-state electron affinity of N_2 is $-1/6D(\mathrm{N_2})$ or -2 eV.

The valence-state electron affinities of the remaining homonuclear diatomic molecules can be estimated from the number of bonding and antibonding electrons and the bond dissociation energy of the neutral and electron affinity of the atom. The group I and group VII elements have a relative bond order of $\frac{1}{2}$ (two net bonding electrons in the neutral and one in the anion); $D_e(X_2^-)/D_e(X_2) =$ the net number of bonding electron in the ion divided by that of the neutral; and $E_a(X_2, M_2) = 1/2D(X_2, M_2) - E_a(X, M)$. Within this approximation the electron affinities of all the homonuclear diatomic molecules in a given family will have the same relative bond order. However, the experimental values for the group I and VII elements range from 0.55 to 1.0. The predicted value for groups III and IV are 1.5 and 1.25 but most experimental values are larger. This is primarily due to the additional correlation energy in the anion.

The simple molecular orbital theory of bonding in homonuclear diatomic molecules can be used to estimate the electron affinities of clusters. In these cases, there can be different geometries. The C_n clusters have been studied most extensively. In the case of the triatomic molecules, there are now two distances and one angle that

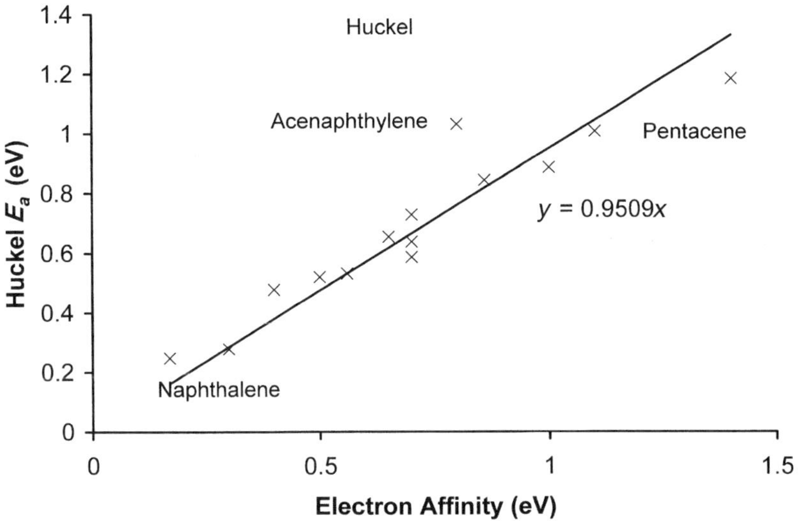

Figure 6.18 Precision and accuracy plot for Huckel E_a for aromatic hydrocarbons. The slope of the zero intercept line is less than 1, indicating systematic uncertainties [79].

can be different in the anion than in the neutral. The electron affinities of the molecules that are linear in the neutral and become bent in the anion illustrate these effects dramatically. The triatomic molecules CO_2, CS_2, COS, and N_2O have been studied in the ECD. Many of these electron affinities are still in question. There are clearly two bound electronic states of the anion, the linear form and the bent form. The E_a can also be determined using CURES-EC. Many of the electron affinities of these molecules have also been calculated using density functional methods. Other small polyatomic molecules studied using the ECD and CURES-EC are nitromethane, biacetyl, and SF_6.

The calculations of the E_a for species containing fewer than six heavy atoms agree with experimental values to within less than 0.1 eV. However, for all but these smallest molecules the more rigorous techniques still give E_a that agree with experiment to better than 0.1–0.2 eV. The use of simple density functional theory for small radicals where the anion is a closed shell species and the neutral is an open shell gives absolute and standard deviations of about 0.12 eV. Where the neutral is a closed shell species and the anion an open shell species, the density functional theory give an average deviation of 0.2 to 0.3 eV. The standard deviations are greater than ± 0.15 eV [8]. Consequently, a need to find a simpler more accurate procedure for calculating the electron affinities of large organic molecules exists. The semi-empirical self-consistent field CURES-EC procedure is such a possibility.

Huckel theory was used to confirm the electron affinities of aromatic hydrocarbons determined with the ECD. Figure 6.18 shows the E_a for several aromatic hydrocarbons calculated from a linear correlation of the Huckel coefficients of the lowest unoccupied molecular orbital (LUMO) versus the EvV from gas phase

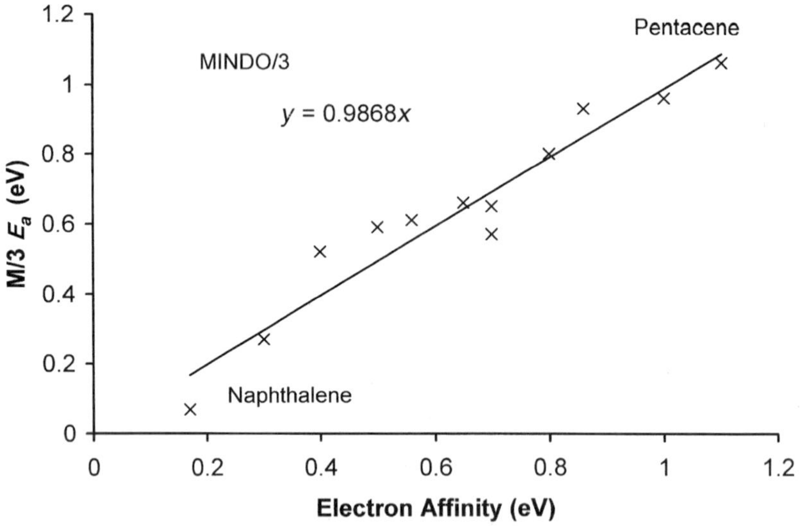

Figure 6.19 Precision and accuracy plot for MINDO/3 E_a for aromatic hydrocarbons. The slope of the zero intercept line is approximately 1, but there are outliers, [89].

measurements. The slope with zero intercept is 0.95. The greatest deviations are ±0.2 eV for pentacene and acenaphthylene. The standard deviation is ±0.12 eV. The MINDO/3 self-consistent field procedures for calculating electron affinities represented a large advance in theoretical calculations. In Figure 6.19 the MINDO/3 semi-empirical self-consistent field calculated values for the above hydrocarbons are plotted against the EvV [39, 85–87]. The slope with a zero intercept becomes 1.00 and the standard deviation is reduced to ±0.1 eV. These results are the basis for the use of CURES-EC to improve the calculated values and experimental electron affinities from half-wave reduction potentials [39]. Modern ab initio and density functional techniques have been applied to the calculation of the E_a of aromatic hydrocarbons, but insufficient values to test the methods exist [8]. The MNDO semi-empirical self-consistent field procedures have been used to calculate electron affinities. In the first calculations using the MNDO procedure it was concluded that the MNDO procedure reproduced the available experimental electron affinities of a variety of "delocalized" radicals with a mean error of ±0.4 eV [87]. There were some notable successes, such as the calculation of the electron affinity of benzoquinone, 1.62, 1.88 eV versus 1.860 eV, and TCNE, 3.06 eV versus 2.95 eV [85, 87]. Extensive calculations of electron affinities using the MNDO, AM1, and PM3 procedures did not occur until the development of CURES-EC.

The electron affinities of a series of substituted quinones have been calculated using the hybrid Hartree Fock/density functional B3LYP method with a 6-311G(3d,p) basis set. The precision and accuracy plot for the E_a obtained from

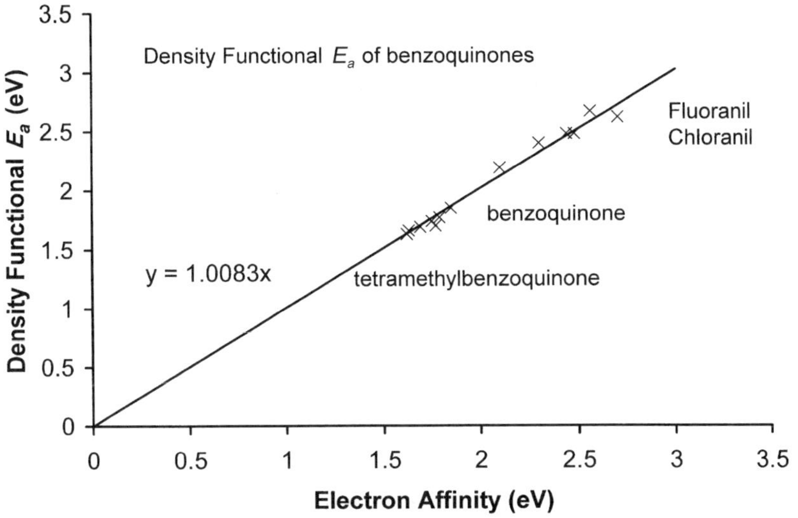

Figure 6.20 Precision and accuracy plot for density functional E_a for the quinones in Figure 6.17. The slope of the zero intercept line is approximately 1 and the deviations are random [9].

these calculations is shown in Figure 6.20. However, these calculations have not been extended to enough molecules to test their predictive ability [9]. Calculations for 20 substituted nitrobenzenes have been carried out using the Hartree Fock procedure and the 6-31+G* basis set to find the optimized geometry for the reactants and products in the thermal charge transfer reactions: $AB(-) + CD \Leftrightarrow CD(-) + AB$, where AB is nitrobenzene. A good correlation between the calculated and experimental energies was obtained [88]. This represents an improvement over the ab initio calculations for hydrocarbons, but the values of the LUMO are still positive, implying a negative E_a. By displacing the LUMO and scaling to the E_a for nitrobenzene of 1.00 eV, the unit slope intercept is 0.0 eV and the standard deviation $\pm$0.13 eV. These are shown in the precision and accuracy plot of Figure 6.21. These values will be discussed further in Chapter 10. The success of this procedure suggests that the LUMO obtained from the semi-empirical SCF calculations could also be correlated to the experimental values. It is important to have theoretical methods that can be used to verify the experimental values because there are so few confirmed experimental E_a. The modified density functional calculations give good results for a limited number of quinones. The ab initio values for aromatic nitro compounds can be scaled to render accurate and precise predictions. However, these still require more computational time than simple SCF semi-empirical calculations. This is especially true for the larger molecules important to organic, biological, and environmental chemistry.

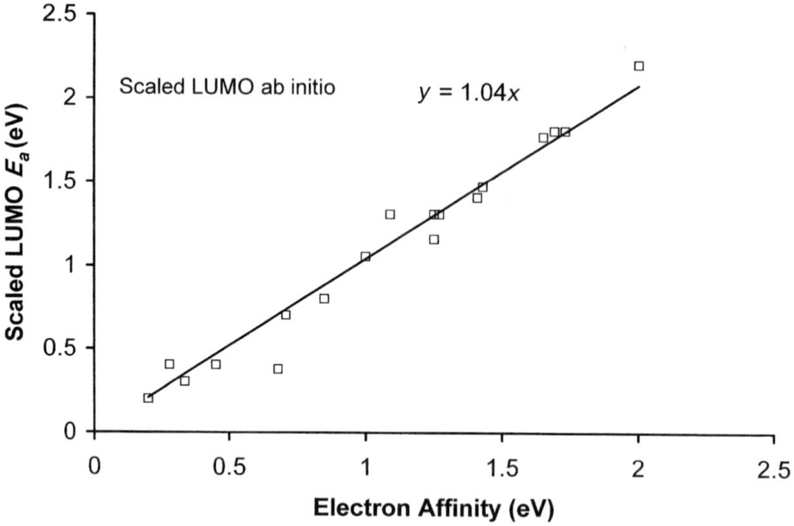

Figure 6.21 Precision and accuracy plot for scaled ab initio LUMO E_a for nitrobenzenes, benzonitriles, benzophenones, and benzaldehydes. The slope of the zero intercept line is approximately 1 and the deviations are random [91].

6.8 RATE CONSTANTS FOR ATTACHMENT, DETACHMENT, AND RECOMBINATION

The specific rate constants of interest to the ECD and NIMS are dissociative and nondissociative electron attachment, electron detachment, unimolecular anion dissociation, and electron and ion recombination. The reactions that have been studied most frequently are electron attachment and electron and ion recombination. To measure recombination coefficients, the electron concentration is measured as a function of time. The values are dependent on the nature of the positive and negative ions and most important on the total pressure in the system. Thus far few experiments have been carried out under the conditions of the NIMS and ECD. However, the values obtained under other conditions suggest that there is a limit to the bimolecular rate constant, just as there is a limit to the value of the rate constant for electron attachment. The bimolecular rate constants for recombination are generally large, on the order of 10^{-7} to 10^{-6} cc/molecule-s or 10^{14} to 10^{15} l/mole-s at about 1 atm pressure. Since the pseudo-first-order rate constants are approximately 100 to 1,000 s^{-1}, the positive-ion concentrations in the ECD and NIMS are about 10^9 ions/cc.

The electron beam and electron swarm experiments [13] can also be used to determine attachment rate constants. However, these are determined as a function of energy and can then be extrapolated to thermal energy. Other techniques used to

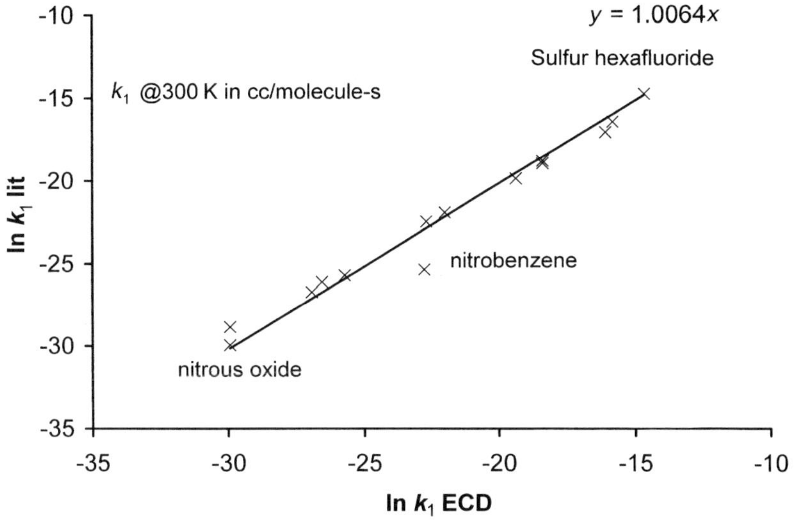

Figure 6.22 Precision and accuracy plot for rate constants determined by other techniques versus those determined by the ECD at room temperature [13, 91].

obtain rate constants for thermal electron attachment include the microwave stationary and flowing afterglow, microwave conductivity method, ion cyclotron method, electron cyclotron resonance method, electron density sampling method, and flowing afterglow Langmuir probe (FALP) method. In general, the rate constants are obtained by measuring the electron concentration as a function of time. A large body of data has been obtained using these techniques. The data are generally only reported at a single temperature [12–14, 89–107]. However, some have been investigated as a function of both electron temperature and the bulk temperature. These studies have been summarized by D. Smith and P. I. Spanel [106].

Figure 6.22 is a log plot of the ECD values versus those obtained by other techniques. If the value of the activation energy has been determined and a value of the rate constant has been measured by a different technique, then a value can be calculated for A_1. Alternatively, a value of A_1 equal to the DeBA could be assumed to calculate activation energies. When this is done, the range of activation energies is 0 kcal/mole to 15 kcal/mole, or 0 eV to 0.65 eV. In Chapter 11 this procedure will be applied to compounds of environmental interest. The activation energies measured in recent studies are equal to those obtained by the ECD [107]. The electron energy has been varied independently from the gross temperature. Significant differences for the types of energy variation were observed, in keeping with the earlier swarm studies of electron attachment cross-sections as a function of average energy [14, 106].

6.9 SUMMARY

The experimental methods for determining electron affinities and rate constants for thermal electron attachment that complement the ECD and NIMS methods of study have been summarized. The nominal uncertainties for each method of measuring electron affinities were established by comparing the values with the current evaluated values. By calibrating to the E_a of benzoquinone and maleic anhydride, the E_a from the general TCT methods are (0.8 eV, benzophenone) < (0.9 eV, C_6F_6) < (1.0 eV, nitrobenzene) < (1.1 eV, p-F-nitrobenzene) < (1.2 eV, phthalic anhydride) < (1.3 eV, m-Cl-nitrobenzene) < (1.4 eV, maleic anhydride) < (1.82 eV, naphthoquinone) < (1.85 eV, benzoquinone), all ±0.1 eV. Some values are lower than the evaluated values and have been assigned to excited states. These are SF_6, CS_2, and NO from AMB, the ICR value for fluoroanil (2.45(5) eV), and the HPMS values for C_6F_6 (0.52(10) eV), CS_2 (0.62(10) eV), azulene (0.70(10) eV), and anthracene (0.60(10) eV). The magnetron values for SF_6, (1.4(2)) and (C_6F_6) (1.2(2)), and the AMB value for fluoranil (2.9(2) eV) are high but included in the weighted average values. The EnCT value of 1.8(3) eV for C_6F_6 is significantly larger than the EvV and included in the average.

The theoretical methods for calculating the electron affinities of atoms, diatomic molecules, and polyatomic molecules have been summarized and compared with the CURES-EC method for molecules. The density functional calculations of the electron affinities of substituted benzoquinones and scaled ab initio LUMO agree with the evaluated values for nitrobenzenes.

The rate constants for thermal electron attachment by alternative techniques are compared to those obtained with the ECD. A method of calculating activation energies for rate constants measured at a single temperature is suggested.

REFERENCES

1. Chen, E. C. M. and Wentworth, W. E. *Mol. Cryst. Liq. Cryst.* **1989**, 171, 271.
2. Page, F. M. and Goode, G. C. *Negative Ions and the Magnetron*. New York: Wiley-Interscience, **1969**.
3. Kebarle, P. and Chowhury, S. *Chem. Rev.* **1987**, 87, 513.
4. Kleyn, A. W. and Moutinho, A. M. C. *J. Phys. B* **2001**, 34, R1.
5. Dillard, J. G. *Chem. Rev.* **1973**, 73, 700.
6. Jordan, K. D. and Burrow, P. D. *Chem. Rev.* **1987**, 87, 557.
7. Reinstra-Kiracofe, J. C.; Tschumper, G. S.; Schaefer, H. F.; Nandi, S.; and Ellison, G. B. *Chem. Rev.* **2002**, 102, 231.
8. Reinstra-Kiracofe, J. C.; Barden, C. J.; Brown, S. T.; and Schaefer, H. F. *J. Phys. Chem. A* **2001**, 105, 524.
9. Boesch, S. E.; Grafton, A. K.; and Wheeler, R. A. *J. Phys. Chem.* **1996**, 100, 10083.

10. Christodoulides, A. A.; McCorkle, D. L.; and Christophorou, L. G. "Electron Affinities of Atoms, Molecules and Radicals" in *Electron-Molecule Interactions and Their Applications*. New York: Academic Press, **1984**.
11. National Institute of Standards and Technology (NIST). *Chemistry WebBook*, **2003**. Available at http://webbook.nist.gov.
12. Caledonia, G. E. *Chem. Rev.* **1975**, 75, 333.
13. Christodoulides, A. A.; McCorkle, D. L.; and Christophorou, L. G. "Electron Attachment Processes" in *Electron-Molecule Interactions and Their Applications*. New York: Academic Press, **1984**.
14. Spanel, P.; Matejcik, S.; and Smith, D. *J. Phys. B* **1995**, 28, 2941.
15. Stemmler, E. A. and Hites, R. A. *Electron Capture Negative Ion Mass Spectra*. New York: VCH, **1988**.
16. Pack J. L. and Phelps, V. *Phys. Rev. Lett.* **1961**, 6, 111.
17. Freeman, R. R. Ph. D. dissertation, University of Houston, **1971**.
18. Van de Wiel, H. J. and Tommasen, H. *J. Chromatogr.* **1972**, 71, 1.
19. Rolla, L. and Picardi, G. *Atti. Accad. Linef.* **1935**, 2VI, 29, 128, 73.
20. Smith, H. and Sugden, T. M. *Proc. Roy. Soc.(Lond.)* **1952**, A211, 31, 58.
21. Sutton, P. and Mayer, J. E. *J. Chem. Phys.* **1935**, 3, 20.
22. Glockler, G. and Calvin, M. *J. Chem. Phys.* **1936**, 4, 492.
23. Christophorou, L. G. and Datkos, P. G. *Int. J. Mass Spectrom. Ion Proc.* **1995**, 149/150, 59.
24. Mock, R. S. and Grimsrud, E. P. *J. Amer. Chem. Soc.* **1989**, 111, 2861.
25. Brinkman, E. A.; Gunther, E.; and Brauman, J. I. *J. Chem. Phys.* **1991**, 95, 6185.
26. Smith, S. J. and Branscomb, L. M. *J. Res. Nat. Bur. Stds.* **1955**, 55, 165.
27. Burch, D. S.; Smith, S. J.; and Branscomb, L. M. *Phys. Rev.* **1958**, 112, 171.
28. Branscomb, L. M. (ed.). *Photodetachment in Atomic and Molecular Processes*. New York: Academic Press, **1962**, p. 130.
29. Berry, R. S. and Reimann, C. W. *J. Chem. Phys.* **1963**, 38, 1540.
30. Scheidt, J. and Weinkauf, R. *Chem. Phys. Lett.* **1997**, 266, 201.
31. Chen, E. S. D. and Chen, E. C. M. *J. Chromatogr. A* **2002**, 952, 173.
32. Oakes, J. M. and Ellison, G. B. *Tetrahedron*, **1986**, 42, 6263.
33. Schiedt, J. and Weinkauf, R. *Z. Natforsch.* **1995**, 50a, 1041.
34. Chen, E. C. M.; George, R.; Carr, S.; Wentworth, W. E.; and Chen, E. S. D. *J. Chromatogr. A* **1998**, 811, 250.
35. Kraus, K.; Muller-Duysing, W.; and Neuert. *Z. Natforsch.* **1961**, 16a, 1385.
36. Hughes, B. M.; Lifschitz, C.; and Tiernen, T. O. *J. Chem. Phys.* **1973**, 59, 3162.
37. Compton, R. N.; Reinhardt, P. W.; and Cooper, C. D. *J. Chem. Phys.* **1975**, 63, 3821.
38. Duncan, M. A. M.; Knight, M.; Negishi, Y.; Nagao, S.; Nakamura, Y.; Kato, A.; Nakajima, A.; and Kaya, K. *Chem. Phys. Lett.* **1999**, 309, 49.
39. Chen, E. S. D.; Chen, E. C. M.; Sane, N.; Talley, L.; Kozanecki, N.; and Shultze, S. *J. Chem. Phys.* **1999**, 110, 9319.
40. Wojnarvits, L. and Foldiak, G. *J. Chromatogr.* **1981**, 206, 511.

41. Crocker, L.; Wang, T. B.; and Kebarle, P. *J. Amer. Chem. Soc.* **1993**, 115, 7818.
42. Chen, E. C. M.; Carr, S.; Wentworth, W. E.; and Chen, E. S. D. *J. Chromatogr. A* **1998**, 827, 91.
43. Desfrancois, C.; Periquet, V.; Lyapustina, S. A.; Lippa, T. P.; Robinson, D. W.; Bowen, K. H.; Nonaka, H.; and Compton. *J. Chem. Phys.* **1999**, 111, 4569.
44. Chen, E. C. M.; Welk, N.; Chen, E. S. D.; and Wentworth, W. E. *J. Phys. Chem. A* **1999**, 103, 9072.
45. Compton, R. N.; Carman Jr.; Desfrancois, C.; Abdoul-Carmine, H.; Schermann, J. P.; and Hendricks, J. H. *J. Chem. Phys.* **1996**, 105, 3472.
46. Lecomte, F.; Carles, S.; Desfrancois, C.; and Johnson, M. A. *J. Chem. Phys.* **2000**, 113, 10973.
47. Lyons, L. E. and Palmer, L. D. *Aust. J. Chem.* **1976**, 29, 1919.
48. Lyons, L. E. and Palmer, L. D. *Int. J. Mass Spectrom. Ion Phys.* **1975**, 16, 431.
49. Lyons, L. E. and Palmer, L. D. *Chem. Phys. Lett.* **1973**, 21, 442.
50. Wenthold, P. G.; Hrovat, D. A.; Borden, W. T.; and Lineberger, W. C. *Science* **1996**, 272, 1456.
51. Scheidt, J.; Weinkauf, R.; Neumark, D. M.; and Schlag, E. W. *Chem. Phys.* **1998**, 123, 511.
52. Bartmess, J. E. and McIver, R. T. "The Gas Phase Acidity Scale," in *Gas Phase Ion Chemistry*, Vol. 2, edited by M. T. Bowers. New York: Academic Press, **1979**, p. 88.
53. Henglein, A. and Muccini, G. A. *J. Chem. Phys.* **1959**, 31, 1426.
54. Kraus, K.; Muller-Duysing, W.; and Neuert, H. *Z. Naturfor.* **1961**, 16A, 1385.
55. Goldan, P. D.; Schmeltekopf, A. L.; Schiff, H. I.; and Ferguson, E. E. *J. Chem. Phys.* **1966**, 44, 4095.
56. Rains, L. J.; Moore, H. W.; and McIver, R. T. *J. Chem. Phys.* **1978**, 68, 3309.
57. Fukuda, E. K. and McIver, R. T. In *Lecture Notes in Chemistry*, edited by H. Hartman and K. P. Wanzek. Berlin: Springer, **1982**, p. 164.
58. Fukuda, E. K. and McIver, R. T. *J. Chem. Phys.* **1982**, 77, 4942.
59. Fukuda, E. K. and McIver, R. T. *J. Amer. Chem. Soc.* **1985**, 107, 2291.
60. Burinsky, D. J.; Fukuda, E. K.; and Campana, J. E. *J. Amer. Chem. Soc.* **1984**, 106, 2270.
61. Chen, G.; Cooks, R. G.; Corpuz, E.; and Scott, L. T. *J. Amer. Soc. Mass Spectrom.* **1996**, 7, 619.
62. Denault, J. W.; Chen, G. D.; and Cooks, R. G. *J. Amer. Soc. Mass Spectrom.* **1998**, 9, 1141.
63. Chen, G. and Cooks, R. G. *J. Mass Spectrom.* **1997**, 32, 333.
64. Chupka, W. A.; Berkowitz, J.; and Gutman, D. *J. Chem. Phys.* **1971**, 55, 2724.
65. Hughes, B. M.; Lifschitz, C.; and Tiernan, T. O. *J. Chem. Phys.* **1973**, 59, 3162.
66. Lifshitz, C.; Tiernan, T. O.; and Hughes, B. M. *J. Chem. Phys.* **1973**, 59, 3182.
67. Laramee, J. A.; Mazurkiewicz, P.; Berkout, V.; and Deinzer, M. L. "Discrete Electron Capture Negative Ion Mass Spectrometry" in *Encyclopedia of Analytical Chemistry*. New York: Wiley, **2000**.
68. Kline, L. E.; Davies, D. K.; Chen, C. L.; and Chantry, P. J. *J. Appl. Phys.* **1979**, 50, 6789.

69. Dressler, R.; Allan, M.; and Haselbach, E. *Chemia* **1985**, 39, 385.
70. Fenzlaff, H. and Illenberger, E. *Int. J. Mass Spectrom. Ion Phys.* **1984**, 32, 185.
71. Spyrou, S. M. and Christophorou, L. G. *J. Chem. Phys.* **1985**, 82, 1048.
72. Wentworth, W. E.; Limero, T.; and Chen, E. C. M. *J. Phys. Chem.* **1987**, 91, 241.
73. Klots, C. E.; Compton, R. N.; and Raaen, V. F. *J. Chem. Phys.* **1974**, 60, 1177.
74. Hiraoka, K.; Mizuno, T.; Eguchi, D.; Takao, K.; Iino, T.; and Yamabe, S. *J. Chem. Phys.* **2002**, 116, 7574.
75. Miller, T. M.; Morris, R. A.; Miller, A. E. S.; Viggiano, A. A.; and Paulson, J. F. *Int. J. Mass Spectrom. Ion Proc.* **1994**, 135, 195.
76. Dillow, G. W. and Kebarle, P. *J. Amer. Chem. Soc.* **1989**, 111, 5592.
77. Chen, E. S. D.; Chen, E. C. M.; Sane, N.; and Shultze, S. *Bioelectrochem. Bioenerget.* **1999**, 48, 69.
78. Chen, E. C. M. and Wentworth, W. E. *J. Chem. Phys.* **1975**, 63, 3183.
79. Streitweiser, A. S. *Molecular Orbital Theory for Organic Chemists.* New York: Wiley, **1961**, pp. 159–172.
80. Dirac, P. *Proc. Roy. Soc. Lond.* **1929**, A123, 714.
81. Lesk, A. M. *Phys. Rev.* **1968**, 171, 7.
82. Hylleras, E. *Z. Phys.* **1930**, 60, 624.
83. Pekeris, C. L. *Phys. Rev.* **1962**, 126, 1470.
84. Lykke, K. R.; Murray, K. K.; and Lineberger, W. C. *Phys. Rev. A* **1991**, 43, 6104.
85. Weis, A. *Phys. Rev.* **1961**, 122, 1826.
86. Clementi, E. *Phys. Rev.* **1964**, 135, A980.
87. Curtis, L. A.; Redfern, P. C.; Raghavachari, K.; and Pople, J. A. *J. Chem. Phys.* **1998**, 109, 42.
88. Younkin, J. M.; Smith, L. J.; and Compton, R. N. *Theoret. Chem. Acta* **1976**, 41, 157.
89. Dewar, M. J. S.; Hashmall, J. A.; and Trinajatic. *J. Amer. Chem. Soc.* **1970**, 92, 5555.
90. Dewar, M. J. S. and Rzepa, H. S. *J. Amer. Chem. Soc.* **1978**, 100, 784.
91. Huh, C.; Kang, C. H.; Lee, H. W.; Nakamura, H.; Mishima, M.; Tsuno, Y.; and Yamataka, H. *Bull. Chem. Soc. Japan* **1999**, 72, 1083.
92. Lee, T. J. *J. Phys. Chem.* **1963**, 76, 360.
93. Mahan, B. H. *J. Chem. Phys.* **1966**, 44, 2182.
94. Chen, E. C. M.; George, R. D.; and Wentworth, W. E. *J. Chem. Phys.* **1968**, 49, 1973.
95. Fehsenfeld, F. C. *J. Chem. Phys.* **1970**, 53, 2000.
96. Fessenden, R. W. and Bansal, K. M. *J. Chem. Phys.* **1970**, 53, 3468.
97. Mothes, K. G. and Schindler. *Ber. Bunsenges Phys. Chem.* **1971**, 75, 936.
98. Davis, F. J.; Compton, R. N.; and Nelson, D. R. *J. Chem. Phys.* **1973**, 59, 2324.
99. Gant, K. S. and Christophorou, L. G. *J. Chem. Phys.* **1976**, 65, 2977.
100. Hildebrandt, G. F.; Kellert, F. G.; Dunning, F. B.; Smith, K. A.; and Stebbings, R. F. *J. Chem. Phys.* **1978**, 68, 1349.
101. Crompton, R. W. and Haddad, G. N. *Aust. J. Phys.* **1983**, 36, 15.
102. Spyrou, S. M. and Christophorou, L. G. *J. Chem. Phys.* **1985**, 82, 1048.

103. Adams, N. G.; Smith, D.; Alge, E.; and Burdon, J. *Chem. Phys. Lett.* **1985**, 116, 460.

104. Chutjian, A. and Alajajian, S. H. *J. Phys. B* **1985**, 18, 4159.

105. Wentworth, W. E.; Limero, T.; and Chen, E. C. M. *J. Phys. Chem.* **1987**, 91, 241.

106. Shimamori, H.; Tatsumi, Y.; Ogawa, Y.; and Sunagawa, T. *J. Chem. Phys.* **1992**, 97, 6335.

107. Shimamori, H.; Tatsumi, Y.; and Sunagawa, T. *J. Chem. Phys.* **1993**, 99, 7787.

108. Miller, T. M.; Miller, S.; and Paulson, J. F. *J. Chem. Phys.* **1994**, 100 8841.

109. Smith, D. and Spanel, P. I. *Adv. At. Mol. Opt. Phys.* **1994**, 32, 307.

110. Matejcik, S.; Mark, T. D.; Spanel, P.; Smith, D.; Jaffke, T.; and Illenberger, E. *J. Chem. Phys.* **1995**, 102, 2516.

111. Vogt, D.; Hauffe, B.; and Neuert, H. *Z. Phys.* **1970**, *232*, 439.

112. Bailey, T. L. and Mahadevan, P. *J. Chem. Phys.* **1970**, 52, 179.

113. Stockdale, J. A. D.; Compton, R. N.; Hurst, G. S.; and Reinhardt, P. W. *J. Chem. Phys.* **1969**, 50, 2176.

114. D'sa, E. D.; Wentworth, W. E.; Batten, C. F.; Shuie, L. R.; and Chen, E. C. M. *J. Chromatogr.* **1987**, 390, 249.

CHAPTER 7

Consolidating Experimental, Theoretical, and Empirical Data

7.1 INTRODUCTION

> The first step in finding a law is to guess at the truth. The second step is to make calculations based on the guess. Then you test the calculations against experimental results. If the calculations do not agree with the experiments, then the guess is wrong.
>
> —Richard Feynman
> *The Pleasure of Finding Things Out*

In this chapter methods of consolidating experimental and theoretical data from various sources will be described and examples given. The semi-empirical SCF procedures developed using the HYPERCHEM software are CURES-EC and READS-TCT. The HYPERCHEM programs were chosen because they are easy to use with standard semi-empirical parameters [1]. The Herschbach classifications and construction of ionic Morse potential energy curves (HIMPEC) are used to characterize negative-ion states [2, 3]. The iso-electronic principle, correlations with charge densities, electronegativities, and substitution and replacement effects are used to predict the electron affinities, ionization potentials, and bond dissociation energies based on values for the parent molecules. A detailed description of the calculations will be presented for selected examples. These techniques will be applied throughout the remainder of the book to atoms, clusters, small molecules, large organic molecules, and environmentally and biologically significant species.

The CURES-EC method was named because it CURES the ELECTRON CORRELATION problem in semi-empirical calculations. This is accomplished by adding multiconfiguration configuration interaction (MCCI) to the wave functions. The acronym CURES-EC stands for configuration interaction or unrestricted orbitals to relate experimental electron affinities to self-consistent field values by estimating electron correlation. Electron affinities of radicals can be obtained from experimental gas phase acidities by calculating C—H bond dissociation

The Electron Capture Detector and the Study of Reactions with Thermal Electrons
by E. C. M. Chen and E. S. D. Chen
ISBN 0-471-32622-4

energies. The charge densities calculated for the anions can be used to classify molecules to improve electron affinities from reduction potentials.

The relative electron affinities of molecules and radicals have been verified by a quantum mechanical procedure simulating the thermal charge transfer between a neutral molecule and an anion. This process is given the acronym READS-TCT, which stands for the determination of RELATIVE ELECTRON AFFINITIES by DIATOMIC SIMULATION OF THERMAL CHARGE TRANSFER reactions. The spin distributions in such simulations reflect electron transfer for nonbonded interactions. The linear acenes, benzene to novacene, are used to illustrate these procedures. The electron affinities of the perfluorinated linear acenes are predicted. The vertical and adiabatic electron affinities of cytosine have been measured by half-wave reduction potentials and PES. These can also be calculated using CURES-EC. The difference between the geometry of two forms of the negative ion of cytosine is illustrated. Such changes result in variations of the spin distribution in the anion.

The procedure for calculation of Morse potentials for negative-ion states from diverse experimental data will be presented. These use the modified classification of negative-ion curves originally proposed by Herschbach [2, 3]. The attractive, repulsive, and frequency portions of the negative-ion curves are related to those for the neutral through three dimensionless constants determined from experimental quantities. The classifications are based on the products of the electron reactions in the Franck Condon region and the VE_a, E_a, and EDEA. Values needed to construct the negative-ion curves can be obtained from quantum mechanical calculations. The Morse potentials consolidated experimental and theoretical data for homonuclear diatomic ions. This resulted in a complete characterization and classification of the states. Extension to polyatomic ions illustrates the effect of geometrical changes in the anions.

Curves for the negative-ion states of H_2 and I_2 are chosen to illustrate the procedures for the homonuclear diatomic molecules. Curves for benzene and naphthalene are examples of excited states for larger molecular negative ions. These illustrate the relationship between gas phase acidities and thermal electron attachment reactions. Such correlation procedures can be applied to systematic predictions for many different problems.

Other empirical concepts have been used to consolidate data. These include the isoelectronic principle, electronegativity correlations, the relationship to absorption spectra, and the effects of single and multiple substitutions of functional groups on molecules. A similar change in the electron affinities is observed when the more electronegative nitrogen atom replaces a CH group in aromatic molecules. Statistical techniques are used to establish these correlations. Because they are empirical and require additional data for their application, the results are less certain.

7.2 SEMI-EMPIRICAL QUANTUM MECHANICAL CALCULATIONS

In calculating ionization potentials or electron affinities, we are comparing two systems with different numbers of electrons. By assuming a constant value of W_c

(the bonding energy of a p_z electron in carbon) we arrive at electron affinities and ionization potentials that are too large. A possibility is that W_c may vary with the formal charge on the molecule. . . . The MOs of a negative ion should be more diffuse than those of the corresponding neutral molecule; our procedure would then overestimate the binding energy in the anion and hence the electron affinity.

—M. J. S. Dewar
The Molecular Orbital Theory of Organic Chemistry

In spite of the development of new methods of measurements, fewer than 300 molecular electron affinities have been measured in the gas phase [5–7]. With the limited number of experimental E_a it is important to have a simple rigorous quantum mechanical procedure for the calculation of electron affinities of large molecules. Electron correlation is the crucial problem in the calculation of E_a. The correlation energies are significant because E_a represent a small difference between two large quantities. Since an electron is added to the system, the effects of geometry changes and correlation reinforce each other rather than cancel out, as in ionization potentials.

Modern desktop computers allow more sophisticated self-consistent field calculations, such as SCF-AM1, MINDO/3, MNDO, PM3, or ZINDO multiconfiguration configuration interaction semi-empirical calculations. Thus, the simple Huckel procedure is no longer used as a "black box" method of calculating quantities such as electron affinities. The MINDO/3 Hamiltonian and parameters reproduce the experimental electron affinities of the aromatic hydrocarbons measured in the gas phase. The largest deviations are 0.12 eV. By adding electron correlation to the calculation of the energies of the ions and neutrals, agreement is improved to 0.05 eV. Application of the same procedure using MINDO/3 for compounds containing O, N, F, S, and Cl as well as carbon and hydrogen was largely unsuccessful. The CURES-EC procedure with the MNDO Hamiltonian and MNDO, AM1, and PM3 parameters gives agreement with experiment, with standard deviations of ± 0.05 eV. Since these procedures reproduce experiment, they can be used to predict values for similar compounds [8, 9].

The quantum mechanical calculations are carried out on a Pentium desktop computer with commercial software. The HYPERCHEM input files (HIN) for the various species contain charge densities and a complete description of the geometric and energy properties of the neutral molecule and anion. These compact files are an efficient way to store and communicate this information [1].

The first step in CURES-EC is to obtain the most appropriate electron affinity, either experimentally or by chemical logic. In the case of aromatic hydrocarbons these are obtained from gas phase experiments or from reduction potentials. The second step is to conduct geometry optimization quantum mechanical calculations for both the neutral and negative ion. If large geometry changes are possible, the structure is annealed. The E_a is the difference in the electronic energies of the neutral and negative ion at the global minimum. Then MCCI is added to minimize the difference between the calculated and selected values. The MINDO/3 Hamiltonian and parameters are best for aromatic hydrocarbons. For molecules containing CHON and X the MNDO Hamiltonian and AM1 parameters are best. Starting from the geometry of the neutral and adding an electron without geometry, optimization yields the VE_a.

Only a few MCCI combinations result in significant changes in the values. These are the UHF and RHF values with up to three filled and three unfilled orbitals, taken pairwise, designated as (UHF, RHF(0000), (0011), (1100), (0022), (2200), (3300), (0033)). The numbers represent the filled and unfilled orbitals used in the RHF configuration interaction for the ion and neutral, respectively. The designation RHF(3300) means that the energy of the neutral is obtained without additional electron correlation (00). For the anion the original parameters did not include enough electron correlation so three filled and three unfilled orbitals (33) were mixed in the MCCI calculation. The RHF(3300) or UHF value will be the maximum value. The minimum value is the RHF(0033) value. This is because the addition of MCCI can only lower the energy of the system so that the greatest difference in energy will occur for the negative ion with the maximum number of filled and unfilled orbitals (three and three) for large molecules. The UHF configuration interaction energy for the anion could be lower than the RHF(33) value and give the maximum E_a. Likewise, the lowest E_a will occur for the anion with no configuration interaction and the neutral with three filled and three unfilled orbitals, RHF(0033). When the neutral is a singlet molecule, the UHF energy is equal to that with no configuration interaction. A standard procedure is to calculate the E_a for the UHF, RHF(0000), (0033), and (3300). For the aromatic hydrocarbons the test value falls between the extremes, and agreement can be optimized. This is done by modifying the MCCI through the use of, for example, (2233). This is summarized as UR3XO, UHF, RHF, CI(33), eXtremes, Optimization. The CURES-EC procedure can only improve on the UHF or RHF(0000) values since these are included in the choices for the optimum. The calculated values have quantum properties since the number of the MCCI orbitals is discrete.

The linear acenes, benzene to pentacene, are used as examples of the CURES-EC procedure. The results obtained utilizing MINDO/3 and AM1 are compared. In addition to calculating the E_a by subtracting the energies of the optimized form, the LUMO of the neutral is compared with the experimental E_a. The electron affinity of hexacene has been estimated from the electronegativity and experimental ionization potential. As a further example of the use of CURES-EC, both the ionization potential and electron affinity of heptacene are estimated. The E_a of octacene and novacene are calculated for comparison to values obtained by using Koopman's theorem and a semi-empirical method based on a variable-parameter modification of the Pariser Parr Pople (PPP) approximation to the Hartree Fock equation [10].

The valence-state electron affinity of benzene is negative from reduction potentials, but the adiabatic electron affinity is positive. The electron affinities of the other linear acenes up to pentacene have been measured in the gas phase. Multiple values are reported for anthracene and naphthalene. The naphthalene E_a are -0.2 eV by electron transmission and others, and 0.16 ± 0.03 eV and 0.13 ± 0.05 eV by ECD. For anthracene the three values are 0.53 ± 0.01 eV by PES, 0.60 ± 0.10 eV by TCT, and 0.54, 0.60, and 0.68 ± 0.02 eV by ECD. In the PES of anthracene there are also peaks at 0.42, 0.54, 0.60, and 0.68 eV

TABLE 7.1 Optimization of CURES-EC for Benzene, Naphthalene, Anthracene, Tetracene, and Pentacene with MINDO/3 and AM1 Parameters

			E_a for MCCI (eV)			opt E_a (eV)	
Molecule	MCCI	UUOO	3300	0033	0000	opt(aabb)	
Benzene	MIN/3	−0.628	−0.483	−1.470	−0.800	−0.803	0033
Naphthalene	MIN/3	−0.058	0.429	−0.458	−0.021	0.139	2233
Anthracene	MIN/3	0.776	0.897	−0.055	0.514	0.655	2300
Tetracene	MIN/3	1.199	1.277	0.412	0.908	1.118	UU21
Pentacene	MIN/3	1.408	1.509	0.753	1.187	1.346	UU22
Benzene	AM1	−0.057	0.429	−0.876	−0.232	−0.848	0022
Naphthalene	AM1	0.049	1.282	−0.165	0.612	0.130	2333
Anthracene	AM1	1.153	1.452	0.733	1.212	0.733	0033
Tetracene	AM1	1.898	1.918	1.089	1.608	1.089	0033
Pentacene	AM1	2.213	2.126	1.442	1.907	1.442	0033

[5–8]. The UR3XO combinations of the MCCI for benzene, naphthalene, anthracene, and pentacene are given in Table 7.1. The calculated E_a for these molecules and about 80 aromatic hydrocarbons have been optimized using the MINDO/3 procedure to give a standard deviation of ±0.03 eV. These will be discussed in Chapter 10 [8].

The experimental and theoretical E_a for the linear acenes are shown in Table 7.2. The optimum CURES-EC values are equal to experiment within the uncertainty. They are compared with MINDO/3 LUMO, the AM1 LUMO, the earlier Pariser Parr Pople E_a, and the calculated VE_a. From the calculation for pentacene the MINDO/3 UHF value at 1.35 eV is the best value. The AM1 results are all larger than those obtained by experiment. The best one, 1.44 eV, is the RHF(0033) or

TABLE 7.2 Experimental and Theoretical Electron Affinities (in eV) for Benzene, Naphthalene, Anthracene, Tetracene, and Pentacene

Molecule	Exp	CEC	PPP	LUMO MIN/3	LUMO AM1	VE_a
Benzene	−0.78	−0.803	−0.87	−1.25	−0.55	−0.9
Naphthalene	0.16	0.139	−0.06	−0.48	0.26	0.0
Anthracene	0.68	0.655	0.49	0.07	0.84	0.5
Tetracene	1.08	1.118	0.84	0.44	1.23	0.96
Pentacene	1.39	1.346	1.06	0.69	1.52	1.31
Hexacene	—	1.6	1.23	0.84	1.73	1.43
Heptacene	—	1.8	1.34	1.02	1.89	1.60
Octacene	—	1.9	1.43	1.13	2.00	1.65
Novacene	—	2.0	1.49	1.21	2.10	1.70

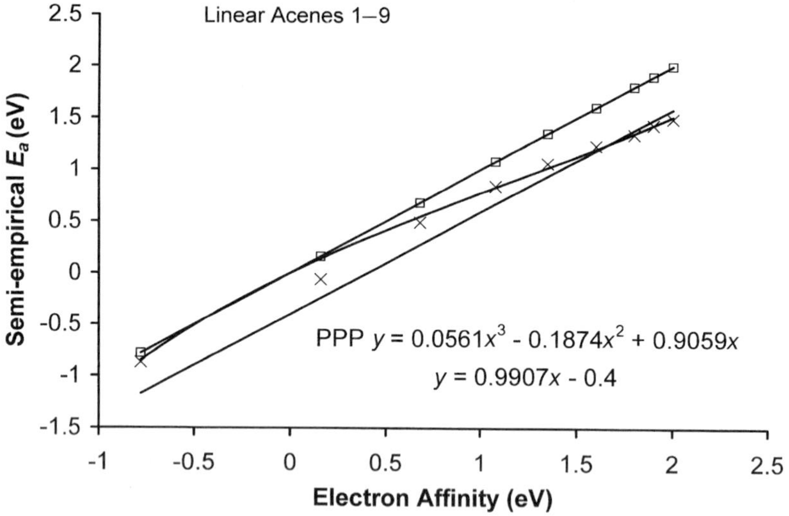

Figure 7.1 Plot of the semi-empirical E_a versus the experimental E_a for the linear acenes 1 to 9. The squares are the CURES-EC values that are equal to the experimental values within the uncertainties [8]. The x's stand for the Pariser Parr Pople calculated values [10]. The latter have systematic uncertainties that vary with magnitude.

minimum value. The use of the AM1 procedure for the higher acenes will overestimate the E_a and the MINDO/3 procedure can underestimate the value. The predicted electron affinity for heptacene is between 1.72 eV and 1.91 eV or 1.8 ± 0.1 eV. The calculated value of the ionization potential for heptacene is 6.14 eV, and when combined with the above E_a, it gives an electronegativity of 3.97 eV, consistent with earlier observations. The PPP value is 1.34 eV [10]. Likewise, the PPP values for octacene and novacene are lower than the CURES-EC values by 0.46 eV. The difference for benzene is 0.2 eV to 0.3 eV. As shown in Figure 7.1, this is an example in which the systematic uncertainty is not constant but depends on magnitude.

The calculated MINDO/3 LUMOs for these molecules are lower than the experimental E_a by 0.70 eV and for AM1 are higher than the experimental E_a by about 0.1 eV. These are shown in Figure 7.2. The systematic uncertainties remain constant. Figure 7.3 provides the optimum values of the VE_a for the linear acenes. The VE_a are between the AE_a and MINDO/3 LUMO values. The calculated VE_a for hexacene to novacene support the AE_a values of 1.65 eV to 2.0 eV. The systematic difference is the rearrangement energy. It is approximately 0.2 eV for anthracene and above.

In both the AM1 and MINDO/3 procedures low-lying excited states are observed. Interestingly, in the case of naphthalene, the splitting is about 0.4 eV, and in the case of the AM1 procedure, one is negative by 0.2 eV, while the other is positive. More significant is the increase in the number of low-lying states that

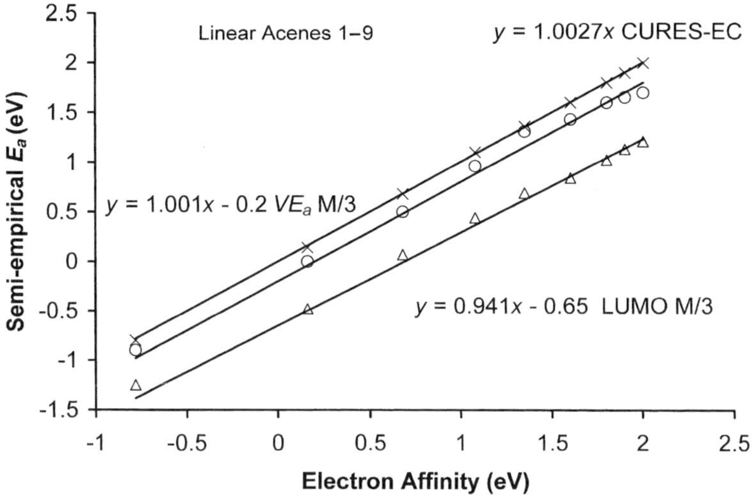

Figure 7.2 Plot of the semi-empirical E_a versus the experimental E_a for the linear acenes 1 to 9. The x's are the CURES-EC values that are equal to the experimental values within the uncertainties. The circles are the calculated VE_a and the triangles the LUMO. These are displaced from the CURES-EC values by a constant amount. This implies that the rearrangement energies are approximately the same for the linear acenes, as determined in [8] and this book.

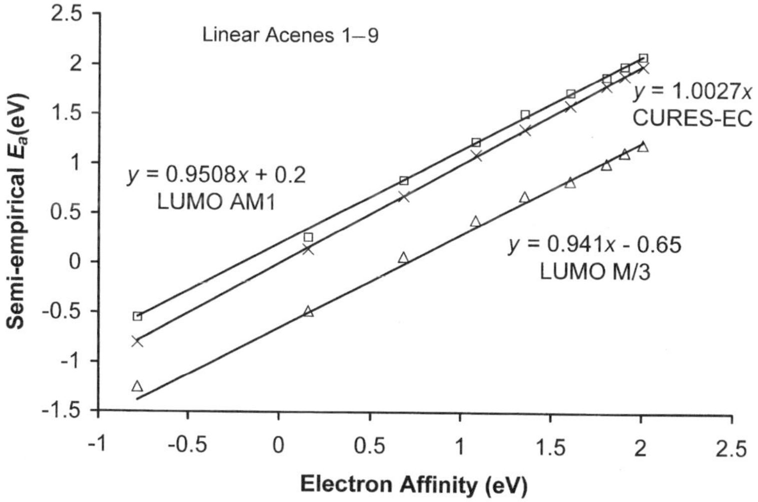

Figure 7.3 Plot of the semi-empirical E_a versus the experimental E_a for the linear acenes 1 to 9. The x's are the CURES-EC values that are equal to the experimental values within the uncertainties. The squares are the AM1 calculated with LUMO, while the triangles are the MINDO/3 LUMO. These are displaced from the CURES-EC values by a constant amount, as determined in [8] and this book.

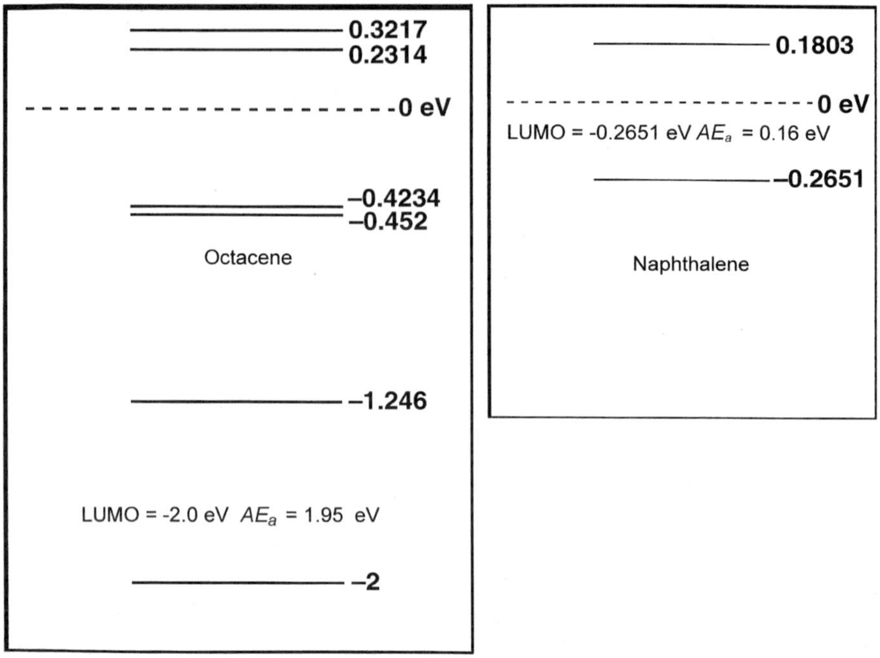

Figure 7.4 Lowest unoccupied molecular orbitals for naphthalene and octacene calculated using AM1. The values are negative, indicating a positive E_a. There are more positive E_a for octacene than naphthalene.

could accommodate the extra electron as the number of rings increased. These are illustrated in Figure 7.4 for naphthalene and octacene using AM1. For octacene there are four unoccupied orbitals with negative energies with a positive electron affinity by Koopman's theorem. The energy of the LUMO is −2.0 eV and gives an E_a of 2.0 eV. The CURES-EC AE_a is 1.90 eV. Several low-lying orbitals have small positive energies and could be involved in configuration interactions. These are related to the different types of C—H bonds with different gas phase acidities. In the case of naphthalene these are the 1 and 2 C—H bonds with gas phase acidities that differ by 3.1 kcal/mole or 0.14 eV [11–13].

The charge densities for the anions can be displayed in HYPERCHEM as illustrated in Figure 7.5, where the charges on C and H are shown for the naphthalene anion. The Values of the linear acenes vary from −0.06 to 0.26q for benzene to −0.05 to 0.15q for naphthalene, to −0.04 to 0.18q for anthracene and tetracene. For pentacene the range is lowered to 0.03 to 0.14q. Thus, the solution energy differences should be smaller for the higher acenes and fullerenes. In predicting the reduction potential value for heptacene, this variation should be included. The reduction potential based on a value of −0.6 V versus Hg pool for hexacene is predicted to be −0.5 V versus the Hg pool. The value, if we assume a constant

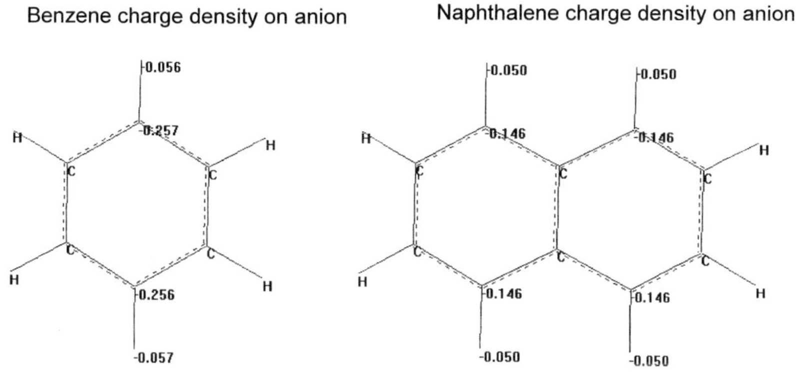

Figure 7.5 Charge densities on the anions of benzene and naphthalene calculated with MINDO/3. The localized charge is lower in naphthalene than benzene.

solution energy difference, is −0.4 V versus Hg pool. The reference energy of the Hg pool electrode is 4.15 V so that for heptacene with an E_a of 1.80 eV and a mdG of 1.85 V, the $E_{1/2} = -4.15 + 1.85 + 1.80 = -0.4$ V.

The application of CURES-EC to CS_2 illustrates a simple molecule where the geometry changes in the anion. The ECD data show two E_a at 0.85 eV and 0.6 eV. The neutral molecule is linear, while the anion is bent. The excited state E_a is the energy difference between the linear neutral and linear anion and is 0.6 eV. The MINDO/3(0000) energy difference between the linear neutral and bent anion is 0.81 eV. The optimum value is 0.88 eV for MINDO/3(2200). The photodetachment energy from the bent anion to the bent neutral is 1.2 eV, compared to the experimental 1.46 ± 0.02 eV.

The linear acenes can also be used to illustrate the READS-TCT procedure. The first step in such a procedure is to establish the charge transfer for two identical molecules when one electron is added to the system. This is done for two anthracene molecules in Figure 7.6. The spin distributions are shown as a three-dimensional plot along with the charge densities. The charge densities are the numbers in the background. The total charge on one of the species can be read from the HYPERCHEM window when one of the molecules is selected. When the species are the same, the charge and spin densities are equally distributed through space without a bridge or bond. For the heteromolecular systems a larger charge and spin distribution is on the molecule with the larger electron affinity. For the anthracene/naphthalene system this is 10 to 1. For higher consecutive acenes the ratio approaches 3 to 1 at heptacene and hexacene. For naphthalene to benzene it is 10 to 1, whereas for anthracene to benzene the charge on benzene is negligible. The READS (phenanthrene/naphthalene) is 3 to 1 for an electron affinity ratio of 0.30/0.17. For coronene/anthracene READS (C/A) is 2 to 1 as shown in Figure 7.7, where anthracene is selected and shows a total charge of 0.34. In Figure 7.8 the relative spin densities for these same two molecules can be seen. By calculating

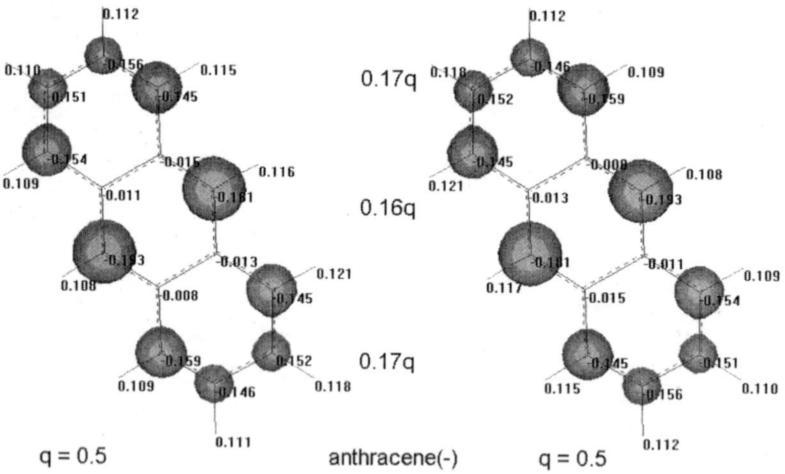

Figure 7.6 Three-dimensional spin densities and charges for one electron distributed on two anthracene molecules calculated using MINDO/3. Both the spin densities and charge densities are equally distributed through space.

the READS (CS_2/coronene) in the bent and linear forms of CS_2, the E_a of coronene falls between the E_a of the bent and linear forms: READS (bent/C) = (0.54/0.46), READS (linear/C) = (0.2/0.8). The electron affinity of coronene is greater than 0.6 eV but less than 0.88 eV. This stands in contrast to the PES and collisional ionization values that are less than 0.6 eV [14–16].

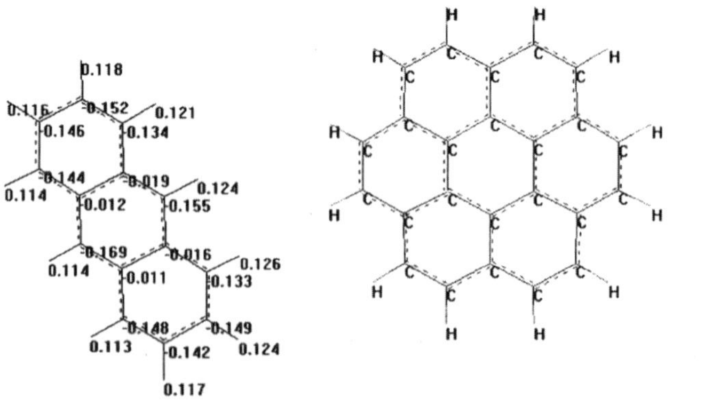

Molecule 1 anthracene(-) selected q = 0.343793 Coronene(-) q = 0.636207

Figure 7.7 Charges for one electron distributed on one anthracene and one coronene molecule calculated using MINDO/3. Seventy percent of the electron is localized on the coronene molecule, as indicated in the window when the anthracene molecule is selected.

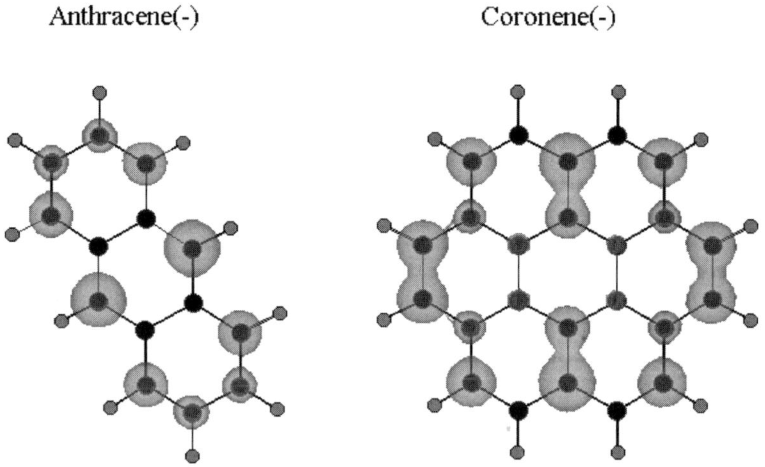

Figure 7.8 Three-dimensional spin densities and charges for one electron distributed on one anthracene and one coronene molecule calculated using MINDO/3.

A very important use of this technique is determining the relative electron affinities of multisubstituted compounds for which little data are available. For hexafluorobenzene and petafluorobenzene the READS (hexaF/pentaF) = (0.611/0.389), while the value for READS (p-diF/FBz) = (0.595/0.405) and READS(s-triF/p-diF) = (0.595/0.405). The sequential addition of a fluorine increases the E_a, and the READS values are relatively constant. The charge distributions are shown in Figure 7.9 for hexafluorobenzene and petafluorobenzene. The pentafluorobenzene value is 0.389q. The value for the hexafluorobenzene is the complement or 0.611q.

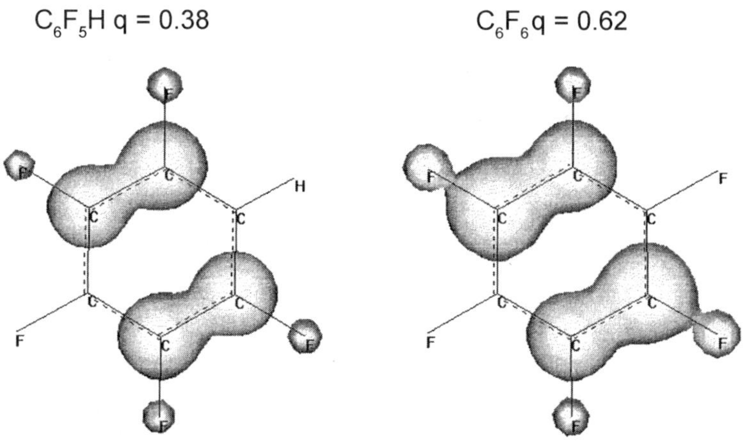

Figure 7.9 Charges for one electron distributed on hexafluorobenzene and pentafluorobenzene calculated using AM1. Sixty percent of the charge is localized on hexafluorobenzene by adding up the charges or selecting one molecule.

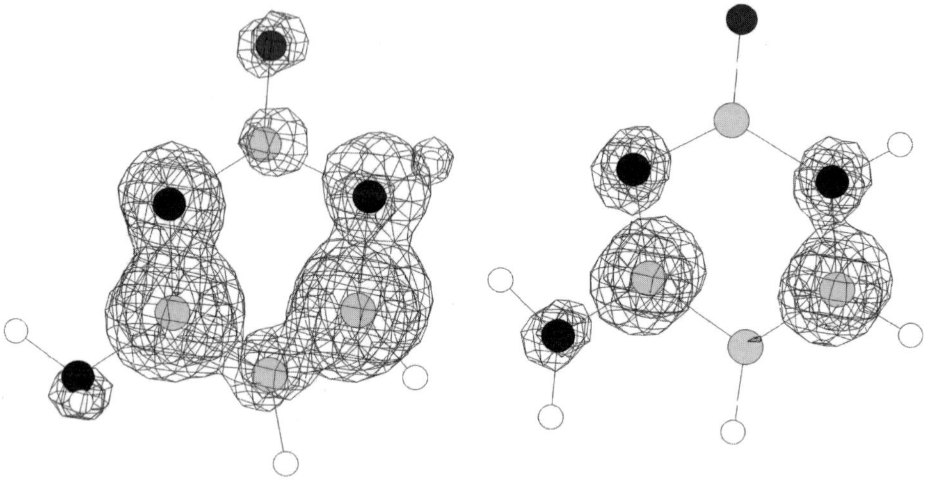

Figure 7.10 Three-dimensional spin densities for planar and distorted anions of cytosine calculated with AM1 [3].

The CURES-EC and READS-TCT procedures can be applied to more complicated biological molecules such as cytosine and thymine. The NH_2 on cytosine is twisted out of the plane in the anion, as shown in Figure 7.10. The AM1(0000) E_a for the planar anion and neutral is 0.65 eV. The minimum AM1(0033) E_a is 0.3 eV. The energy of the planar form is −1,355.8 kcal/mole, whereas that in the global minimum is −1,360.73 kcal/mole. In order to go from the planar form that is a local minimum in the energy surface to the global minimum, the sample must be annealed. For the global minimum in the energy with the twisted NH_2 group, the AM1(0033) E_a is 0.56 eV, which is equal to the values obtained in experiment. The two geometries can be seen in Figure 7.10, where the differences in the spin distributions may be easily visualized. In the global minimum there is spin density on the carbonyl group, whereas in the case of the planar anion none exists for the carbonyl group, but there is spin density on the C—H group opposite the carbonyl group [17]. These changes are especially important for the GC hydrogen-bonded base pairs, as will be discussed in Chapter 12.

7.3 MORSE POTENTIAL ENERGY CURVES

In 1963 negative-ion Morse parameters for the ground-state anions of Br_2 and I_2 were obtained by estimating D, r_e, and ν from the VE_a measured from charge transfer spectra and properties of the excited states of the neutral. Multiple excited states of $I_2(-)$ were characterized by D. R. Herschbach in 1966. He presented general forms for ionic Morse potential energy curves (HIMPEC). Nine total groups were based on dissociation or molecular ion formation in the Franck Condon region, the repulsive nature of the anion curves, the sign of the VE_a, and the sign of $-\text{EDEA} = D - E_a(X)$. Herschbach described the HIMPEC as "a convenient qualitative classification of the types of XY(−) potential energy curves which may occur is given [see Figure 7.11 here]. . . . The category I or II is determined from the sign of the EDEA. In both cases A and B, the free anion is bound, but for B as well as the unbound case C, the vertical transition leads to dissociation" [2]. The "C" curves are slightly bound due to polarization forces. These are stronger than the attractions in He_2, but weaker than the interactions responsible for dipole bound anions of molecules. However, there are several spaces where no curves can exist. Thus, the parameters for the classification were modified in 2001. For convenience the original classifications given in Chapter 2 are compared to the updated ones in Figure 7.11. The characterization of these Herschbach Morse groups is the first objective of this section.

The second objective will be to describe in detail the procedure for obtaining Morse potentials from diverse data using H_2, I_2, benzene, and naphthalene as examples. Hydrogen is chosen because it is the simplest homonuclear diatomic molecule and its adiabatic electron affinity corresponds to the polarization curve. Quantum mechanical calculations represent this as a free electron and molecule. Iodine is chosen because the electron impact distributions yield 12 curves. This splitting of the states is predicted theoretically for both $I_2(-)$ and $Xe_2(+)$. The bond dissociation energies of the $Xe_2(+)$ have been determined for these 10 states. Negative-ion states not observed in the $Xe_2(+)$ have been observed in $I_2(-)$ [18–25]. By using the split curves and absorption spectra of the negative ions, 10 excited-state curves are defined. Benzene is chosen because the polarization anion is in the ground state like H_2. Naphthalene is the first aromatic hydrocarbon with a positive valence-state electron affinity. Multiple low-lying negative-ion states of naphthalene have been observed, one of which has a negative electron affinity. Naphthalene has two distinct types of C—H bonds for which two different gas phase acidities have been measured [11–13].

7.3.1 Classification of Negative-Ion Morse Potentials

In the updated HIMPEC the sign of three independent metrics—the electron affinities E_a, the vertical electron affinities VE_a, and the energy for dissociative electron attachment EDEA—gives $2^3 = 8$ possible curves. These are described in a symmetrical notation: $M(m)$ and $D(m)$, $m = 0$ to 3, where m is the number of positive metrics and M and D signify molecular anion formation and dissociation in a vertical transition. The EDEA is positive when the electron affinity is larger than the bond dissociation energy since it is $\text{EDEA} = E_a(X) - BDE_{XY}$. The E_a is the difference in energy between the individual anions and ground state of the neutral. By convention E_a are positive for exothermic reactions. The VE_a are the energy differences for each anion in the geometry of the neutral. By definition the ground-state adiabatic E_a, AE_a, is positive, but excited-state E_a and VE_a can be negative or positive. By considering the $r(e-)$-XY separation as a third dimension, eight subclasses

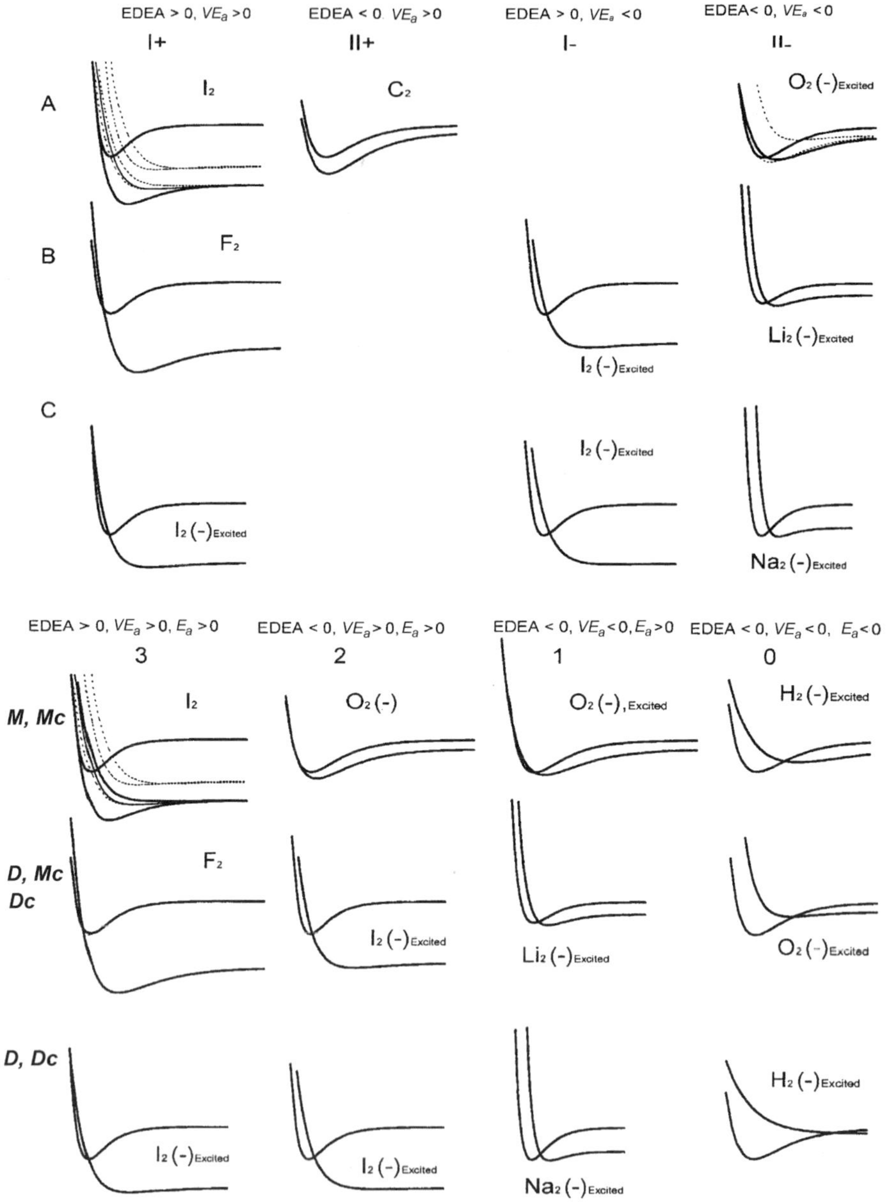

Figure 7.11 Original Hershbach ionic morse potential energy curves and the modified HIMPEC [2, 3]. The curves are calculated for the current best available data. The multiple curves for $O_2(-)$ and $I_2(-)$ are given to illustrate the relative positions of the curves. The specific example is the curve that is solid.

can be defined as $Mc(m)$ and $Dc(m)$ in which the ground state neutral crosses the anion curves to yield molecular ions or dissociation.

The electron affinities of many homonuclear diatomic molecules have been measured by anion photoelectron spectroscopy or photodetachment, but few Morse potential energy curves have been calculated. The AE_a for H_2 and N_2 are positive but small since the ground-state anions correspond to long-range ion polarization states. The anions are as follows: H_2 [M(0), Mc(0); D, Dc(0)]; Li_2 [M(2), Mc(2), D(1)]; C_2 [M(2), Mc(2)]; O_2 [M(2, 1), Mc(2, 1, 0); D(0), Dc(0)]; F_2 [D(3, 2), Dc(3, 2)]; Na_2 [M(2), Mc(2); D(1), Dc(1)]; and I_2 [M(3), Mc(3); D(3, 2), Dc(3, 2)] illustrate the classes as shown in Figure 7.11.

The Morse potentials for the neutral and HIMPEC as referenced to zero energy at infinite separation are given by

$$U(X_2) = -2D_e(X_2)\exp(-\beta(r-r_e)) + 2D_e(X_2)\exp(-2\beta(r-r_e)) \tag{7.1}$$

$$U(X_2^-) = -2k_AD_e(X_2)\exp(-k_B\beta(r-r_e)) + k_RD_e(X_2)\exp(-2k_B\beta(r-r_e)) - E_a(X) + E(X^*) \tag{7.2}$$

where $D_e(X_2)$ is the spectroscopic dissociation energy; r is the internuclear separation; $r_e = r$ at the minimum of $U(X_2)$; $E_a(X)$ is the electron affinity of X; $E(X^*)$ is the excitation energy; $\beta(X_2) = \nu_e(2\pi^2\mu(X_2)/D_e(X_2))^{1/2}$; μ is the reduced mass; k_A, and k_B, and k_R are dimensionless constants. The variation in the reduced mass of the anion occurs in the k_B term since $k_B = \beta(X_2(-))/\beta(X_2)$. The HIMPEC are also Morse potentials, and the negative-ion parameters can be calculated from those of the neutral and dimensionless constants using these equations:

$$D_e(X_2(-)) = [k_A^2/k_R]D_e(X_2) \tag{7.3}$$

$$r_e(X_2(-)) = [\ln(k_R/k_A)]/[k_B\beta(X_2)] + r_e(X_2) \tag{7.4}$$

$$\nu_e(X_2(-)) = [k_Ak_B/k_R^{1/2}]\nu_e(X_2) \tag{7.5}$$

$$-VE_a = D_e(X_2)(1-2k_A+k_R) - EA(X) + E(X^*) - 1/2h\nu_e(x_2) \tag{7.6}$$

The negative-ion properties have not been measured directly. Therefore, it is necessary to combine data from different sources to obtain the dimensionless parameters to define the curves. Three data points on the curve will define a negative-ion curve. The Herschbach metrics EDEA, E_a, and VE_a give k_A and k_R from equations 7.3 and 7.6 since the $D_e(X_2(-))$ are directly related to the E_a and EDEA. This only gives two constants since the EDEA is a displacement of the negative-ion curve from the neutral, and a third point is required. The $\nu_e(X_2(-))$ or $r_e(X_2(-))$ gives k_B from equation 7.4 or 7.5 to define the anion Morse potentials.

7.3.2 The Negative-Ion States of H_2

The first step in assigning the experimental E_a to electronic states is to determine the number of theoretical states. The H_2 anion has a polarization ground state and

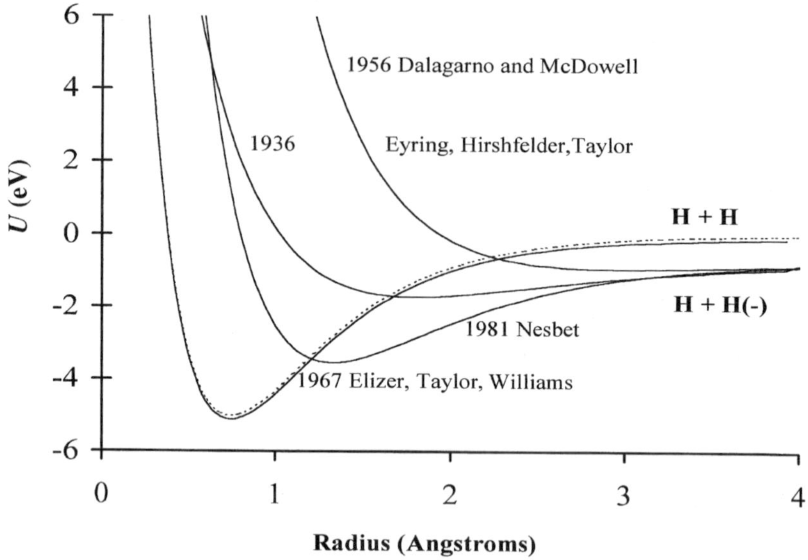

Figure 7.12 Historical Morse potential energy curves for H_2 and its anions, dating back to 1936 [25], 1956 for the excited state [26], 1967 for the polarization ground state [27], and 1981 for the valence excited state [28].

two covalent states. The ground state is a "neutral molecule plus free electron." Because of the simplicity of the $H_2(-)$ ions, many attempts have been made to characterize them. Morse parameters D, r, and ν) (0.9 eV, 190 nm, 1,127 cm^{-1} and 0.16 eV, 300 nm, 600 cm^{-1}) were calculated for the bonding and antibonding states in 1936 and 1956. In 1967 a curve for the polarization state was calculated, but the bond dissociation energy was less than that of the neutral, giving it a negative E_a. In 1967 and 1981 improved bonding curves were calculated; however, the dissociation energy remained too small. One net bonding electron is present in $H_2(-)$, as in $He_2(+)$ and $H_2(+)$. The neutral has two bonding electrons, and the dissociation energy of the anion is one-half that of the neutral or $D_e = 2.65$ eV. The bond dissociation energy of the polarization state should be slightly larger than that for the neutral and the internuclear distance and frequencies should be about the same. When we use these bond dissociation energies and the calculated frequencies and internuclear distances, the Morse potentials for the anion states of $H_2(-)$ can be calculated. These are illustrated in Figure 7.12, along with the 1936 curve formulated by Eyring Hirschfelder and Taylor. They can be compared with the current "best" curves shown in Figure 7.13 [26–29].

Until recently, the only experimental data that were available for $H_2(-)$ were electron impact distributions for the formation of H(−). The H(−) peaks were observed with an abrupt onset at −EDEA = 3.75 eV, an onset at 7 eV, and a peak at 10 eV [30]. These data are shown in Figure 7.14 with the calculated distributions.

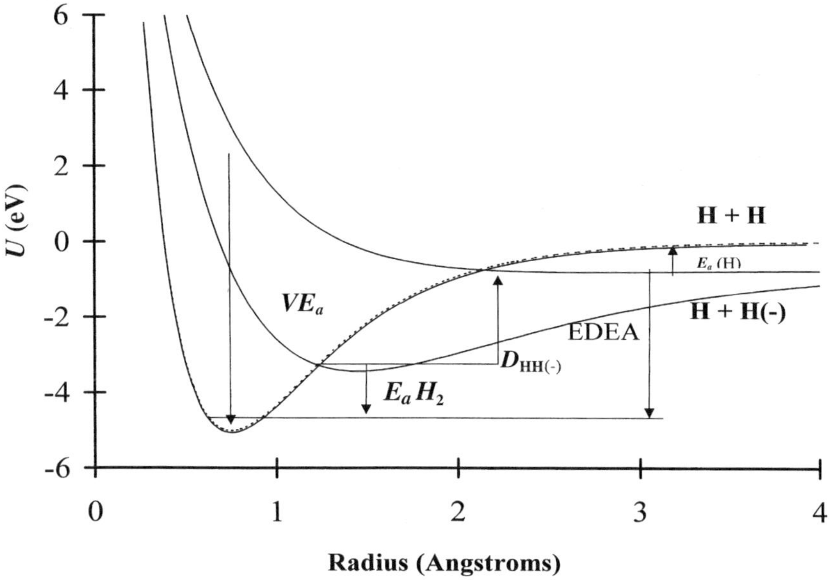

Figure 7.13 Current "best" Morse potential energy curves for H_2 and its anions, from [30–32].

Electron scattering data give a VE_a of −3.3 eV for the bonding state [31]. The ESR spectrum of the valence-state anion has been observed in irradiated solid H_2 [32]. The combination of these data yields $r_e = 153$ pm and ν_e 1,700 cm^{-1} for the bonding valence-state anion. The H(−) distribution and an assumed dissociation energy

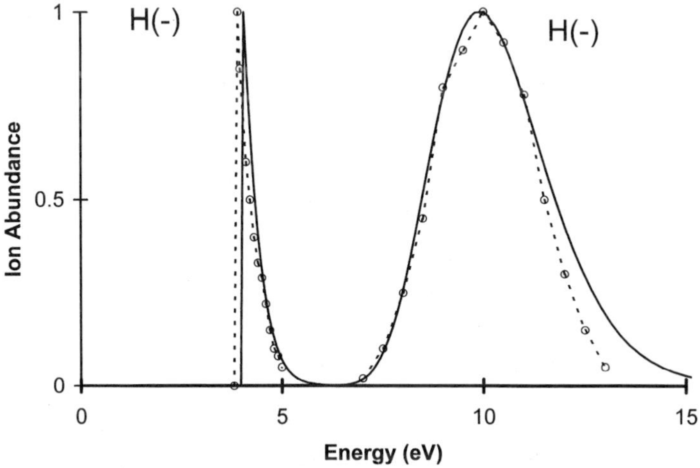

Figure 7.14 Electron impact spectrum calculated from the current "best" Morse potential energy curves for H_2 and $H_2(-)$ shown in Figure 7.13, from [30].

TABLE 7.3 Morse Parameters, Dimensionless Constants, and Experimental Data for Neutral and Ionic H_2 and $He_2(+)$

		k_A	k_B	k_R	D_0 (eV)	r_e (pm)	ν_e (cm^{-1})
$He_2(+)$	$^2\Sigma_u^+$	1.11	0.62	2.40	2.58	108	1,964
	$^2\Sigma_g^+$	0.16	0.49	1.38	0.08	190	300
$H_2(+)$	$^2\Sigma_u^+$	0.84	0.69	1.26	2.79	106	2,250
H_2 Neutral		1.00	1.00	1.00	5.01	75.4	4,395
$H_2(-)$	Polar	1.00	1.00	1.00	5.1	75.4	4,395
	$^2\Sigma_u^+$	1.07	0.53	2.14	2.58	145	700
	$^2\Sigma_g^+$	0.12	0.56	0.99	0.05	280	285

of 0.05 eV were used to calculate the antibonding Morse potential [31]. The bonding curve is an $M(0)$ curve, while the antibonding curve is a $D(0)$ curve. The Morse parameters and dimensionless constants for the neutral and anions of H_2 and the cations of H_2 and He_2 are in presented Table 7.3. The attractive term $k_A = 1$, is about the same in the bonding valence-state anion, but the repulsive term k_R is doubled. In the antibonding state the attractive term is diminished, while that for the repulsive term is the same as for the neutral. The radius of H(−) is obtained from $r_e(H_2(-) - r(H) = 145 - 38 = 107$ pm.

7.3.3 The Negative-Ion States of I_2

Twelve negative-ion states for I_2 have been predicted [25]. The ground-state curves have been completely characterized by photon experiments. This is the only $X_2(-)$ ground-state curve so well characterized [23]. The bond dissociation energy is the same for the ground state of iso-electronic $Xe_2(+)$ [3, 18, 21]. Electron impact data define the vertical electron affinities of 10 states, while the bond dissociation energies of the excited states of isoelectronic $Xe_2(+)$ define excited-state E_a. One excited-state curve with a much lower bond dissociation energy than measured for $Xe_2(+)$ has been observed in photon experiments [24]. The third point on some curves is the peak in the absorption spectra of the ground-state anion. The E_a of one state can be calculated from the observed bond dissociation energy. For all the VE_a the I(−) distributions have been measured [19]. This defines the slope of the excited-state curves in the Franck Condon region. Finally, the activation energies of the thermal electron attachment to two excited states have been measured in ECD experiments [8, 22]. Thus, some of these curves are overdetermined with five data points. The values of k_A, k_B, and k_R could be calculated using the analytical expressions 7.3 to 7.6. However, this is not done. Instead, the electron impact curves are fit to the E_a and VE_a to give the ion distribution. Then the calculated absorption peak is adjusted to determine the experimental values. An isobestic point is assumed for several curves since not all the absorption peaks have

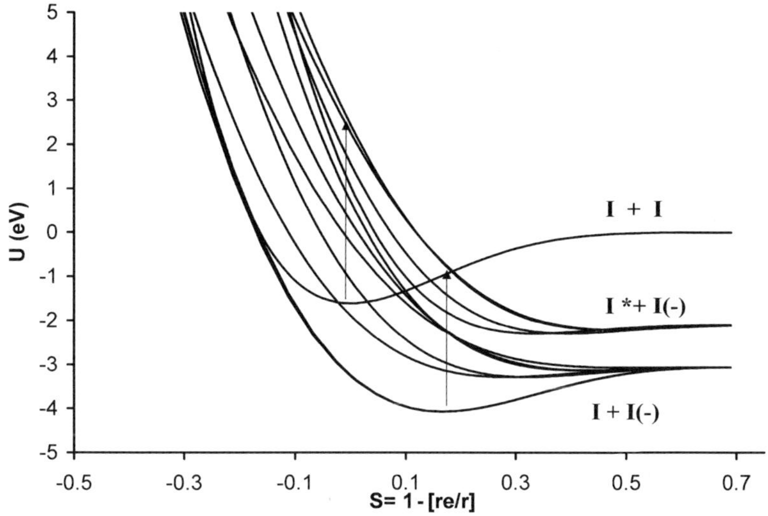

Figure 7.15 Current "best" Morse potential energy curves for I_2 and its anions. The X axis is the reduced internuclear distance $S = 1 - [r_e/r]$, where r is the internuclear distance and r_e is the equilibrium internuclear distance. The data are taken from [18]. The 12 curves were predicted in [23].

been measured. Figure 7.15 illustrates the Morse potential energy curves for I_2 and its anions using the S dimensionless variable, where $S = 1 - (r_e/r)$. This type of plot spreads out the curves in the region where the experimental data are available and makes it easier to see the HIMPEC classifications of the curves. The two ground-state curves are $M(3)$ and $Mc(3)$ curves since all the Herschbach metrics are positive. The next curve is $D(3)$ because the vertical process leads to dissociation. It is $Mc(3)$ because it crosses the polarization state on the "backside." These three states could lead to the formation of the molecular anion on electron impact. The remaining curves are all $D(1)$ and $Dc(1)$ curves because the EDEA is positive, but the E_a and VE_a are negative and dissociative electron attachment occurs in the Franck Condon region. The calculated and experimental electron impact data for I_2 appear in Figure 7.16. The good fit and the need for multiple negative-ion states are clear. The Morse parameters and dimensionless constants used to calculate these curves are given in Table 7.4.

7.3.4 The Negative-Ion States of Benzene and Naphthalene

The negative-ion curves for benzene are very similar to those for H_2. Five negative-ion states should exist. The X polarization ground-state curve is an $M(2)$ curve. There should be two bonding and two antibonding curves corresponding to the two dissociation limits, Ph + H(−) and Ph(−) + H. The two bonding curves are $M(0)$ curves. The aromatic C—H bond dissociation energy and electron affinities of the H atom and the phenyl radical have been measured experimentally, giving

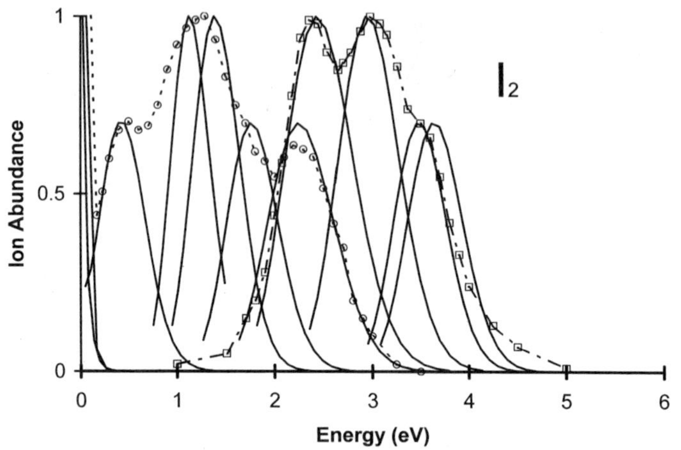

Figure 7.16 Electron impact spectrum calculated from the current "best" Morse potential energy curves for I_2 and $I_2(-)$ shown in Figure 7.15. The data are taken from [19].

the EDEA. The vertical electron affinities are the same for the two bonding states, but the E_a are different, at -0.74 eV determined from reduction potential data and at -1.15 eV based on electron transmission spectra. The internuclear distances and frequencies for the anions are close to the values for the neutral. The Morse parameters are given in Table 7.5. Figure 7.17 illustrates the curves. The two antibonding $D(0)$ and $Dc(0)$ curves are not shown. They have larger frequencies and internuclear distances than the bonding curves.

TABLE 7.4 Morse Parameters, Dimensionless Constants, and Experimental Data for Neutral and Ionic I_2

	k_A	k_B	k_R	D_0 (eV)	r_e (pm)	ν_e (cm^{-1})	VE_a	E (abs)
Neutral	1.00	1.00	1.00	1.54	267	215	—	–
A1(1/2)	1.187	0.645	2.242	1.007	320.5	110	−1.67	—
A1(1/2)	1.194	0.651	2.264	1.007	320.5	111	−1.61	—
B1(3/2)	0.491	0.649	1.702	0.225	371	52.5	−0.30	0.87
B1(3/2)	0.491	0.649	1.702	0.225	392	49.5	0.50	1.01
B1(1/2)	0.125	0.557	1.992	0.012	537	10.6	1.35	**1.68**
B1(1/2)	0.173	0.601	2.297	0.020	501	14.7	1.69	**1.65**
C1(3/2)	0.376	0.523	2.789	0.080	475	25.3	1.83	**1.68**
C1(3/2)	0.383	0.587	3.080	0.075	460	27.5	2.27	**1.69**
C2(1/2)	0.628	0.654	3.118	0.194	400	50	2.35	**2.11**
C2(1/2)	0.663	0.524	3.491	0.194	439	40	2.82	2.45
D2(1/2)	0.501	0.439	3.569	0.108	510	25	3.45	**3.03**
D2(1/2)	0.509	0.461	3.691	0.108	500	26.3	3.61	**3.03**

TABLE 7.5 Morse Parameters, Dimensionless Constants, and Experimental Data for Neutral and Ionic Benzene and Naphthalene

	D_0	r_e	n	k_A	k_B	k_R	E_a (R)	E_a (RH)
Benzene	5.16	1.09	3350	1.000	1.000	1.000	0.000	—
Polariz(−) gs	5.20	1.09	3350	1.008	0.996	1.008	0.000	0+
M(0)(−) B	3.20	1.11	3000	0.695	1.137	0.779	0.750	−1.10
M(0)(−) A	3.20	1.14	3000	0.824	1.137	1.095	1.100	−0.78
Naphthalene	5.06	1.09	3350	1.000	1.000	1.000	0.000	—
Polariz(−) B	5.10	1.09	3350	1.008	0.996	1.008	0.000	0+
M(0)(−) C	4.10	1.12	3300	0.955	1.094	1.126	0.750	−0.2
M(2)(−) gs	3.80	1.10	3300	0.795	1.136	0.842	1.390	0.16
M(2)(−) A	3.75	1.10	3310	0.805	1.147	0.874	1.370	0.13

For naphthalene there should be nine negative-ion states: the polarization state, four bonding states, and four antibonding states. The ground state has a positive electron affinity of 0.16 eV and dissociates via the 1—C-H bond. The polarization state and a state dissociating via the 2-C-H bond have positive electron affinities of 0+ and 0.13 eV respectively. The excited-valence-state curve dissociating to H(−)

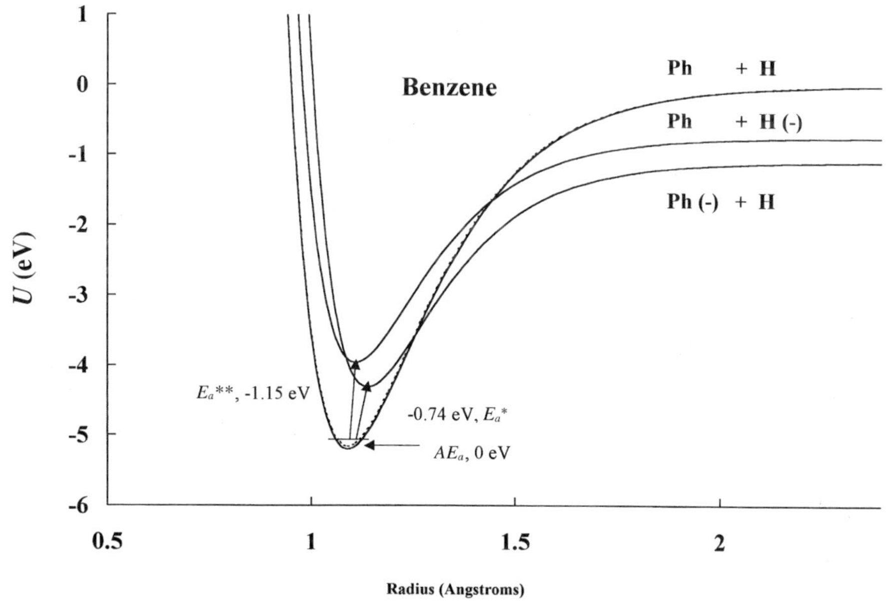

Figure 7.17 Current "best" Morse potential energy curves for benzene and its anions. There are two additional "antibonding" curves going to each dissociation limit that are not shown. The adiabatic electron affinity corresponds to the polarization curve.

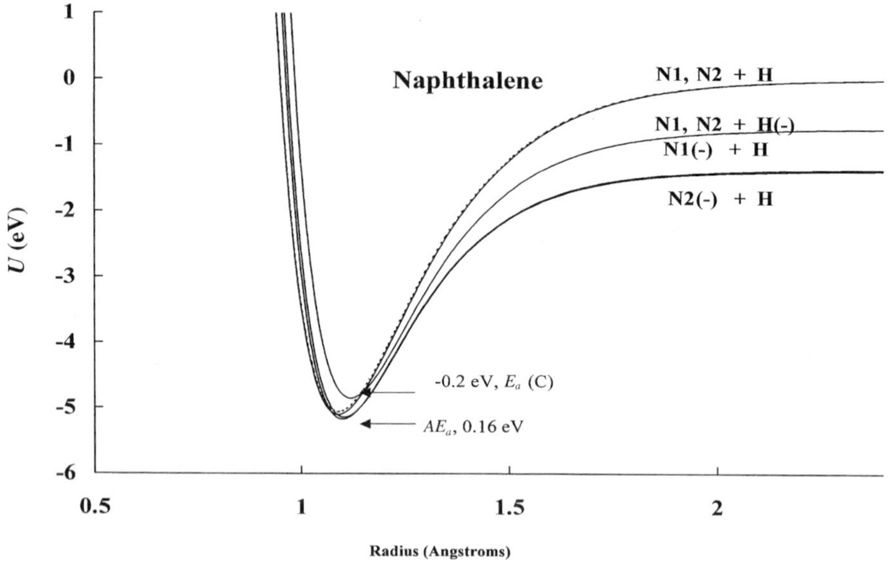

Figure 7.18 Current "best" Morse potential energy curves for naphthalene and its anions. There are four additional "antibonding" curves that are not shown, giving a total of eight valence-state curves. The adiabatic electron affinity corresponds to the valence state with an E_a of 0.16 eV. The polarization curve has an E_a of about zero.

has a negative electron affinity, -0.2 eV. The electron affinity of the 2-naphthalenide radical is 1.34(2) eV, while that of 1-naphthalenide is 1.39(2) eV. The experimental gas phase acidities differ by 0.05 eV, making the two curves very close since the 2-C-H and 1-C-H bond dissociation energies are only slightly different. The internuclear distances and frequencies for the anion should be similar to those of the neutral. Three anion states have been observed experimentally. The two lower valence-state curves are $M(2)$ curves with a negative EDEA and positive E_a and VE_a. Two bonding curves leading to H($-$) are $M(0)$ since molecular ion formation takes place in the Franck Condon regions and all three Herschbach parameters are negative. All the bonding curves are Mc(2 or 0) curves. The four antibonding curves are D and $Dc(0)$. Figure 7.18 shows four of these curves.

The dimensionless constants for the ground state for $I_2(-)$ correspond to an increase in the attractive term but a larger increase in the repulsive term to give a smaller bond dissociation energy. In the case of bonds for aromatic hydrocarbons the attractive terms are decreased and the repulsive terms increased, but by a smaller amount than for $I_2(-)$ or $H_2(-)$. Notice that the relative bond order is given by $\mathrm{BO} = D_e(X_2(-))/D_e(X_2) = [k_A^2/k_R]$. Thus, the bond orders for the aromatic hydrocarbons are larger than for the diatomic molecules. If there are no experimental data for the construction of negative-ion states of aromatic hydrocarbons, then these values can be used as first approximations or the theoretical values could be used.

7.4 EMPIRICAL CORRELATIONS

The isoelectronic equivalence is the simplest procedure for estimating electron affinities. It was applied to H_2 and I_2 and to the atomic electron affinities. Species with the same outer electronic configuration should have similar electron affinities and bond dissociation energies. This results in the relative constancy of the electron affinities of a given family of atoms. The equivalence of the bond dissociation energies for the $X_2(-)$ and $Rg_2(+)$ ions is also based on this principle. The systematic variation of the electron affinities of the homonuclear diatomic molecules is another example.

Related procedures for estimating electron affinities make use of the concept of electronegativity (EN). These use the Mulliken [33] definition of "absolute" electronegativity, the average of the first ionization potential and the first electron affinity, $EN = (IP + EA)/2$. With an estimate of the Mulliken electronegativity and the experimental value of the ionization potential, the electron affinities can be calculated. When both the electron affinity and ionization potential are measured, the relationship between the EN and experiment can be examined. This has been accomplished for aromatic hydrocarbons and will be discussed in Chapter 10.

In the case of atomic species Linus Pauling defined a thermochemical scale of electronegativities and reported the correlation between the absolute scale and thermochemical scales as follows: "It is seen that the values of x [$x = (IP + EA)/125$] are closely proportional to those of the sum of the EA and IP except for hydrogen, which, with its unique electronic structure, might be expected to misbehave" [34]. There are other definitions of electronegativity related to these two definitions. Other properties such as the work functions of metals are also related to electronegativity. With these estimates the Mulliken EN for all elements can be obtained. The electronegativities of the alkali metals decrease slightly going down the Periodic Table from 3 eV for Li to 2.2 eV for Cs. The ionization potentials decrease from 5.4 eV for Li to 3.9 eV for Cs. The electron affinities calculated from these data are 0.6 eV for Li and 0.5 eV for Cs. These are only 0.03 eV different from those obtained in experiment. In Chapter 8 we will correlate the electron affinities for all elements with different measures of electronegativity.

The substitution and replacement rules have their origin in correlations between thc thermodynamic and kinetic properties of chemical reactions that form part of traditional physical organic chemistry. The Hammett relations consolidate much experimental information based on properties of substituents. These rules are empirical. However, the effects have been attributed to conjugation, mesomeric, or resonance effects; inductive effects; and geometric effects. The substitution of a halogen atom will increase the electron affinity by an inductive effect. The substitution of a vinyl group will increase the electron affinity by conjugation. The substitution of a methyl group will lower the electron affinity by an inductive effect, but with multiple substitutions can increase the electron affinity by geometric effects. The electron affinity of biacetyl is expected to be higher than that of butadiene because of the increased electronegativity of the oxygen atom. The electron affinity of pyridine should be larger than that of benzene, and that of pyrimidine should be larger than that of pyridine because of the greater electronegativity of the nitrogen atom. Multiple substitutions will generally go in the same direction but will be attenuated. These rules have also been incorporated in the simple Huckel theory in the modification of parameters upon substitution on a parent molecule. The replacement rules also have their foundation in simple perturbation theory [4, 35–37].

The substitution rules were originally described in 1963 for the estimation of electron affinities of organic charge transfer complexes. The origin of the rules and their uncertainty are reflected in the following passage:

> The relatively great inaccuracy in the electron affinity values precludes discussion of the more refined effects induced by the structure of the compounds involved. Alkyl substituents tend to reduce the electron affinity, the effect becoming more pronounced as the number increases. Electrophillic (halogen, CN, NO_2) groups tend to augment the electron affinity of a compound, this effect increasing with the number of substitutions. The cyano and nitro groups tend to increase the electron affinity to a larger extent than do the halogens, which do not differ appreciably in this respect. Molecules with two adjacent electrophillic carbonyl groups are relatively good electron acceptors. The relatively high electron affinity of CCl_4 is noteworthy.... Carbon tetrachloride and chloroform are σ acceptors which, despite their high electron affinity, have an EDA (electron donor acceptor) interaction which is lower than that of π acceptors with comparable electron affinities. [38]

In 1969 the effect of substitutions on electron affinities determined using the magnetron method was thus described: "While it is tempting to produce a set of numbers for group contributions to π electron affinities which may be used to correlate all of the experimental data, and to predict unknown affinities, the data are far too scanty to do this or to do more than put forward some tentative values." The contribution of an F atom to the electron affinity was assigned a value of 0.2 eV, whereas that for Cl was 0.15 eV. The effect of a single cyano group was 0.8 eV, whereas that for a nitro group was 1 eV. The addition of a second cyano group was only 0.05 lower than the addition of a single cyano group. The passage continues, "While this leads to a conflict between the magnetron values and those of Compton, it must be remembered that the latter are vertical values and that therefore, they are upper limits. Other work by Compton suggests that the adiabatic attachment energy for benzene may be 0.65 eV lower than the vertical energy, and this difference would account for the supposed ring affinity." The value of 0.65 eV is thus the rearrangement energy. If the modern value for the vertical electron affinity of benzene, -1.15 eV, is used, this gives an E_a of -0.5 eV, as compared to the value of -0.74 eV obtained from reduction potentials. If we use the group contribution of 0.2 eV, the AE_a of C_6F_6 is $-0.5 + 6 \times 0.2 = 0.7$ eV. The value predicted by Compton was 0.9 eV. These compare favorably with 0.86(2), the ECD experimental value. The accuracy of the values for benzene and hexafluorobenzene is notable since they were reported in 1969 [39].

By 1988 sufficient gas phase data were available to define substitution effects for the NO_2 group and multiple substitutions. The new data were obtained from alkali

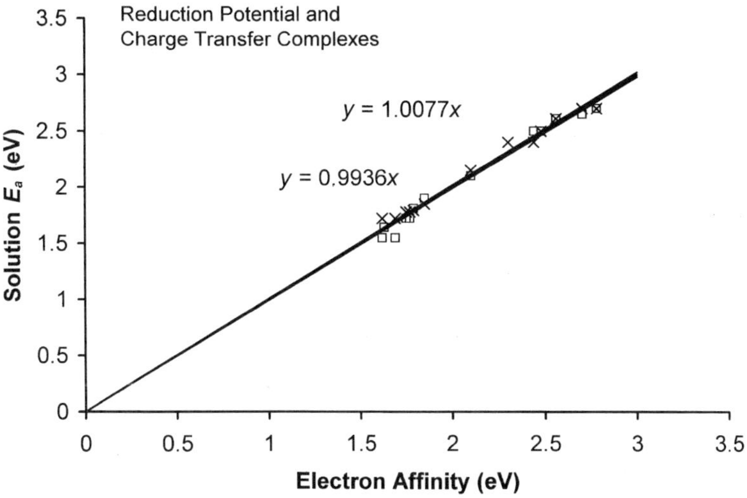

Figure 7.19 Precision and accuracy plot of the corrected values of the electron affinities of halogenated and methylated benzoquinones. These should be compared to the parallel lines in Figure 6.17. The compounds are listed in Table 6.3.

metal beam and thermal charge transfer experiments. The leveling effect of the number of substituents was noted. In addition, a brief look at the effect of substituents on electron affinities obtained from half-wave reduction potentials was considered. This procedure will be extended to additional molecules in Chapter 10.

The CURES-EC values have been calculated for the methyl- and chloro-substituted benzoquinones discussed in Chapter 6. They agree with the experimental and density functional calculated E_a. Based on the charge distributions of the C=O group, the solution energy differences for the halogenated compounds should be smaller by approximately 0.20 eV than for the parent compound, while the methyl substituted compounds should be about 0.05 eV larger. The charge transfer complex values are also systematically lower than the gas phase values for the chlorinated compounds. This implies that the constants used for those compounds should be larger by about 0.20 eV for the chlorocompounds and 0.10 eV smaller for the methyl compounds. By adding a constant amount to each E_a in the different categories, the deviations are significantly reduced. Since only two additional parameters are used, correlation is made with three intercepts and a unit slope instead of the typical slope and intercept. As shown in the modified precision and accuracy plot in Figure 7.19, the zero intercept slopes for the two sets of data are essentially unity. This procedure will be extended in Chapter 10, and the values from reduction potential and charge transfer complex data will be tabulated for compounds with E_a measured in the gas phase and predicted for other compounds.

TABLE 7.6 Theoretical Electron Affinities (in eV) for Benzene, Naphthalene, Anthracene, Tetracene, and Pentacene and the Fully Fluorinated Compounds

	F	non-F	$\Delta E_a/F$
Benzene	0.86	0.00	0.14
Naphthalene	1.69	0.16	0.19
Anthracene	2.30	0.68	0.16
Tetracene	2.70	1.10	0.13
Pentacene	3.08	1.39	0.12
Hexacene	3.26	1.60	0.10
Heptacene	3.40	1.80	0.09
Octacene	(3.50)	1.95	0.08
Novacene	(3.60)	2.10	0.07

The effect of multiple substitutions can be determined by examining the perfluorinated acenes. The electron affinity of hexafluorobenzene has been measured using the ECD, PES, and TCT. It is 0.86 ± 0.02 eV. When a fluorine atom is added, the E_a is increased 0.16 eV. The AM1(0033) minimum values for the perfluorinated acenes are provided in Table 7.6 with the experimental value for the nonfluorinted acenes. Also shown are the increments in the electron affinity per fluorine atom. The largest is 0.19 eV/F for perfluoronaphthalene. The E_a of the linear acenes are compared to those of the fluorinated compounds in Figure 7.20. The change in the increment is also shown. The electron affinities of the perfluorinated benzenes are predicted to vary from about 0.1 eV to 0.86 eV. This agrees with experiment, as will be shown in Chapter 11. The increment can be applied to the value for perfluorobenzoquinone to predict an electron affinity of $1.85 + 4(0.19) = 2.61$ eV, the same as the value obtained from reduction potential data.

The linear acenes illustrate the effect of extended conjugation. As the number of rings is extended in both the hydrocarbons and perfluorinated hycrocarbons, the E_a increases. However, for a nonlinear extension the E_a may or may not increase. For example, the E_a of phenanthrene is 0.30 ± 0.02 eV, while that for anthracene is 0.68 ± 0.02 eV. Likewise, the E_a of benzanthracene is less than that of tetracene. The curves for the linear acenes are an upper limit for the electron affinities of polycyclic aromatic hydrocarbons with the same number of six membered rings. The inclusion of five or seven membered rings will increase the electron affinity above those with solely six rings. The simplest example is the E_a of azulene, 0.8 eV, versus that of naphthalene, 0.16 eV. Interestingly, the calculated electron affinity of the perfluorinated azulene is 2.6 eV or about 0.2 eV per fluorine atom.

Another systematic variation of the E_a is observed for the replacement of a CH by a nitrogen atom. The electron affinity of pyridine is expected to be larger than that for benzene and that for quinoline should be greater than that for naphthalene.

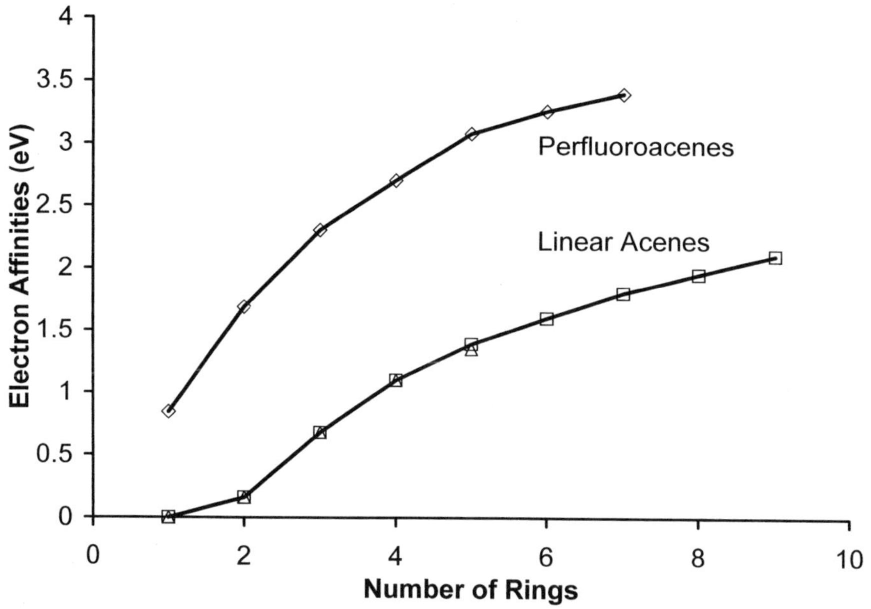

Figure 7.20 Plot of the experimental electron affinities of the linear acenes and the calculated electron affinities of the perfluorinated linear acenes versus number of rings. The values are calculated using AM1. The values are given in Table 7.6.

Reduction potential data support this prediction. Also, the inclusion of multiple substitutions of CH by N in benzene is expected to increase electron affinity. On the basis of reduction potentials the E_a of pyridine is about zero; and that for the diazines ranges from 0.2 to 0.36 eV. The triazines vary from 0.5 to 0.9 eV. The predicted value for hexazine is 2.8 eV. These substitution and replacement effects can be used to predict electron affinities. Indeed, the first attempt at estimating the electron affinities of AGCUT was made using such correlations. These will be discussed in more detail in Chapter 12.

7.5 SUMMARY

Speedy and accurate desktop computers and modern programs such as HYPERCHEM place quantum mechanical calculations within the reach of any experimental chemist. The CURES-EC procedure simulates equilibrium methods of measuring electron affinities by calculating the difference between the optimized forms of the anion and neutral. The READS-TCT determination of charge densities in anion complexes simulates thermal charge transfer experiments. The effect of geometry on energies can be observed by calculating vertical electron affinities or electron affinities for different forms, such as the linear and bent anions of CS_2. The charge densities calculated using HYPERCHEM can be utilized to classify molecules so that different solution energy differences may be used to determine more accurate electron affinities from reduction potentials and/or charge transfer complex energies. Experimental values of electron affinities can be tested against periodic trends that are predicted by simple quantum mechanical concepts. The classification of ionic Morse potential energy curves and their construction from experimental and theoretical data have been described. These can be used to assign experimental electron affinities to excited electronic states.

REFERENCES

1. HYPERCHEM. Available at http://www.hyper.com.
2. Herschbach, D. R. *Adv. Chem. Phys.* **1966**, 10, 250.
3. Chen, E. S. and Chen, E. C. M. *J. Phys. Chem. A* **2002**, 106, 6655.
4. Dewar, M. J. S. *The Molecular Orbital Theory of Organic Chemistry.* New York: McGraw-Hill, **1969**, p. 275.
5. National Institute of Standards and Technology (NIST). *Chemistry WebBook*, **2003**. Available at http://webbook.nist.gov.
6. Christodoulides, A. A.; McCorkle, D. L.; and Christophorou, L. G. "Electron Affinities of Atoms, Molecules and Radicals" in *Electron-Molecule Interactions and Their Applications.* New York: Academic Press, **1984**.
7. Reinstra-Kiracofe, J. C.; Tschumper, G. S.; Schaefer, H. F.; Nandi, S.; and Ellison, G. B. *Chem. Rev.* **2002**, 102, 231.
8. Chen, E. S. D.; Chen, E. C. M.; Sane, N.; Kozenecki, N.; and Schultz, S. *J. Chem. Phys.* **1999**, 110, 9319.
9. Chen, E. S. D.; Chen, E. C. M.; Sane, N.; and Schultz, S. *Bioelectrochem. Bioenerget.* **1999**, 48, 69.
10. Younkin, J. M.; Smith, L. J.; and Compton, R. N. *Theoret. Chem. Acta* **1976**, 41, 157.
11. Ervin, K. M.; Ramond, T. M.; Davico, G. E.; Schwartz, R. L.; Casey, S. M.; and Lineberger, W. C., *J. Phys. Chem. A* **2001**, 105, 10822.
12. Lardin, H. A.; Squires, R. R.; and Wenthold, P. G. *J. Mass Spectrom.* **2001**, 36, 607.
13. Reed, D. R. and Kass, S. R. *J. Mass Spectrom.* **2000**, 35, 534.
14. Chen, E. C. M.; George, R.; Carr, S. D.; Wentworth, W. E.; and Chen, E. S. D. *J. Chromatogr. A* **1998**, 811, 250.
15. Duncan, M. A.; Knight, A. M.; Negishi, Y.; Nagao, S.; Nakamura, Y.; Kato, A.; Nakajima, A.; and Kaya, K. *Chem. Phys. Lett.* **1999**, 309, 49.
16. Chen, G.; Cooks, R. G.; Corpuz, E.; and Scott, L. T. *J. Amer. Soc. Mass Spectrom.* **1996**, 7, 619.
17. Chen, E. C. M. and Chen, E. S. D.; *J. Phys. Chem. B* **2000**, 104, 7835.
18. Chen, E. S. D. and Chen, E. C. M. *J. Phys. Chem. A* **2003**, 107, 169.

19. Azria, R.; Abouaf, R.; and Tellet-Billy, D. *J. Phys. B: At. Mol. Opt. Phys.* **1988**, 21, L213.
20. Chen, E. C. M. and Wentworth, W. E. *J. Phys. Chem.* **1985**, 89, 4099.
21. Chen, E. S. and Chen, E. C. M. *Chem. Phys. Lett.* **1998**, 293, 491.
22. Ayala, J. A. Chen, E. C. M.; and Wentworth, W. E. *J. Phys. Chem.* **1981**, 85, 768.
23. Maslen, P. E.; Papanikolas. J. M.; Faeder, J.; Parson, R.; and O'Neil, S. V. *J. Chem. Phys.* **1994**, 101, 5731.
24. Zanni, M. T.; Taylor, T. R.; Greenblatt, B. J.; Miller, W. H.; and Neumark, D. M. *J. Chem. Phys.* **1997**, 107, 7613.
25. Zanni, M. T.; Batista, V. S.; Greenblatt, B. J.; Soep, B.; and Neumark, D. M. *J. Chem. Phys.* **1999**, 110, 3748.
26. Eyring, H.; Hirshfelder, J. O.; and Taylor, H. S. *J. Chem. Phys.* **1936**, 4, 479.
27. Dalgarno, A. and McDowell, M. R. C. *Proc. Phys. Soc.* **1956**, A69, 617.
28. Elizer, I.; Taylor, H. S.; and Williams, J. K. *J. Chem. Phys.* **1967**, 47, 2165.
29. Nesbet, R. K. *Comm. Atom. Molec. Phys.* **1981**, 11, 25.
30. Schulz, G. J. *Phys. Rev.* **1959**, 113, 816.
31. Allan, M. *J. Phys. B* **1985**, 18, L451.
32. Ichikawa, T.; Tachikawa, H.; Kumada, T.; Kumagai, J.; and Miyzaki, T. *Chem. Phys. Lett.* **1999**, 307, 283.
33. Mulliken, R. S. *J. Chem. Phys.* **1934**, 2, 782.
34. Pauling, L. *The Nature of the Chemical Bond*, 3rd ed. Ithaca, NY: Cornell University Press, **1960**.
35. Streitweiser, A. S. *Molecular Orbital Theory for Organic Chemists.* New York: Wiley, **1961**.
36. Coulson, C. A.; O'Leary, B.; and Mallion, R. B. *Huckel Theory for Organic Chemists.* New York: Academic Press, **1978**.
37. Hine, J. *Physical Organic Chemistry.* New York: McGraw-Hill, **1962**.
38. Briegleb, G. *Angew. Chem. Internat. Edit.* **1964**, 3, 617.
39. Page, F. M. and Goode, G. C. *Negative Ions and the Magnetron.* New York: Wiley, **1969**.

CHAPTER 8

Selection, Assignment, and Correlations of Atomic Electron Affinities

8.1 INTRODUCTION

The electron affinities E_a of the main group atoms are the most precisely measured values. Recall that the E_a is the difference in energy between the most stable state of the neutral and a specific state of a negative ion. It was once believed that only one bound anion state of atoms and molecules could exist. However, multiple bound states for atomic and molecular anions have been observed. This makes it necessary to assign the experimental values to the proper state. The random uncertainties of some atomic E_a determined from photodetachment thresholds occur in parts per million. These are confirmed by photoelectron spectroscopy, surface ionization, ion pair formation, and the Born Haber cycle. Atomic electron affinities illustrate the procedure for evaluating experimental E_a.

Random and systematic errors are characteristics of the method, not the values. Random errors can be determined by repeating the experiment. Systematic errors can only be determined by comparisons of values determined by different methods. Uncertainties can be estimated from precision and accuracy plots if it is assumed that there are only random errors. The E_a of the d and f block elements, electronegativities, and work functions of the elements will be evaluated in this chapter using this procedure.

The extrapolation of values in the Periodic Table for the main group of elements can be examined and applied to the transition elements. The evaluated values can be combined with ionization potentials to obtain Mulliken electronegativities. These may be correlated with other electronegativities and the work functions of the elements. This provides another example of the use of the Periodic Table to evaluate and predict values.

The Electron Capture Detector and the Study of Reactions with Thermal Electrons
by E. C. M. Chen and E. S. D. Chen
ISBN 0-471-32622-4

An application of the electron affinities of the elements and the experimental work functions involves the prediction of the electron affinities of clusters. The C_n molecules are an important type of cluster studied experimentally and theoretically. With experimental data the CURES-EC method of calculating electron affinities can be evaluated. The READS-TCT procedure can also be used to determine relative electron affinities. The clusters of C, Si, and Ge involve covalent bonds, while the bonds in the Sn and Pb clusters are partially metallic. With available electron affinities the relationship between the electron affinities and work functions of these anion clusters can be investigated.

The precision of a metric is determined by the random uncertainties of a method and the number of replications. The equipment, ability of the investigator, and material investigated affect the random uncertainties. It is important to know the "best" precision that has been attained and the number of replications used to attain that precision. In establishing the precision, it is assumed there are no systematic uncertainties. In the case of atomic electron affinities the largest systematic uncertainty is the state assignment.

The statistical evaluation of experimental data involves the following questions: "Do the results agree with previous results within the random and systematic uncertainties?" and "Is this value assigned to the correct state?" For the main group elements the data are the E_a obtained by different techniques. The largest precise value is assumed to be the ground state E_a. If this value has been confirmed by measurement with another technique, the assumption is validated. The weighted average is the "best" value. If the E_a has not been determined by an alternative technique, but is confirmed by theoretical calculations or the expected variation within the Periodic Table, the assumption may be validated. If there are no validations, the assumption can only remain an assumption.

8.2 EVALUATION OF ATOMIC ELECTRON AFFINITIES

As new values were obtained, atomic electron affinities were reviewed periodically beginning in 1953 [1–13]. All the available experimental, extrapolated, and theoretical values were tabulated in 1984 [7]. Presently, experimental values are available at the NIST website [12]. Prior to 1970 the majority of the values for the main group elements were determined by the Born Haber cycle, electron impact, or relative and absolute equilibrium surface ionization techniques. However, values for C, O, and S had been measured by photodetachment [1–3]. By the mid-1970s virtually all the E_a of the main group elements in the first three rows had been measured by photon methods [4–7]. By the early 1980s values were obtained for the transition elements by photon techniques [7, 8]. In the 1990s the values of Ca, Sr, and Ba were measured [9–13]. Recently, experimental values have been reported for Ce, Pr, Tm, and Lu [14–17].

In 1971 all the experimental atomic E_a in the literature were evaluated and compared with extrapolated values [3]. None of the experimental values were eliminated. Considered were 123 values for 23 elements. All but four of the elements

TABLE 8.1 Atomic Electron Affinities: Current Best Averages to 1970, Weighted Averages to 1970, and Weighted Average of Photon Values to 1975 [3–5]

		E_a (eV)			
A	*N*	Current	Average	Weighted Average	1975
H	1	0.754	0.77	0.77	0.754
Li	3	0.618	0.50	0.50	0.62
C	6	1.263	1.31	1.31	1.263
O	8	1.462	1.71	1.465	1.462
F	9	3.401	3.53	3.41	3.400
Na	11	0.548	0.3	0.3	0.548
S	16	2.077	2.22	2.1	2.077
Cl	17	3.613	3.75	3.6	3.613
K	19	0.502	0.5	0.5	0.502
Cr	24	0.676	1.2	1.2	0.77
Cu	29	1.236	1.5	1.5	1.226
Br	35	3.364	3.58	3.47	3.365
Rb	37	0.486	0.6	0.6	0.486
Mo	42	0.747	1.0	1.0	1.0
Ag	47	1.303	1.95	1.95	1.303
Sb	51	1.047	1.5	1.5	1.15
I	53	3.059	3.19	3.12	3.07
Cs	55	0.472	0.6	0.5	0.472
W	74	0.815	0.5	0.5	0.5
Re	75	0.2	0.15	0.15	0.15
Au	79	2.309	2.8	2.8	2.309
Tl	81	0.4	1.5	1.5	0.5
Pb	82	1.1	1.1	1.1	1.2
Bi	83	0.942	1.76	1.76	1.0

were from the main group. Ninety-seven of the measurements were for the halogens and oxygen. The data are summarized in Table 8.1. The first column is the current "best" value. The simple average is given in the second column and the weighted average in the third column. Given in the fourth column are the weighted average of the values determined by photon methods up to 1975 [1–6]. Figure 8.1 is a precision and accuracy (P and A) plot for the E_a values given in Table 8.1. The "simple average" values have a slightly larger zero intercept slope than do the weighted averages primarily because of the E_a of the halogens. The weighted average of the values determined up to 1975 is equal to the current best values within the experimental errors. The outliers in the earlier values are indicated by the large deviations from the unit slope line. For subsequent averages these values will automatically be weighted out by their uncertainties.

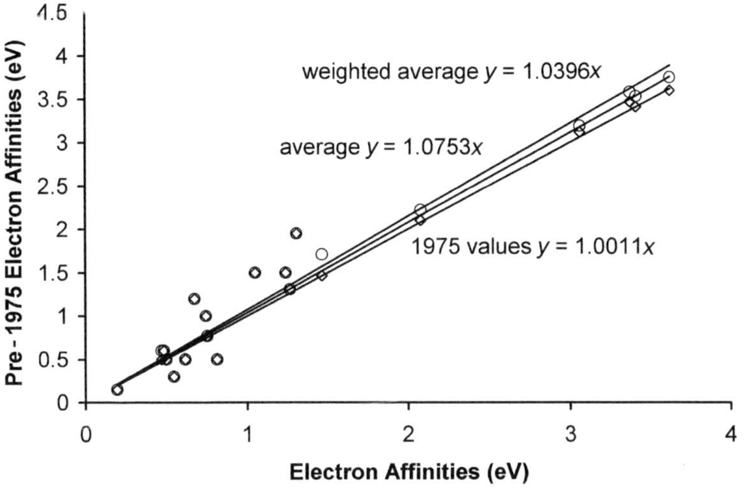

Figure 8.1 Precision and accuracy plot of atomic electron affinities measured up to 1971. The simple average and weighted average are shown. The simple average has a zero intercept slope of 1.08, indicating systematic errors. The random uncertainties are large, indicating possible outliers. The weighted average values have a slope of 1.04 because some photodetachment values are included. The weighted average values reported up to 1975 indicate no systematic or significant random uncertainties.

Table 8.2 lists the best confirming E_a for the main group elements determined by the photon and surface ionization methods for selected atoms. These are the weighted averages of the values determined by the method up until that time [1–13, 18–29]. Many of the current "best" accurate and precise values have been determined from photodetachment thresholds. The random uncertainties range from 50 parts per billion (ppb) for 0 to 10 parts per million (ppm) for Ir. The AE_a of N, Be, and Mg and the rare gases are slightly positive and have not been determined by photon methods. A complete list of the E_a and random uncertainties for the elements is given in the appendix.

The photodetachment threshold values for Ga, As, and Pb (0.30(15), 0.81(3), and 1.10(5) eV, respectively) were determined by an older photodetachment technique [30]. The PES values for Ga, 0.43(3), and As, 0.814(8), are more precise, but are supported by the PD values [31, 32]. The PES value for Pb, 0.364(8), is significantly lower than the PD value, 1.10(5) eV. The photodetachment values for the majority of the other main group elements have also been confirmed by photoelectron spectroscopy. The values for the halogens have been determined by most methods: photodetachment thresholds, photoabsorption, ion pair photodissociation, relative and absolute surface ionization methods, and the Born Haber cycle. Values for H, Li, C, F, Cl, Cu, Ge, Br, Nb, Ag, Sn, I, W, Re, Au, and Pb have also been determined by relative and absolute surface ionization methods, as shown in Table 8.2.

TABLE 8.2 Atomic Electron Affinities (in eV) Determined by Photodetachment, Photoelectron Spectroscopy and Surface Ionization Techniques [4, 7–12]

		PD	PES	SI
H	1	0.7542	0.754	0.8(1)
Li	3	0.6182	0.620(7)	0.8(2)
B	5	0.2797	0.28(1)	—
C	6	1.2621	1.268(5)	1.24(2)
O	8	1.4611	1.462(3)	—
F	9	3.4012	3.399(3)	3.40(2)
Na	11	0.5479	0.546(5)	—
Al	13	0.4328	0.46(3)	—
Si	14	1.3895	1.385(5)	—
P	15	0.7465	0.743(10)	—
S	16	2.0771	2.077	2.09(7)
Cl	17	3.6127	3.615(4)	3.67(4)
K	19	0.5015	0.5015(5)	—
Ca	20	0.0245	—	—
Cu	29	1.2358	1.226(10)	1.18(6)
Ga	31	0.30(15)	0.43(3)	—
Ge	32	1.2327	1.2(1)	1.14(3)
As	33	0.814	0.80(5)	—
Se	34	2.0207	2.0206(3)	—
Br	35	3.364	3.364(4)	3.49(2)
Rb	37	0.4859	0.486	—
Sr	38	0.0521	—	—
Ag	47	1.3045	1.303	1.38(10)
In	49	0.404	0.3	—
Sn	50	1.1121	1.15	1.16(5)
Sb	51	1.0474	1.07	—
Te	52	1.9709	1.9708	—
I	53	3.0590	3.059	3.07(2)
Cs	55	0.4716	0.4716	—
Ba	56	0.1446	—	—
Au	79	2.3086	2.309	2.34(10)
Tl	81	0.38(1)	0.3(2)	—
Pb	82	1.10(5)	0.364(8)	1.05(8)
Bi	83	0.9424	1.1(2)	—
Fr	87	0.491(5)	0.46	—

In 1957 a thermal charge transfer reaction showed that the $E_a(\text{Pb}) > E_a(\text{Sb})$. Since it is now known that the $AE_a(\text{Sb}) = 1.05$ eV, the value for Pb is 1.1(1) eV. In 1971 a surface ionization procedure established that $E_a(\text{Ag}) - E_a(\text{Pb}) = 0.1(2)$ eV so $E_a(\text{Pb}) = 1.2(2)$ eV. In 1981 an electron beam dissociative electron attachment experiment gave a value of 1.2(1) eV. These three determinations confirm the photodetachment value of 1.10(5) [24, 33, 34]. An E_a for Pb was determined to be 0.364(8) eV by photoelectron spectrometry in 1981 [35]. The weighted average of the four higher E_a is: $(A = \sum[a/s^2]/N; N = \sum[1/s^2]; S^2 = 1/N) : A = 1.1/0.01 + 1.2/0.01 + 1.2/0.04 + 1.10/.0025)/N$ and $N = (100 + 100 + 25 + 400) = 1.10 \pm 0.04$, where the weights are given by $w_i = (1/s_i)^2$. The weighted average of the three thermodynamic values is 1.10(5) eV. The ground-state electron affinity for lead is assigned the value 1.10(5) eV. The PES value does not overlap at the 14σ level $n = d/s = 0.7/0.04 = 18$. There are two possible explanations. The PES value may be in error by more than normal for photoelectron spectroscopy data, or it could be for an excited state. Because there are four independent determinations of the higher value that agree within the random uncertainty, the possibility of two states must be considered. This is especially likely since the other Group IV elements have low-lying bound excited states.

Bound excited-state E_a have been measured for Al, P, C, Si, Ge, and Sn. The ground state for all the anions of the Group IVA elements is the $^4S_{3/2}$ state. For C(-) the ^2D_m state is observed with an E_a of 0.035 eV. Two excited states are observed for Si(-): The ^2D state E_a for is 0.523(5) eV and the ^2P state is 0.029(5) eV. For Ge(-) and Sn(-) the ^2D_m state E_a is about 0.4 eV. Thus, for Pb(-) it is reasonable that the E_a of the $^4S_{3/2}$ state is 1.10(5) eV and the E_a of the ^2D_m state is 0.381(8) eV. The ground state for Al(-) is $^3P_0$ with an E_a of 0.428 eV and the excited state $^1D_2$ (m) has an E_a of 0.12(3) eV. The E_a for P(-) are a ground state of 0.7465(3) eV and a $^1D_2$ of about 0 eV. Many other possible excited valence states with negative E_a have been observed. These have been discussed extensively in the literature [11, 36].

Other reported values of atomic E_a are larger than the selected values. For indium the 1998 laser PD value is 0.404(9) eV, an earlier PD value is 0.3(2) eV, and an electron impact value is 0.85(30). The three values overlap at the 2σ level. The weighted average is $A = [0.4040/(0.009)^2 + 0.3/(0.2)^2 + 0.85/(0.3)^2]/(N)$, where $N = 1/(0.009)^2 + 1/(0.2)^2 + 1/(0.3)^2$ and $A = 0.4042 \pm 0.009$ or 0.404(9). None of the data are discarded, but the "best" estimate of the E_a is the properly weighted average [8–13, 39, 40].

It is generally agreed that nitrogen and the rare gases will not form bound valence-state anions because of their closed-shell configurations. The "best" adiabatic electron affinity for the elements for which E_a is a slightly positive value due to the polarization attractions is 0+. This is consistent with the precise definition of adiabatic electron affinity. It is more accurate than the statements "less than zero," "does not exist," or "is unstable." It was once believed that the Group IIA and IIB elements did not possess bound anion states. Subsequently, small positive E_a were measured for Ca, Sr, and Ba.

The electron affinities of atoms are presented in Figure 8.2 in the form of a Periodic Table. This format will be used to concisely present the data, whereas the complete values will be given in the appendix. The experimental values are given with the proper number of significant figures, or with the random error in the last figure specified in parentheses. Because chemical accuracy and precision are often considered to be 1 meV, the values are only given to within one-tenth of meV. The values are offered with the date of their determination or the date of their selection as the "best" value. Figure 8.3 plots the E_a for a given family versus the period. The values for the members of a given family are different by about 25 to 30% except for the Group VA elements. The variation is systematic, with the only major discontinuity occurring between the first and second period. Figure 8.4 illustrates the trends in the rows. It is easier to visualize the relative constancy of

O	I	II	Ib	Iib	III	IV	V	VI	VII
	1 H 0.7542 1975								
2 He [0+] 2002	3 Li 0.6180 1975	4 Be [0+] 2002			5 B 0.2797 1998	6 C 1.2621 1998	7 N [0+] 2002	8 O 1.4611 1999	9F 3.4012 2001
10 Ne [0+] 2002	11Na 0.5479 1975	12 Mg [0+] 2002			13 Al 0.4328 1998	14 Si 1.3895 2001	15 P 0.7465(3) 1990	16 S 2.0771 1975	17 Cl 3.6127 1995
18 Ar [0+] 2002	19 K 0.5015 1975	20 Ca 0.0245 1996	29 Cu 1.2358 1998	30 Zn [0+] 2002	31 Ga 0.43(3) 1998	32 Ge 1.2327 1998	33 As 0.814(8) 1990	34 Se 2.0207 1975	35 Br 3.3636 1989
36 Kr [0+] 2002	37 Rb 0.4859 1975	38 Sr 0.0521 1997	47 Ag 1.3045 1975	48 Cd [0+] 2002	49 In 0.404(9) 1998	50 Sn 1.1121 1998	51 Sb 1.0474 1997	52 Te 1.9709 1990	53 I 3.0590 1992
54 Xe [0+] 2002	55 Cs 0.4716 1975	56 Ba 0.1446 1995	79 Au 2.3086 1975	80 Hg [0+] 2002	81 Tl 0.38 1999	82 Pb 1.1 ± 0.05 2002	83 Bi 0.9424 1999	84 Po [1.9(3)] 1975	85 At [2.8(3)] 1975
86 Rn [0+] 2002	87 Fr 0.491(5) 2002	88 Ra 0.17 2002							

21 Sc 0.19 1990	22 Ti 0.08 1990	23 V 0.53 1990	24 Cr 0.6758 1998	25Mn [0+] 2002	26 Fe 0.151(3) 1990	27 Co 0.663 1990	28 Ni 1.1572 1990
39 Y 0.31 1990	40 Zr 0.43 1990	41 Nb 0.89(3) 1990	42 Mo 0.747 1990	43 Tc 0.6(2) 1990	44 Ru 1.046 1999	45 Rh 1.1429 1998	46 Pd 0.5621 1998
57 La 0.47(2) 1998	72 Hf 0.1 1998	73 Ta 0.32(1) 1990	74 W 0.82 1990	75 Re 0.2(2) 1975	76 Os 1.0778 1999	77 Ir 1.5644 1999	78 Pt 2.1251 1999
58 Ce 0.96(3) 2002	59 Pr 0.96(2) 2002	60 Nd 0.05+ 1997	61 0+ 1997	62-68 0.1+ 1997	69Tm 1.03(3) 2001	70Yb 0.01+ 1997	71 Lu 0.34(1) 2001
89 Ac 0+ 1997	90 Th 0.05+ 1997	91 Pa 0.05+ 1997	92 U 0.05+ 1997	93 Np 0+ 1997	94 Pu 0.05+ 1997		

Figure 8.2 Electron affinities of the elements in the form of a Periodic Table. The values in parentheses are the uncertainties in the last figure. The other statistics are given with their proper number of significant figures. The dates are those of the determination or selection as the evaluated values [10–17].

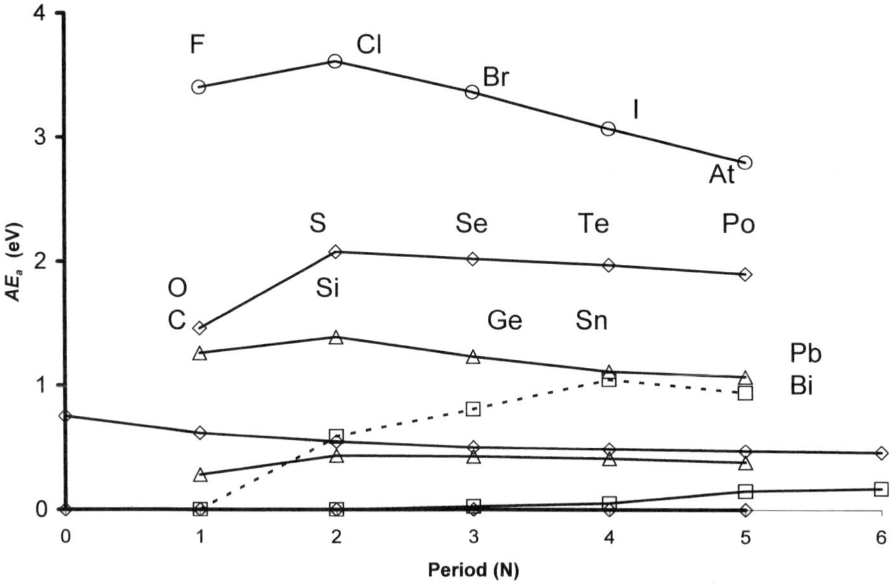

Figure 8.3 Plots of the E_a of the main group elements versus the period number to illustrate the consistency of the electron affinities of a given family.

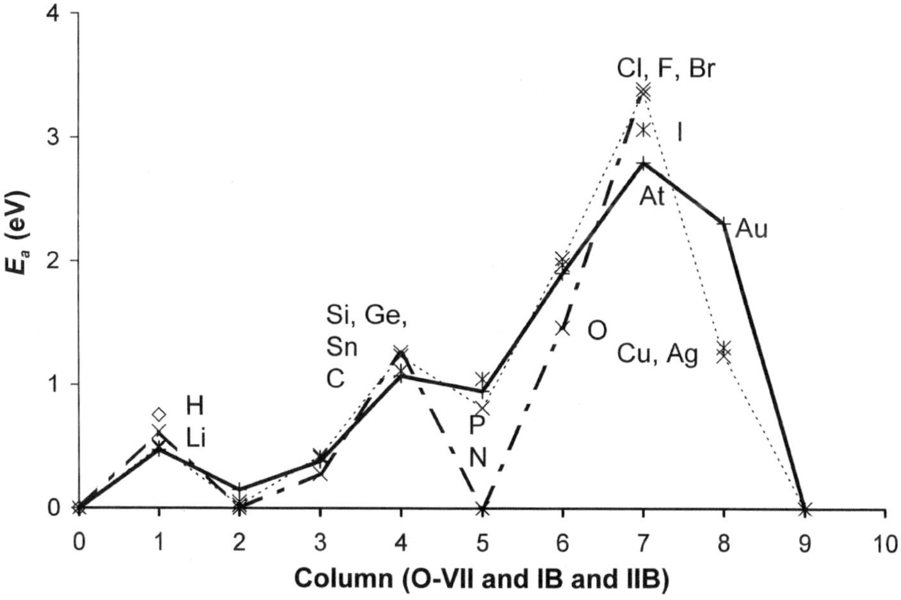

Figure 8.4 Plot of the E_a of the main group elements versus the column number to illustrate the consistency of the electron affinities of a given family and across a period.

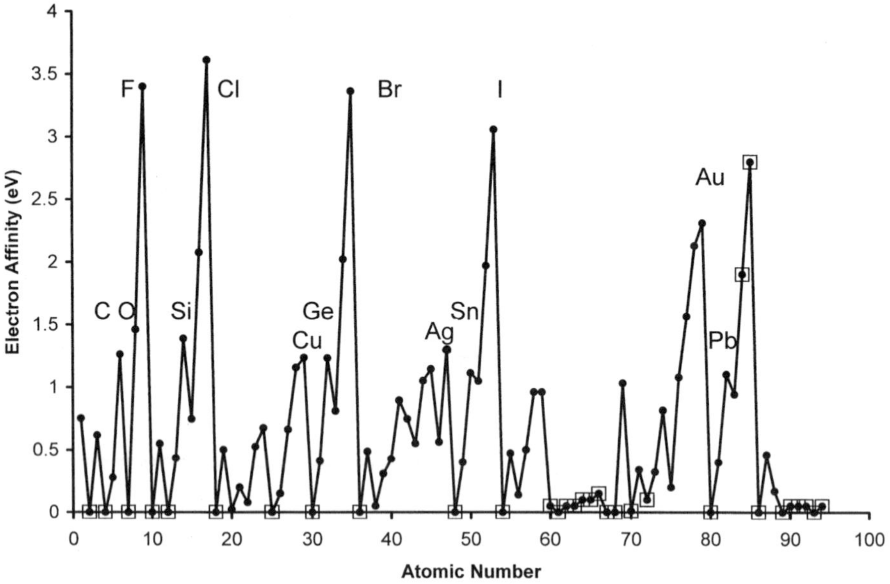

Figure 8.5 Plot of the E_a of the elements versus the atomic number to illustrate the consistency of the electron affinities of a given family and across a period.

the values for a given family in this plot. The only unusual data point is the very low AE_a of the nitrogen atom. This can be explained by its small size. Figure 8.5 is a plot of the selected values versus atomic number for all the elements. This shows the consistency in horizontal variations. The importance of using the value of 0+ for the closed-shell elements is emphasized.

A complete review of the theoretical calculations for the electron affinities of atoms is beyond the scope of this book. The quantum mechanically calculated electron affinities of the first and second row elements—the alkali metals, Ca, Ba, and Sr—support experimental results within their mutual uncertainties. 5 meV has been determined to be the best precision and accuracy of theoretical methods for atoms [13]. For example, the calculated values for Li, Na, K, Rb, and Cs agree with the experimental values to within 5 meV. Thus, the AE_a of Fr is 0.491(5) eV calculated theoretically. By the same method, the predicted value for eka-francium (element 119) is 0.663(5) eV [41]. The predicted E_a for Ra is also larger than the experimental value for Ba, 0.145 eV.

The higher value for the AE_a of Pb plotted in Figures 8.3 through 8.5 fits the expected trends. The average E_a for C, Si, Ge, and Sn is 1.25 eV so that a change in the E_a of 0.7 eV from Sn to Pb is unexpected. The horizontal trend from Tl to Pb to Bi is the same as for In, Sn, Sb and Ga, Ge, As. This is clearly shown in Figures 8.4 and 8.5. The electronegativities and adiabatic ionization potentials of Pb, Sn, and Bi are virtually the same so the adiabatic electron affinities should also be the same. Finally, the higher value gives a relative bond order of 1.3 for Pb_2 to Pb_2 (-), the same as predicted by simple molecular orbital theory (i.e., 5/4),

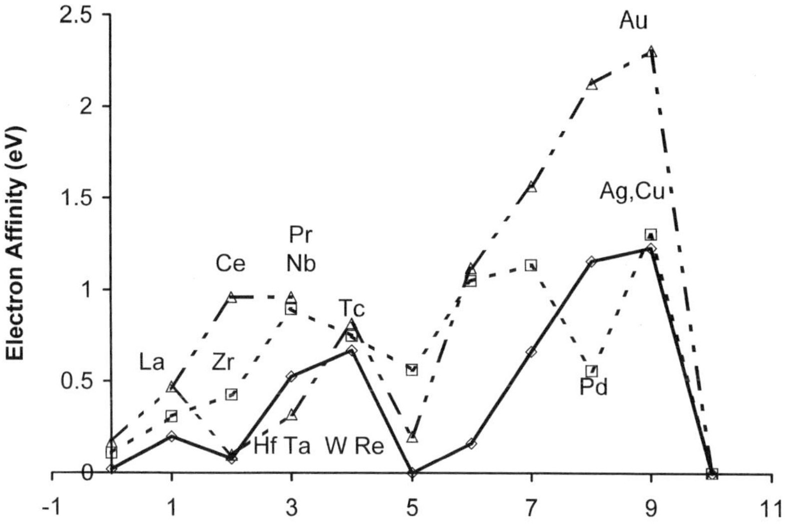

Figure 8.6 Plot of the E_a of the transition elements versus the horizontal number to illustrate the inconsistencies of the electron affinities for Pd and Re.

while the lower value gives 2, which is larger than observed for any other main group element.

The AE_a of the d block elements should generally increase across a period with some tendency to increase as one goes down a column. This can be observed in Figure 8.6, where the electron affinities of the elements from IIA (Ca, Sr, Ba) to IIB (Zn, Cd, Hg) are plotted. Two values are plotted for the third block: Ce and Pr approaching from the IIA elements and Re, Ta, W, and Hf approaching from the IIB elements. The inconsistencies suggest interesting characteristics. The "high" value for Nb, Zr, and Tc and the "low" values for Hf, Pd, Ta, and Re based on both horizontal and vertical trends are especially noted. Some of the horizontal inconsistencies in the 5d block are removed if the values of Ce and Pr are considered instead of the values for Hf and Ta. This leaves only the low values for Re and Pd as inconsistencies in both the horizontal and vertical trends. Since these inconsistencies have not been resolved, the periodic trends cannot be used to support complete assignment to the AE_a.

The E_a for the d block elements can be supported by theoretical estimates. However, the experience with lanthanides suggests that calculations presently give lower limits to the actual values because of the difficulties of relativity and electron correlation effects. The determination of more precise and accurate experimental values should improve these calculations. For the present the only conclusion that can be reached for the transition elements is that lower limits to the adiabatic electron affinity have been measured and that some of them are equal to the adiabatic electron affinities based on periodic trends.

8.3 MULLIKEN ELECTRONEGATIVITIES

Now that experimental values for the E_a of almost all the elements have been measured, the Mulliken electronegativities can be calculated from $\mathrm{EN} = (\mathrm{IP} + \mathrm{EA})/2$. These can be correlated with other electronegativities and work functions. Figure 8.7 gives the ionization potentials, electron affinities, and EN tabulated in

O	I	II	Ib	IIb	III	IV	V	VI	VII
	1 H 0.7542 13.6 7.18		Ea IP EN						
2 He [0+] 24.6 12.3	3 Li 0.6180 5.39 3.00	4 Be [0+] 9.32 4.66			5 B 0.2797 8.30 4.29	6 C 1.2621 11.26 6.26	7 N [0+] 14.53 7.27	8 O 1.4611 13.62 7.54	9F 3.4012 17.42 10.41
10 Ne [0+] 21.56 10.78	11Na 0.5479 5.15 2.84	12 Mg [0+] 7.65 3.83			13 Al 0.4328 5.99 3.21	14 Si 1.3895 8.15 4.77	15 P 0.7465(3) 10.49 5.62	16 S 2.0771 10.36 6.22	17 Cl 3.6127 12.97 8.29
18 Ar [0+] 15.76 7.88	19 K 0.5015 4.34 2.42	20 Ca 0.0245 6.11 3.08	29 Cu 1.2358 7.73 4.48	30 Zn [0+] 9.39 4.7	31 Ga 0.43(3) 6 3.2	32 Ge 1.2327 7.9 4.57	33 As 0.814(8) 9.81 5.31	34 Se 2.0207 9.75 5.89	35 Br 3.364 11.81 7.59
36 Kr [0+] 14 7	37 Rb 0.4859 4.18 2.33	38 Sr 0.0521 5.7 2.88	47 Ag 1.3045 7.58 4.44	48 Cd [0+] 8.99 4.5	49 In 0.404(9) 5.79 3.1	50 Sn 1.1121 7.34 4.23	51 Sb 1.0474 8.64 5.02	52 Te 1.9709 9.01 5.49	53 I 3.0590 10.45 6.75
54 Xe [0+] 9.01 5.49	55 Cs 0.4716 3.8 6.75	56 Ba 0.1446 5.21 6.06	79 Au 2.3086 9.23 5.77	80 Hg [0+] 10.4 5.2	81 Tl 0.38 6.11 3.24	82 Pb 1.1 ± 0.05 7.42 4.26	83 Bi 0.9424 7.29 4.12	84 Po [1.9(3)] 8.48 5.19	85 At [2.8(3)] 9.4 6.1
86 Rn [0+] 4 2.23	87 Fr 0.491(5) 5.28 2.73	88 Ra 0.17 6.9 3.45							

21 Sc 0.19 6.54 3.36	22 Ti 0.08 6.82 3.45	23 V 0.53 6.74 3.63	24 Cr 0.6758 6.77 3.73	25Mn [0+] 7.44 3.72	26 Fe 0.151(3) 7.87 4.01	27 Co 0.663 7.86 4.26	28 Ni 1.1572 7.64 4.39
39 Y 0.31 6.38 3.34	40 Zr 0.43 6.84 3.63	41 Nb 0.89(3) 6.88 3.89	42 Mo 0.747 7.1 3.92	43 Tc 0.6(2) 7.28 3.94	44 Ru 1.046 7.37 4.21	45 Rh 1.1429 7.46 4.3	46 Pd 0.5621 8.34 4.45
57 La 0.47(2) 5.589 3.03	72 Hf 0.1 7 3.55	73 Ta 0.32(1) 7.98 4.15	74 W 0.82 7.98 4.34	75 Re 0.2(2) 7.88 4.04	76 Os 1.0778 8.7 4.89	77 Ir 1.5644 9.1 5.33	78 Pt 2.1251 9 5.56
58 Ce 0.96(3) 5.47 3.23	59 Pr 0.96(2) 5.42 2.76	60 Nd 0.05+ 5.49 2.77	61 0+ 5.55 2.78	62-68 0.1+ 6.1 3.05	69Tm 1.03(3) 6.18 3.11	70Yb 0.01+ 6.25 3.13	71 Lu 0.34(1) 5.43 2.77
89 Ac 0+ 6.9 3.45							

Figure 8.7 Electron affinities, ionization potentials, and Mulliken electronegativities of the elements in the form of a Periodic Table. The ionization potentials were taken from [12] and the electronegativity calculated as $\mathrm{EN} = (\mathrm{IP} + E_a)/2$.

the form of a Periodic Table [12]. In Figure 8.8 the scaled Pauling and Allen Rochow electronegativities and EN are plotted against the atomic number. The Pauling and Allen Rochow electronegativities are scaled to give the minimum deviation from the Mulliken values. The larger values of the Pauling electronegativities of the transition metals, as compared to the Mulliken electronegativities and smaller Allen Rochow values, should be noted. However, the trends in all three scales are consistent. The general trends for the main group elements in all the figures support the assignments of the adiabatic electron affinities of the main group elements. The electronegativities of the d block elements show consistent trends with all three measures of electronegativities and support the assignments for some of the E_a to the ground state.

Ionization potentials are plotted against the atomic number in Figure 8.9. When we compare the ionization potential plot in Figure 8.9 with the electron affinity plot, the most striking difference is the range of values. For the electron affinities the values go from 0+ for many elements to 3.61 eV for Cl. The ionization potentials range from 3.89 eV for Cs to 24.59 eV for He. The E_a for Cs is 0.47 eV, while that for He is 0+. The periodic trends for the electronegativities, ionization potentials, and work functions are apparent. In Figure 8.10 the work functions and Mulliken EN are plotted against the atomic number, with EN replacing the work function for those elements with unmeasured work functions. Figure 8.10 easily visualizes the correspondence between the work functions and Mulliken electronegativities.

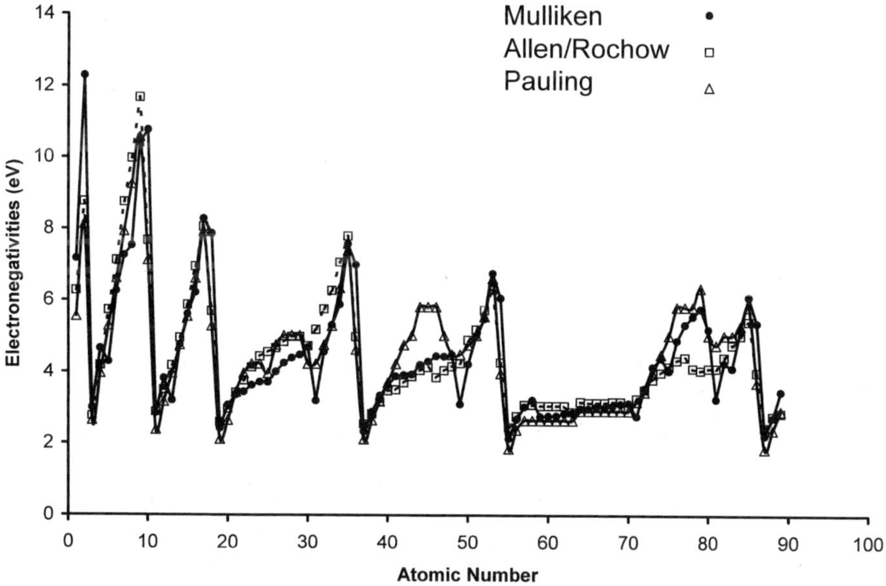

Figure 8.8 Plots of the Mulliken, scaled Pauling, and scaled Allen Rochow electronegativities of the elements versus the atomic number to illustrate the consistency and differences between the measures of electronegativity [41, 42].

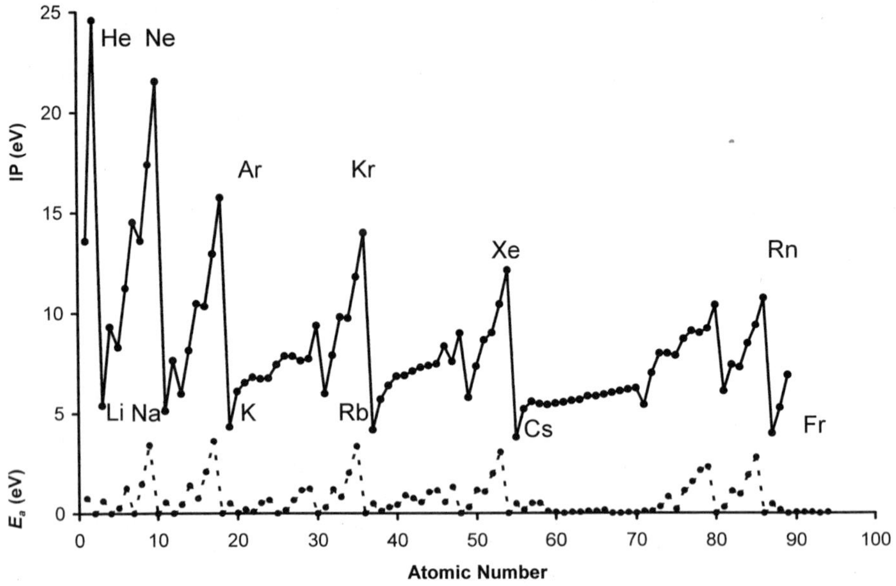

Figure 8.9 Plots of the E_a and IP of the elements versus the atomic number to illustrate the consistency of the IP of a given family and across a period (compare with Figure 8.8).

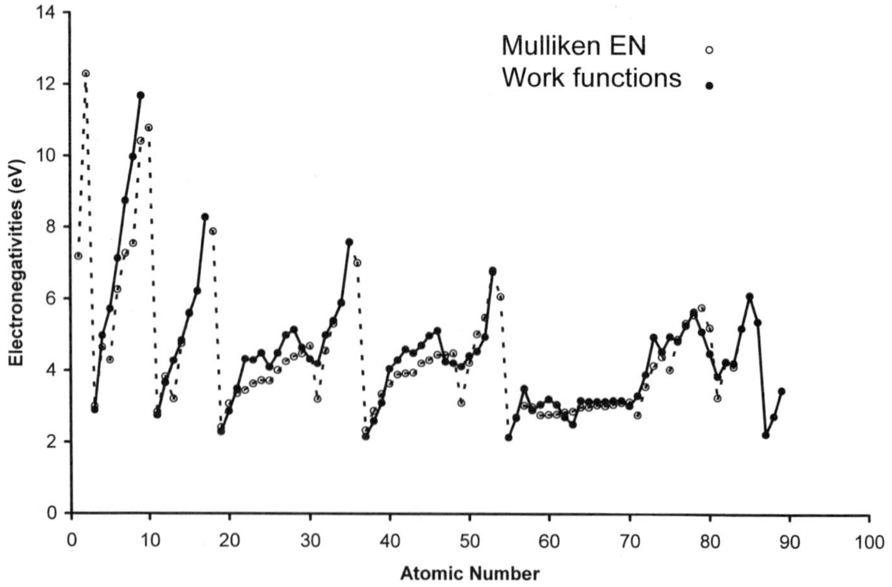

Figure 8.10 Plots of the Mulliken, electronegativities, and work functions of the elements versus the atomic number to illustrate their consistency and differences [43, 44].

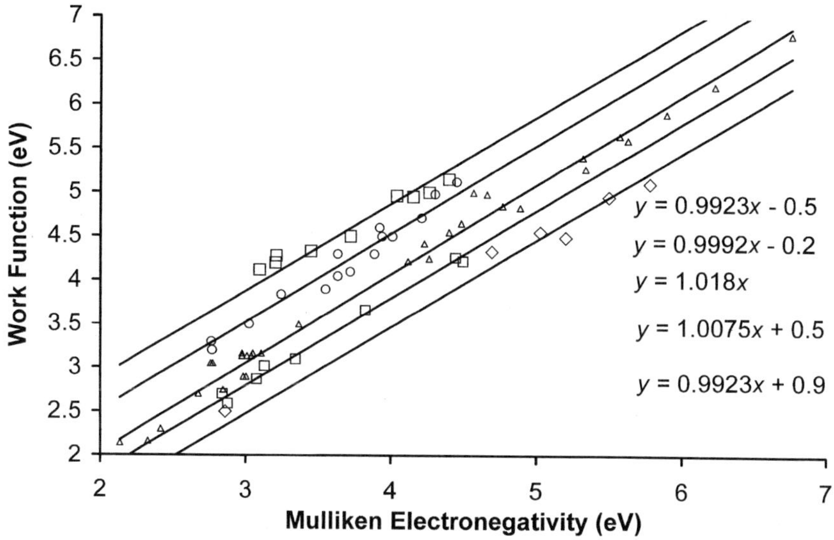

Figure 8.11 Plots of the work function versus the Mulliken electronegativities of the elements to illustrate the constant displacements of groups to determine $P = \text{EN-WF}$ [43, 44].

The work functions of the metals are directly related to Mulliken electronegativity. Figure 8.11 is a plot of the experimental work functions of a series of metals against Mulliken electronegativity. In 1978 a systematic investigation of the relationship between electronegativities and work functions was carried out using a constant displacement of the members of different groups of elements [44, 45]. The equation was $\text{EN} = \text{WF} + P$. The quantity P was defined as the periodicity parameter and assigned to 1 of 10-values: 0.3, 0.2, 0.1, 0, −0.2, −0.3, −0.4, −0.6, −0.7, and −1.0 eV. By examining Figure 8.11, which is based on additional data for both electron affinities and work functions, five values for P can now be determined: −0.9, −0.5, 0.5, 0.2, and 0.0. The intercepts are fixed at 0.2, 0.5, 0, −0.5, and −0.9 and the slopes determined by a linear regression. The slopes are all approximately 1. Twenty-eight of the elements have $P = 0$. The Group IIIA elements and several 3d elements have $P = -0.9$. The $P = -0.5$ group consists of transition elements, Hf, and Tl. The $P = 0.2$ group consists of Mg, Ca, Rb, Sr, Cd, Y, Ag, Sm, and Yb. Zn, Sb, Te, Eu, Au, and Hg are in the $P = 0.5$ group. These P values can be used to predict the work function of Tc as 4.6(1) eV. The Mulliken electronegativities are an approximation to the work functions. The corrected values of the work function are plotted against the EN values in Figure 8.12. This is an example of how species can be classified to obtain additional fundamental information. It is similar to the classification of molecules to estimate solution energy differences to obtain E_a from half-wave reduction potentials.

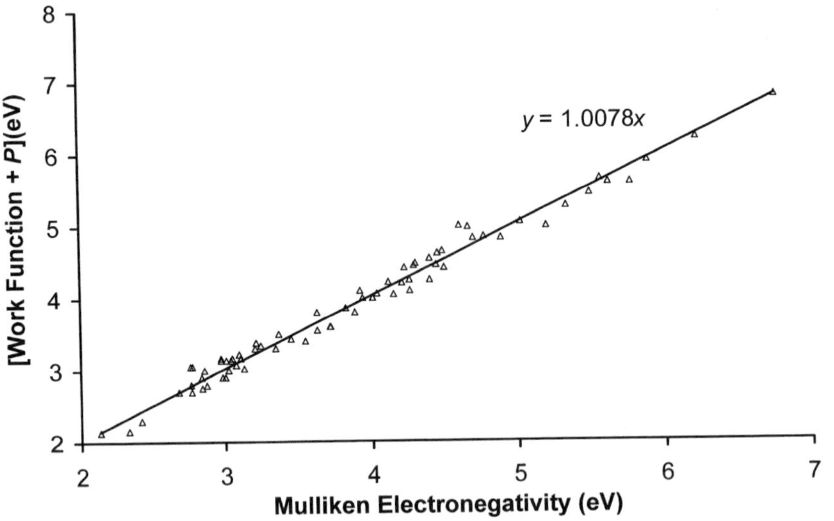

Figure 8.12 Plot of the work function $+P$ versus the Mulliken electronegativities of the elements [43, 44].

The P values for the larger set of data are compared to the earlier ones in Table 8.3. Figure 8.13 is a plot of the P values versus the atomic number. Except for Li, Be, and B, there are no data plotted for the first row atoms. The abnormalities can be noted in Table 8.3. The P for Be is 0.0 eV, but the other Group IIA

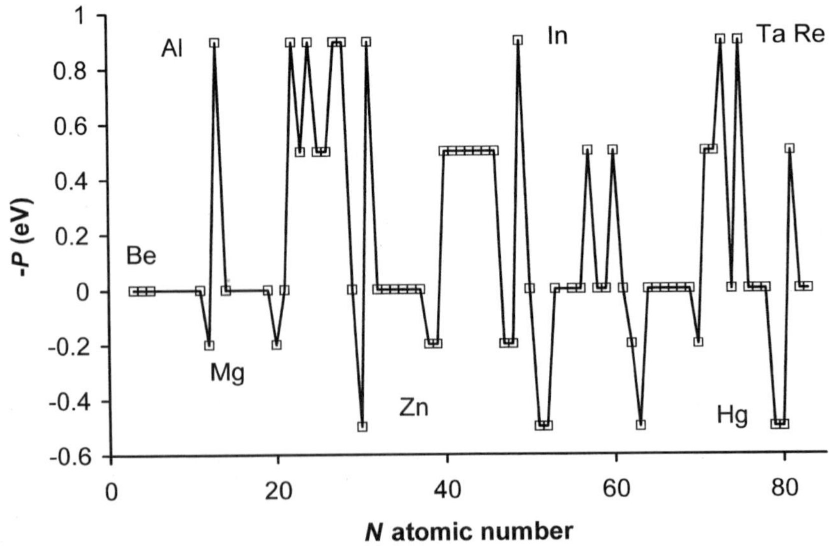

Figure 8.13 Plot of -P versus atomic number to illustrate the consistency of the values for families of elements.

TABLE 8.3 Values of *P* (in eV) to Relate Work Functions and Mulliken Electronegativities [12, 45]

Atom	*N*	*P*	*P* [45]	Atom	*N*	*P*	*P* [44]
Zn	30	0.50	0.20	Tb	65	0.00	—
Sb	51	0.50	—	Dy	66	0.00	—
Te	52	0.50	0.50	Ho	67	0.00	—
Eu	63	0.50	—	Er	68	0.00	—
Au	79	0.50	0.20	Tm	69	0.00	—
Hg	80	0.50	0.20	W	74	0.00	−0.60
Mg	12	0.20	0.00	Os	76	0.00	−0.40
Ca	20	0.20	0.00	Ir	77	0.00	−0.40
Sr	38	0.20	0.00	Pt	78	0.00	−0.40
Y	39	0.20	−0.20	Pb	82	0.00	−0.20
Ag	47	0.20	0.20	Bi	83	0.00	0.10
Cd	48	0.20	0.20	V	23	−0.50	−0.30
Sm	62	0.20	—	Mn	25	−0.50	−0.70
Yb	70	0.20	—	Fe	26	−0.50	−0.40
Li	3	0.00	0.10	Nb	41	−0.50	−0.30
Be	4	0.00	0.00	Mo	42	−0.50	−0.60
B	5	0.00	0.10	Tc	43	−0.50	−0.70
Na	11	0.00	0.10	Ru	44	−0.50	−0.40
Si	14	0.00	—	Rh	45	−0.50	−0.40
K	19	0.00	0.10	Pd	46	−0.50	−0.40
Sc	21	0.00	−0.20	La	57	−0.50	−0.20
Ge	32	0.00	−0.20	Nd	60	−0.50	—
Cu	29	0.00	0.20	Lu	71	−0.50	—
As	33	0.00	0.10	Hf	72	−0.50	−0.6
Se	34	0.00	0.30	Tl	81	−0.50	−1.0
Rb	37	0.00	0.10	Al	13	−0.90	−1.0
Sn	50	0.00	−0.20	Ti	22	−0.90	−0.6
I	53	0.00	—	Cr	24	−0.90	−0.6
Cs	55	0.00	0.10	Co	27	−0.90	−0.4
Ba	56	0.00	0.00	Ni	28	−0.90	−0.4
Ce	58	0.00	—	Ga	31	−0.90	−1.0
Pr	59	0.00	—	In	49	−0.90	−1.0
Pm	61	0.00	—	Ta	73	−0.90	−0.3
Gd	64	0.00	—	Re	75	−0.90	−0.7

$WF = 0.5(IP + E_a) - P$

elements have $P = 0.2$. Even with a different *P* value the calculated EN of Be is more than 0.3 eV lower than the "best" value. The Group IIB elements belong to two groups. The electron affinity of Be, Zn, Cd, and Hg was set to zero in calculating the Mulliken electronegativity. Among the Group IVA elements the value of $P = 0$ for Ge gives a larger deviation of 0.2 eV than for the other Group IVA elements. The work function of carbon must be modified by the band gap energy to be equated to the Mulliken EN. For the most part the value of *P* is constant for elements in a given column. This could possibly be the physical significance of the P parameter if there is in fact one [44].

8.4 ELECTRON AFFINITIES OF ATOMIC CLUSTERS

One of the major applications of the CURES-EC procedure has been in calculating the electron affinities of the carbon clusters C_n [46]. Carbon clusters are common species in nature, since they are a product of combustion and contribute to environmental problems. They have been observed in the interstellar medium. The fullerenes represent one of the most recent discoveries of a new form of carbon. Carbon clusters have many forms—linear chains, cyclic compounds and the fullerenes—because carbon can form covalent bonds. The NIST tables give 54 electron affinities values for C_n; 14 for Si_n; 17 for Ge_n; 45 for Sn_n; and 55 for Pb_n. Only the structures for the carbon compounds from $n = 3$ to 30 are identified [12].

The linear forms of C_n are more stable for $n = 1$ to 12. Above C_{10}, the monocyclic forms are more stable. The observed E_a for the linear species range from 2.0 eV to about 4.8 eV, while the values for the monocyclic compounds range from 2.0 eV to about 3.9 eV. The uncertainty of the E_a for the linear $n = 2$ to 9 and 11 is 0.01 to 0.02 eV. The other values are less precise, with uncertainties of ± 0.1 eV. Both the linear and cyclic forms show an even odd alternation with the even-numbered clusters having the larger electron affinity for the linear form and the odd-numbered clusters having the larger value for the ring structures. The differences are much larger than the uncertainty in the experimental values. This is partially due to the fact that the ground state for the even-numbered linear clusters from 4 to 20 is the $^3\Pi_g$ state, while that for the odd numbered clusters is the $^1\Sigma_g^+$ state. The bond alternation is observed in Figure 8.14, where the experimental values are shown as X's. The linear forms are at the top of the graph, while the cyclic forms appear at the bottom [47–49].

The experimental and theoretical data for the anions of carbon clusters were reviewed in 1999 [50]. Earlier theoretical studies concentrated on the linear clusters up to C_{10}. There were limited studies of the cyclic structures. During 2002 two studies focused on the density functional calculations for $n = 3$ to 13 and 2 to 20 [51, 52]. In 1999 the CURES-EC method was applied to the linear clusters for $n = 2$ to 30 and cyclic structures for $n = 4$, 10 to 30. Table 8.4 gives the results of these calculations and additional calculated values for the cyclic C_{6-9}. These are compared to the experimental values, density functional values, and earlier theoretical values. In Figure 8.14 the dotted lines represent the density functional values and the solid lines the CURES-EC values. The CURES-EC values are plotted in Figure 8.15 with the experimental values for the linear chains. The cyclic structures would merge with the data for Si, Ge, Sn, and Pb. The very good agreement between the experimental and CURES-EC values is easily seen. The agreement between the CURES-EC values for the cyclic clusters is not as good as for the linear clusters, but it is better than for the density functional values, as shown in Figure 8.14. The

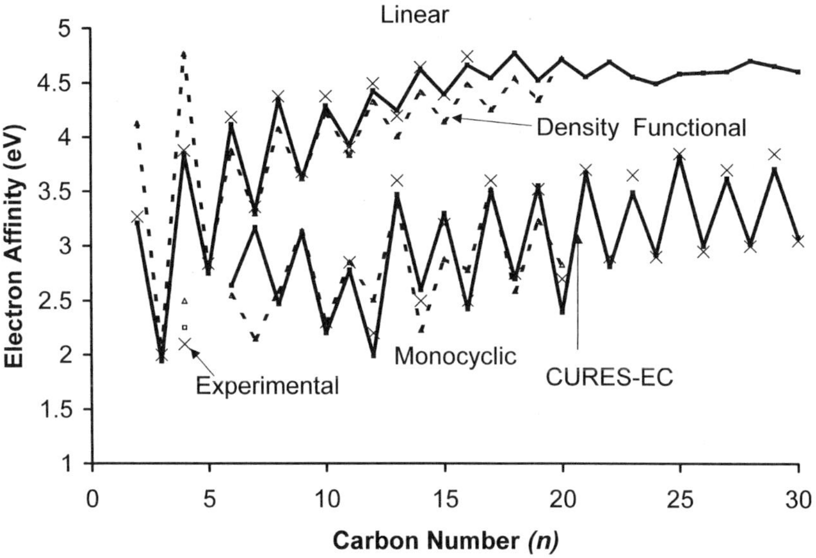

Figure 8.14 Plot of experimental and theoretical electron affinities of carbon clusters for cyclic and linear molecules versus number of carbon atoms. The experimental data are from [45–49], the theoretical values from [45, 50, and 51].

largest absolute deviation in the values is 0.3 eV for cyclic C_{20}. It is possible that this large difference results from the isomers of this compound. In addition, the uncertainty in the experimental values is much larger, partially due to the use of the maximum in the photoelectron spectroscopy peaks as the electron affinity.

The density functional values for cyclic C_2, C_4, and C_8 are significantly different from those of the CURES-EC. The PM3 values, not shown, give good agreement with the experimental values up to C_{10} but the deviations for C_{12} and C_{13} are about 0.5 eV. The use of modified electron correlation improves the agreement. The standard deviation between the CURES-EC and experimental values is ±0.06 eV. Without the clear outliers for density functional values the standard deviation is ±0.2 eV. The CURES-EC values cover a wider range of complexes and agree better with experiment.

The values for the cyclic C_5 to C_9 and linear C_{18} to C_{30} complexes are predictions. Even numbered carbon clusters containing from 32 to 92 carbon atoms have been observed with electron affinities between 3.0 eV and 3.6 eV, the same as the range for the cyclic clusters. The structure of these compounds is unspecified. The E_a of the C_{92} cluster is 3.0 eV, while the maximum value of 3.6 eV lies in the C_{50} to C_{60} range. The odd-numbered clusters reach a maximum of 3.85 eV at C_{29}. The only other E_a reported for an odd-numbered cluster is 3.40 eV for C_{59}. Regardless, the limiting value for the linear clusters (4.75–5 eV) is higher than that for the cyclic clusters (3.0–3.6 eV). The work function of graphite (4.6 eV) is about the same as the limiting value for the linear chains, and less than the value for the Mulliken electronegativity, 6.26 eV.

TABLE 8.4 Experimental, CURES-EC, and Density Functional Electron Affinities (in eV) of Carbon Clusters, n = 2 to 30 [45–51]

	Linear			Monocyclic		
n	Exp	CEC	DF	Exp	CEC	DF
2	3.27	3.21	4.13	—	—	—
3	2.00	1.94	2.03	—	—	—
4	3.88	3.85	4.77	2.1	2.25	2.5
5	2.84	2.75	2.82	—	—	—
6	4.19	4.12	3.88	—	2.64	2.55
7	3.36	3.32	3.30	—	3.17	2.15
8	4.38	4.35	4.08	—	2.47	2.59
9	3.68	3.62	3.62	—	3.13	3.14
10	4.38	4.29	4.22	2.30	2.20	2.30
11	3.91	3.93	3.84	2.85	2.78	2.85
12	4.5	4.43	4.33	2.20	1.99	2.51
13	4.20	4.25	4.01	3.60	3.47	3.39
14	4.65	4.63	4.42	2.50	2.60	2.23
15	4.40	4.40	4.15	3.20	3.30	2.88
16	4.75	4.67	4.49	2.50	2.42	2.78
17	—	4.55	4.26	3.60	3.51	3.52
18	—	4.78	4.55	2.75	2.7	2.59
19	—	4.53	4.35	3.52	3.55	3.23
20	—	4.72	4.73	2.70	2.39	2.83
21	—	4.56	—	3.70	3.67	—
22	—	4.70	—	2.90	2.81	—
23	—	4.56	—	3.65	3.49	—
24	—	4.50	—	2.90	2.92	—
25	—	4.59	—	3.85	3.82	—
26	—	4.60	—	2.95	3.00	—
27	—	4.61	—	3.70	3.62	—
28	—	4.71	—	3.00	3.02	—
29	—	4.66	—	3.85	3.71	—
30	—	4.61	—	3.05	3.08	—

The trends for the electron affinities of carbon might also be expected for the clusters of Si, Ge, Sn, and Pb. Table 8.5 lists the electron affinities for some of the lower clusters of these elements. These are plotted versus the number of atoms in Figure 8.15 [53–55]. The general shape of these curves is similar to that for the cyclic C_n compounds, but with less regularity in the alternation. The values for the other Group IV elements level off at about 2.5 to 3.0 up to $n = 20$. At higher n the electron affinities generally remain the same for Ge to Pb. At $n = 45$ the Sn value is 3.010 eV, while the Pb value is 2.630 eV. None of the values reach the Mulliken electronegativity or experimental work functions. The Pb value is 3.18(14) eV for $n > 200$. It has been extrapolated to the work function of Pb,

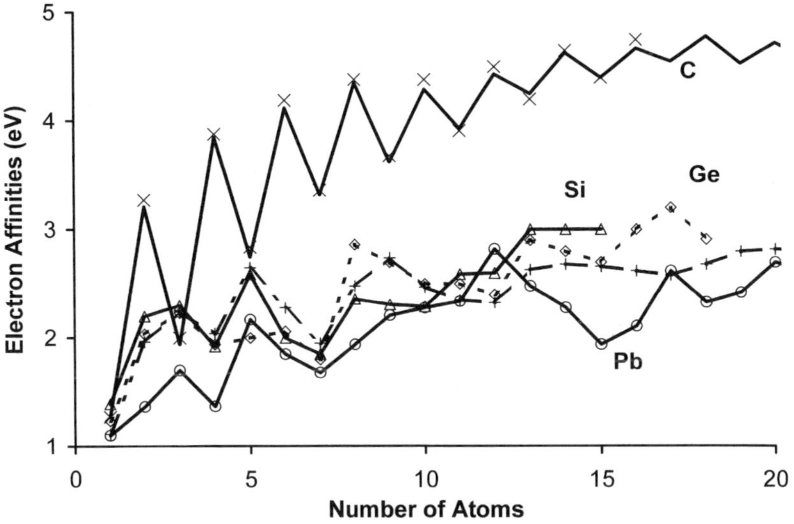

Figure 8.15 Plot of experimental and theoretical electron affinities of carbon silicon, germanium tin, and lead clusters versus number of atoms. The experimental data are from [12, 45–54].

TABLE 8.5 Electron Affinity (in eV) of Linear Carbon Clusters and Si, Ge, Sn, and Pb Clusters, $n = 2$ to 20 [12]

N	C	Si	Ge	Sn	Pb
1	1.27	1.39	1.23	1.11	1.10
2	3.27	2.20	2.04	1.97	1.37
3	2.00	2.30	2.23	2.24	1.70
4	3.88	1.92	1.94	2.04	1.37
5	2.84	2.59	2.00	2.65	2.17
6	4.19	2.00	2.06	2.28	1.85
7	3.36	1.85	1.80	1.95	1.68
8	4.38	2.36	2.86	2.48	1.94
9	3.68	2.31	2.70	2.74	2.21
10	4.38	2.29	2.50	2.47	2.28
11	3.91	2.59	2.50	2.35	2.34
12	4.50	2.60	2.40	2.33	2.82
13	4.20	3.00	2.90	2.63	2.48
14	4.65	3.00	2.80	2.68	2.28
15	4.40	3.00	2.70	2.66	1.94
16	4.75	—	3.00	2.62	2.11
17	—	—	3.20	2.58	2.62
18	—	—	2.91	2.68	2.33
19	—	—	—	2.80	2.42
20	—	—	—	2.82	2.70

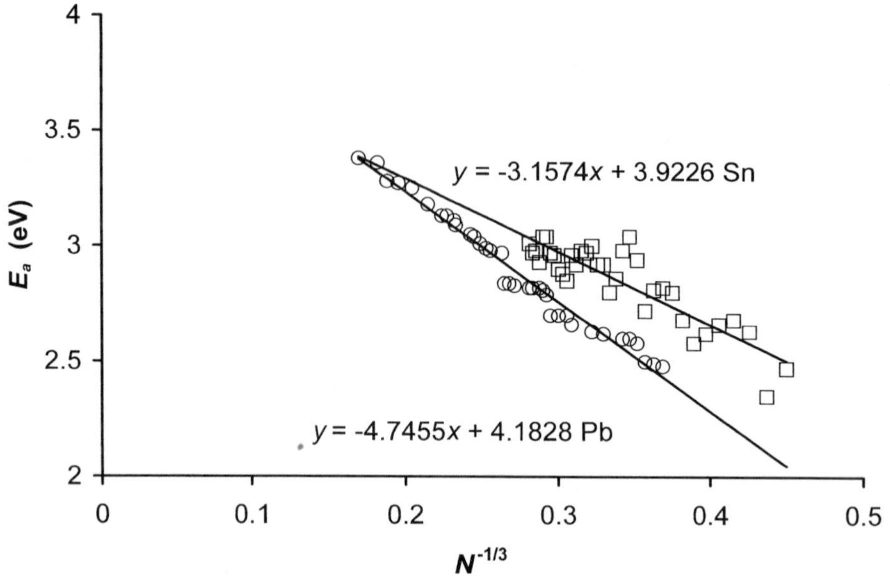

Figure 8.16 Plot of experimental and theoretical electron affinities of tin and lead clusters versus $N^{-1/3}$, where N is the number of atoms. This plot should extrapolate to the work functions for metals. The extrapolated value for lead is approximately the same as the work function, but that for tin is lower than the work function. The experimental data are from [12].

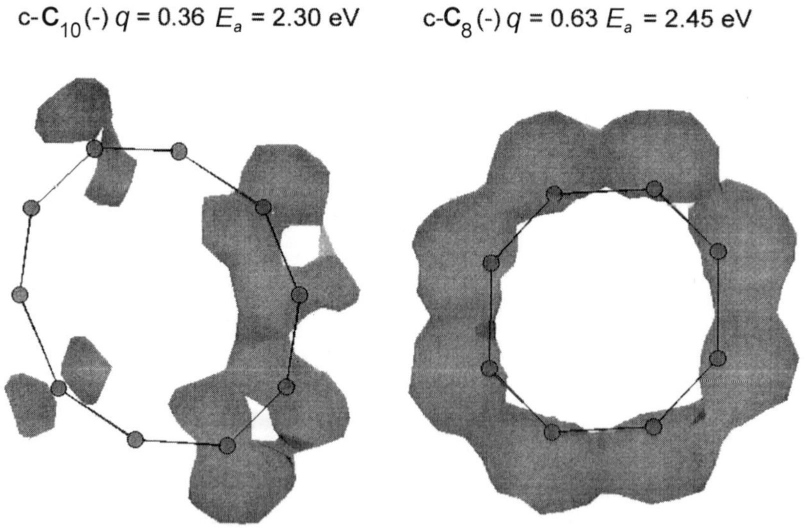

Figure 8.17 Three-dimensional spin densities of the negative ions of cyclic C_8 and cyclic C_{10} calculated for the addition of one electron to the combined system using AM1. The values of the charge densities indicate that the electron affinity of c-C_8 > c-C_{10} = 2.30 eV.

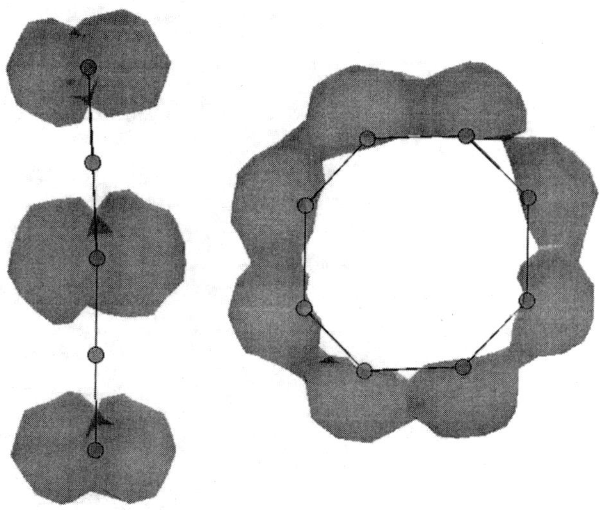

Figure 8.18 Three-dimensional spin densities of the negative ions of cyclic C_8 and linear C_5 calculated for the addition of one electron to the combined system using AM1. The CURES-EC value for c-C_8 is equal to 2.45 eV, while the density functional value is 2.57 eV. The READS-TCT data shown in Figures 8.17 and 8.18 are consistent with a value of 2.5 ± 0.1 eV for the E_a of c-C_8.

4.25 eV, using a metallic model where E_a is related to $N^{-1/3}$. The extrapolated value for Sn is less than the work function, indicating some covalent rather than metallic bonding. These plots can be observed in Figure 8.16 [54].

In Figures 8.17 and 8.18 the results of a READS-TCT calculation are given. The three-dimensional spin densities of the negative ions of the cyclic C_8 and C_{10} are shown in Figure 8.17, while those for the cyclic C_8(-) and linear C_5(-) are shown in Figure 8.18. The experimental electron affinity of C_{10} is 2.30 eV, whereas that for the cyclic C_8 has not been reported. The CURES-EC and density functional values are 2.47 eV and 2.59 eV for the cyclic C_8. The READS TCT calculation shows electron transfer from C_{10} ($q = 0.36$) to C_8 ($q = 0.63$). This indicates the E_a of $C_8 > 2.30$ eV. The electron affinity of the cyclic C_8 is about the same as that for the linear C_5, as are the q values. The experimental E_a for the linear C_5 is 2.57 eV. This is consistent with an E_a of cyclic $C_8 = 2.5 \pm 0.1$ eV. The spin densities reflect the charge distribution and illustrate electron transfer through space.

8.5 SUMMARY

Three groups of atomic AE_a can be discerned: those that are precisely and accurately measured to one meV; those that have been precisely measured but have not been confirmed; and those that have only been estimated. If a value has been measured by two or more methods and those values agree within the random uncertainty, the weighted average is the "best" value. When this is the largest value, it is the AE_a. A single very precise value will dominate the weighted average and the uncertainty will be approximately equal to that determination. Precise positive electron affinities for the Group 0 to Group VII elements can be assigned to the ground state. Some other positive values are assigned to excited states. Experimental values for the electron affinities of all the d elements and some of the f elements are available. However, based on the abnormalities in periodic trends of the d and f group elements and the lack of verification of the experimental values, these are not necessarily the AE_a.

The experimental work functions of the metals are related to the Mulliken electronegativities by a constant displacement, $P = -0.9$, -0.5, 0, 0.2, and 0.5 eV, for the majority of the elements. The Pauling and Allen Rochow electronegativities have systematic variations similar to the Mulliken electronegativities for the main group of elements, but the values are different in magnitude for the d and f block elements.

The electron affinities of the linear and monocyclic C_n clusters have been calculated using the CURES-EC method and agree with the experimental values. The electron affinities of the Si_n, Ge_n, Sn_n, and Pb_n vary with N in a manner similar to those of the cyclic C_n compounds. Up to $N = 15$ to 20, the values are considerably lower than the bulk work function. This behavior is different from that for metallic clusters, where the electron affinity is a function of $N^{-1/3}$. However, the data for Pb taken up to $N = 204$ can be extrapolated to the work function.

REFERENCES

1. Pritchard, H. O. *Chem. Rev.* **1953**, 52, 529.
2. Berry, R. S *Chem. Rev.* **1969**, 69, 533.
3. Zeforov, Y. V. *Zh. Strukturnoi Khimi* **1973**, 14, 762.
4. Chen, E. C. M. and Wentworth, W. E. *J. Chem. Educ.* **1975**, 52, 486.
5. Hotop, H. and Lineberger, W. C. *J. Phys. Chem. Ref. Data* **1975**, 4, 539.
6. Massey, H. S. W. *Negative Ions*. New York: Cambridge University Press, **1976**.
7. Christodoulides, A. A.; McCorkle, D. L.; and Christophorou, L. G. "Electron Affinities of Atoms, Molecules and Radicals" in *Electron-Molecule Interactions and Their Applications*. New York: Academic Press, **1984**.
8. Hotop, H. and Lineberger, W. C. *J. Phys. Chem. Ref. Data* **1985**, 14, 731.
9. Wheeler, J. C. *J. Chem. Educ.* **1997**, 74, 123.
10. Nadeau, M.-J.; Garwan, M. A.; Zhao, X.-L.; and Litherland, A. E. *Nucl. Instr. Meth. Phys. Res. B* **1997** 123, 521.
11. Andersen, T., Haugen, H. K.; and Hotop, H. *J. Phys. Chem. Ref. Data* **1999**, 28, 1511.
12. National Institute of Standards and Technology (NIST). *Chemistry WebBook*, **2003**. Available at http://webbook.nist.gov.

13. Rienstra-Kiracofe, J. C.; Tschumper, G. S.; Schaeffer, H. F.; Nandi, S.; and Ellison, G. B. *Chem. Rev.* **2002**, 102, 231.
14. Davis, V. T. and Thompson, J. S. *Phys. Rev. A* **2001**, 65, 10501R.
15. Davis, V. T. and Thompson, J. S. *J. Phys. B* **2001**, 354, L433.
16. Davis, V. T. and Thompson, J. S. *J. Phys. B* **2002**, 35, L11.
17. Davis, V. T. and Thompson, J. S. *Phys. Rev. Lett.* **2002**, 88, 073003.
18. Page, F. M. and Goode, G. C. *Negative Ions and the Magnetron.* New York: Wiley-Interscience, **1969**.
19. Khvostenko, V. I. and Dukel'skil, V. M. *Zh. Eksp. Teor. Fiz.* **1958**, 37, 651.
20. Sheer, M. D. and Fine, J. *J. Chem. Phys.* **1967**, 46, 3998.
21. Sheer, M. D. and Fine, J. *J. Chem. Phys.* **1969**, 50, 4343.
22. Sheer, M. D. *J. Res. Nat. Bur. Stds.* **1970**, A74, 37.
23. Bailey, T. L. *J. Chem. Phys.* **1958**, 28, 792.
24. Zandberg, E. Y.; Kamenev, A. G.; and Paleev, V. I. *Sov. Phys.-Tech Phys.* **1971**, 16, 832.
25. Zandberg, E. Y.; Kamenev, A. G.; and Paleev, V. I. *Sov. Phys.-Tech Phys.* **1972**, 16, 1657.
26. Ansdell, D. A. and Page, F. M. *Trans. Faraday Soc.* **1962**, 58, 1084.
27. McCullum, K. J. and Mayer, J. E. *J. Chem. Phys.* **1943**, 11, 56.
28. Doty, P. M. and Mayer, J. E. *J. Chem. Phys.* **1944**, 12, 323.
29. Sutton, J. and Mayer, J. E. *J. Chem. Phys.* **1935**, 3, 20.
30. Feldman, D. In *Proc. 8th Internat. Conf. on Electr. Atom. Collis.* Belgrade, **1973**.
31. Williams, W. W.; Carpenter, D. L.; Covington, A. M.; Koepnick, M. C.; Calabrese, D.; and Thompson, J. S. *J. Phys. B* **1998**, 31, L341.
32. Lippa, T. P.; Xu, S. J.; Lyapustina, S. A.; Nilles, J. M.; and Bowen, K. H. *J. Chem. Phys.* **1998**, 109, 10727.
33. Dukel'skil, V. M.; and Sokolov, V. M. *Zh. Eksp. Teor. Fiz.* **1957**, 32, 394.
34. Brunot, A.; Cottin, M.; Gotchiguian, P.; and Muller, J. C. *Int. J. Mass Spectrom. Ion Phys.* **1981**, 41, 31.
35. Feigerle, C. S.; Corderman, R. R.; and Lineberger, W. C. *J. Chem. Phys*, **1981**, 74, 1513.
36. Buckman, S. J. and Clark, C. W. *Rev. Mod. Phys.* **1994**, 66, 539.
37. Williams, W. W.; Carpenter, D. L.; Covington, A. M.; Thompson, J. S.; Kvale, T. J.; and Seely, D. G. *Phys. Rev. A* **1998**, 58, 3582.
38. Brunot, A.; Cottin, M.; Donnart, M. H.; and Muller, J. C. *Int. J. Mass Spectrom. Ion Phys.* **1980**, 33, 417.
39. Carpenter, D. L.; Covington, A. M.; and Thompson, J. S. *Phys. Rev. A* **2000**, 61, 42501.
40. Landau, A.; Eliav, E.; Ishikawa, Y.; and Kaldor, U. *J. Chem. Phys.* **2001**, 115, 2389.
41. Pauling, L. *The Nature of the Chemical Bond*, 3rd ed. Ithaca, NY: Cornell University Press, **1960**.
42. Porterfield, W. W. *Inorganic Chemistry.* New York: Academic Press, **1993**.
43. Chen, E. C. M.; Wentworth, W. E.; and Ayala, J. A. *J. Chem. Phys.* **1977**, 67, 2642.
44. Michaelson, H. B. *IBM J. Res. Dev.* **1978**, 22, 72.
45. Chen, E. C. M.; Schulze, Welk, S.; Brown, T.; and Chen E. S. D. *The Chemist* **1999**.
46. Yang, S.; Taylor, K. J.; Craycraft, M. J.; Conceicao, J.; Pettiette, C. L.; Cherhnovsky, O.; and Smalley, R. E. *Chem. Phys. Lett.* **1989**, 144, 431.

47. Kohno, M.; Suzuki, S.; Shiromaru, H.; Moriwaki, T.; and Achiba, Y. *Chem. Phys. Lett.* **1998**, 282, 330.
48. Arnold, D. W.; Bradforth, S. E.; Kitsopoulos, T. N.; and Neumark. *J. Chem. Phys.* **1991**, 95, 8753.
49. Van Orden, A. and Saykally, R. J. *Chem. Rev.* **1998**, 98, 2313.
50. Lepine, F. Allouche, A. R.; Baguenard, B.; Bordas, C.; and Aubert-Frecon, M. *J. Phys. Chem. A* **2002**, 106, 7177.
51. Giuffreda, M. G.; Deluze, M. S.; and Francois, J. P. *J. Phys. Chem. A* **2002**, 106, 8569.
52. Liu, Y.; Zhang, Q.-L.; Tittel, F. K.; Curl, R. F.; and Smalley, R. E. *J. Chem. Phys.* **1986**, 85, 7434.
53. Negishi, Y.; Kawamata, H.; Nakajima, A.; and Kaya, K. *J. Elec. Spectrosc. Rel. Phenom.* **2000**, 106, 117.
54. Luder, C. and MeiwesBroer, K. H. *Chem. Phys. Lett.* **1998**, 294, 391.

CHAPTER 9

Diatomic and Triatomic Molecules and Sulfur Fluorides

9.1 INTRODUCTION

In Chapter 8 the electron affinities of atoms were evaluated. In this chapter the electron affinities of diatomic and triatomic molecules and SF_n ($n = 1$ to 6) will be considered. The ECD has been used to study Cl_2, Br_2, I_2, NO, O_2, CO_2, COS, CS_2, N_2O, NO_2, SO_2, SF_6. All the E_a for these molecules have been calculated by the CURES-EC method. The comparison of the relative electron affinities of COS, CS_2, and N_2O will be illustrated by READS-TCT calculations.

The homonuclear diatomic molecules are the simplest closed set of molecules. Many of the electron affinities of the main group diatomic molecules have been measured by anion photoelectron spectroscopy (PES), but only a few have been confirmed. These E_a can be examined by their systematic variation in the Periodic Table. Calculating Morse potential energy curves for the anions and comparing them with curves for isoelectronic species confirm experimental values. The homonuclear diatomic anions of Group IA, IB, VI, VII, and 3d elements and NO are examined first.

Next, electron affinities and Morse potential energy curves for triatomic molecules will be considered. Electron affinities have been reported for CO_2, COS, CS_2, SO_2, N_2O, NO_2, O_3, and N_3. The experimental E_a for CO_2, COS, CS_2, and N_2O are uncertain. These are linear in the neutral form and bent in the anion. The valence-state electron affinities are very different, ranging from a negative value for CO_2 to 0.89(2) eV for CS_2. The energy of these ions is calculated as a function of the bonding angle using CURES-ES. Morse potential energy curves are constructed for the bent and linear forms of these anions from theoretical and experimental data.

Anions of SF_n ($n = 1$ to 6) provide a transition to larger molecules. The anions of SF_6 and SF_4 are among the most frequently studied in the gas phase. The ECD, atmospheric pressure ionization, and low-pressure chemical ionization mass spectrometry have been used to study SF_6. The precise and accurate adiabatic electron

The Electron Capture Detector and the Study of Reactions with Thermal Electrons
by E. C. M. Chen and E. S. D. Chen
ISBN 0-471-32622-4

affinity of SF_6 was measured independently using high-pressure and atmospheric pressure negative-ion mass spectrometry. Values that differ from this largest precise value have been assigned to excited anion states. Morse potential energy curves were constructed for the ground state and excited states using these assignments, electron affinities, and electron impact ion distributions. The ground-state curve of $SF_6(-)$ is compared with an ab initio calculation. By analogy the Morse curves for other $SF_n(-)$ are calculated. These illustrate the use of excited states to explain the different E_a reported for these compounds. The anion curves provide examples of the Herschbach classifications, D(0–2) and M(0–3).

9.2 DIATOMIC MOLECULES

9.2.1 Electron Affinities and Periodic Trends of Homonuclear Diatomic Molecules

The electron affinities of the main group homonuclear diatomic anions have been measured by PES. A few experimental values for the transition metal dimers are also available. The electron affinities of all the 3d homonuclear diatomic molecules have been calculated using density functional methods [1–4]. Only the AE_a of I_2, 2.52_4 eV; C_2, 3.2_7; Si_2, 2.2_0; S_2, 1.6_7; F_2, 3.0_8; Cl_2, 2.4_5; Br_2, 2.5_6; and O_2, 1.0_7 have been measured by more than one method [1–3]. CURES-EC calculations confirm these to within ± 0.1 eV. Positive excited states E_a have been measured for O_2, C_2, and I_2 and are inferred for other X_2 [5–8]. Just as in the case of the atomic E_a, the trends in the Periodic Table can support the assignments of AE_a for the other elements.

The establishment of accurate and precise E_a for molecules is the ultimate goal of experimental and theoretical studies. The E_a and bond dissociation energies for the Group IA, IB, and IIIA–VIIA homonuclear diatomic molecules are shown in the form of a Periodic Table in Figure 9.1. These are the second entry in each block below the E_a for the atoms [1–3]. The specific references for these data are given in Appendix I, where the E_a of the homonuclear diatomic molecules are listed. The values for the rare gases are 0+ because they only result from the polarization attractions of the dimers.

The E_a of the main group homonuclear diatomic molecules are consistent with molecular orbital predictions of the bond order: BO $= D_e(X_2(-))/D_e(X_2) =$ the net number of bonding electrons in the ion divided by that of the neutral, as shown in Figure 9.2. The Group I and VII dimers have two net bonding electrons in the neutral and one in the anion such that the predicted bond order is 0.5. The experimental values range from 0.55 to about 1.0. The predicted bond order for Groups III and IV are 1.5 and 1.25, but the experimental values are larger. The change in the bond dissociation energy $D_e(X_2) - D_e(X_2(-))$ is approximately given by $E_a(X_2) - E_a(X)$ and obtained by subtracting the first two entries in each block in Figure 9.1. The largest increase in dissociation energy for the ground state is for $C_2(-)$, $3.273 - 1.261 = 2.012$ eV, while the largest decrease occurs for $Cl_2(-)$, $2.45 - 3.6127 = -1.16$ eV.

O	I	II	Ib	IIb	III	IV	V	VI	VII
	1 H 0.7542 [0+] 4.51			Element Ea(X) eV Ea(X2) eV BDE (eV)					
2 He [0+] [0+]	3 Li 0.6180 0.509 1.13	4 Be [0+]	Element Ea(X) eV Ea(X2) eV BDE (eV)		5 B 0.2797 1.30 3.08	6 C 1.2621 3.273 6.3	7 N [0+] [0+] 9.8	8 O 1.4611 1.07 5.16	9F 3.4012 3.05 1.69
10 Ne [0+] [0+]	11Na 0.5479 0.430 0.77	12 Mg [0+]		Element Ea(X) eV Ea(X2) eV BDE (eV)	13 Al 0.4328 1.46 1.93	14 Si 1.3895 2.200 3.4	15 P 0.7465(3) 0.61 5.07	16 S 2.0771 1.69 4.42	17 Cl 3.6127 2.45 2.56
18 Ar [0+] [0+]	19 K 0.5015 0.497 0.6	20 Ca 0.0245	29 Cu 1.2358 0.84 2.01	30 Zn [0+]	31 Ga 0.43(3) 1.6 1.43	32 Ge 1.2327 2.035 2.73	33 As 0.814(8) 0.739 3.96	34 Se 2.0207 1.94 3.45	35 Br 3.3636 2.57 1.99
36 Kr [0+] [0+]	37 Rb 0.4859 0.498 0.47	38 Sr 0.0521	47 Ag 1.3045 1.10 1.7	48 Cd [0+]	49 In 0.404(9) 1.27 1.04	50 Sn 1.1121 1.962 2.03	51 Sb 1.0474 1.282 3.1	52 Te 1.9709 1.92 2.9	53 I 3.0590 2.524 1.56
54 Xe [0+] [0+]	55 Cs 0.4716 0.469 0.47	56 Ba 0.1446	79 Au 2.3086 1.94 2.34	80 Hg [0+]	81 Tl 0.38(1) 0.95 0.66	82 Pb 1.10(5) 1.366 0.91	83 Bi 0.9424 1.271 2.07	84 Po [1.9(3)]	85 At [2.8(3)]
86 Rn [0+] [0+]	87 Fr 0.491(5)	88 Ra 0.17							

21 Sc 0.19 0.89 1.65(22)	22 Ti 0.08 0.63 1.54(18)	23 V 0.53 0.54 2.753(1)	24 Cr 0.6758 0.51 1.44(5)	25Mn [0+] 0.68 0.8	26 Fe 0.151(3) 0.9 1.15(9)	27 Co 0.663 1.11 1.69(26)	28 Ni 1.1572 0.95 2.068(1)
39 Y 0.31	40 Zr 0.43	41 Nb 0.89(3)	42 Mo 0.747	43 Tc 0.6(2)	44 Ru 1.046	45 Rh 1.1429	46 Pd 0.5621
57 La 0.47(2)	72 Hf 0.1	73 Ta 0.32(1)	74 W 0.82	75 Re 0.2(2)	76 Os 1.0778	77 Ir 1.5644	78 Pt 2.1251
58 Ce 0.96(3)	59 Pr 0.96(2)	60 Nd 0.05+	61 0+	62-68 0.1+	69Tm 1.03(3)	70Yb 0.01+	71 Lu 0.34(1)
89 Ac 0+	90 Th 0.05+	91 Pa 0.05+	92 U 0.05+	93 Np 0+	94 Pu 0.05+		

Figure 9.1 Electron affinities of the elements, electron affinities, and bond dissociation energies of the homonuclear diatomic molecules in the form of a Periodic Table [1].

The bond order is $BO = 1 + [D_e(X_2) - D_e(X_2(-))]/D_e(X_2) = 1 + [E_a(X_2) - E_a(X)]/D_e(X_2)$. This is obtained from the data in Figure 9.1 by adding 1 to the difference in the electron affinities divided by the bond dissociation energy given as the third entry for each element. For C_2 $BO = 1 + 2/6.3 = 1.32$; for Cl_2 $BO = 1 - 1.16/2.56 = 0.55$. The value for the rare gases is 1. The bond orders for the Group IA and VIA elements approach 1 going down the table. The values for the Group IIIA and IVA elements are all above unity and increase down the table. The bond orders for the Group VA elements go from less than 1 to greater than 1 from P to Bi. The vertical trends can be observed in Figure 9.2. The smooth

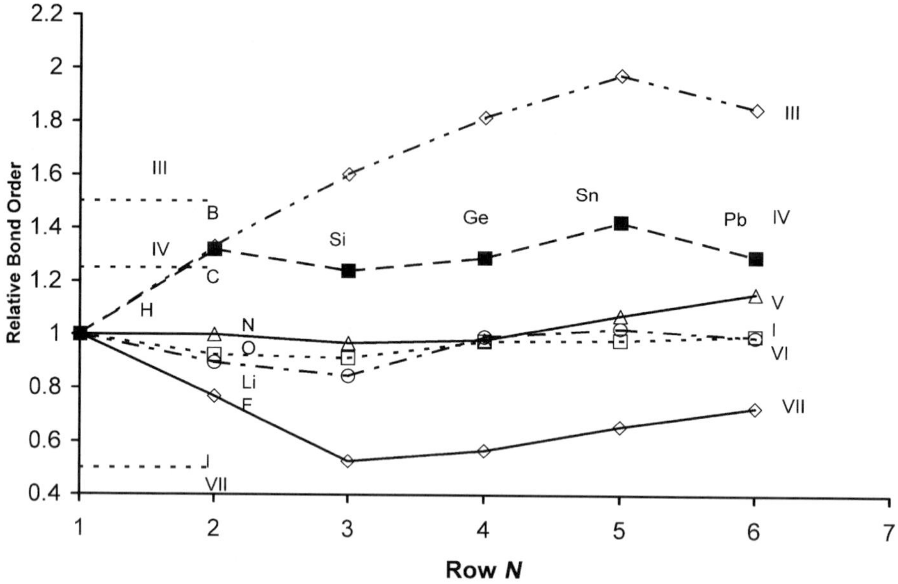

Figure 9.2 Bond energies of the homonuclear diatomic anions divided by bond energies of the neutral versus the row number. This illustrates the consistency of the values for a given family, from Figure 9.1.

changes down the Periodic Table support the assignments of the AE_a of the main group elements.

The horizontal trends in the bond order are shown in Figure 9.3, where the relative bond orders are plotted against the atomic number. The values for the 3d homonuclear diatomic molecules are a combination of experimental and theoretical values. The bond dissociation energies for the anions are calculated from the average of the density functional electron affinities for the elements where experimental data are not available. The values for the 4d and 5d elements should follow the same trend as for the 3d elements and are shown with dotted lines. The relative bond orders not determined experimentally are set to unity to plot the complete graph. The extreme values are identified. The values for O_2 and Pb_2 are pointed out to illustrate their consistency with the periodic trends. For Pb the $BO = 1 + 0.27/0.91 = 1.30$ and supports the assignment of the 1.10(5) eV value for the AE_a of Pb. For the lower atomic E_a of Pb with a value of 0.364(8) eV the bond order is $1 + 1/0.9 = 2.11$, which would be the largest value in the graph. For O_2 the $BO = 1 - 0.39/5.2 = 0.92$, consistent with the other Group VIA elements. If the lower electron affinity of O_2 was used, the bond order would be lower than Li_2 and slightly higher than the value for F_2. The horizontal trends also support the assignments for the AE_a of the main group diatomic anions and elements made in Chapter 8. These are the major support for the assignment of the experimental E_a of the homonuclear diatomic molecules to the AE_a. The horizontal trend in the relative bond orders for 3d homonuclear diatomic anions and

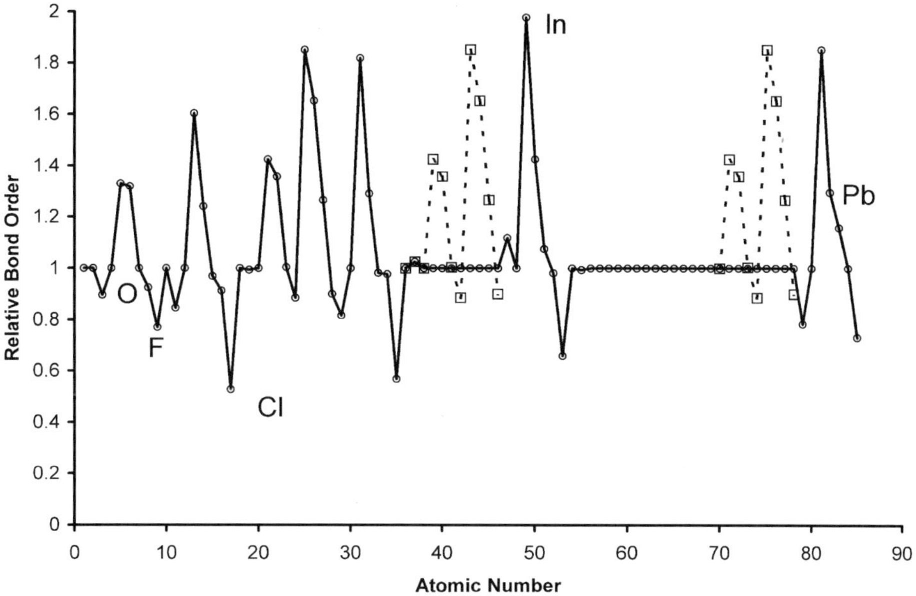

Figure 9.3 Bond energies of the homonuclear diatomic anions divided by bond energies of the neutral versus atomic number. The values for the 3d elements are taken from [4]. The 4d and 5d elements are estimated to be the same as for the 3d elements.

data support the experimental and theoretical electron affinities and provide targets for the 4d and 5d elements.

9.2.2 Electron Affinities and Morse Potential Energy Curves: Group VII Diatomic Molecules and Anions

In 1985 we began a series of papers devoted to Morse potential energy curves for the halogen anions, $X_2(-)$, the rare gas positive-ion dimers, $Rg_2(+)$, and the Group VI homonuclear diatomic anions. These curves were updated periodically based on new experimental data [8–11]. Curves for the anions of the alkali metal dimers and coinage metals dimers, $M_2(-)$, have been constructed but not published. Originally, the ground-state electron affinities for the X_2 were available from gas phase experiments. The ECD had been used to measure the activation energies for thermal electron attachment. The electron impact ion distributions had been measured for all the X_2 except for I_2. For some of the anions the frequencies had been measured in the solid state and the absorption spectra in solution.

As early as the 1960s D. R. Herschbach calculated six Morse curves for $I_2(-)$. The data used to calculate these curves were described as follows:

> Electron impact experiments only show that that the curves for some states of $X_2(-)$ must cross the ground state of the parent near its minimum.... The dissociation energy of $I_2(-)$ should be one half of the neutral or 0.7 eV (this gives an E_a of

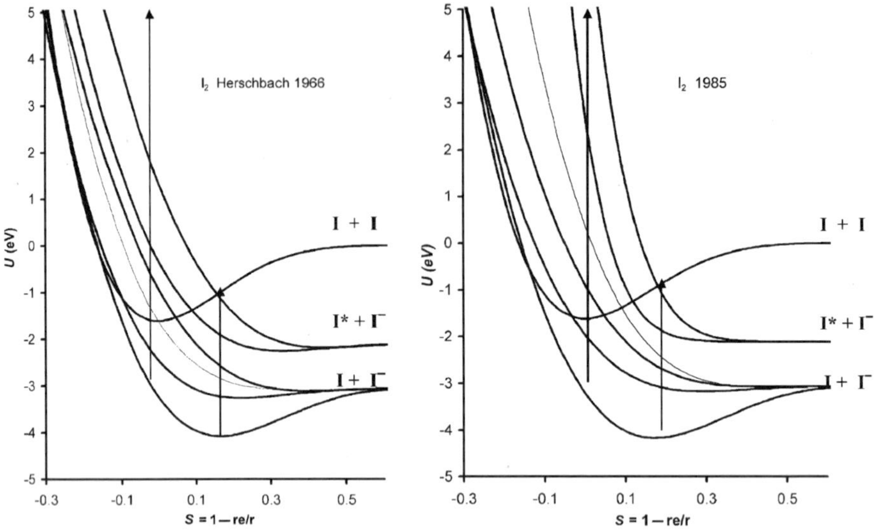

Figure 9.4 Historical Morse potential energy curves for I_2 and $I_2(-)$: 1966 [12] and 1985 [8].

> 2.3 ± 0.5 eV). The VE_a of I_2 is 1.7 ± 0.5 eV and 1.1 ± 0.5 eV for Br_2. . . . The location of the excited state curves above the minimum for the ground state has been derived from a study of color centers in doped alkali halide crystals. . . . Flash photolysis of aqueous and ethanoic solutions of alkali halides gives rise to spectra that have been assigned to the $X_2(-)$. [12]

These curves were the prototypes for the original HIMPEC classifications.

In 1985 six curves for all the X_2 were obtained from updated data and ECD E_1. No electron impact data existed for the two highest $I_2(-)$ curves. The original Herschbach curves and curves obtained in 1985 are compared in Figure 9.4. The normal plot versus internuclear distance r is replaced by a plot of U versus $S = 1 - [r/r_e]$. R. G. Parr suggested this variable as a technique for spreading out the curves in the region of the experimental data [13]. The VE_a, absorption energies, activation energies, and crossings are more easily visualized with the S variable. The absorbance data used in 1985 are the same as those used in 1966. The major difference in the 1985 curves resulted from the assumed electron impact distribution for I_2. The Herschbach curves are closer to the present-day curves, except that 6 curves are drawn instead of 12 [8, 12].

Figure 9.5 gives the original absorbance data used by Herschbach to draw the $I_2(-)$ curves. The energies represented by dotted lines in the 1966 data were approximated. Figure 9.6 offers the most recent data. These data are important for constructing the excited-state curves. However, before we are able to use these data, good curves for the ground-state anion must be available.

In Table 9.1 the optimum Morse parameters for the ground states of $X_2(-)$ are compared with the 1985 values, AMB and PES and theoretical calculations

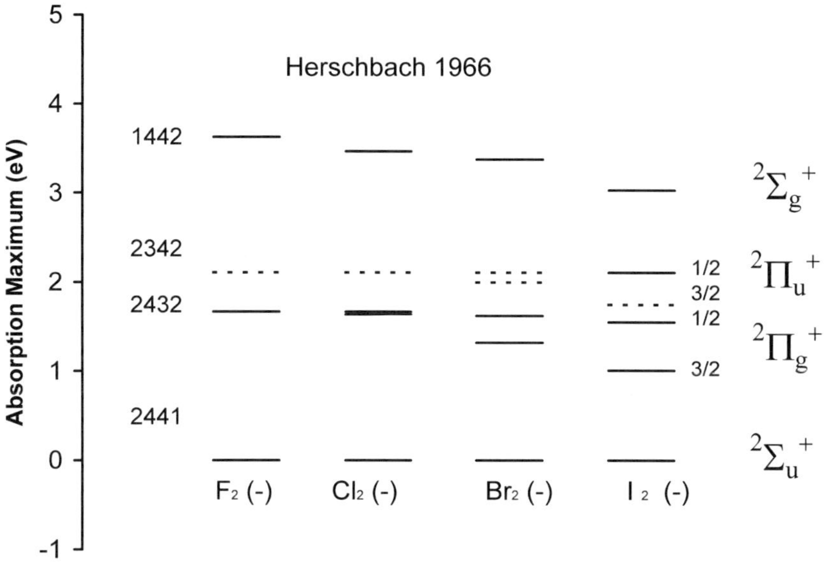

Figure 9.5 Historical absorbance data for the construction of the Morse potential energy curves for the Halogen diatomic anions. The energies indicated by dotted lines were predicted in 1966 [12].

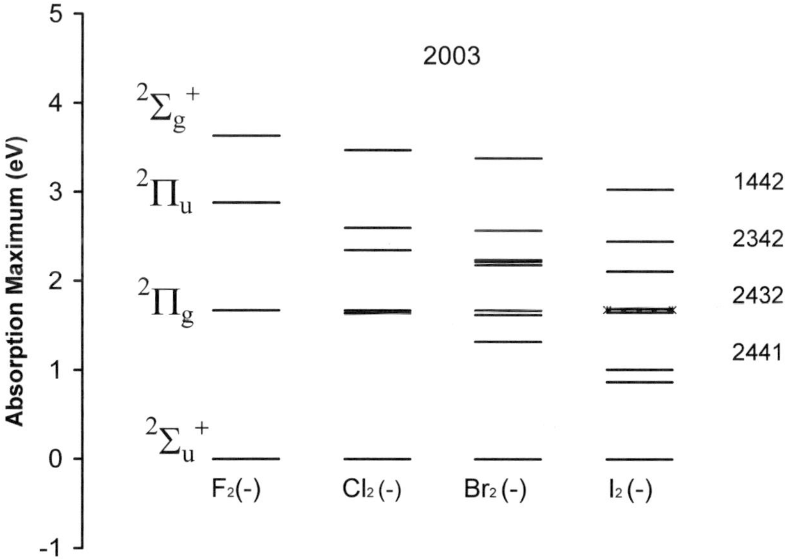

Figure 9.6 Recent absorbance data for the construction of the Morse potential energy curves for the Halogen diatomic anions. The calculated energies agree with the experimental values when available and predict experimental values which have not been measured.

TABLE 9.1 Morse Parameters for Ground State $X_2(-)$

	D_0 (eV)	r_e (pm)	ν (cm^{-1})	AE_a (eV)	References
$F_2(-)$	1.28(5)	181(5)	525(30)	3.08(5)	[5]
	1.23	192	462	3.0	[10]
	1.21(10)	179(2)	580(30)	3.08	[14]
	—	189	510	3.08	[1]
$Cl_2(-)$	1.32(2)	256(5)	255(4)	2.45(2)	[5]
	1.35	262	249	2.46	[10]
	1.30(15)	255(9)	2.4	—	[14]
$Br_2(-)$	1.18(2)	283(5)	160(4)	2.56(2)	[5]
	1.20	285	158	2.57	[10]
	1.15(10)	—	160(5)	2.6	[14]
$I_2(-)$	1.01(1)	321(0.5)	110(2)	2.52(1)	[5]
	1.01(1)	321(0.5)	111(2)	2.52(1)	[5]
	1.07	323	113	2.57	[10]
	1.02(5)	—	109	2.55	[14]
	1.01(1)	321(.5)	110(2)	—	[15]

available in the literature that best agree with experiment. The best $I_2(-)$ values derive from PES. The internuclear distances for $Cl_2(-)$ and $Br_2(-)$ are determined from the experimental VE_a. The dissociation energies are defined from the AE_a and $Rg_2(+)$ values. The frequencies are experimental values measured in solids. Five or six data points define these curves, and systematic uncertainties should be smaller than random uncertainties. The parameters for F_2 are less certain. The dissociation energy from the experimental AE_a is 3 meV lower than the value for $Ne_2(+)$. The internuclear distance is higher than the one obtained via AMB, but less than the NIST value. The frequency chosen is the NIST value with AMB random uncertainty. There are no ECD data for F_2, but other studies give a low E_1, as will be shown in simulated ECD data [1–3, 5, 8, 9, 14–19].

The majority of excited-state curves used the dissociation energy of the excited states of $Rg_2(+)$ [9, 16–23]. The parameters and data used to calculate these curves are given in Table 9.2. The lower bond dissociation energy observed in the PES of $I_2(-)$ is used to calculated the $^2\Pi_{g1/2}$ curves for $I_2(-)$ [7]. By analogy two new curves for $Br_2(-)$ are calculated. Figure 9.7 presents the calculated and experimental distributions. Their agreement with experiment is apparent [20–23]. The highest curve for $Cl_2(-)$ shows two states, one of which could be for an ion pair state. The curves were then adjusted to fit the measured absorption maxima indicated in bold in Table 9.2. Four curves for $I_2(-)$ were adjusted to a common absorption maximum at 1.68 eV, as shown in Table 9.2. No attempt was made to fit the absorption distributions. The calculated values agree with experiment within the uncertainties. Each of these curves can have four points: the dissociation energy, absorbance maxima, electron impact maxima, and electron impact distribution. Some absorbance maxima are estimated by the systematic variation shown in Figure 9.6 and predicted

TABLE 9.2 Morse Parameters and Dimensionless Constants for the Neutral and Anion Morse Potentials for X_2

Species		k_A	k_B	k_R	D_o (eV)	r_e (pm)	ν (cm^{-1})	$-VE_a$ (eV)	E(abs) (eV)
F_2	Neutral	1.00	1.00	1.00	1.60	141	917	—	—
$F_2(-)$	A $^2\Sigma_u^+$	1.634	0.625	3.372	1.28	180	510	−1.62	—
	B $^2\Pi_{g3/2}$	0.555	0.525	2.840	0.170	245	159	1.07	**1.67**
	B $^2\Pi_{g1/2}$	0.390	0.581	2.439	0.095	247	133	0.95	**1.67**
	C $^2\Pi_{u3/2}$	0.379	0.390	3.692	0.060	337	71	3.06	2.88
	C $^2\Pi_{u1/2}$	0.517	0.380	4.011	0.105	322	90	3.14	2.86
	D $^2\Sigma_g^+$	0.430	0.460	5.596	0.050	328	76.8	6.05	**3.63**
Cl_2	Neutral	1.00	1.00	1.00	2.48	199	560	—	—
$Cl_2(-)$	A $^2\Sigma_u^+$	1.082	0.641	2.328	1.31	256	255	−1.01	—
	B $^2\Pi_{g3/2}$	0.431	0.638	2.370	0.191	331	101	2.66	**1.64**
	B $^2\Pi_{g1/2}$	0.251	0.780	2.016	0.074	331	78	2.67	**1.64**
	C $^2\Pi_{u3/2}$	0.324	0.655	3.251	0.077	373	67	5.41	2.35
	C $^2\Pi_{u1/2}$	0.462	0.619	3.826	0.135	368	83	6.28	2.60
	D $^2\Sigma_g^+$	0.419	0.653	5.441	0.077	393	66	10.56	**3.47**
Br_2	Neutral	1.00	1.00	1.00	1.98	228	323	—	—
$Br_2(-)$	A $^2\Sigma_u^+$	1.180	0.641	2.328	1.18	282	160	−1.45	—
	B $^2\Pi_{g3/2}$	0.438	0.598	1.926	0.195	354	61	0.70	**1.32**
	B $^2\Pi_{g1/2}$	0.134	0.744	1.653	0.020	400	25	1.37	**1.62**
	B $^2\Pi_{g1/2}$	0.147	0.728	1.671	0.024	398	27	1.35	**1.62**
	C $^2\Pi_{u3/2}$	0.344	0.628	3.253	0.070	410	39	3.72	2.18
	C $^2\Pi_{u3/2}$	0.383	0.632	3.546	0.080	407	42	3.72	2.24
	C $^2\Pi_{u1/2}$	0.627	0.651	4.380	0.180	380	63	5.28	2.57
	D $^2\Sigma_g^+$	0.494	0.692	5.875	0.080	410	46	8.79	**3.38**
I_2	Neutral	1.00	1.00	1.00	1.54	267	215	—	—
$I_2(-)$	A 1(1/2)	1.187	0.645	2.242	1.007	320.5	110	−1.67	—
	A 1(1/2)	1.194	0.651	2.264	1.007	320.5	111	−1.61	—
	B 1(3/2)	0.491	0.649	1.702	0.225	371	53	−0.30	0.87
	B 1(3/2)	0.491	0.649	1.702	0.225	392	50	0.50	1.01
	B 1(1/2)	0.125	0.557	1.992	0.012	537	11	1.35	**1.68**
	B 1(1/2)	0.173	0.601	2.297	0.020	501	15	1.69	**1.65**
	C 1(3/2)	0.376	0.523	2.789	0.080	475	25	1.83	**1.68**
	C 1(3/2)	0.383	0.587	3.080	0.075	460	28	2.27	**1.69**
	C 2(1/2)	0.628	0.654	3.118	0.194	400	50	2.35	**2.11**
	C 2(1/2)	0.663	0.524	3.491	0.194	439	40	2.82	2.45
	D 2(1/2)	0.501	0.439	3.569	0.108	510	25	3.45	**3.03**
	D 2(1/2)	0.509	0.461	3.691	0.108	500	26	3.61	**3.03**

by Herschbach. The curves without these data are still defined by three points so these energies are experimental values to be compared to others.

Four sets of curves are easily compared for X_2, as shown in Figure 9.8. The overall similarity of the curves is striking and supports the isoelectronic principle. When we compare the values of the dimensionless constants for a given state in Table 9.2,

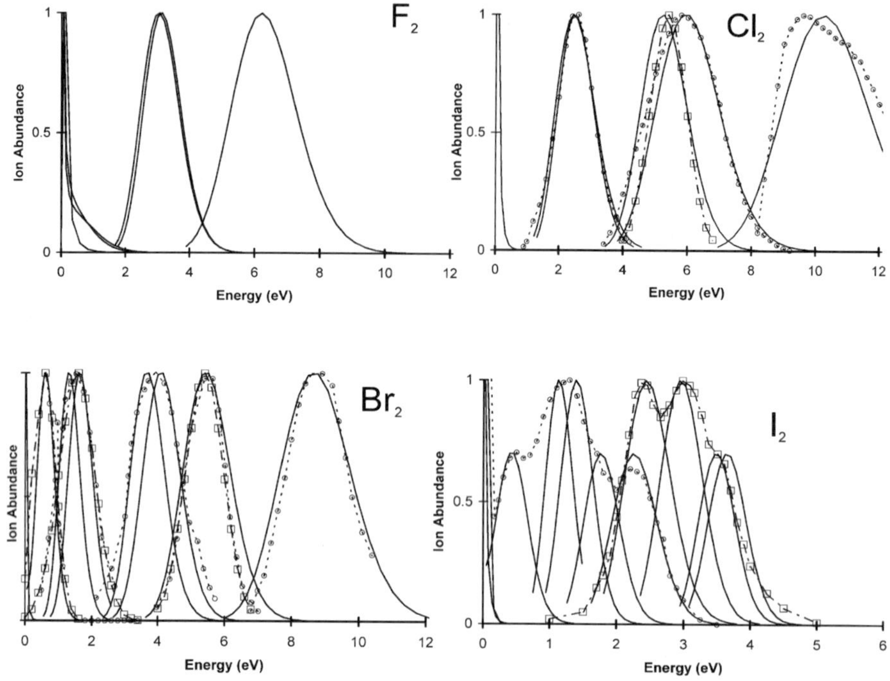

Figure 9.7 Experimental and calculated ion distributions for the electron impact of the halogen diatomic molecules [20–23].

this is emphasized. All the ground-state k_A are greater than 1, while the k_R are greater than 2. Likewise, the dimensionless constants for the D states are similar. The major differences occur in the spin orbital coupling of the atomic species. The ground-state curves are all $Mc(3)$, but the ground-state curves for $F_2(-)$ and $Cl_2(-)$ are $D(3)$, while those for $Br_2(-)$ and $I_2(-)$ are $M(3)$. The excited-state curves for all but the first excited state of $I_2(-)$ are $D(2)$. The first excited state of $I_2(-)$ is $D(3)$. The first excited-state curves for all but $Cl_2(-)$ are $Mc(2)$ since they cross the neutral in the vicinity of the internuclear distance of the neutral. The remaining curves are $D(2)$ and $Dc(2)$. On the basis of this information the formation of the parent negative ions of $I_2(-)$ and $Br_2(-)$ in electron impact spectra can occur at an electron energy sufficient to overcome the activation energy for crossing to the negative-ion curve. Such ions have been observed but not explained [18, 19].

Figure 9.9 shows all the extant ECD data for homonuclear diatomic molecules [5, 10, 24, 25]. The parameters for O_2 will be discussed in the next section. Originally, only one state was assumed for the halogens. If we use two states, two activation energies can be obtained from the ECD data. This requires E_a, Q, and A_1 measured using other techniques or assumed. These two activation energies can be used to define two potential energy curves. The parameters utilized to calculate the ECD curves in Figure 9.9 are given in Table 9.3. The crossings of the ground

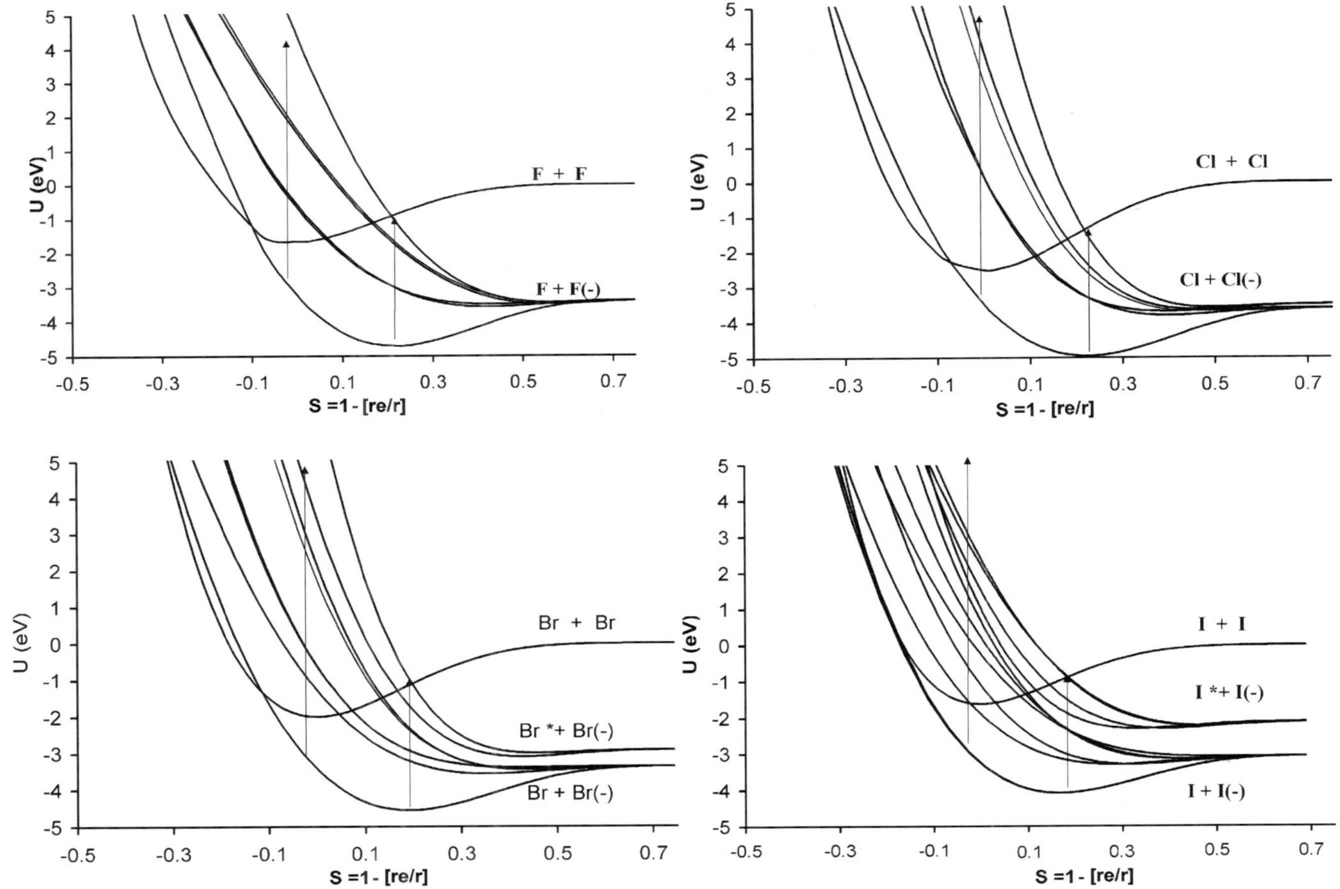

Figure 9.8 Morse Potential energy curves for the halogen molecules and their anions [5]. The data are given in Table 9.2.

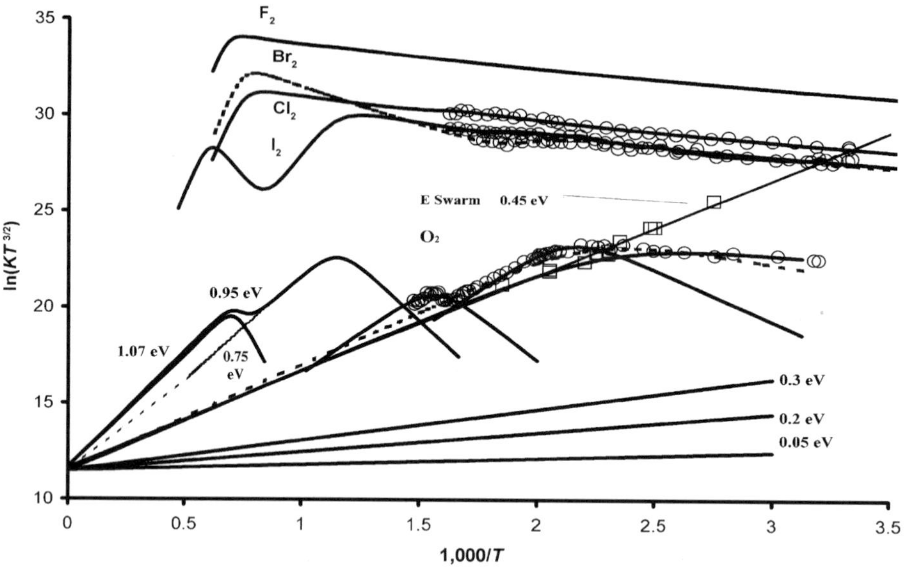

Figure 9.9 ECD data for homonuclear diatomic molecules plotted as ln $KT^{3/2}$ versus 1,000/T. The curves drawn through the data are calculated using the parameters given in Table 9.3.

TABLE 9.3 ECD Parameters for Homonuclear Diatomic Molecules

Species	ln (A_1)	E_1 (eV)	Q	E_a (eV)
F_2	(35.5)	(0.05)	(1)	(3.05)
F_2	(33)	(0.03)	(1)	(1.7)
Cl_2	(33)	0.06	(1)	(2.45)
Cl_2	(35.5)	0.30	(1)	(1.1)
Br_2	(35.5)	0.28	(1)	(2.56)
Br_2	(33)	0.03	(1)	(1.4)
I_2	(35.5)	0.45	(1)	(2.52)
I_2	(33)	0.05	(1)	(1.5)
O_2	(24.9)	(0.05)	(1)	(0.45)
	(24.9)	(0.05)	(0.5)	(0.43)
	(24.9)	0.1	(0.8)	(0.5)
	(34.7)	0.4	0.01	0.7
	(35.2)	0.8	0.02	0.75
	(35.5)	0.9	(0.8)	(0.95)
	(35.5)	(1.9)	(0.8)	(1.07)

The values in parentheses are experimental values from other methods or have been estimated.

state and/or the first excited states with the neutral curve agree with the E_1 from ECD data. The activation energy for the backside crossing of the curves for $I_2(-)$ and $Br_2(-)$ is higher than that for the frontside crossing. Thus, dissociative thermal electron attachment will occur via the frontside curves. In the case of $Cl_2(-)$ the activation energy for the backside crossing is lower than that for the frontside crossing so dissociation occurs via the ground-state curve. These curves are prototypes for the organic halides.

9.2.3 Electron Affinities and Morse Potential Energy Curves: Group VI Diatomic Molecules and Anions

The electron affinity of the oxygen molecule is very important. In 1970 an excited-state electron affinity for O_2 was obtained from ECD data, but an unexplained upturn at higher temperatures was also present. Recent ECD results have confirmed these data, but show a second downturn at higher temperatures. These data are shown in Figure 9.9 [5, 24, 25]. Using the higher E_a from other experiments in the ECD data analysis allowed us to determine additional kinetic and thermodynamic parameters for excited states. The E_a now range from 0.15 eV to 1.07 eV: photodetachment, 0.1_5 (eV), 1958; CTC, 0.7_5, 1961 to 1971; electron swarm, 0.4_5, 1961; EB, 0.6, $1_{.1}$, 1961 to 1971; AMB, 0.2, 0.3, 0.5, 0.8, $1_{.3}$, 1970 to 1977; IB, 1.0_7, 1970; PES, 0.43, 0.45, 1971 to 1995; and ECD, 0.4_5, 0.5, 0.9, 0.75, 0.70, 1970 to 2002. This range is much larger than that obtained for the halogens using the same techniques, indicating significant differences [24–36]. The VE_a determined from electron impact data range from -0.2 eV to -17 eV. The larger negative values are assigned to D curves [37].

In 1942 D. R. Bates and H. S. W. Massey drew potential energy curves for four bound superoxide anion states leading to $O(^3P_g) + O(-)\ ^2P_u$ [38]. Later 24 total anion states were postulated: $^2\Sigma_g^-(2)$, $^2\Sigma_u^-(2)$, $^4\Sigma_g^-(2)$, $^4\Sigma_u^-(2)$, $^2\Sigma_g^+$, $^2\Sigma_u^+$, $^4\Sigma_g^+$, $^4\Sigma_u^+$; $^2\Pi_g(2)$, $^2\Pi_u(2)$, $^4\Pi_g(2)$, $^4\Pi_u(2)$, and $^2\Delta_g$, $^2\Delta_u$, $^4\Delta_g$, $^4\Delta_u$ [39]. In 1981 H. H. Michels presented curves for these states and calculated dissociation energies and E_a for the bound states of $O_2(-)$ from -1.5 eV to -3.7 eV [40]. Assigning the experimental and theoretical E_a and VE_a, we obtain 12 M and 12 D HIMPEC for the 24 states [5, 41–43]. The relative bond orders of the ground state agree with simple molecular orbital predictions. The bond orders for the excited states are reasonable compared to the predicted values. Two anion curves for S_2, Se_2, and Te_2 have been constructed from experimental data [44–47].

The experimental E_a for O_2 are assigned to bound states. The VE_a for these states range from 0.9 eV to -0.75 eV. The activation energies for thermal electron attachment determined from the ECD data range from 0.05 eV to 1.9 eV. The frequencies observed in solids and the Morse parameters obtained from PES, ECD, electron swarm, and other techniques can be used to approximate the internuclear distance and frequency of the predicted states. These give six $M(2)$ curves (EDEA < 0, and E_a and $VE_a > 0$) and six $M(1)$ curves (only $E_a > 0$). The PES E_a, with values of 0.430, 0.450 ± 0.002 eV, is the most precise E_a and the Born

TABLE 9.4 Electron Affinities for O_2 and Morse Parameters for $O_2(-)$

States	D_e (eV)	r_e (pm)	ν_e (cm^{-1})	E_a (eV)	VE_a (eV)	E_1 (eV)
			Quartet States			
$^4\Sigma_u^-$	4.6	129	1,250	0.9_5	0.75	0.9
$^4\Delta_g$	4.4	130	1,210	0.7_5	0.5	0.8
$^4\Sigma_g^+$	4.4	130	1,210	0.7_5	0.5	0.8
$^4\Pi_u$	3.9	138	1,000	0.2_5	−0.4	0.2
$^4\Sigma_g^+$	3.8	136	1,089	0.1_5	−0.45	0.2
$^4\Pi_u$	3.7	140	1,089	0.0_5	−0.9	0.5
$^4\Sigma_u^-$	1	169	700	−2.65	−9.5	3.5
$^4\Delta_g$	0.8	183	510	−2.85	−9.5	3.5
$^4\Sigma_g^+$	0.8	170	650	−2.85	−9.6	3.5
$^4\Pi_u$	0.3	220	260	−3.35	−10.8	4
$^4\Sigma_g^+$	0.1	220	190	−3.55	−10.9	4
$^4\Pi_u$	0.1	230	180	−3.55	−11.9	4
			Doublet States			
$^2\Pi_g$	4.7	129	1,250	1.0_7	0.86	1.9
$^2\Delta_g$	4.3_5	130	1,200	0.7	0.46	0.8
$^2\Sigma_u^+$	4.1_5	132	1,125	0.5	0.17	0.1
$^2\Pi_u(1/2)$	4.1	134	1,125	0.45	−0.03	0.05
$^2\Pi_u(3/2)$	4.1	134	1,125	0.43	−0.03	0.05
$^2\Sigma_g^+$	4.0_5	135	1,089	0.36	−0.16	0.1
$^2\Sigma_u^-$	3.8_5	136	1,089	0.3	−0.39	0.2
$^2\Pi_g$	1.0	169	670	−2.65	−8.16	3
$^2\Delta_u$	0.9	169	620	−2.75	−7.65	3
$^2\Sigma_u^+$	0.82	169	610	−2.82	−7.77	3
$^2\Pi_u$	0.77	169	592	−2.88	−7.55	3
$^2\Sigma_g^+$	0.3	230	235	−3.35	−10.7	4
$^2\Sigma_u^-$	0.2	250	180	−3.15	−11.1	4

Haber cycle E_a at 0.9 ± 0.1 eV is the least precise [19, 32]. The ECD values have no greater uncertainties since a data reduction procedure cannot degrade quality.

Michels calculated dissociation energies for $O_2(-)$ ranging from 0 to 2.11 eV for nine bound states. These give negative E_a and VE_a but dissociate in the Franck Condon region. Therefore, they are all $D(0)$ curves. The only M curve presented was determined from PES data with an E_a of 0.45 eV [31]. Between the ground state and first excited $^2\Pi_u$ state there are three quartet states and two doublet states separated by 1 eV. By assigning the experimental values to the predicted order, the E_a shown in Table 9.4 were obtained. The E_a of the $^4\Sigma_u^-$ state is assigned to 0.95 eV, and the electron affinities of the $^4\Delta_g$ state and accidentally degenerate $^4\Sigma_g^+$ state are assigned $E_a = 0.75$ eV. The next two doublet states are assigned these E_a: 0.7 eV and 0.5 eV [28–30]. The five other bound states are assigned E_a of 0.4, 0.3, 0.2_5, 0.1_5, and 0.0_5 eV, yielding 12 states with positive E_a that lead to molecular anions, 6 $M(2)$ and 6 $M(1)$ states [5, 26, 31–36].

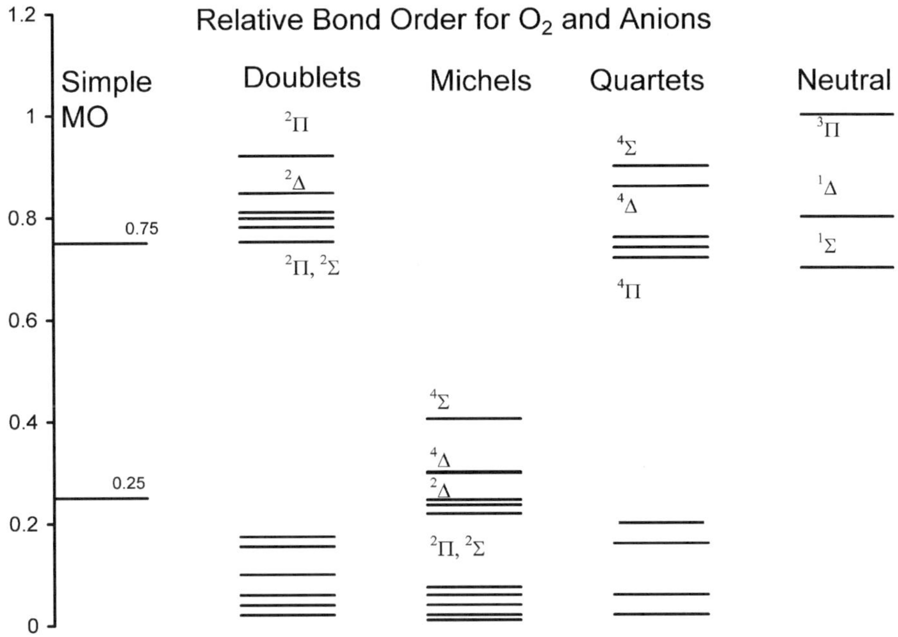

Figure 9.10 Predicted and experimental relative bond orders for $O_2(-)$. The predicted values were taken from [40].

The experimental bond orders for the states of the neutral and negative ions are shown in Figure 9.10. Also shown are the bond orders of the anions predicted by simple MO theory and calculated by Michels. The bond orders of the dissociative states are zero. The predicted bond orders of the excited states can be used to define the 12 D curves relative to the 12 M curves. The bond orders for the M and D states calculated by Michels are all much lower than the values predicted by simple MO theory: 0.75 for the M states and 0.25 for the D states. The experimental bond orders for the M states are all larger than the simple MO value. The D states should have bond orders less than 0.25 but greater than zero to preserve symmetry. In Figure 9.10 we can observe the relative positions of the experimental bond orders for the M states and assumed bond orders for the D states.

The remaining 12 states are $D(0)$ because they all lead to dissociation and all three Herschbach metrics are negative. The E_a are negative, but the potential energy curves have nonzero dissociation energies. Since $E_a = \text{EDEA} + D_0 \pm 0.1$ eV, the state with a dissociation energy of 1 eV has an E_a of $-2.6(2)$ eV, and that with a dissociation energy of 0.1 eV, $-3.5(1)$ eV. The observed $-VE_a$ from electron impact and scattering or dissociative electron attachment on solids or liquids are 0.2, 1, 7.8, up to 11.9 eV [41]. It is important to note that the observed electron impact data can be modeled using only one upper negative-ion state. The gap between the VE_a of $(8 - 1) = 7$ eV is the approximate splitting of the states at the internuclear distance of the neutral. This gap has been confirmed by dissociative

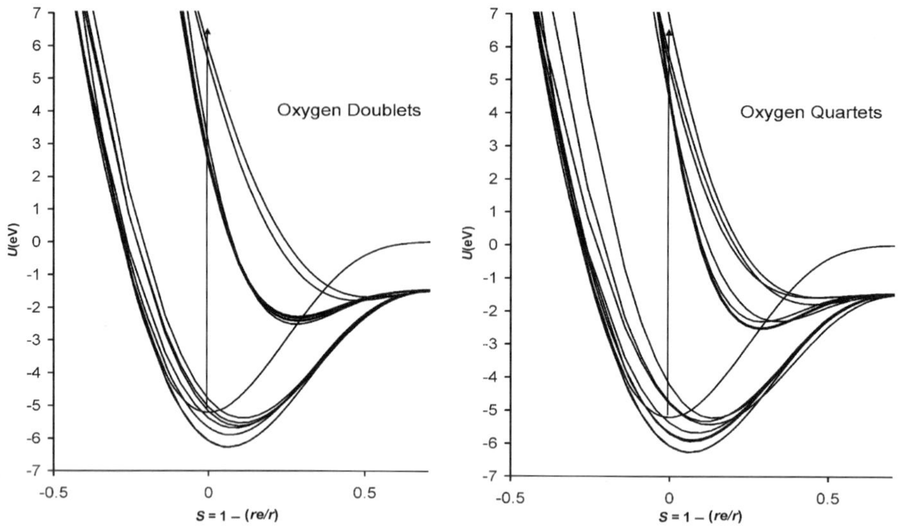

Figure 9.11 Morse potential energy curves for 24 negative-ion states of O_2. The parameters used to calculate the curves and assignments are given in Table 9.4 [5].

electron attachment to the excited state of the neutral formed by application of microwave energy. The peak in the O(−) distribution for electron attachment to the excited state of the neutral is shifted by about 1 eV and a second excited state dissociating to the next highest limit is identified approximately 2 eV higher [42].

The 12 D peaks leading to the lowest limit have $-VE_a$ values in the range of 8 eV to 12 eV. The lowest state in the Franck Condon region is assigned to $^2\Pi_u$, but several states are assigned larger dissociation energies and cross this state at larger internuclear distances to yield the same vertical energy. The published Morse parameters of the $^2\Pi_u$ state guide the other curves [43]. The D states are assigned in the same order as the M states, but these curves are highly speculative. If the dissociative electron impact peaks were more definitively assigned the curves could be improved. The curves for all 12 states are illustrated in Figure 9.11. The doublet states are shown on one plot, while the quartet states are shown on another. The spin orbital coupling states that have been identified in both ECD and PES data are not shown. The state designations, Morse parameters, E_a, VE_a, and E_1 values are given in Table 9.4.

The negative-ion states of S_2, Se_2, and Te_2 have much less data. The E_a of the Group VI molecules have been calculated using CURES-EC. The ground states for the neutrals are all triplet states. The ZINDO method gives the best agreement with experiment for O_2 with a σ constant of 1.0 and a π constant of 0.15. The ZINDO(7100) and ZINDO(5100) values are 1.08 eV and 0.5 eV, respectively. The higher value corresponds to the AE_a. The PM3(7100) and PM3(0000) values are 0.7 eV and 0.4 eV. The PM3(7100) values for the others are S_2, 1.7 eV; Se_2, 1.9 eV, and Te_2, 2.1 eV; these are in agreement with experiment. Several photoelectron spectra have been obtained for S_2. The most recent has the highest resolution.

The relative bond orders for the remaining Group VI homonuclear diatomic molecules are larger than 0.75 and approach 1 for $Se_2(-)$ and $Te_2(-)$. The Morse parameters for the ground-state anions are very similar to those for $O_2(-)$. The $^2\Pi_u$ anion states characterized by electronic spectra for the higher members of the Group VI family are M states. These are $M(0)$ states for S_2 and Se_2, but the curve for Te_2 is an $M(2)$ curve because only the EDE_a is negative. It also crosses the neutral on the backside. The dimensionless constants are different from those for $O_2(-)$. In addition, their relative bond orders are different going from about 0.2 in O_2 to 0.6 in Te_2. This increase mirrors that for the ground state since the relative bond orders for $Se_2(-)$ and $Te_2(-)$ are approximately 1, while that for $O_2(-)$ is about 0.9. There should be as many states for the isoelectronic ions as for $O_2(-)$. The consistency of the dimensionless constants for the ground-state curves, as shown in Table 9.5, supports the isoelectronic principle for the ground state, but the differences in the values for the one excited state do not justify the use of the $O_2(-)$ parameters for the other diatomic anions. The parameters for all the curves are given in Table 9.5. Since no other data are available, only two curves will be calculated for each of these anions, as shown in Figure 9.12 [44–47].

9.2.4 Electron Affinities and Morse Potential Energy Curves: Group IA and IB Homonuclear Diatomic Molecules and Anions

The Morse potentials for the Group IA and IB homonuclear diatomic anions can be constructed for the ground state and one excited state from experimental data. However, there is a low-lying excited state for the alkali metal and coinage metal anions. This gives two dissociation limits so there should be four low-lying states. Theoretical calculations have been carried out for these states in $Li_2(-)$ [48]. Adiabatic electron affinities have been measured by photoelectron spectroscopy. The values of the Morse parameters for the neutral have also been measured [48–53]. The $Li_2(-)$ ground-state curve is $M(2)$ with positive values for the E_a and VE_a. The lower antibonding curve is $D(0)$. The bonding curve dissociating to the higher limit is $M(0)$, while the complementary antibonding curve is $D(0)$ [48].

For the higher-atomic-weight IA and IB homonuclear diatomic molecules two "bonding" and two antibonding curves going to two limits can be constructed by analogy to $Li_2(-)$. These are shown in Figure 9.13 for Li_2 and Na_2 and in Figure 9.14 for Cu_2 and Au_2. The Morse parameters for all the IA and IB homonuclear diatomic molecules and anions and the relative anion bond orders are given in Table 9.6. The anion curves for K_2, Rb_2, and Ag_2 are similar to those for Na_2. The ground-state curves for all but Au_2 are $M(2)$ since all the EDEA are negative. They all cross on the backside of the neutral. All the bonding curves dissociating to the higher limit are $M(0)$. All the complementary nonbonding anion curves are $D(0)$. For all of the first excited-state curves dissociative thermal electron attachment is possible and has been observed in the case of Li_2 and Na_2. For $Li_2(-)$ the dissociative curve crosses the neutral between 3 and 4 Å at $v = 10$ to 13 [53]. The anion curves for Au_2 are unique. The two states dissociating to the lowest

TABLE 9.5 Morse Parameters and Dimensionless Constants for Diatomic Group VA Molecules and Anions

Species	D_o (eV)	r_e (pm)	ν (cm^{-1})	k_A	k_B	k_R
O_2 Neutral	5.11	121	1,580	1.000	1.000	1.000
$O_2(-)$ $^4\Sigma_u^-$	4.6	129	1,250	1.080	0.835	1.298
$^4\Delta_g$	4.4	130	1,210	1.054	0.826	1.292
$^4\Sigma_g^+$	4.4	130	1,210	1.054	0.826	1.292
$^4\Pi_u$	3.9	138	1,000	1.062	0.726	1.482
$^4\Sigma_g^+$	3.8	136	1,089	1.028	0.800	1.422
$^4\Pi_u$	3.7	140	1,089	1.096	0.810	1.660
$^4\Sigma_u^-$	1	169	700	0.713	0.990	2.537
$^4\Delta_g$	0.8	183	510	0.608	0.808	2.311
$^4\Sigma_g^+$	0.8	170	650	0.617	1.024	2.356
$^4\Pi_u$	0.3	220	260	0.353	0.668	2.055
$^4\Sigma_g^+$	0.1	220	190	0.189	0.825	1.664
$^4\Pi_u$	0.1	230	180	0.205	0.780	1.972
$^2\Pi_g$	4.7	129	1,250	1.105	0.824	1.325
$^2\Delta_g$	4.3_5	130	1,200	1.041	0.824	1.276
$^2\Sigma_u^+$	4.1_5	132	1,125	1.027	0.791	1.302
$^2\Pi_u(1/2)$	4.0_7	134	1,125	1.057	0.798	1.404
$^2\Pi_u(3/2)$	4.1_0	134	1,125	1.062	0.796	1.409
$^2\Sigma_g^+$	4.0_5	135	1,089	1.050	0.780	1.412
$^2\Sigma_u^-$	3.8_5	136	1,089	1.039	0.795	1.435
$^2\Pi_g$	1.0	169	670	0.675	0.948	2.276
$^2\Delta_u$	0.9	169	620	0.582	0.940	1.942
$^2\Sigma_u^+$	0.82	169	610	0.558	0.951	1.890
$^2\Pi_u$	0.77	169	592	0.525	0.952	1.780
$^2\Sigma_g^+$	0.3	230	235	0.350	0.605	2.026
$^2\Sigma_u^-$	0.2	250	180	0.283	0.566	1.973
S_2 Neutral	4.46	190	726	1.000	1.000	1.000
$S_2(-)$ $^2\Pi_g$	4.11	202	601	1.128	0.858	1.367
$^2\Pi_u$	1.60	230	364	0.677	0.831	1.258
Se_2 Neutral	3.44	217	387	1.000	1.000	1.000
$Se_2(-)$ $^2\Pi_g$	3.43	226	330	1.153	0.853	1.331
$^2\Pi_u$	1.50	250	217	0.730	0.845	1.210
Te_2 Neutral	2.69	256	247	1.000	1.000	1.000
$Te_2(-)$ $^2\Pi_g$	2.63	266	223	1.141	0.911	1.325
$^2\Pi_u$	1.40	270	180	0.661	1.007	0.833

limit are $M(3)$ since EDEA is also positive due to the large electron affinity of the gold atom. The other curves dissociating to the higher limit are $M(0)$ and $D(0)$, respectively.

The internuclear distances for the ground-state anions are 10% larger than the neutral, while the frequencies are 30% lower than the neutrals. The bond dissociation energies are only slightly lower in the anion than in the neutral [49–52]. The bond order varies from 0.8 to 1.0 for the bonding curves, as seen in Table 9.6. The

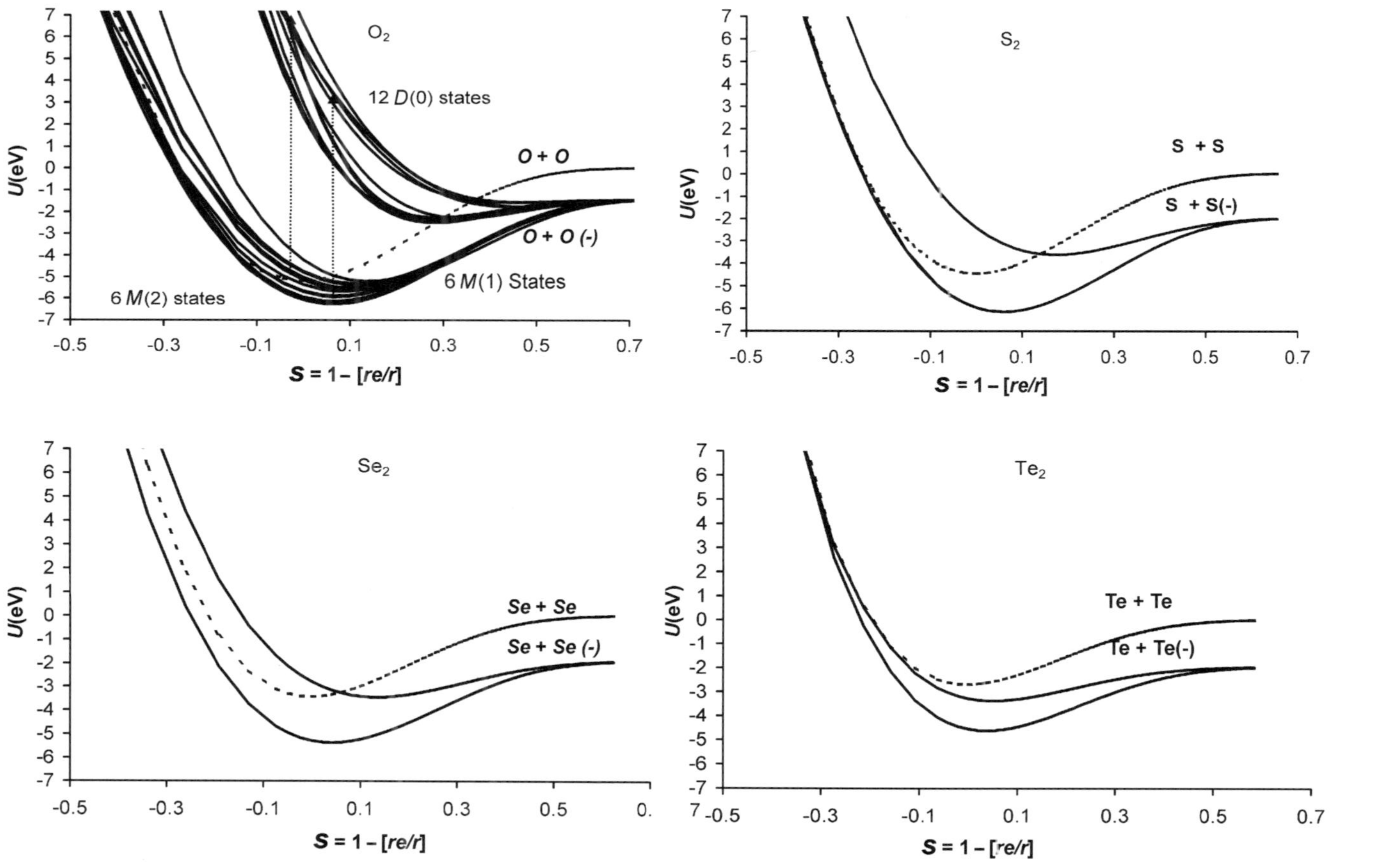

Figure 9.12 Morse potential energy curves for the neutral and negative-ion states of O_2, S_2, Se_2, and Te_2. The parameters used to calculate the curves are given in Table 9.5 [5].

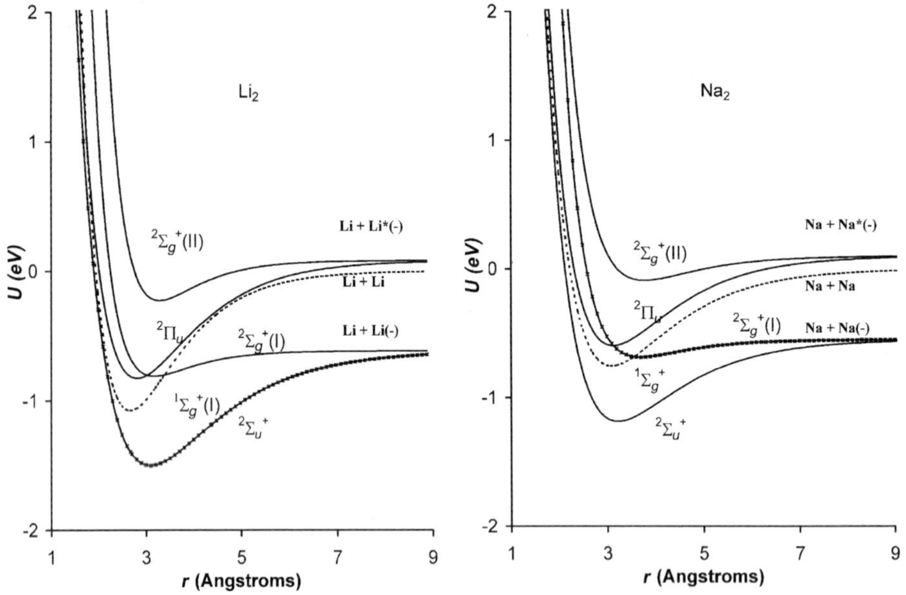

Figure 9.13 Morse potential energy curves for the neutral and negative-ion states of Li_2 and Na_2. The parameters used to calculate the curves are given in Table 9.6.

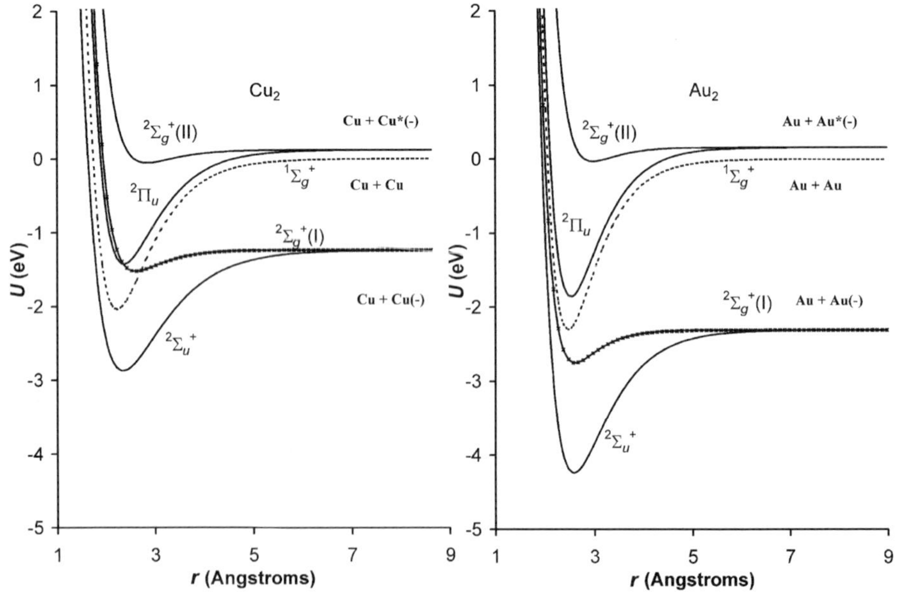

Figure 9.14 Morse potential energy curves for the neutral and negative-ion states of Cu_2 and Au_2. The parameters used to calculate the curves are given in Table 9.6.

TABLE 9.6 Morse Parameters and Bond Order for Homonuclear Diatomic Group IA and IB Neutral and Anions

Species		D_e (eV)	r_e (pm)	ν (cm^{-1})	$E_a + E^*$	Bond Order
Li_2	$^1\Sigma_g^+$	1.054	267	351	0.00	1.00
$Li_2(-)$	$^2\Sigma_u^+$	0.869	309	232	0.618	0.82
$Li_2(-)$	$^2\Pi_u$(I)	0.182	318	186	0.618	0.17
$Li_2(-)$	$^2\Sigma_g$(II)	0.869	282	280	−0.080	0.84
$Li_2(-)$	$^2\Sigma_g^+$	0.292	329	230	−0.080	0.27
Na_2	$^1\Sigma_g^+$	0.746	308	159	0.000	1.00
$Na_2(-)$	$^2\Sigma_u^+$	0.628	322	136	0.548	0.84
$Na_2(-)$	$^2\Pi_u$(I)	0.130	369	84	0.548	0.17
$Na_2(-)$	$2\Sigma_g$(II)	0.685	312	147	−0.100	0.92
$Na_2(-)$	$2\Sigma_g^+$	0.184	377	88	−0.100	0.25
K_2	$^1\Sigma_g^+$	0.520	391	92	0.000	1.00
$K_2(-)$	$^2\Sigma_u^+$	0.512	404	75	0.502	0.98
$K_2(-)$	$^2\Pi_u$(I)	0.091	456	49	0.502	0.18
$K_2(-)$	$^2\Sigma_g$(II)	0.511	397	81	−0.120	0.98
$K_2(-)$	$^2\Sigma_g^+$	0.107	470	51	−0.120	0.21
Rb_2	$^1\Sigma_g^+$	0.490	418	57	0.000	1.00
$Rb_2(-)$	$^2\Sigma_u^+$	0.475	440	46	0.486	0.96
$Rb_2(-)$	$^2\Pi_u$(I)	0.064	475	31	0.486	0.13
$Rb_2(-)$	$^2\Sigma_g$(II)	0.470	428	60	−0.140	0.96
$Rb_2(-)$	$^2\Sigma_g^+$	0.085	490	44	−0.140	0.17
Cs_2	$^1\Sigma_g^+$	0.452	464	42	0.000	1.00
$Cs_2(-)$	$^2\Sigma_u^+$	0.437	491	37	0.689	0.97
$Cs_2(-)$	$^2\Pi_u$(I)	0.031	525	18	0.689	0.07
$Cs_2(-)$	$^2\Sigma_g$(II)	0.455	485	42	−0.160	1.00
$Cs_2(-)$	$^2\Sigma_g^+$	0.047	542	25	−0.160	0.10
Cu_2	$^1\Sigma_g^+$	2.020	222	266	0.000	1.00
$Cu_2(-)$	$^2\Sigma_u^+$	1.620	234	196	1.235	0.80
$Cu_2(-)$	$^2\Pi_u$(I)	0.280	261	125	1.235	0.14
$Cu_2(-)$	$^2\Sigma_g$(II)	1.531	236	230	−0.120	0.76
$Cu_2(-)$	$^2\Sigma_g^+$	0.168	284	91	−0.120	0.09
Ag_2	$^1\Sigma_g$	1.650	248	192	0.000	1.00
$Ag_2(-)$	$^2\Sigma_u^+$	1.370	260	145	1.488	0.83
$Ag_2(-)$	$^2\Pi_u$(I)	0.288	279	103	1.488	0.18
$Ag_2(-)$	$^2\Sigma_g$(II)	1.350	257	185	−0.140	0.82
$Ag_2(-)$	$^2\Sigma_g$	0.175	293	75	−0.140	0.10
Au_2	$^1\Sigma_g^+$	2.290	247	190	0.000	1.00
$Au_2(-)$	$^2\Sigma_u^+$	1.920	258	149	2.309	0.84
$Au_2(-)$	$^2\Pi_u$(I)	0.434	270	104	2.309	0.18
$Au_2(-)$	$^2\Sigma_g$(II)	2.010	253	175	−0.160	0.88
$Au_2(-)$	$^2\Sigma_g^+$	0.188	298	65	−0.160	0.08

bond orders of the antibonding curves are 0.1 to 0.2. In the anions of the halogens and O_2 the bond orders of the excited states vary from 0.025 to 0.11.

9.2.5 Electron Affinities and Morse Potential Energy Curves: NO and NO(−)

The electron affinities for NO listed in the NIST table cluster around 0.02, 0.6, and 0.85 eV. All are confirmed by more than one technique. In 1971 the ECD response for both O_2 and NO were measured as a function of temperature [1–3, 5, 23, 24, 54–62]. The low-temperature data for O_2 gave an E_a that is consistent with the excited state E_a. In 2002 the data for O_2 and NO were analyzed in terms of two states. However, the number of predicted states for each is larger. Additional ECD data for O_2 indicated more excited states, as discussed above. Therefore, the original data for NO have been reanalyzed in terms of multiple states. Three correspond to the three states observed for the O_2 molecule, which is isoelectronic with NO(−). The ground state is the $^3\Sigma$ state, $AE_a = 0.86(5)$ eV, while the intermediate state occurs at the $^1\Delta$ state, $E_a = 0.60(5)$ eV or 0.40(5) eV. Figure 9.15 is a combined plot of the calculated ECD curves for four valence-state anions of NO. Table 9.7 gives the parameters used to calculate these curves. The ground state and first excited state have a larger A_1 value than the two lower states. The Q_{an} of the first excited state is lower than that for the other three, all with an approximate value of 1. The activation energies increase from about zero for the 0.2 eV state to 0.7 eV for the ground state. With these electron affinities and activation energies Morse

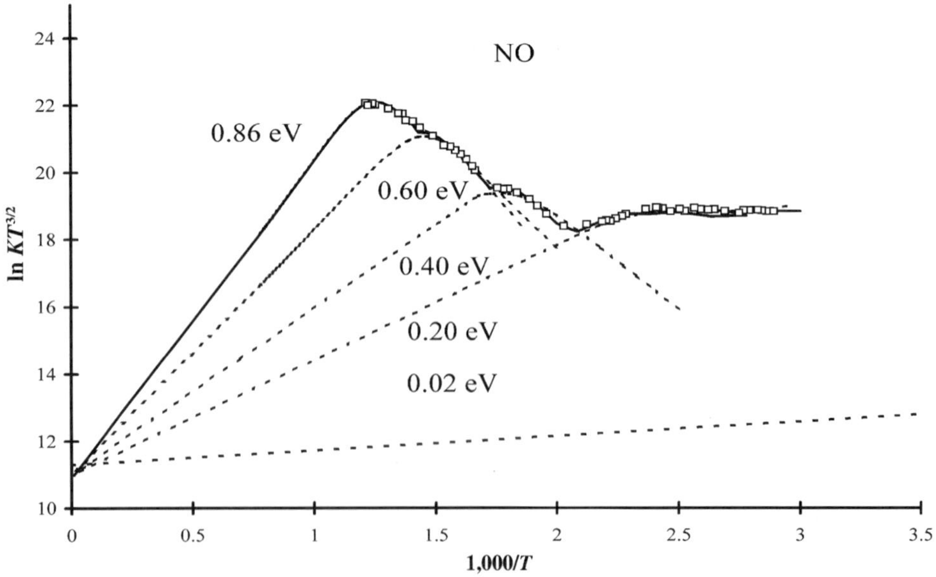

Figure 9.15 ECD data for NO plotted as ln $KT^{3/2}$ versus 1,000/T. The curves drawn through the data are calculated using the parameters given in Table 9.7.

TABLE 9.7 ECD Parameters for NO

Species	ln (A_1)	E_1 (eV)	Q	E_a (eV)	E_{-1} (eV)
Ground state	34.3	0.70(2)	0.93	0.86(5)	1.57(3)
A	34.9	0.60(2)	0.82	0.60(5)	1.21(3)
B	29.3	0.41(2)	0.86	0.40(5)	0.81(3)
C	28.0	0.02(2)	0.88	0.21(5)	0.23(3)
D (not observed)	—	—	1.00	0.02	0.02

potential energy curves for four $M(2)$ states and a dipole bound state can be calculated. These are illustrated in Figure 9.16. The state with an E_a of 0.02 eV could lead to the N(−) + O limit or be a dipole bound state. All the $M(2)$ valence-state curves cross on the "backside" of the neutral curve and have an activation energy that corresponds to one of those measured in the ECD. The lowest activation energy will exist for the dipole bound state. When this state is accessed, it could be metastable with respect to transitions to the lower states and the ground state will not be populated. In addition, rapid autodetachment to the neutral might not take place because of the similar geometry in the anion and neutral. Such could be the reason this ion is so frequently observed experimentally. The experimental data for the electron attachment to NO can be rationalized using these curves [54–62].

The CURES-EC for the E_a of the singlet and triplet states of NO(−) are 0.9, 0.5, and 0.3 eV. The triplet ground state of the anion is 0.6 eV lower than that for the singlet state. The PM3(t6100) AE_a is 0.90 eV. The PM3(s4100) E_a is 0.5 eV, while the PM3(s6100) E_a is 0.3 eV. The internuclear distances for the anions are about 0.07 Å longer than the neutral, whereas the frequencies are about 10% lower in the anions. The Morse curves are consistent with these data. The importance of

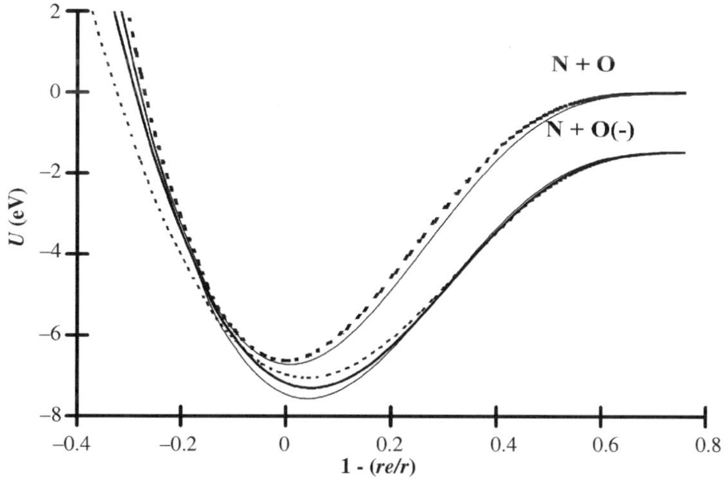

Figure 9.16 Morse potential energy curves for the neutral and negative-ion states of NO.

the negative-ion states of O_2 and NO to biological processes will be discussed in Chapter 12.

9.3 TRIATOMIC MOLECULES AND ANIONS

The electron affinities of triatomic molecules are important because they are the simplest species with a third geometrical dimension. Those of many of the homonuclear triatomic molecules have already been measured. The E_a of the Group I, III, and VII trimers are larger than those for the monomer and dimer. Those of C, N, and O are particularly important. For C_3 the electron affinity is lower than that of the dimer, while for N_3 and O_3 the opposite is true. The azide radical is the simplest species containing only nitrogen with a positive valence-state electron affinity, 2.76(4) eV, determined by PD and PES. Ozone is an important form of oxygen in the upper atmosphere. Its electron affinity of 2.103(3) eV has been confirmed by nine determinations using three techniques [63–68]. The heteronuclear triatomic molecules SO_2 and NO_2 are environmental hazards. The electron affinity of SO_2 is well established at 1.107(7) eV. The weighted average of nine different values equals the PES value. This includes the first value determined in 1957 by equilibrium bracketing. The E_a of SO_2 has been used as a reference point for TCT measurements. The electron affinities of NO_2 range from 1.8 eV to 3.9 eV and have been determined by many different techniques. Ten values cluster around the PES value of 2.273(5) eV, including one determined by dissociative electron attachment to CH_3NO_2 in the ECD. However, three values are reported at 2.5 eV, two at about 3.0 eV, and two at 3.9 eV. In addition, excited-state electron affinities have been postulated for NO_2, one at 0.0 eV [1–3, 69–89].

The azide radical is linear, as is its anion. Ozone, sulfur dioxide, and NO_2 are bent in the neutral form and in the anion. The neutrals of CO_2, CS_2, COS, and N_2O are linear and the anions bent, as predicted by simple molecular orbital theory. The electron affinities of these species are uncertain [90–102]. In Figure 9.17 the ECD data for these compounds are shown. The diversity in the temperature dependence of the isoelectronic molecules is remarkable. The ECD response of CO_2 is unexplained. The temperature dependence for CS_2 is typical of nondissociative electron attachment to two states. The low-temperature dependence of COS is unusual, but can be attributed to attachment to two states with a low A_1. Dissociative electron attachment and molecular ion formation are observed in the ECD data for N_2O. Consequently, only an upper limit to the electron affinity may be obtained [90–93].

Photoelectron spectra and electron impact data are available for N_2O and CS_2. Figures 9.18 and 9.19 provide the photoelectron spectra for both. The comparison of the PES for these two compounds is striking. The broad peak at the higher energies and the low-energy onset of photodetachment in the spectra of N_2O are very different from the vibrational fine structure observed for CS_2. The onset for the PES of CS_2 occurs at 0.6 eV, but the well-resolved progression has its origin at about 0.9 eV. The peak at 1.6 eV is the vertical detachment energy [94]. The onset in the PES for N_2O occurs at 0.0 eV and corresponds to the linear anion. The PES

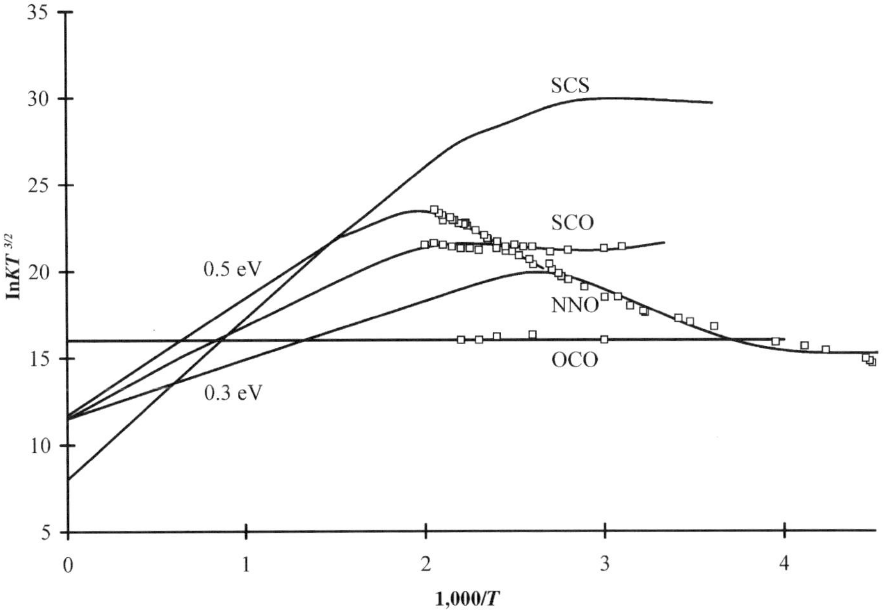

Figure 9.17 ECD data for N_2O, CO_2, CS_2, and COS plotted as ln $KT^{3/2}$ versus 1,000/T, [90–92].

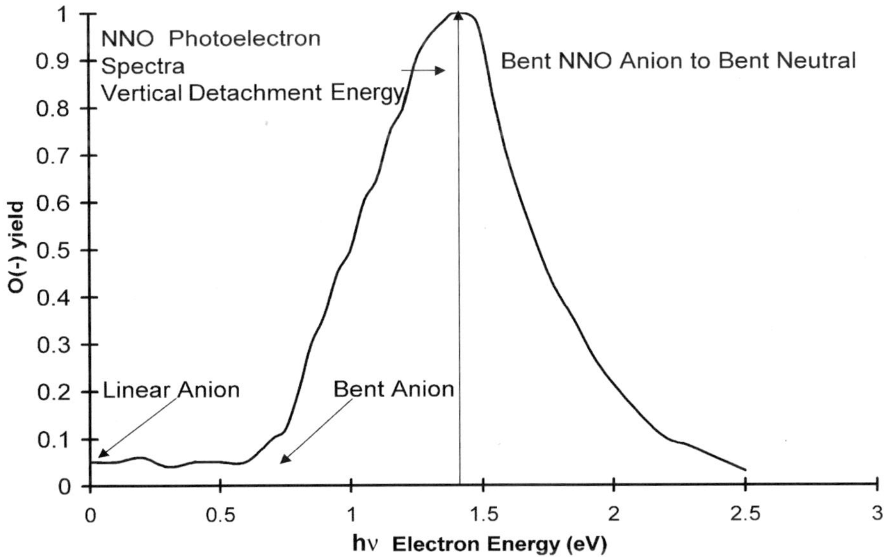

Figure 9.18 Photoelectron spectra of N_2O, replotted from [100].

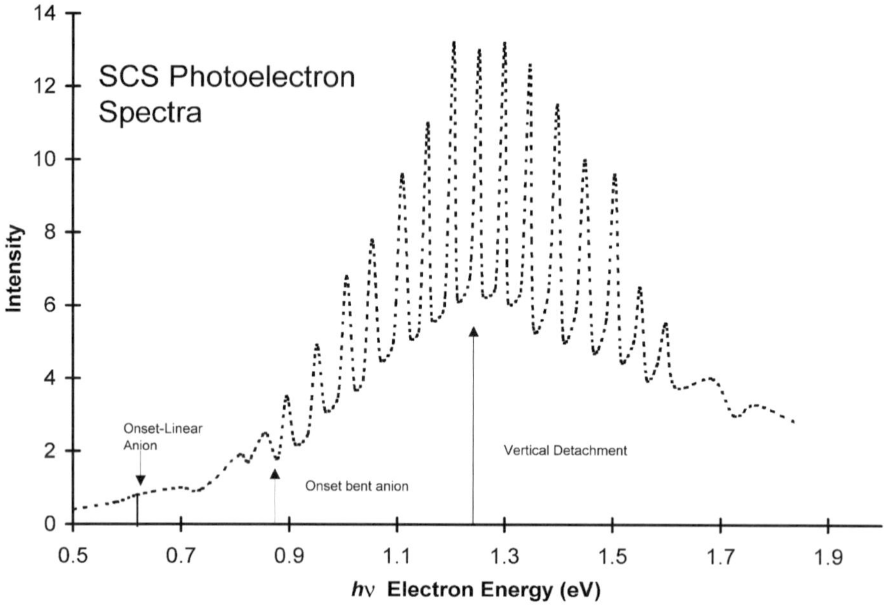

Figure 9.19 Photoelectron spectra of CS_2, replotted from [94].

onset for the bent form of the anion appears at 0.5 eV. The vertical detachment energy for the bent anion is 1.4 eV [100].

Thermal electron attachment to nitrous oxide has been studied for more than 75 years. In spite of this extensive experimental and theoretical work the adiabatic electron affinity of N_2O remains uncertain. The reported electron affinities are 0.0 ± 0.1 eV, 0.22 ± 0.2 eV; an upper limit of 0.76 ± 0.1 eV is determined by PES [92, 98–103]. By fitting the ECD data to an expanded kinetic model, the data can be attributed to two states. The E_a obtained from the ECD data are -0.17 ± 0.05 eV for the linear anion and 0.40 ± 0.15 eV for the bent anion. The larger uncertainty in the ECD value results because the transition temperature to dissociative electron attachment cannot be determined. If the calculated curve is extended to the highest temperature, the E_a is 0.5 eV and the E_1 value is 0.5 eV. If it is terminated as shown in the second curve at approximately 350 K, the E_a is 0.3 eV and the E_1 is 0.4 eV. The "best" value of the AE_a is thus 0.40(15) eV.

Figure 9.20 is the most recent electron impact spectrum for N_2O [98]. The broad peak at 2.2 eV was observed earlier, but that at 0.5 eV was not. The peak at 0.0 eV gives the activation energy for the formation of O(−) of 0.21 eV, which is equal to −EDEA [99]. From earlier electron impact data that goes to higher temperatures, the activation energy is 0.5 eV, as observed for the ECD data. This is attributed to the bending of the neutral so that dissociative electron attachment can take place. Consequently, the lower-temperature data could result from unimolecular dissociation, whereas the higher activation energy might be due to the crossing of the bent

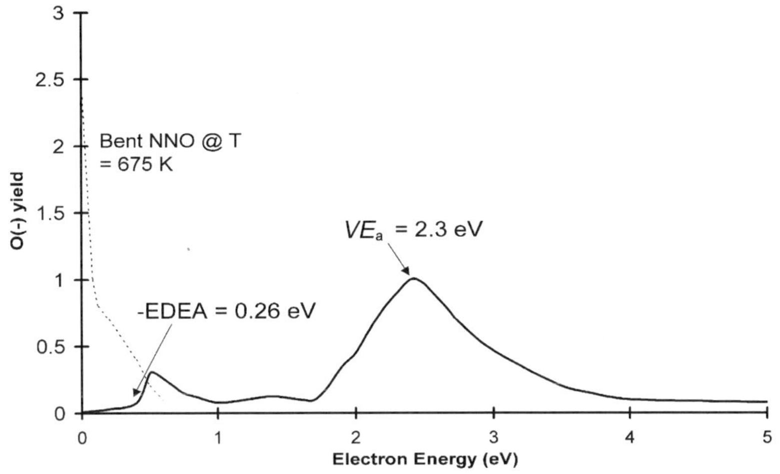

Figure 9.20 Electron impact spectra of N_2O, replotted from [98].

anions and neutrals. Some of these crossings could lead to the molecular ion, which could then undergo thermal detachment.

The CURES-EC E_a are -0.10 eV for the formation of linear anion and 0.50 eV for the bent anion of N_2O. In Figure 9.21 the relative positions of the vertical electron affinities can be seen at constant angles. At 180° the neutral is above the anion,

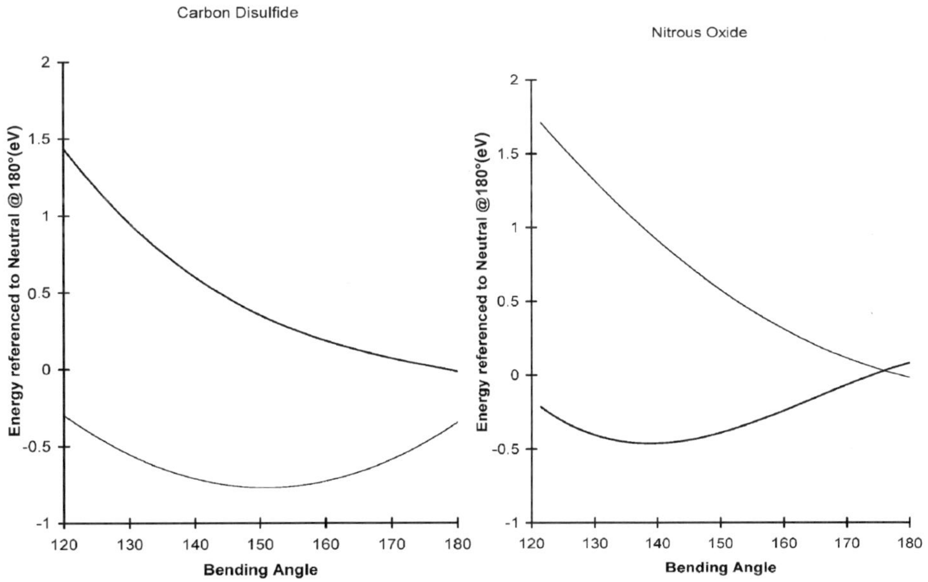

Figure 9.21 The energy of the neutral and negative ions of CS_2 and N_2O as a function of the bending angle calculated using CURES-EC and AM1.

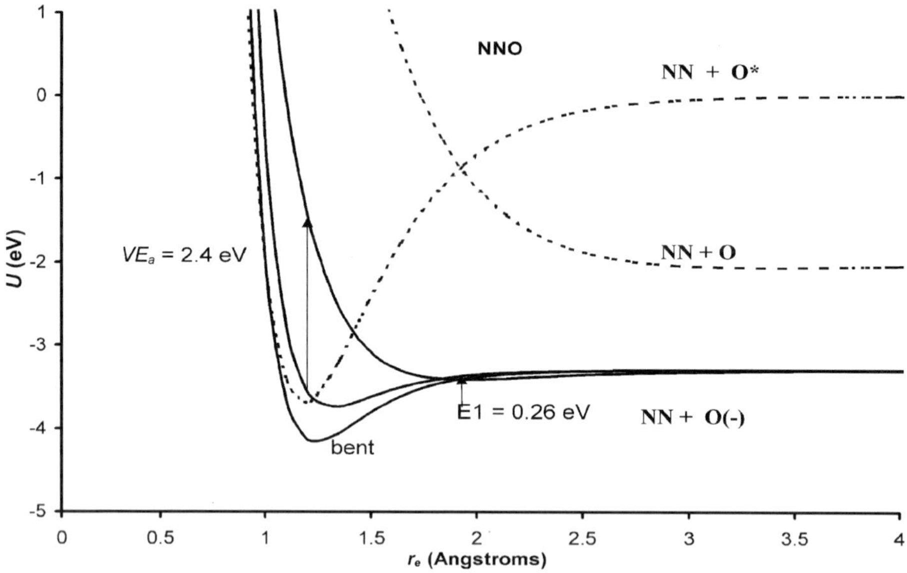

Figure 9.22 Morse potential energy curves for the neutral and negative-ion states of N_2O.

but at 140° the energy of the anion is below that of the linear neutral. The calculated vertical detachment energy is 1.4 eV and agrees with experiment, as shown in Figure 9.21.

With these data Morse potential energy curves can be calculated for the linear and bent forms of the anion. These are shown in Figure 9.22. The dissociation of the neutral to the lowest level is spin-forbidden, giving a maximum in the neutral curve. Two linear anion curves are drawn. The bond dissociation energy of one is lower than the other. The lower curve is responsible for the 0.5 eV vertical attachment peak, while the higher curve with less bonding is responsible for the 2.3 eV vertical attachment peak. These curves are both $D(0)$ curves since they lead to dissociation in the Franck Condon region. However, the lower curve is a $Mc(0)$ curve since it crosses the long-range anion curve below the dissociation limit, while the upper curve is $Dc(0)$ since it crosses that curve above the dissociation limit. The ground-state $M(2)$ curve is bent and will intersect the ground-state neutral on the backside. The PES of $N_2O(-)$ arises primarily from transitions of the bent anion to the bent neutral. Improved photoelectron spectra could resolve many of the questions that still exist for N_2O.

Of these four molecules the electron affinities of CS_2 are the most precise and accurate. Two thresholds are observed in the PES for CS_2, but the electron affinity obtained from the lower-energy onset was not reported. These can be seen in the high-resolution PES for CS_2 in Figure 9.19. Shown in Figure 9.23 is an expanded plot of the ECD data with the slopes drawn in the two regions. The electron affinities of these two states are 0.6 ± 0.05 eV and 0.89 ± 0.02 eV. Figures 9.17, 9.19,

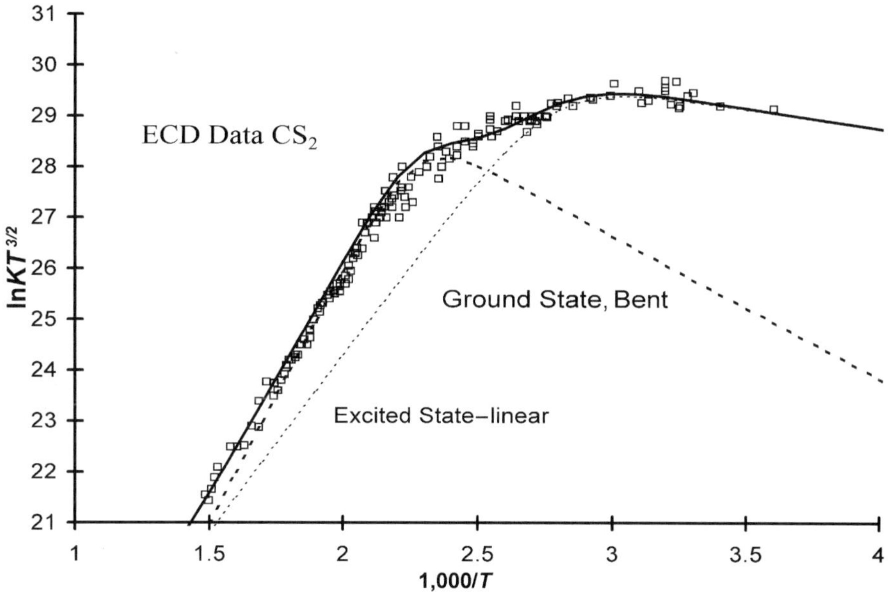

Figure 9.23 ECD data for CS_2 plotted as ln $KT^{3/2}$ versus 1,000/T. The resolved curves are associated with the linear and bent anions.

and 9.23 clearly show that two electronic states of $CS_2(-)$ can be observed in the ECD and PES experiments. Two negative-ion states have also been observed in the same Rydberg electron transfer experiment [95]. In addition, different values have been reported using the same technique. Two TCT values, 0.53(10) and 0.9(2), and two AMB values, 0.6(2) and 1.0(2) eV, have been reported. If the lower values are assigned to the linear anion, the data are reconciled.

The CURES-EC calculations support the two electron affinities. The calculations of the energies as a function of bond angle give additional data for the calculation of potential energy curves, as shown in Figure 9.22 along with those of N_2O. The calculated VE_a for the bent-to-bent transition is in agreement with the PES data. The lowest energy for dissociative electron attachment is more than 2 eV so only molecular ion formation can take place with thermal electrons.

The dissociation in the C—S dimension has two limits: one leading to CS + S(−) and the other leading to CS(−) + S. Figure 9.24 gives the Morse potentials for the linear and bent neutrals and anions for CS_2 leading to these two dissociation limits. The electron affinities used to construct the curves in the CS + S(−) dimension are the two experimental values. The crossing of the linear anion with the linear neutral occurs approximately at the internuclear distance of the neutral. The "backside" crossing of the ground state requires activation energy, as shown in the resolved curves for the ECD data in Figure 9.23. Vertical electron affinity is obtained from the PES spectra and CURES-EC calculation. The vertical transition for the bent anion to the bent neutral occurs at about 1.25 eV, as shown in Figure 9.20.

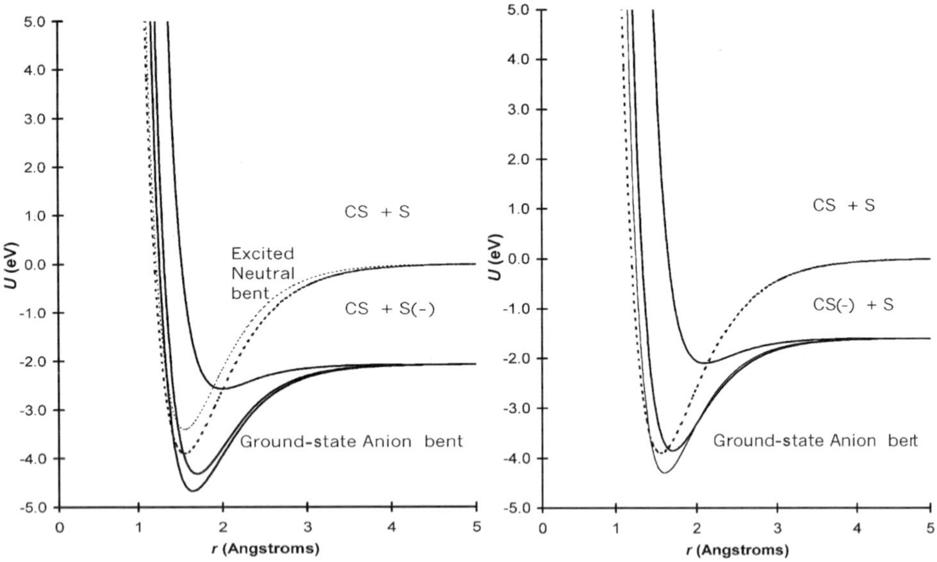

Figure 9.24 Morse potential energy curves for the neutral and negative-ion states of CS_2 for the linear and bent forms.

The transition for the linear anion to the linear neutral has a lower VE_a. The anion curves are both $M(2)$ and $Mc(2)$ curves. There is also a $D(0)/Dc(0)$ curve, which leads to S(−) at about 3.6 eV. Similar curves can be drawn to the dissociation limit of CS(−) + S. The electron affinity of CS is uncertain with one value at 0.20(2) eV and the other at 1.6(3) eV. Using 1.6 eV and the same bond orders as utilized for the two lower negative-ion curves, we obtain two low-lying curves: a $M(1)$ and $M(0)$. The $D(0)$ curve gives CS(−) at an electron energy of 6.2 eV.

The reported electron affinities of COS are about 0.5 eV determined from ECD and alkali metal beam data. Two electron affinities of COS have been obtained from ECD data by fitting to the two state models. The ground state is assigned a value of 0.40 ± 0.1 eV and the excited state one of 0.25 ± 0.1 eV. Except for the A_1 the parameters used to calculate the curves are similar to those for CS_2. The Q_{an} for the ground state is less than 1, and the A_1 values are less than the Debroglie A_1. The CURES-EC E_a are 0.5 eV and 0.3 eV. The angular dependence of the energies of the anions and neutrals is illustrated in Figure 9.25, along with those for CO_2. The negative ions have a lower energy than the neutrals in the calculated range of angles. The minimum for the anion data is not as deep as for the other curves, which might be due to the two different types of bonds. The dissociation in the C—O dimension has two limits below the dissociation limit of the neutral: one leading to CS + O(−) and the other leading to CS(−) + O. In the C—S dimension only one limit leading to CO + S(−) is below the dissociation limit. Since the C—S bond should be weaker than the C—O bond, this should be the least endothermic dissociation. In this dimension two bonding curves, one for the bent form and the other

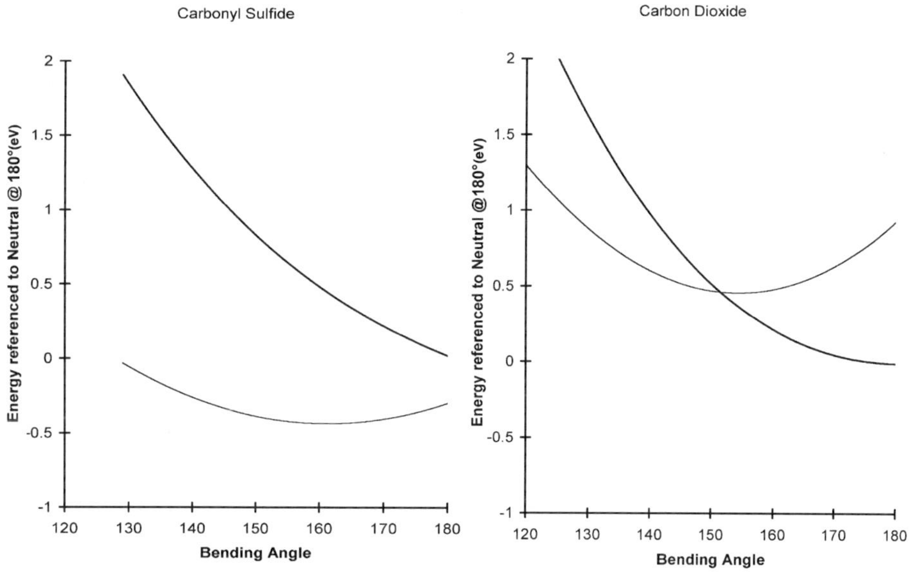

Figure 9.25 CURES-EC AM1 calculations of the energy for the neutral and negative ions of CO_2 and COS as a function of the bending angle.

for the linear form of the anion, can be constructed. In addition, a dissociative curve may be drawn. In the other two dimensions five curves could be drawn, as in the case of $CS_2(-)$. This gives nine M states and three D states.

In the case of CO_2 the linear anion is less stable than the bent anion. The valence-state E_a of CO_2 has been reported as -0.6 eV to -0.8 eV and -1.1 eV to -2.2 eV. The CURES-EC calculations give E_a of -0.5 eV and -0.9 eV for the bent and linear forms. The variation of the energies with the bending angle can be calculated using CURES-EC. These curves are shown in Figure 9.25 along with curves for COS. The reference point for the energy scale is the linear neutral at 180°. The negative-ion curve crosses the neutral curve at an angle of about 150°. At smaller angles the anion will be stable to autodetachment, but still higher than the ground-state neutral. The $CO_2(-)$ anion has been observed in the electron impact spectra of anhydrides.

The relative electron affinities of these four compounds in the linear and bent forms were evaluated using READS-TCT and geometry restraints. The charge density on the bent CS_2 anion is greater than on the linear CS_2 anion in agreement with experiment. In a comparison of the bent COS anion with the linear CS_2 anion the larger charge density resides on linear $CS_2(-)$. The bent COS anion is more stable than the linear or bent N_2O anion. Thus, the order of the electron affinities is CS_2 (b) $>$ CS_2 (l) $>$ COS (b) $>$ N_2O (b) $=$ COS (l) $>$ N_2O (l) $>$ CO_2 (b) $>$ CO_2 (l). This is the same order as experiment and the CURES-EC calculated E_a. Figure 9.26 presents an example of the CS_2 (l) $>$ COS (l) spin density difference of 0.63 versus 0.37.

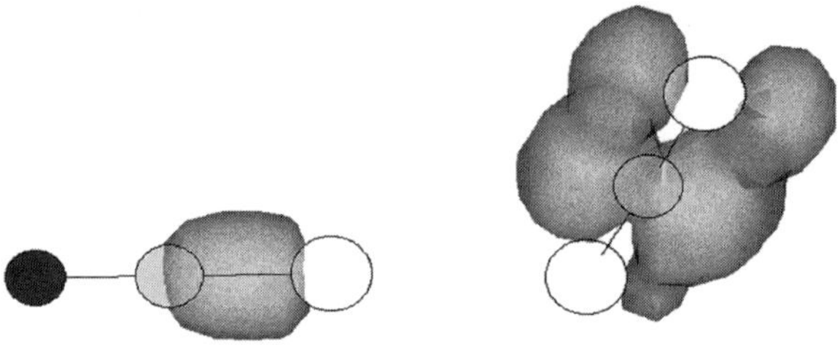

COS $q = 0.37$ SCS $q = 0.63$

Figure 9.26 Three-dimensional spin densities for electron added to CS_2 (l) and COS (l). The spin density difference is 0.63 versus 0.37, indicating that the E_a of linear CS_2 is greater than the E_a of linear COS.

The charge density is automatically calculated when one of the molecules is selected in HYPERCHEM.

9.4 ELECTRON AFFINITIES AND MORSE POTENTIAL ENERGY CURVES: SULFUR FLUORIDES AND ANIONS

The reactions of electrons with the SF_n compounds provide a nice transition between the smaller molecules and larger ones [103–126]. In this section the E_a for the SF_n will be assigned. The only electron impact data available are for SF_6. Using the curves for SF_6 as a prototype, we can utilize the electron affinities measured for the other SF_n to construct qualitative Morse potentials. Curves are drawn dissociating to each of the two lowest limits. In Figures 9.27 and 9.28 Morse potentials for SF, SF_4, SF_5, and SF_6 and their anions are given. The bonding and antibonding curves are easily distinguished in both illustrations.

The electron affinities of all the SF_n from $n = 0$ to 7 have been measured or calculated theoretically. The NIST tables contain the E_a for many of these compounds, but do not assign or evaluate them. A. E. S. Miller and coworkers evaluated the available ground-state values but did not consider excited states [106]. The assignment of E_a for SF_6 to excited states was used in Chapter 5 as an example. The ground-state value is 1.07 ± 0.07 eV, and excited-state values are 0.75, 0.5, 0.3, and 0.0 eV.

The E_a of S, F, SF_3, SF_5, and SF_6 were determined using the magnetron method. The differences between the values obtained by the magnetron method and the current "best" values are -0.2, $+0.2$, -0.2, -0.14, and $+0.3$ eV, respectively, indicating no systematic uncertainties and a random uncertainty on the order of ± 0.2 eV. Except for SF_6 these E_a were determined by dissociative reactions. Two measures of the E_a of SF_5 were obtained: one by the initial dissociation of S_2F_{10} and the other by the determination of the energy of dissociation of $SF_6(-)$ to $F + SF_5(-)$. This energy of 0.0 ± 0.1 eV combined with the current S—F bond dissociation energy

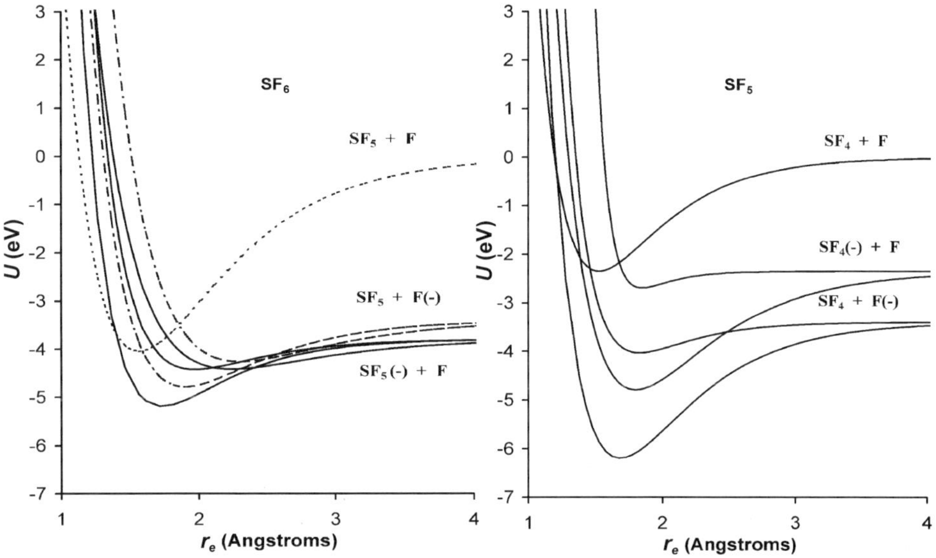

Figure 9.27 Morse potential energy curves for the neutral and negative-ion states of SF_5 and SF_6.

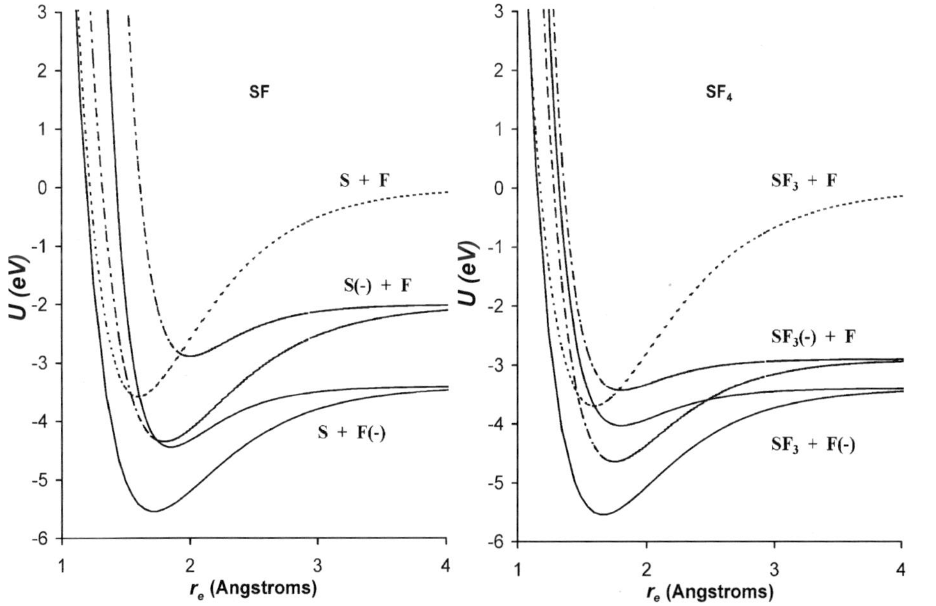

Figure 9.28 Morse potential energy curves for the neutral and negative-ion states of SF and SF_4.

yields $E_a(SF_5) = D(S{-}F) = 4.0 \pm 0.2$ eV. More recently, this energy has been determined to be 0.15 eV by NIMS and electron attachment reactions. Dissociative electron attachment to SF_4 determined the magnetron E_a of SF_3 to be 2.71(15) eV. By using the current best S—F dissociation energy, the E_a is 2.91(15) eV, the same as the more recent experimental and theoretical values. With these modifications the differences between the magnetron values and most recent values are reduced and the random uncertainties are ± 0.15 eV [109, 117].

The electron affinity of SF, the difference in the bond energies of the neutral and anion, and the internuclear distance and vibrational frequency of the ground-state anion were determined using PES and confirmed by theoretical calculations [104]. One larger value of the E_a is listed in the NIST table. Since it is a lower limit of 4.0 eV and is larger than the E_a of F, the value is not used here. Such a large value would require a relative bond order of 1.2 for the addition of an electron to an antibonding orbital. Two excited states with positive E_a have been predicted but not observed.

The E_a of SF_2 is the least certain of any of the values, bracketed between 0.2 eV and 1.6 eV. The NIST values are 0.9(6) eV or 3.1 eV. The latter value was determined using the same technique that resulted in an E_a of greater than 4.0 eV for SF. Therefore, this value is not used in the current analysis. Theoretical values are about 2.0 eV and two nondetaching anion states are predicted. The CURES-EC value is 1.45 eV. The weighted average of the NIST value for the E_a of SF_3 is 2.9(1) eV. Four E_a of SF_4 have been reported: 2.35(10), 1.5(2), 1.26(10), and 0.8(2) eV. The reported E_a of SF_5 are 3.80(3) eV and 3.0(1) eV.

Thermal electron attachment to SF_6 has been studied in the ECD and NIMS at atmospheric pressure and in a chemical ionization mass spectrometer at lower pressures. From these data rate constants for thermal electron attachment and detachment to SF_6, and the Q_{an} values and electron affinities of SF_6 and SF_5 have been determined. The Morse potential energy curves for multiple negative-ion states were calculated using electron impact data and electron affinities.

The Morse potential energy curve for the ground state of $SF_6(-)$ has been calculated by ab initio methods. The recommended electron affinity is 1.39 eV, but values ranging from 0.49 eV to 1.39 eV were reported. The theoretical curve differs from the semi-empirical one in that the internuclear distances and frequencies are smaller for the ground-state anion [126]. By using the theoretical ground-state curve, an additional excited-state curve is required to account for the electron impact data and reactions with thermal electrons. This is not a problem since multiple negative-ion states have been observed for SF_6. The curves shown in Figure 9.27 have been drawn with the E_a, the vertical E_a, and the ion distributions. There can also be many other curves since $SF_5(-)$ has a large number of excited states. The ground-state curve is $M(2)$ with positive E_a and VE_a and negative EDEA. The upper bonding curve is $M(1)$ since the curve crosses the neutral at the zero point vibrational level, making VE_a negative. The excited-state antibonding curves are $D(1)$ since E_a is positive, but the other metrics are negative and the vertical transition leads to dissociation.

The Morse curves for $SF_5(-)$ shown in Figure 9.27 are drawn with positive E_a. The ground-state curve is $M(3)$ and has a large activation energy. The first excited

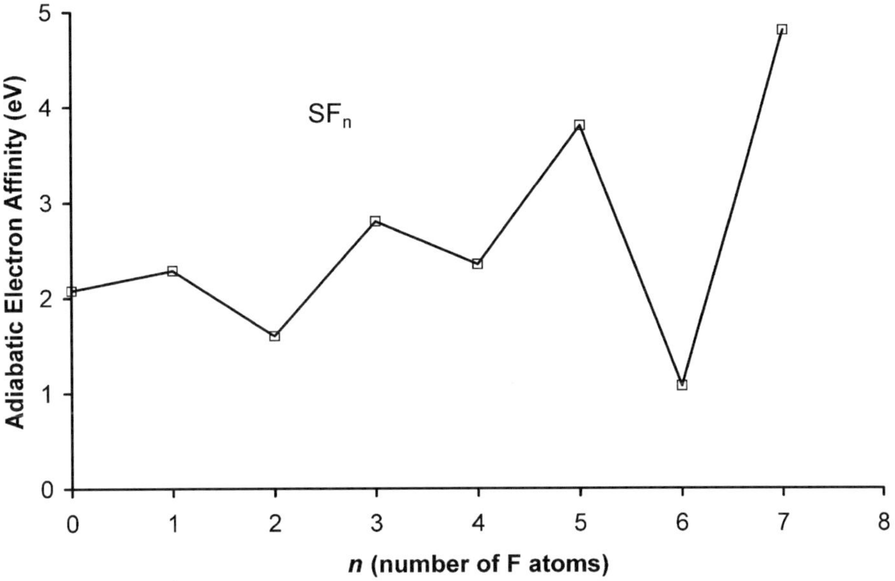

Figure 9.29 Adiabatic electron affinities of SF_n. Data from [1].

state is also $M(3)$ but dissociates to a higher limit, leading to $SF_4(-)$. The two antibonding curves are $D(3)$ and $D(2)$ curves since the E_a and EDEA are positive and the VE_a is negative for the curve, leading to the higher limit. The vertical transition leads to the fluoride ion for one curve and $SF_4(-)$ for the other. The anion curves in Figure 9.27 are examples of five of the eight possible HIMPEC.

The curves for $SF_4(-)$ and $SF(-)$ shown in Figure 9.28 provide additional example HIMPEC. The ground-state curve for $SF(-)$ is $M(3)$. The bonding curve leading to $S(-) + F$ is $M(2)$, with positive E_a and VE_a but negative EDEA. The higher antibonding curve is $D(0)$, while the lower antibonding curve is $D(2)$. The two excited-state curves for $SF_4(-)$ are $M(2)$ and $M(0)$ curves since they lead to molecular ion formation in the vertical transition. The EDEA, E_a, and VE_a are negative for the antibonding curve, while only the EDEA is negative for the bonding curve. The ground-state curve is $M(2)$, while the complementary antibonding curve is $D(2)$. The only HIMPEC not illustrated by the anion curves in Figures 9.28 and 9.29 is the $D(3)$ curve.

The systematic trends in the E_a, D(SF), and relative bond orders were used in these constructions. The quantities are plotted against the number of fluorine atoms in Figures 9.29 and 9.30 and given in Table 9.8. The CURES-EC values reproduce the experimental E_a. The first S—F bond dissociation energy in SF_5 is the weakest of all the SF_n. This accounts for the relative bond order greater than 1 for the anion. The dissociation energies show minima at $n = 3$, 5, and 7 and maxima at 2, 4, and 6. The experimental E_a illustrated in Figure 9.29 were used

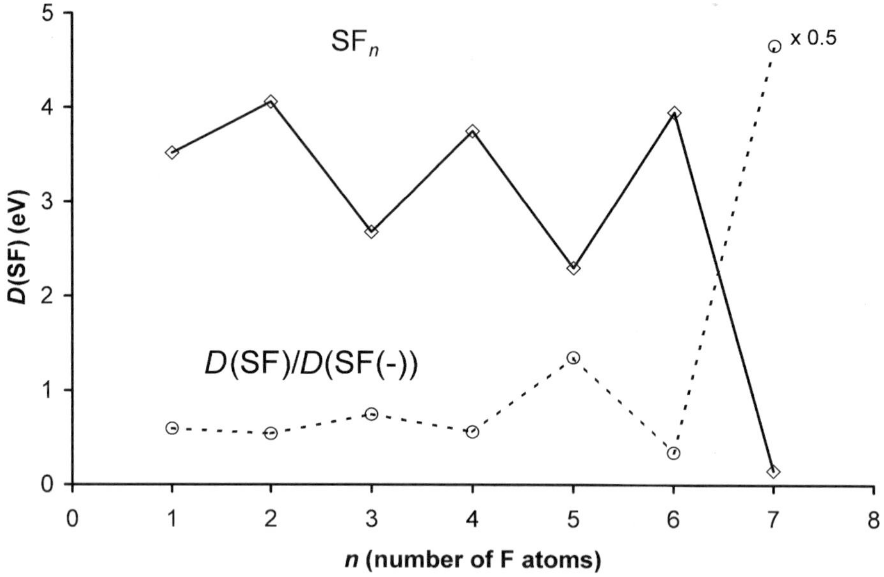

Figure 9.30 Dissociation energies and relative bond order of SF_n.

to calculate the curves for the excited states when available. The relative bond orders for the ground state given in Figure 9.30 are upper limits to the bond orders for the excited states [106–124].

The limitations of the two-dimensional representations must be emphasized. For example, for SF_5 the geometry of the anion is very different from that of the neutral. Therefore, by showing only the S—F stretching dimension, the representation is only a cut through the surface. In the case of triatomic molecules this was taken into account by showing the manner in which the neutral changes with the bond angle. In the case of these polyatomic molecules the geometric changes cannot be thus represented. In the case of more rigid molecules, such as those involving benzene rings, the dissociations in one dimension can be very effective in representing negative-ion states.

TABLE 9.8 Electron Affinities, Dissociation Energies, and Anion Bond Order SF_n

Species	E_a (eV)	Source	CEC (eV)	*D*(S—F)	Bond Order
SF	2.285(6)	PES	2.23	3.52(5)	0.60
SF_2	1.6(5)	TCT	1.38	3.98(19)	0.54
SF_3	2.91(15)	MGN	2.81	2.68(10)	0.75
SF_4	1.50(20)	TCT	1.58	3.75(10)	0.56
SF_5	3.85(3)	NIMS	3.80	2.30(26)	1.34
SF_6	1.07(7)	NIMS	1.02	3.95(20)	0.15
SF_7	4.8	CALC	4.7	0.15(10)	9.33

9.5 SUMMARY

The electron affinities of the homonuclear diatomic molecules are supported by periodic trends and simple molecular orbital concepts. The negative-ions states of the Group IA, IB, VI, and VII homonuclear diatomic anions are supported by Morse potential energy curves. Six curves for $F_2(-)$ and $Cl_2(-)$, eight curves for $Br_2(-)$, and 12 curves for $I_2(-)$ are obtained. The curves for the $^2\Pi_g$ (1/2) state of $I_2(-)$ and $Br_2(-)$ are calculated using a lower excited-state dissociation energy (0.02 eV). The curves for $F_2(-)$ are the least certain for the halogens. The other curves should remain the same as new data are obtained. The AE_a are as follows: F_2, 3.08; Cl_2, 2.45; Br_2, 2.56; and I_2, 2.50 (all ±0.02 eV) obtained by assuming $D_0(Rg_2(+)) = D_0(X_2(-))$ and $AE_a(X_2) = D_0(X_2(-)) + AE_a(X) - D_0(X_2)$. They agree with those measured directly. A new interpretation of ECD data for Cl_2, Br_2, and I_2 gives activation energies for thermal electron attachment E_1 to two states. Recent ECD data for $O_2(-)$ confirm multiple electron affinities previously observed and define new activation energies for thermal electron attachment. Small peaks observed in published high-resolution PES for $O_2(-)$ have been assigned to closely spaced negative-ion states based on experimental E_a and the prediction of 24 states. Twelve curves leading to molecular ion formation and 12 curves leading to dissociation in a Franck Condon process were constructed for $O_2(-)$ from experimental data. Two anion curves are presented for $S_2(-)$, $Se_2(-)$, and $Te_2(-)$. The anion curves for the Group IA and IB homonuclear diatomic molecules have been constructed using experimental E_a. The relative bond order for these ground-state anions is approximately 1. The excited-state anion curves are based on theoretical curves for $Li_2(-)$.

The electron affinities of several triatomic molecules and the azide radical have been evaluated and are supported by the CURES-EC method. The molecules that are linear in the neutral and bent in the anion have been emphasized. The ECD data support two negative-ion states. Morse potential energy curves for N_2O and CS_2 have been constructed for the linear and bent ions. The relative energies of the anion and neutrals for CO_2, COS, CS_2, and N_2O were presented to explain the electron attachment data. The electron affinities of the SF_n molecules $n = 1$ to 6 were evaluated and the largest values assigned to the ground state.

All the curves discussed in this chapter have been assigned to the general Herschbach ionic Morse potential curves classification. While many of these curves are speculative, they reflect the data. As stated by Robert Sanderson Mulliken in offering curves for I_2 in 1971, "While the curves shown cannot possibly be quantitatively correct, they should be useful as forming a sort of zeroth approximation to the true curves." [128].

REFERENCES

1. National Institute of Standards and Technology (NIST). *Chemistry WebBook*, **2003**. Available at http://webbook.nist.gov.

2. Christodoulides, A. A.; McCorkle, D. L.; and Christophorou, L. G. "Electron Affinities of Atoms, Molecules and Radicals" in *Electron-Molecule Interactions and Their Applications*. New York: Academic Press, **1984**.
3. Reinstra-Kiracofe, J. C.; Tschumper, G. S.; Schaefer, H. F.; Nandi, S.; and Ellison, G. B. *Chem. Rev.* **2002**, 102, 231.
4. Gutsev, G. L. and Bauschlicher, C. W. *J. Phys. Chem. A* **2003**, 107, 4755.
5. Chen, E. S. D. and Chen, E. C. M. *J. Phys. Chem. A* **2003**, 107, 169.
6. Jones, P. L.; Mead, R. D.; Kohler, B. E.; Rosner, S. D.; and Lineberger, W. C. *J. Chem. Phys.* **1980**, 73, 4419.
7. Zanni, M. T.; Batista, V. S.; Greenblatt, B. J.; Soep, B.; and Neumark, D. M. *J. Chem. Phys.* **1999**, 110, 3748.
8. Chen, E. C. M. and Wentworth, W. E. *J. Phys. Chem.* **1985**, 89, 4099.
9. Chen, E. S. and Chen, E. C. M. *Chem. Phys Lett.* **1998**, 293, 491.
10. Chen, E. S. and Chen, E. C. M. *J. Phys. Chem. A* **2002**, 106, 6655.
11. Ayala, J. A.; Chen, E. C. M.; and Wentworth, W. E. *J. Phys. Chem.* **1981**, 85, 768.
12. Herschbach, D. R. *Adv. Chem. Phys.* **1966**, 10, 250.
13. Parr, R. G. and Ayres, P. W. *J. Phys. Chem. A* **2002**, 106, 5060.
14. Kleyn, A. W. and Moutinho, A. M. C. *J. Phys. B: At. Mol. Opt. Phys.* **2001**, 34, R1.
15. Zanni, M. T.; Taylor, T. R.; Greenblatt, B. J.; Miller, W. H.; and Neumark, D. M. *J. Chem. Phys.* **1997**, 107, 7613.
16. Leininger, T. and Gadea, F. X. *J. Phys. B: At. Mol. Opt. Phys.* **2001**, 33, 735.
17. Paidarova, I. and Gadea, F. X. *Chem. Phys.* **2001**, 274, 1.
18. Caledonia, G. E. *Chem. Rev.* **1975**, 75, 333.
19. Pritchard, H. O. *Chem. Rev.* **1953**, 52, 529.
20. Kurepa, M. V.; Babic, D. S.; and Belic, D. S. *J. Phys. B: At. Mol. Opt. Phys.* **1978**, 11, 3719.
21. Kurepa, M. V.; Babic, D. S.; and Belic, D. S. *J. Phys. B: At. Mol. Opt. Phys.* **1981**, 14, 375.
22. Azria, R.; Abouaf, R.; and Tellet-Billy, D. *J. Phys. B: At. Mol. Opt. Phys.* **1982**, 16, L569.
23. Azria, R.; Abouaf, R.; and Tellet-Billy, D. *J. Phys. B: At. Mol. Opt. Phys.* **1988**, 21, L213.
24. Freeman, R. R. Ph. D. dissertation, University of Houston, **1971**.
25. Chen, E. S.; Wentworth, W. E.; and Chen, E. C. M. *J. Mol. Struct.* **2002**, 606, 1.
26. Burch, D. S.; Smith, S. J.; and Branscomb L. M. *Phys. Rev.* **1958**, 112, 171.
27. Pack J. L. and Phelps, V. *Phys. Rev. Lett.* **1961**, 6, 111.
28. Vogt, D.; Hauffe, B.; and Neuert, H. *Z. Phys.* **1970**, 232, 439.
29. Bailey, T. L. and Mahadevan, P. *J. Chem. Phys.* **1970**, 52, 179.
30. Stockdale, J. A. D.; Compton, R. N.; Hurst, G. S.; and Reinhardt, P. W. *J. Chem. Phys.* **1969**, 50, 2176.
31. Celotta, R. J.; Bennett, R. A.; Hall, J. L.; Siegel, M. W.; and Levine, J. *Phys. Rev. A: At. Mol. Opt. Phys.* **1972**, 6, 631.
32. Schiedt, J. and Weinkauf, R. *Z. Natforsch.* **1995**, 50a, 1041.
33. Baeda, A. P. M. *Physica* **1972**, 59, 541.
34. Nalley, S. J. and Compton, R. N. *Chem. Phys. Lett.* **1971**, 9, 529.
35. Tiernan, T. O.; Hughes, B. M.; and Lifschitz, C. *J. Chem. Phys.* **1971**, 55, 5692.
36. Lacmann, K. and Herschbach, D. R. *Chem. Phys. Lett.* **1970**, 6, 106.

37. Sambe, H. and Ramaker, D. E. *Phys. Rev. A: At. Mol. Opt. Phys.* **1989**, 40, 3651.
38. Bates, D. R. and Massey, H. S. W. *Proc. Roy. Soc. A* **1943**, 239, 33.
39. Gilmore, F. R. *J. Quant. Spectr. Rad. Trans.* **1965**, 5, 369.
40. Michels, H. H. *Adv. Chem. Phys.* **1981**, 45, 227.
41. Bass, A. D.; Parenteau, L.; Weik, F.; and Sanche, L. *J. Chem. Phys.* **2001**, 115, 48.11.
42. Jaffke, T.; Meinke, M.; Hashemi, R.; Christrophorou, L. G.; and Illenberger, E. *Chem. Phys. Lett.* **1992**, 193, 62.
43. Bailey, C. J.; Lavrich, D. J.; Serxner, D.; and Johnson, M. A. *J. Chem. Phys.* **1996**, 105, 1807.
44. Celotta, R. J.; Bennett, R. A.; and Hall, J. L. *J. Chem. Phys.* **1986**, 60, 1740.
45. Hunsicker, S.; Jones, R. O.; and Gantefor, G. *J. Chem. Phys.* **1995**, 102, 5917.
46. Moran, N. R. and Ellison, G. B. *J. Phys. Chem.* **1988**, 92, 1794.
47. Snodgrass, J. T.; Coe, J. V.; McHugh, K. M. ; Friedhoff, C. B.; and Bowen, K. H. *J. Phys. Chem.* **1989**, 93, 1249.
48. Hobbs, R. H.; Michels, H. H.; and Wright, L. A. *Chem. Phys. Lett.* **1985**, 118, 67.
49. McHugh, K. M.; Eaton, J. G.; Lee, G. H.; Sarkas, H. W.; Kidder, L. H.; Snodgrass, J. T.; Manaa, M. R.; and Bowen, K. H. *J. Chem. Phys.* **1989**, 91, 3792.
50. Ho, J.; Ervin, K. M.; and Lineberger, W. C. *J. Chem. Phys.* **1990**, 93, 6987.
51. Gantefor, G. F.; Cox, D. M.; and Kaldor, A. *J. Chem. Phys.* **1990**, 93, 8395.
52. Taylor, K. J.; Pettiettehall, C. L.; Cheshnovsky, O.; and Smalley, R. E. *J. Chem. Phys.* **1992**, 96, 3319.
53. McGeoch, M. W. and Schlier, R. E. *Phys. Rev. A: At. Mol. Opt. Phys.* **1985**, 33, 1708.
54. Travers, M. J.; Cowles, D. C.; and Ellison, G. B. *Chem. Phys. Lett.* **1989**, 164, 449.
55. Siegel, M. W.; Celotta, R. J.; Hall, J. L.; Levine, J.; and Bennett, R. A. *Phys. Rev. A: At. Mol. Opt. Phys.* **1972**, 6, 607.
56. Tiernan, T. O. and Wu, R. L. C. *Adv. Mass Spectrom.* **1978**, 7A, 136.
57. Burrow, P. D. *Chem. Phys. Lett.* **1974**, 26, 265.
58. Parkes, D. A. and Sugden, T. M. *J. Chem. Soc. Farad. Trans. 2*, **1972**, 68, 600.
59. Refaey, K. M. A. *Int. J. Mass Spectrom. Ion Phys.* **1976**, 21, 21.
60. Berkowitz, J.; Chupka, W. A.; and Gutman, D. *J. Chem. Phys.* **1971**, 55, 2733.
61. Williams, J. M. and Hamill, W. H. *J. Chem. Phys.* **1968**, 49, 4467.
62. Farragher, A. L.; Page, F. M.; and Wheeler, R. C. *Farad. Disc. Chem. Soc.* **1964**, 37, 203.
63. Arnold, D. W.; Bradforth, S. E.; Kitsopoulos, T. N.; and Neumark, D. M. *J. Chem. Phys.* **1991**, 95, 8753.
64. Oakes, J. M. and Ellison, G. B. *Tetrahedron* **1986**, 42, 6263.
65. Yang, S.; Taylor, K. J.; Craycraft, M. J.; Conceicao, J.; Pettiette, C. L.; Cherhnovsky, O.; and Smalley, R. E. *Chem. Phys. Lett.* **1989**, 144, 431.
66. Illenberger, E.; Comita, P.; Brauman, J. I.; Fenzlaff, H-P.; Heni, M.; Heinrich, N.; Koch, W.; and Frenking, G. *Ber. Bunsen-Ges. Phys. Chem.* **1985**, 89, 1026.
67. Jackson, R. L.; Pellerite, M. J.; and Brauman, J. I. *J. Amer. Chem. Soc.* **1981**, 103, 1802.
68. Engleking, P. C. and Lineberger, W. C. *J. Chem. Phys.* **1976**, 65, 4323.
69. Franklin, J. L.; Dibeler, V. H.; Reese, R. M.; and Krauss, M. *J. Amer. Chem. Soc.* **1958**, 80, 298.

70. Arnold, D. W.; Xu, C. S.; Kim, E. H.; and Neumark, D. M. *J. Chem. Phys.* **1994**, 101, 912.
71. Novich, S. E.; Engelking, P. C.; Jones, P. L.; Futrell, J. H.; and Lineberger, W. C. *J. Chem. Phys.* **1979**, 70, 2652.
72. Wong, S. F.; Vorburger, T. V.; and Woo, S. B. *Bull. Am. Phys. Soc.* **1971**, 16, 213.
73. Smith, G. P.; Lee, L. C.; Cosby, P. C.; Peterson, J. R.; and Moseley, J. T. *J. Chem. Phys.* **1978**, 68, 3818.
74. Dotan, I.; Davidson, J. A.; Streit, G. E.; Albritton, D. L.; and Fehsenfeld, F. C. *J. Chem. Phys.* **1977**, 67, 2874.
75. Wu, R. L. C.; Tiernan, T. O.; and Lifschitz, C. *Chem. Phys. Lett.* **1977**, 51, 211.
76. Rothe, E. W.; Tang, S. Y.; and Reck, G. P. *J. Chem. Phys.* **1975**, 62, 3829.
77. Nimlos, M. R. and Ellison, G. B. *J. Phys. Chem.* **1986**, 90, 2574.
78. Chowdhury, S.; Heinis, T.; Grimsrud, E. P.; and Kebarle, P. *J. Phys. Chem.* **1986**, 90, 2747.
79. Feldman, D. *Z. Naturfor.* **1970**, 25A, 621.
80. Grabowski, J. J.; Van Doren, J. M.; De Puy, C. H.; and Bierbaum, V. M. *J. Chem. Phys.* **1984**, 80, 575.
81. Refaey, K. M. A. and Franklin, J. L. *J. Chem. Phys.* **1976**, 65, 1994.
82. Ervin, K. M.; Ho, J.; and Lineberger, W. C. *J. Phys. Chem.* **1988**, 92, 5405.
83. Woo, S. B.; Helmy, E. M.; Mauk, P. H.; and Paszek, A. P. *Phys. Rev. A* **1981**, 24, 1380.
84. Herbst, E.; Patterson, T. A.; and Lineberger, W. C. *J. Chem. Phys.* **1974**, 61, 1300.
85. Smith, G. P.; Lee, L. C.; and Moseley, J. T. *J. Chem. Phys.* **1979**, 71, 4034.
86. Richardson, J. H.; Stephenson, L. M.; and Brauman, J. I. *Chem. Phys. Lett.* **1974**, 25, 318.
87. Pearson, P. K.; Schaefer, H. F.; Richardson, J. H.; Stephenson, L. M.; and Brauman, J. I. *J. Amer. Chem. Soc.* **1974**, 96, 6778.
88. Ferguson, E. E.; Dunkin, D. B.; and Fehsenfeld, F. C. *J. Chem. Phys.* **1972**, 57, 1459.
89. Chen, E. C. M.; Welk, N.; Chen, E. S.; and Wentworth, W. E. *J. Phys. Chem. A* **1999**, 103, 9072.
90. Chen, E. C. M. and Wentworth, W. E. *J. Phys. Chem.* **1983**, 87, 45.
91. Chen, E. C. M.; George, R.; Carr, S. D.; Wentworth, W. E.; and Chen, E. S. D. *J. Chromatogr. A* **1998**, 811, 250.
92. Wentworth, W. E.; Chen, E.; and Freeman, R. *J. Chem. Phys.* **1971**, 55, 2075.
93. Compton, R. N.; Reinhardt, P. W.; and Cooper, C. D. *J. Chem. Phys.* **1975**, 63, 3821.
94. Scheidt, J. and Weinkauf, R. *Chem. Phys. Lett.* **1997**, 266, 201.
95. Compton, R. N.; Dunning, F. B.; and Nordlander, P. *Chem. Phys. Lett,* **1996**, 253, 8.
96. Tsukuda, T.; Hirose, T.; and Nagata, T. *Chem. Phys. Lett.* **1997**, 279, 179–184.
97. Kraus, K.; Muller-Duysing, W.; and Neuert, H. *Z. Naturfor.* **1961**, 16A, 1385.
98. Bruning, F.; Matejcik, S.; Illenberger, E.; Chu, Y.; Senn, G.; Muigg, D.; Denifi, G.; and Mark, T. D. *Chem. Phys. Lett.* **1998**, 292, 177.
99. Chantry, P. J. *J. Chem. Phys.* **1969**, 51, 3369.
100. Knapp, A.; Echt, O.; Kreisle, D.; Mark, T. D.; and Recknagel, E. *Chem. Phys. Lett.* **1986**, 126, 225.
101. Coe, J. V.; Snodgrass, J. T.; Freidhoff, C. B.; McHugh, K. M.; and Bowen, K. H. *Chem. Phys. Lett.* **1986**, 124, 274.
102. Tiernan, T. O. and Wu, R. L. C. *Adv. Mass Spectrom.* **1978**, 7A, 136.

103. Hopper, D. G.; Wahl, A. C.; Wu, R. L. C.; and Tiernan, T. O. *J. Chem. Phys.* **1976**, 65, 5474.
104. Polak, M. L.; Gilles, M. K.; and Lineberger, W. C. *J. Chem. Phys.* **1992**, 96, 7191.
105. Langford, M. L.; Almeida, D. P.; and Harris, F. M. *Int. J. Mass Spectrom. Ion Proce.* **1990**, 98, 147.
106. Miller, A. E. S.; Miller, T. M.; Viggiano, A. A.; Morris, R. A.; Vandoren, J. M.; Arnold, S. T.; and Paulson, J. *J. Chem. Phys.* **1995**, 102, 8865.
107. Harland, P. W. and Thynne, J. C. J. *J. Phys. Chem.* **1971**, 75, 351.
108. Compton, R. N.; Reinhardt, P. W.; and Cooper, C. D. *J. Chem. Phys.* **1978**, 68, 2023.
109. Page, F. M. and Goode, G. C. *Negative Ions and the Magnetron.* New York: Wiley, **1969**.
110. Babcock, L. M. and Streit, G. E. *J. Chem. Phys.* **1981**, 75, 3864.
111. Donovan, R. J.; Harland, P. W.; Knox, J. H.; Makowski, J. A.; and Thynne, J. C. J. *Int. J. Mass Spectrom. Ion Phys.* **1974**, 13, 464.
112. Chen, E. C. M.; Shuie, L. R.; D'Sa, E. D.; Batten, C. F.; and Wentworth, W. E. *J. Chem. Phys.* **1988**, 88, 4711.
113. Larson, J. W. and McMahon, T. B. *J. Amer. Chem. Soc.* **1983**, 105, 2944.
114. Compton, R. N.; Reinhardt, P. W.; and Cooper, C. D. *J. Chem. Phys.* **1978**, 68, 2023.
115. Hubers, M. M. and Los. *J. Chem. Phys.* **1975**, 10, 235.
116. Compton, R. N. and Cooper, C. D. *J. Chem. Phys.* **1973**, 59, 4140.
117. Kay, J. and Page, F. M. *Trans. Farad. Soc.* **1964**, 60, 1042.
118. Chen, E. C. M.; Wiley, J. R.; Batten, C. F.; and Wentworth, W. E. *J. Phys. Chem.* **1994**, 98, 88.
119. Grimsrud, E. P.; Chowdhury, S.; and Kebarle, P. *J. Chem. Phys.* **1985**, 83, 1059.
120. Datskos, P. G.; Carter, J. G.; and Christophorou, L. G. *Chem. Phys. Lett.* **1995**, 239, 38.
121. Drzaic, P. S. and Brauman, J. I. *J. Amer. Chem. Soc.* **1982**, 104, 13.
122. Refaey, K. M. A. and Franklin, J. L. *Int. J. Mass Spectrom. Ion Phys.* **1978**, 26, 125.
123. Fehsenfeld, F. C. *J. Chem. Phys.* **1971**, 54, 438.
124. Compton, R. N.; Christophorou, L. G.; Hurst, G. S.; and Reinhardt, P. W. *J. Chem. Phys.* **1966**, 45, 4634.
125. Tachikawa, H. *J. Phys. B: At. Mol. Opt. Phys.* **2002**, 35, 55.
126. Ziegler, T. and Gutsev, G. L. *J. Chem. Phys.* **1992**, 96, 7623.
127. King, R. A.; Galbraith, J. M.; and Schaefer, H. F. *J. Phys. Chem.* **1996**, 100, 6061.
128. Mulliken, R. S. *J. Chem. Phys.* **1971**, 55, 288.

CHAPTER 10

Negative Ions of Organic Molecules

10.1 INTRODUCTION

The negative ions of organic molecules are more difficult to characterize because of the many degrees of freedom that they have. Organic molecules are defined as closed-shell species that contain C, H, O, S, N, and halogens. The profusion of vibrations and rotations also leads to multiple negative-ion states. In this chapter the negative-ion states of nitromethane and nitrobenzene will be characterized by Morse potential energy curves. The electron affinities of the organic molecules will be evaluated by comparisons with experimental and theoretical determinations. Substitution and replacement rules will then be established from the data.

Fewer than 300 E_a for organic molecules have been determined in the gas phase. The majority of the E_a have been determined by the ECD and/or TCT methods. The direct capture magnetron, AMB, photon, and collisional ionization methods have produced fewer than 40 values. Only the E_a of p-benzoquinone, nitrobenzene, nitromethane, azulene, tetracene, and perylene have been determined by three or more methods. Excited-state E_a have been obtained by each of these methods. Half-wave reduction potentials have determined the electron affinities of 50 aromatic hydrocarbons. The electron affinities of another 50 organic compounds have been determined from half-wave reduction potentials and the energies of charge transfer complexes. It is a manageable task to evaluate these 300 to 400 E_a.

The electron affinities of organic halides and environmental pollutants are evaluated in Chapter 11 and those of biological molecules in Chapter 12. Many of the E_a measured in the gas phase are tabulated in the NIST tables [1]. These are listed in the appendices according to molecules containing CHX, CHNX, CHOX, and CHONX with references. The ECD values for some of the aromatic hydrocarbons in NIST have not been updated. L. G. Christophorou's compilation includes E_a from half-wave reduction potentials and charge transfer complexes [2]. Some of these E_a will be revised and evaluated in this chapter based on gas phase measurements.

The Electron Capture Detector and the Study of Reactions with Thermal Electrons
by E. C. M. Chen and E. S. D. Chen
ISBN 0-471-32622-4

The E_a are considered according to the methods used. Those for hydrocarbons determined by other methods are compared with the ECD values. The E_a for benzaldehydes, acetophenones, benzophenones, bezonitriles, esters, and nitrobenzenes have been determined by the ECD, TCT, and/or NIMS methods. These compounds and related molecules obtained from reduction potentials and charge transfer complex energies will be evaluated. Based on these data, the substitution effects for $CH_3C{=}O$, NO_2, $HC{=}O$, NH_2, F, Cl, and CF_3 and the alkyl groups are determined. The multiple substitution effects of NO_2 and CN on benzene and ethylene are also established.

10.2 ELECTRON AFFINITIES AND POTENTIAL ENERGY CURVES FOR NITROBENZENE AND NITROMETHANE

Nitromethane makes a nice transition from small molecules to large organic molecules. The E_a of NO_2 was determined using the ECD (above 425 K) from the activation energy for dissociative thermal electron attachment to CH_3NO_2. The E_a of CH_3NO_2 was determined from data in the lower-temperature region. The data were presented in Chapter 5 as an example of the determination of a Q_{an} much smaller than 1. The ECD E_a of 0.50 ± 0.02 eV for nitromethane is supported by AMB and TCT values. The PES value is 0.26 ± 0.08 eV and could exist for an excited state. A dipole bound state is observed at 0.012 eV [1–13].

For nitromethane there are two dissociation directions: the C—H dimension and C—N dimension. There are two known excited states for NO_2(-). The dissociation in the C—H dimension is related to the gas phase acidity measurement. There are considerable gas phase acidity, electron impact, absorption, and emission data for nitromethane. Morse potential energy curves have been calculated for seven states in the C—N dimension and six states in the C—H dimension [3–13]. At least two dipole bound curves can also be constructed. Curves in the C—H dimension and C—N dimension are shown in Figures 10.1 and 10.2. The ground-state curve is drawn to give the AE_a and crosses on the backside of the neutral, making it $M(2)$ and $Mc(2)$. The first excited-state curve, $M(1)$, is drawn to give a positive E_a and crosses on the frontside of the neutral. The bonding excited-state curve leading to the H(−) limit is $D(0)$ and $Mc(0)$. It is possible for the parent negative ion to be formed in a vertical process if the ion is stabilized to the dipole bound state before dissociation takes place. The two highest excited states lead to dissociation, giving the H(−) or CH_2NO_2(−) ions.

In the C—N dimension anion curves are drawn to four different dissociation limits: the two complementary limits of NO_2(−) and CH_3 and two limits leading to excited states of NO_2(−) and CH_3. Dipole bound states could also be drawn in these dimensions. The ground-state curves and dipole bound curves are $M(2)$. The lower excited-state curve A is $M(1)$ since the EDEA and VE_a are negative but the E_a is positive. More than three data points from PES, ECD, EI, or anion absorption and emission spectra define the ground state and the first excited valence state. The B state is $D(0)$ but could lead to molecular anion formation via the C, D,

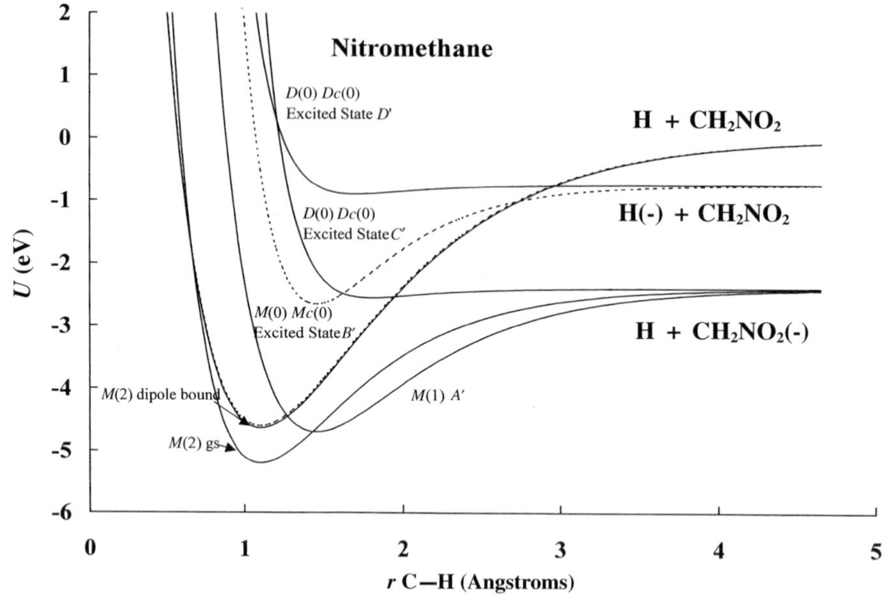

Figure 10.1 Morse potential energy curves for CH_3NO_2 in the C—H dimension. The curves are drawn from experimental data for the E_a and gas phase acidity. The details are given in [3].

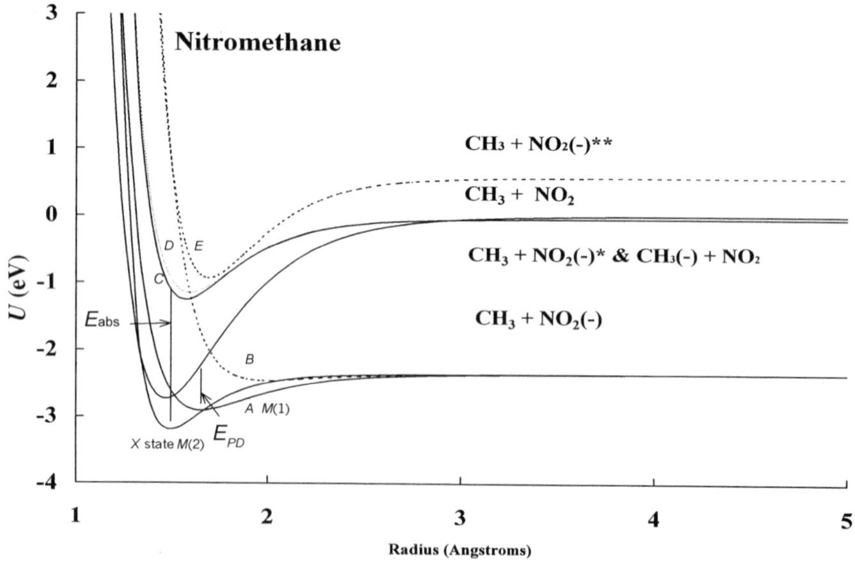

Figure 10.2 Morse potential energy curves for CH_3NO_2 in the C—N dimension. The curves are drawn from experimental data for the E_a and electron impact and negative-ion absorption data. The details are given in [3].

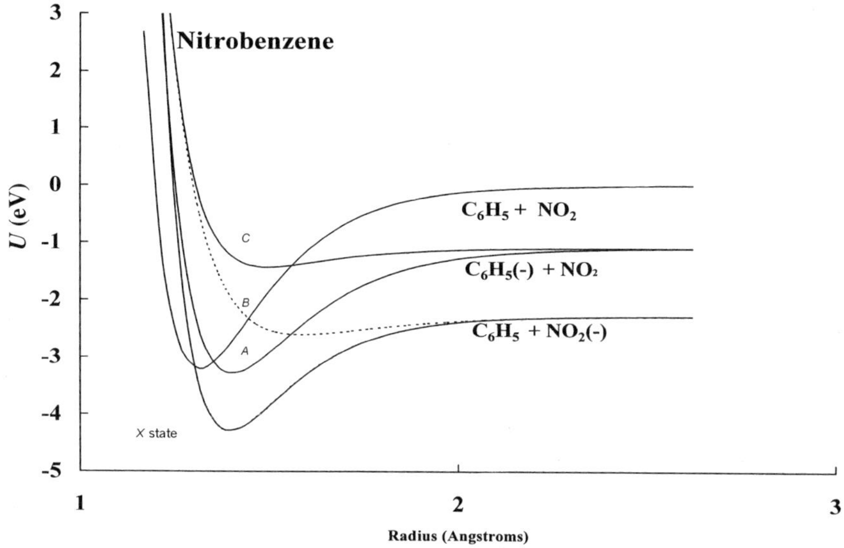

Figure 10.3 Morse potential energy curves for $C_6H_5NO_2$ in the C—N dimension. The curves are drawn from experimental data for the E_a and electron impact data [14–19].

or *E* states. The higher excited-state curves are *M*(0) and *Mc*(0) curves. This stands in contrast to the C' and D' curves shown in Figure 10.1 that are *D*(0) and *Dc*(0). Only electron impact data are available for the higher curves so they are constructed using assumed bond orders or internuclear distances [3].

The electron affinity of nitrobenzene has been determined by the ECD, NIMS, PES, and TCT methods [14–18]. The more recent values are the most precise, but agree in magnitude with the earliest value determined in 1957. The electron impact spectrum for nitrobenzene shows peaks for the molecular ion at 0.03, 3.3, and 6.9 eV. The most intense ion is $NO_2(-)$ with peaks at 1.2 eV and 3.5 eV. The phenyl anion is observed at 3.6 eV and 6.0 eV. Based on these data, curves—two "bonding" and two "antibonding"—are drawn, $C_6H_5 + NO_2(-)$ and $C_6H_5(-) + NO_2$ [19]. These are shown in Figure 10.3 as examples in Chapter 5. The first excited-state curve, designated *A*, has been drawn to cross the neutral curve close to the minimum. The ground-state curve is *M*(2), while the *A* curve is *M*(1) and the higher states are *D*(0). The calculated distributions for the curves reproduce experiment, as shown in Figure 10.4. The peak at 1.2 eV is a vertical onset since the −EDEA is 3.1 − 2.3 eV = 0.8 eV. The onset for the formation of phenyl(−) is 3.1 − 1.1 eV = 2.0 eV, as shown in Figure 10.4. The origin of the peak at approximately 6.0 to 6.5 eV is unknown, but could be a curve leading to the excited state of the phenyl anion or NO_2 anion. These excited-state curves are given as qualitative indications to be compared to those for nitromethane. These curves are good representations of the dissociation in the C—N dimension because the geometry of the phenyl ring does not change significantly upon the dissociation of $NO_2(-)$.

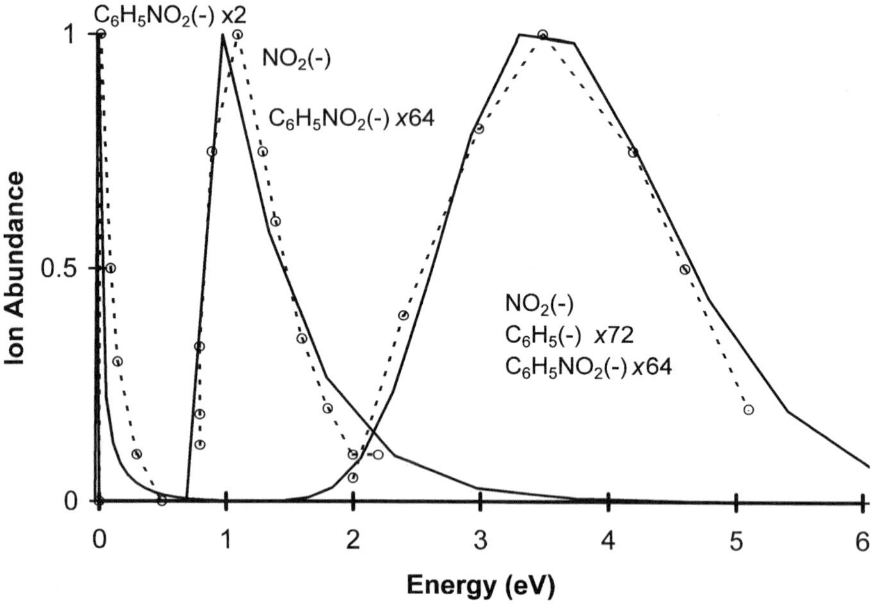

Figure 10.4 Calculated and experimental ion distributions for the electron impact of $C_6H_5NO_2$, from [19] and this book.

10.3 ELECTRON AFFINITIES DETERMINED USING THE MAGNETRON, ALKALI METAL BEAM, PHOTON, AND COLLISIONAL IONIZATION METHODS

10.3.1 Electron Affinities Determined Using the Magnetron Method

The electron affinities of organic charge transfer complex acceptors in the gas phase were first determined using the magnetron method. This method is regarded as unreliable due to the lack of mass identification. However, the MGN E_a for H, O, S, F, Cl, Br, I, CCl_4, NO, tetracyanoethylene, dicyanoethylene, dicyanobenzene, and tetracyanoquinodomethane are equal to the confirmed EvV to ± 0.15 eV. The magnetron values for benzoquinone, anthraquinone, fluoranil, and chloranil are assigned to excited states. The values for C_6F_6 and SF_6 are higher than the EvV by about 0.3 eV. The E_a of some cyanocarbons and s-trinitrobenzene have not been determined by any other technique. They have been confirmed using CURES-EC and reduction potentials. The regularity of the substitution effects of the cyano and nitro groups on ethylene and benzene shown in Figure 10.5 further supports the magnetron values. Table 10.1 lists the magnetron E_a. The regular change in the electron affinities of the chloromethanes with the number of chlorine atoms provides support for these values. The CURES-EC values reproduce these

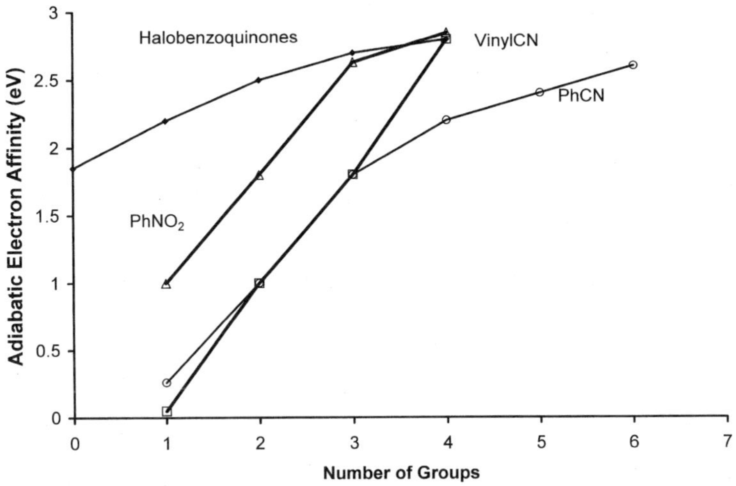

Figure 10.5 The electron affinities of halogenated benzoquinones, cyanoethylene, cyanobenzenes, and nitrobenzenes versus the number of substituents, data from [20].

TABLE 10.1 Evaluated Electron Affinities (in eV) Using the Magnetron [20, 58]

Molecule	Selected	NIST	MGN	$E_a(E_{1/2})$	C-EC
Hexacyanobutadiene	3.29(15)	3.291	3.24(15)	3.30(10)	3.3
TCNE	2.95(10)	3.166	2.87(15)	2.90(10)	3.0
TCNQ	2.80(6)	2.800	2.85(15)	2.80(10)	2.8
Hexacyanobenzene	2.54(15)	2.541	2.48(15)	—	2.7
Chloranil-p	2.77(5)	2.775	2.40(15)	2.65(10)	2.8
Fluoranil-p	2.70(10)	2.702	2.16(15)	2.60(10)	2.7
Fluorobenzoquinone	2.20(15)	—	2.20(15)	2.22(10)	2.2
s-Tetracyanobenzene	2.20(15)	2.203	2.15(15)	2.20(10)	2.2
s-Tetracyanopyridine	2.17(15)	2.173	2.11(15)	—	2.5
s-Trinitrobenzene	2.63(15)	2.628	2.30(15)	2.60(10)	2.6
p-Benzoquinone	1.860(5)	1.860	1.34(15)	1.80(5)	1.9
9,10 Anthraquinone	1.59(6)	1.591	1.15(15)	1.60(10)	1.6
o-Dicyanobenzene	1.03(10)	0.954	1.11(15)	1.00(10)	1.0
t-Dicyanoethylene	1.00(10)	1.249	0.75(15)	—	1.0
SF_6	1.07(5)	1.070	1.39(15)	—	1.0
NO	0.86(10)	0.026	0.83(15)	—	0.8
CCl_4	2.04(10)	0.805	2.06(15)	—	2.2
CBr_4	2.06 (10)	2.060	2.03(15)	—	2.2
$CHCl_3$	1.75(15)	0.622	1.75(15)	—	1.6
CH_2Cl_2	1.30(15)	—	1.30(15)	—	1.25
C_2Cl_6	1.80(15)	1.479	1.48(15)	—	2.0
C_6F_6	0.86(2)	0.520	1.20(15)	0.90(10)	0.85
C_6F_5	3.18(10)	3.183	3.2(1)	—	3.20
SF_5	3.80(3)	3.800	3.7(1)	—	3.80

TABLE 10.2 Evaluated Electron Affinities (in eV) Using the AMB Method [2, 21]

Molecule	AE_a	NIST	AMB	$E_{1/2}$	C-EC
TCNQ	2.80(8)	2.800	2.80(20)	2.80(10)	2.8
Bromanil-o	2.60(20)	2.440	2.44(20)	2.70(10)	2.7
Chloranil-p	2.77(5)	2.775	2.76(20)	2.65(10)	2.8
Fluoranil-p	2.70(10)	2.702	2.91(20)	2.60(10)	2.7
p-Benzoquinone	1.860(5)	1.990	1.89(20)	1.80(5)	1.9
NO	0.86(10)	0.026	0.10	—	0.8
CH_3NO_2	0.50 (2)	0.486	0.45	—	0.50
Cl_2	2.45(2)	2.400	2.45	—	2.43
Br_2	2.57(2)	2.550	2.57	—	2.58
I_2	2.524(5)	2.524	2.52	—	2.33
O_2	1.07(10)	0.450	0.45	—	1.1
CS_2	0.89(2)	0.512	1.00	—	0.85
NO_2	2.273(5)	2.273	2.25	—	2.25
Maleic anhydride	1.44(9)	1.440	1.40	1.42(5)	1.4
CCl_4	2.06(15)	0.80	2.06	—	2.20
CH_3I	1.30(10)	—	0.30	—	0.00
CF_2Cl_2	1.3(2)	—	0.40	—	—
CF_3Br	1.8(3)	—	0.91	—	—
CF_3I	2.20(20)	—	2.20	—	2.3

The uncertainties in the AMB values are nominally quoted as 0.20 eV, but the agreement is often better.

systematic changes in the E_a, as shown in Table 10.1. These will be discussed further in Chapter 11.

Only the E_a for SF_5 and C_6F_5 listed in Table 10.1 are dissociative capture values. These E_a were first determined by the magnetron method in the 1960s. The ECD and NIMS values are the most recent and stand in agreement with the earlier magnetron values ± 0.15. The MGN E_a for the C_6F_5 radical uses more recent bond dissociation energies to update the original value. Other magnetron E_a for phenyl radicals with three or more halogens on the ring are reasonable since they fall between 2.0 eV and 3.0 eV. Some of these have not been determined by any other method [1, 2, 20].

10.3.2 Electron Affinities Determined Using the AMB Method

The AMB E_a are listed in Table 10.2 [21–32]. The AMB method confirmed the magnetron value for tetracyanoquinodimethane and CCl_4. It obtained higher values for the quinones that are equal to the EvV. The AMB values of maleic anhydride, Cl_2, Br_2, I_2, NO_2, and CS_2 are also equal to the current EvV. The AMB E_a for CCl_4 and CF_3I of about 2 eV are for the ground state. However, two lower values for CF_3I have also been reported using AMB studies. The uncertainty in the AMB values is often quoted as ± 0.20 eV, but their agreement with the EvV is often better than this. The values for the other alkyl halides could exist for an excited state. These will be discussed further in Chapter 11.

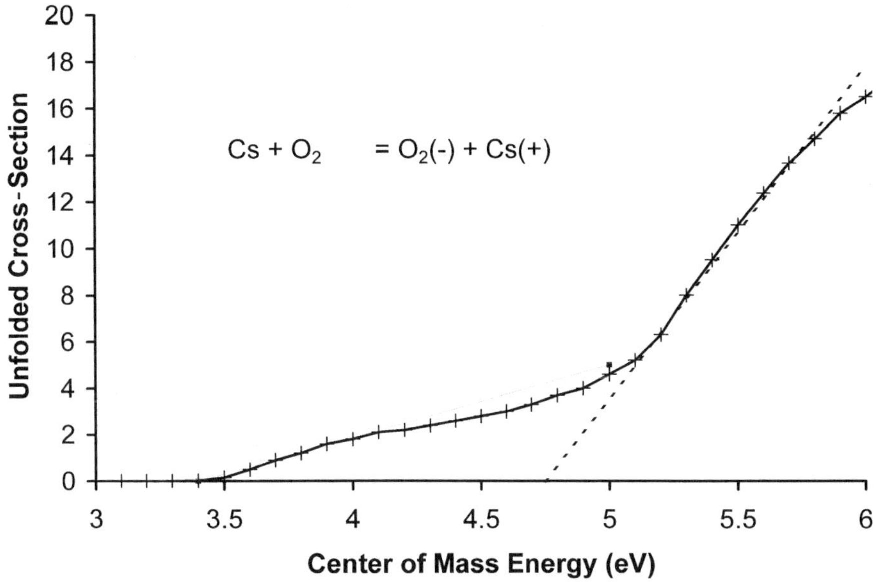

Figure 10.6 Unfolded cross-section for Cs beam reaction with O_2, from [33]. The higher extrapolated appearance potential is AP = IP + E_1. With the 3.89 eV ionization potential of Cs this gives $E_1 = 4.75 - 3.89$ eV = 0.86 eV.

The AMB E_a for O_2 and NO are for excited states. In 1971 Freeman postulated that structure observed in the unfolded cross-section for the reaction of a Cs beam with O_2 could give a measure of the activation energy for electron attachment to form the ground state of $O_2(-)$. In this case the extrapolated appearance potential is AP = IP + E_1. As shown in Figure 10.6, the two AP are 3.44 eV and 4.75 eV. With the 3.89 eV ionization potential of Cs, the first gives $E_a = 3.89 - 3.44 = 0.45$ eV for the excited state and the second $E_1 = 4.75 - 3.89 = 0.86$ eV. This illustrates another interpretation of AMB data to give fundamental experimental information for thermal electron attachment reactions. This E_1 has been subsequently measured and confirmed by isotopic exchange experiments and is incorporated in the ECD data analysis [33].

10.3.3 Electron Affinities Determined Using Photon Methods

Very few electron affinities of large organic molecules have been determined using photoelectron spectrometry or photodetachment techniques. These are summarized in Table 10.3 [34–47]. Limits to the electron affinities of a large number of nitrobenzenes have been determined using the ECD photodetachment technique. Because these are known to be upper limits, they are only assigned to the AE_a in the absence of other measurements. The evaluated E_a of the dinitrobenzenes, nitrotoluenes, and halogenated nitrobenzenes are in agreement with the upper limits

TABLE 10.3 Evaluated Molecular Electron Affinities (in eV) Using the Photon Method

Molecule	AE_a	NIST	PES	PD	C-EC
Tetracyanoethylene	2.90(10)	3.166	—	2.0	2.90
p-Benzoquinone	1.860(5)	[1.990]	1.86	1.99	1.9
SF_6	1.07(5)	1.070	—	3.16	1.0
SF_6	1.07(5)	1.070	—	0.54	1.0
NO	0.86(10)	0.026	0.026	—	0.8
C_6F_6	0.86(2)	0.520	0.80	—	0.85
CH_3NO_2	0.50 (2)	0.486	0.26	—	0.50
I_2	2.524(5)	2.524	—	—	2.33
O_2	1.07(10)	0.450	0.450	0.15	1.10
CS_2	0.89(2)	0.512	0.895	—	0.85
NO_2	2.273(5)	2.273	2.273	—	2.25
Nitrobenzene	1.00(1)	1.006	1.00	1.05	1.0
c-C_8H_8	0.80(10)	0.550	0.65	0.83	0.80
Perylene	0.973(5)	0.973	0.973	—	1.00
Tetracene	1.08(4)	1.067	1.067	—	1.08
Naphthalene	0.16(2)	−0.20	−0.20	—	0.15
Coronene	0.80(5)	0.470	0.470	—	0.68
Azulene	0.84(5)	0.694	0.790	—	0.90
Anthracene	0.68(2)	0.530	0.530	—	0.68

established by the ECD photodetachment technique. Since many have only been determined in the gas phase by TCT methods, these are important confirming values [35]. The photodetachment and PES E_a of p-benzoquinone and nitrobenzene are equal to the EvV. Two photodetachment values are reported for SF_6, one larger than the AE_a due to the change in geometry of the anion and the other lower than the AE_a due to an excited state. The PD E_a reported for tetracyanoethylene is significantly lower than the AE_a. The excited-state E_a of O_2 and NO have been determined using PES. The PES values for C_6F_6, CS_2, SO_2, NO_2, and other small molecules agree with the EvV [36–40].

The electron affinities of several aromatic hydrocarbons, including naphthalene, azulene, anthracene, tetracene, perylene, and coronene, have been measured using PES, and the E_a of c-C_8H_8 has been measured by both PES and photodetachment. The E_a for azulene, perylene, and tetracene agree with the AE_a [34, 43, 44]. The values for naphthalene, anthracene, and coronene are lower than the AE_a by more than the experimental uncertainty and have been assigned to excited states [42, 43, 46]. The E_a for naphthalene is obtained by extrapolating the PES peaks found in anion hydrates to zero water molecules. This gives a negative electron affinity. In the PES of naphthalene, anthracene, and coronene there are indications of a higher electron affinity. In the case of naphthalene and anthracene the intensities of the higher-energy peaks are lower than for the excited state. In the case of coronene the two sets of peaks are of equivalent intensity, as shown in Figure 6.8.

TABLE 10.4 Evaluated Molecular Electron Affinities (in eV) of Selected Hydrocarbons Using the Collisional Ionization Method [51–53]

Molecule	AE_a	NIST	C-Ion	$E_{1/2}$	C-EC
Biphenylene	0.40(5)	(0.89)	0.89(10)	0.40(5)	0.45
Anthracene, 1-CH_3	0.65(2)	0.550	0.55(10)	0.65(10)	0.65
Benz[a]anthracene	0.72(2)	0.390	0.39(10)	0.72(5)	0.74
Coronene	0.80(10)	0.470	0.47(10)	0.74(5)	0.80
1,3,5,7-c-C_8H_8	0.80(10)	0.550	0.58(10)	—	0.80
Benzo[ghi]perylene	0.89(10)	0.420	0.42(10)	0.90(10)	0.90

The photodetachment and PES studies of c-C_8H_8 are interesting because the E_a determined from the first observed peak in the PES is larger than the adiabatic E_a. This occurs because the geometry of the neutral formed by photoionization is different from that of the ground state of the neutral. The anion is planar, while that of the neutral is bent. The lowest-energy peak at 1.1 eV corresponds to the vertical transition from the planar anion to the planar neutral. By estimating the energy to form the planar neutral, an E_a of 0.65 eV was obtained. Thermal charge transfer and kinetic fragmentation values are 0.55(2) eV and 0.60(10) eV, respectively. The photodetachment value is less than 0.82 eV. The E_a for c-C_8H_8 was reported to be 0.58(2) eV from the single-state ECD model. However, the highest-temperature data point indicates a higher-energy state. The lower value is assigned to an excited state of the anion and the ground-state E_a is closer to the PD onset. The CURES-EC value is 0.8 eV. The uncertainty in the E_a, ± 0.1 eV, encompasses all the values at twice the uncertainty. It is larger than the nominal value because of the approximations [47–51].

10.3.4 Electron Affinities Determined Using Collisional Ionization Methods

The relative electron affinities of a number of aromatic hydrocarbons have been measured by the collisional ionization of electron bound dimers. Table 10.4 gives the E_a of some aromatic hydrocarbons determined by the collisional ionization method listed in NIST [51–53]. These values are lower than the AE_a and assigned to excited states. The E_a of biphenylene is significantly larger than the E_a determined from reduction potentials and cannot be explained. One possibility is that the ion is an isomer of biphenylene and acenaphthylene. To illustrate this point, a READS-TCT calculation is made for the two isomers. The three-dimensional spin densities are shown in Figure 10.7. When an electron is added to the system, the charge density on the acenaphthylene is 0.8q, while that for the biphenylene is 0.2q. This indicates that the E_a of acenaphthylene is greater than that of biphenylene. The CURES-EC calculations agree with this order. The ions observed in the collisional ionization could be either of the two isomers. The observation of excited-state E_a by collisional ionization of aromatic hydrocarbons with fullerenes and with bromosubstituted aromatic hydrocarbons has been observed [52, 53].

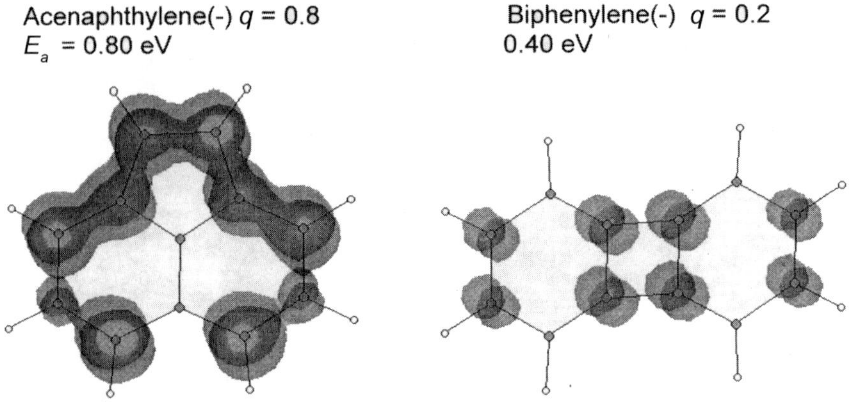

Figure 10.7 READS-TCT calculation three-dimensional spin densities for acenaphthylene and biphenylene with one electron added. Eighty percent of the spin resides on acenaphthalene, indicating that the E_a of acenaphthylene is greater than that of biphenylene.

10.4 ELECTRON AFFINITIES DETERMINED USING THE ECD, NIMS, AND TCT METHODS

10.4.1 Electron Affinities of Aromatic Hydrocarbons by the ECD Method

Many of the E_a determined in the ECD are for molecules with only an α region. A large number of these compounds have the same intercept, designated the common intercept, from which the E_a is calculated. The alternate approach is to estimate the temperature range of the α region and to obtain the "best" slope and intercept through the data. Some of these intercepts are higher than the weighted average intercept and some of them are lower. The lower values can be explained by a Q_{an} value lower than 1. The higher values imply a Q_{an} value greater than 1. This is not likely. For these compounds the common intercept value is chosen. For the compounds with intercepts significantly lower than the weighted average the selected value is determined by the fit through the data. The random uncertainties are larger than those obtained using a fixed intercept. For compounds with only two regions the variable intercept values were obtained using a least-squares fit to the high-temperature data. In 1995 the ECD E_a for 20 hydrocarbons were verified with reduction potential data. The E_a were compared to E_a determined from other gas phase techniques and reduction potentials for 7 fullerenes and 15 metallophorphyrins and metal complexes [54]. At that time the E_a of anthracene, tetracene, perylene, and pentacene had been measured by TCT [55]. This led to the postulate of two states for tetracene and anthracene and the assignment of some of the ECD values to excited states [56–59].

In 1999 the E_a were further verified using CURES-EC calculations and electronegativity considerations. The E_a for another 50 aromatic hydrocarbons were

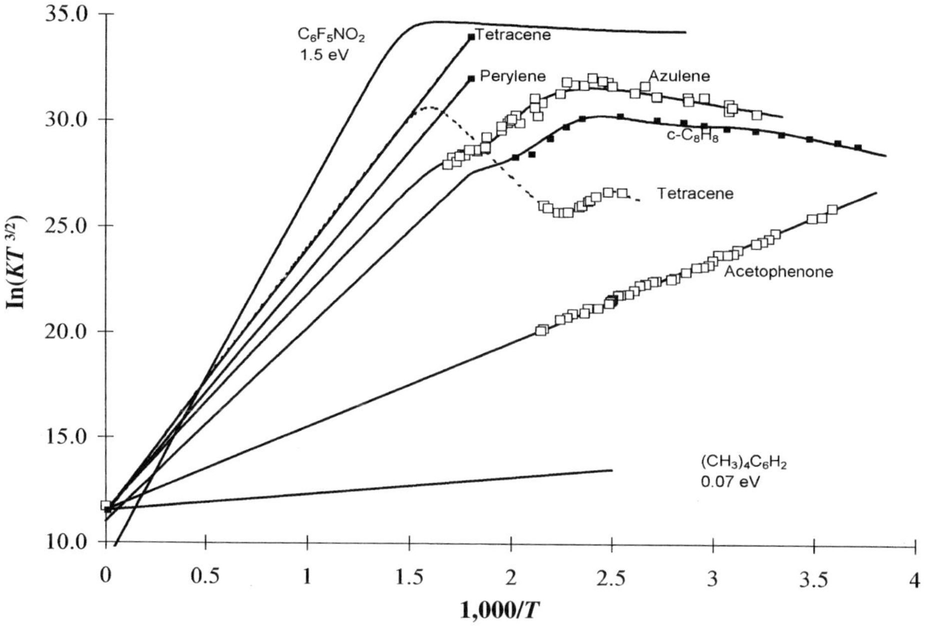

Figure 10.8 Plots of ECD data as ln $KT^{3/2}$ versus 1,000/T for azulene, tetracene, and C_8H_6 with multiple α and β regions. Acetophenone, tetramethylbenzene, $C_6F_5NO_2$, tetracene, and perylene are given for comparison. Table 10.5 provides the parameters. From [39, 49, 57, 59].

determined from a set of reduction potentials published by I. Bergman. M. J. S. Dewar and co-workers had calculated the E_a for all these molecules using MINDO/3. These are referred to as the Bergman Dewar set [60–62]. A complete listing of the AE_a for the Bergman Dewar set is included in the appendix. Excited-state electron affinities have been observed for anthracene, cyclooctaterene, acenaphthylene, fluoranthene, azulene, and tetracene in the ECD.

In Figure 10.8 ECD data for acetophenone, azulene, c-C_8H_8, and tetracene and calculated responses for perylene, tetramethylbenzene, and $C_6F_5NO_2$ are shown. Table 10.5 gives the parameters used to calculate the curves for azulene, c-C_8H_8, and tetracene. The lower values are assigned to excited states. Table 10.6 provides the AE_a of the hydrocarbons reported in the literature determined using the ECD, the NIST values, reduction potentials, and CURES-EC. The NIST values are those returned from a sequential search for molecules containing only C and H. The ECD values are confirmed by the $E_{1/2}$ and CURES-EC values. The largest differences between the NIST E_a and the assigned AE_a exist for naphthalene and the E_a determined by PES and collisional ionization methods. The NIST ECD E_a for acenaphthylene is known to occur for an excited state based on the use of the constant current mode of detection and comparisons with the fixed frequency data [57–59]. The PES E_a for anthracene and coronene, and the collisional ionization E_a for benzanthracene and benzo[g,h,i]perylene are lower than the EvV by more

TABLE 10.5 Experimental Kinetic and Thermodynamic Properties of Hydrocarbons from ECD Data [57]

Compound	E_a (eV)	E_1 (eV)	ln A_1	Q_{an}
c-C_8H_8	0.80(10)	0.36	36.8	1
c-$C_8H_8^*$	0.70(2)	0.20	36.8	1.0
c-$C_8H_8^{**}$	0.57(2)	0.10	34.6	0.1
Azulene	0.84(10)	0.22	35.6	1
Azulene*	0.78(2)	0.13	35.6	1
Azulene**	0.70(2)	0.13	35.6	1
Azulene***	0.60(5)	0.06	35.6	1
Tetracene	1.08(4)	0.95	37.0	1
Tetracene*	0.90(4)	0.80	36.0	1×10^{-4}
Tetracene**	0.53(4)	0.65	36.0	1

* First excited state.
** Second excited state.
*** Third excited state.

than the uncertainty and are assigned to excited states, as shown in Tables 10.3 through 10.6. The remaining values are the same as the ECD values.

Table 10.7 gives the AE_a for hydrocarbons determined from reduction potentials, electronegativities, and CURES-EC [56]. The MINDO/3 E_a originally reported by Dewar are given for comparison [60]. Also shown are the values of mddG and the group classification used to obtain the E_a from the $E_{1/2}$ values. The nominal random uncertainty in the E_a is 0.03 eV. This is one-half the differences between each group. The CURES-EC values were adjusted to the $E_{1/2}$ E_a and have a deviation of less than 0.03 eV. The average deviation between the $E_{1/2}$ E_a and the MINDO/3 values is 0.07 eV. The E_a obtained from electronegativity considerations is determined from $\mathrm{EN} = (\mathrm{IP} + E_a)/2$ so $E_a = 2\mathrm{EN} - \mathrm{IP}$, where IP is the ionization potential. The EN of the alternate aromatic hydrocarbons is nominally 4.00 eV. The E_a calculated in this manner agree with the $E_{1/2}$ E_a with an average deviation of 0.08 eV. Some of the IP have not been measured. The process can be reversed and the IP obtained from $\mathrm{IP} = 8.00 - E_a$. For the molecules containing five membered rings the value of EN is larger, 4.37 eV. The IP for fluoranthenes can be calculated from $\mathrm{IP} = 8.74 - E_a$. These are especially interesting values because they are precursors to the fullerenes. The electronegativity of the fullerenes is 4.5 eV so $\mathrm{IP} = 9.0 - E_a$ [57].

10.4.2 Electron Affinities of Organic Carbonyl Compounds by the ECD Method

Besides the aromatic hydrocarbons the E_a of a number of aromatic carbonyl compounds have been determined using the ECD. These include substituted acetophenones, benzaldehydes, benzophenones, benzoates, phthalates, acetonaphthone, naphthaldehydes, and anthracene and phenanthrene aldehydes. Like the aromatic hydrocarbons, the majority of these compounds only undergo nondissociative

TABLE 10.6 Evaluated Electron Affinities (in eV) of Hydrocarbons Using the ECD Method [57]

Molecule	AE_a	NIST	ECD	$E_{1/2}$	C-EC
1,2,4,5-tetraMeBz	0.07(2)	0.048	0.07(2)	—	0.05
Styrene	0.10(5)	—	0.10(5)	0.12(5)	0.10
1,2,3,5-tetraMeBz	0.10(5)	0.108	0.10(5)	—	0.08
Benzene, hexamethyl-	0.12(5)	0.121	0.12(5)	—	0.10
Biphenyl	0.13(2)	0.130	0.13(5)	0.10(3)	0.14
2-Me-naphthalene	0.14(5)	0.143	0.14(5)	—	0.15
1-Me-naphthalene	0.16(5)	0.147	0.16(5)	—	0.17
Naphthalene	0.16(2)	−0.200	0.16(5)	0.17(3)	0.15
Diphenylmethane	0.16(5)	0.156	0.16(5)	—	0.18
2,6-Di-Me-naphthalene	0.16(5)	0.160	0.16(5)	—	0.15
2,3-Di-Me-naphthalene	0.17(5)	0.173	0.17(5)	—	0.18
Indene	0.17(5)	0.173	0.17(5)	—	0.18
Benzene, pentamethyl-	0.18(5)	0.182	0.18(5)	—	0.17
Naphthalene, 2-ethyl	0.19(5)	0.195	0.19(5)	—	0.19
1,4-Di-Me-naphthalene	0.25(5)	0.247	0.25(5)	—	0.23
Fluorene	0.28(5)	0.278	0.28(5)	—	0.30
Triphenylene	0.29(2)	0.285	0.29(2)	0.29(3)	0.27
Phenanthrene	0.30(2)	0.307	0.30(2)	0.31(3)	0.31
Diphenylethyne	0.32(5)	0.321	0.32(5)	—	0.35
Ethylene-1,1 diphenyl	0.39(5)	0.390	0.39(5)	—	0.36
Stillbene	0.39(5)	0.390	0.39(5)	—	0.40
Chrysene	0.42(4)	0.397	0.42(4)	0.42(3)	0.43
Picene	0.54(3)	0.542	0.50(3)	0.49(3)	0.50
Benz[e]pyrene	0.55(3)	0.534	0.55(3)	0.56(3)	0.56
Benzo[c]phenanthrene	0.58(1)	0.545	0.58(1)	0.58(3)	0.60
Pyrene	0.61(2)	0.500	0.61(2)	0.63(3)	0.62
Dibenz[a,j]anthracene	0.67(3)	0.591	0.67(3)	0.65(3)	0.70
Dibenz[a,c]anthracene	0.69(5)	—	0.69(5)	0.69(3)	0.68
Anthracene	0.68(1)	0.530	0.68(1)	0.72(3)	0.70
Dibenz[a,h]anthracene	0.69(3)	0.595	0.69(3)	0.65(3)	0.66
Benz[a]anthracene	0.72(2)	0.390	0.72(2)	0.72(3)	0.74
Benz[a]pyrene	0.80(3)	0.815	0.80(3)	0.79(3)	0.80
1,3,5,7-c-C_8H_8	0.80(10)	0.550	0.80(10)	0.83(5)	0.8
Acenaphthylene	0.80(2)	0.403	0.80(2)	0.80(10)	0.8
Fluoranthene	0.82(4)	0.630	0.82(4)	0.83(3)	0.81
Azulene	0.84(10)	0.694	0.90(10)	0.78(5)	0.78
Perylene	0.973(5)	0.973	0.98(10)	1.01(3)	1.00
Tetracene	1.08(4)	1.067	1.08(4)	1.09(3)	1.08

capture. The exceptions are the chloroacetophenones. Some of these have been used as reference points for calibrating the TCT scale. Others have been measured by both the TCT and ECD methods. The NIST tables list the majority of the E_a for these compounds. However, the NIST E_a are single-state values. In the data for the F-acetophenones structure is observed so the analysis has been revisited. The

TABLE 10.7 Electron Affinities (in eV) of Aromatic Hydrocarbons from Reduction Potentials [56]

Molecule	$(E_{1/2})^a$	mddG	Group	C-EC	MINDO/3^b	EN
p-Terphenyl	0.27	2.00	A	0.30	0.06	0.05
p-Quarterphenyl	0.41	2.00	A	0.41	0.57	0.42
Tribenzanthracene	0.61	2.00	A	0.62	0.61	0.60
Dibenz[b,g]phenanthrene	0.78	2.00	A	0.78	0.76	—
Benz[j]fluoranthene	1.01	2.00	A	1.00	1.02	—
Dibenz[b,e]pyrene	1.10	2.00	A	1.08	0.93	0.93
Benz[a]perylene	1.24	2.00	A	1.25	1.25	1.29
Benz[g]chrysene	0.58	1.95	A−	0.56	0.60	—
Naphtho[2,3,b]pyrene	0.73	1.95	A−	0.70	0.52	0.65
Benz[b]fluoranthene	0.95	1.95	A−	0.96	0.94	—
Dibenz[a,c]tetracene	1.02	1.95	A−	1.03	0.95	1.04
Benz[a]tetracene	1.06	1.95	A−	1.06	1.00	1.00
Naphtho[2,3,j]fluoranthene	1.17	1.95	A−	1.19	1.15	—
Benz[a]fluoranthene	1.25	1.95	A−	1.27	1.23	—
Dibenz[a,c]pentacene	1.30	1.95	A−	1.29	1.25	1.38
s-indenofluoranthene	1.33	1.95	A−	1.30	1.37	—
Dibenz[a,f]fluoranthene	1.58	1.95	A−	1.60	1.60	—
Dibenz[e]pyrene	0.59	1.90	mAF	0.59	0.56	0.61
Dibenz[c,m]pentaphene	0.76	1.90	mAF	0.77	0.78	—
Benz[p]pentaphene	0.78	1.90	mAF	0.77	0.75	0.80
Dibenz[a,e]pyrene	0.88	1.90	mAF	0.86	0.85	0.89
Benz[k]fluoranthene	0.89	1.90	mAF	0.88	0.83	—
Naphto[2,3]fluoranthene	0.90	1.90	mAF	0.89	0.85	—
Dibenz[a,h]tetracene	1.03	1.90	mAF	1.02	0.96	1.01
Hexaphene	1.06	1.90	mAF	1.05	0.97	1.08
Dibenz[a,n]perylene	1.28	1.90	mAF	1.30	1.30	1.36
Dibenz[a,h]pentacene	1.43	1.90	mAF	1.43	1.24	1.36
Dibenzphenanthrene	0.58	1.90	F+	0.58	—	0.53
Benz[c]chrysene	0.60	1.85	F+	0.60	0.50	0.50
Pentaphene	0.75	1.85	F+	0.73	0.71	0.73
Benz[b]chrysene	0.89	1.85	F+	0.90	0.80	0.86
Dibenz[def,mno]chrysene	1.14	1.85	F+	1.14	1.18	1.10
Dibenz[a,h]pyrene	1.16	1.85	F+	1.16	1.19	1.18
Tetrabenz[a,cd, j,lm]perylene	1.18	1.85	F+	1.21	1.19	1.07
Benz[a]pentacene	1.38	1.85	F+	1.37	1.29	1.39
Dibenz[bc,hl]coronene	1.41	1.85	F+	1.38	1.39	1.50
Dibenz[a, j]perylene	1.45	1.85	F+	1.44	1.46	1.50
Dinaphtho[a,h]pyrene	1.55	1.85	F+	1.53	1.52	1.58
Benzindenofluoranthene	1.73	1.85	F+	1.72	1.70	—
Dibenz[pg,p]chrysene	0.79	1.80	F	0.77	0.78	0.80
Dibenz[b,pqr]perylene	0.86	1.80	F	0.83	0.82	0.88
Benz[g,h,I]perylene	0.89	1.80	F	0.89	0.75	0.85
Dinaphth[p]pentaphene	0.91	1.80	F	0.90	0.80	0.86
Benznaphthopentacene	1.03	1.80	F	1.01	0.96	—
Tribenz[a,e,i]pyrene	1.03	1.80	F	1.02	0.96	1.01

TABLE 10.7 *(Continued)*

Molecule	$(E_{1/2})^a$	mddG	Group	C-EC	MINDO/3[b]	EN
Dibenz[b,k]chrysene	1.05	1.80	F	1.06	0.96	1.03
Dibenz[fq,qr]tetracene	1.10	1.80	F	1.08	1.05	1.04
Naphtho[2,3,e]pyrene	1.23	1.80	F	1.21	1.15	1.28
Pyranthene	1.32	1.80	F	1.32	1.38	—
Tetrabenzpentacene	1.68	1.80	F	1.66	1.60	1.73
Tetrabenzpentaphene	1.70	1.80	F	1.68	1.67	1.78
Benztetrabenzpentaphene	1.90	1.80	F	1.89	—	1.94
Diindeneotetracene	2.04	1.80	F	2.05	2.04	—
Dibenz[cd,lm]perylene	1.25	1.75	F−	1.26	1.25	1.28
Benz[x,y,z]heptaphene	1.28	1.75	F−	1.29	1.16	1.28
Tetracenotetracene	1.65	1.75	F−	1.64	1.64	1.60

structure is apparent in a plot of ECD data for several substituted acetophenones, as shown in Figure 10.9. For the trifluoromethylacetophenones recent TCT E_a can be used in the data analysis to improve the ECD parameters. The E_a of the TCT reference compounds, acetophenone, benzonitrile, methylbenzoate, and benzophenone are supported by reduction potentials.

The electron affinities of acetophenone and benzaldehyde have also been used as internal standards for ECD work. Many of the values of the substituted compounds were determined using acetophenone as an internal standard. Acetophenone was chosen because its temperature dependence was so well established, as shown in Figures 10.8 and 10.9. The value for acetophenone in the NIST table is

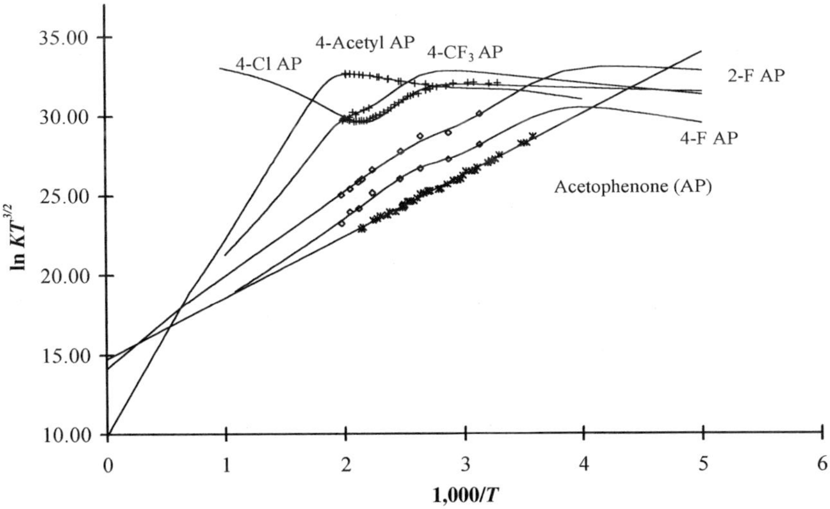

Figure 10.9 Plots of ECD data as ln $KT^{3/2}$ versus 1,000/T for acetophenones that show structure. From [58, 69, 72].

TABLE 10.8 Molecular Electron Affinities (in eV) of Acetophenones and Benzaldehydes Using the ECD Method

Molecule	AE_a	NIST	ECD	C-EC
Acetophenone	0.338(2)	0.334	0.338(2)	0.34
Acetophenone-o-F	0.49(3)	0.442	0.49(3)	0.45
Acetophenone-m-F	0.58(3)	0.577	0.58(3)	0.55
Acetophenone-p-F	0.52(5)	0.395	0.52(5)	0.50
Acetophenone-p-Cl	0.64(5)	0.585	0.64(5)	0.62
Acetophenone-m-Cl	0.67(5)	0.616	0.67(5)	0.66
Acetophenone-o-CF_3	0.79(5)	0.642	0.79(5)	0.83
Acetophenone-m-OH	0.77(2)	—	0.77(2)	0.76
Acetophenone-m-CF_3	0.79(5)	0.768	0.79(5)	0.80
Acetophenone-p-CF_3	0.89(9)	0.898	0.89(10)	1.00
Acetophenone-p-acetyl	1.06(9)	1.062	1.06(10)	1.15
Benzaldehyde	0.457(5)	0.429	0.457(5)	0.45
Benzaldehyde-m-F	0.67(3)	0.668	0.66(4)	0.64
Benzaldehyde-o-F	0.66(3)	0.637	0.66(4)	0.62
Benzaldehyde-p-F	0.57(5)	0.486	0.57(5)	0.55
Benzaldehyde-m-OCH_3	0.48(4)	0.429	0.48(4)	0.45
Benzaldehyde-m-CH_3	0.41(2)	0.408	0.43(2)	0.45
Benzaldehyde-p-CH_3	0.39(2)	0.373	0.39(2)	0.40
Benzaldehyde-2,4,6 triMe	0.44(4)	0.442	0.44(3)	0.47
Propiophenone	0.36(2)	0.351	0.36(1)	0.36

0.334 eV. This is slightly lower than the average of all the determinations for acetophenone, 0.338(2). Likewise, for benzaldehyde the average of multiple determinations is 0.457(5), while the NIST value is 0.429 eV. The NIST ECD values for benzonitrile and methylbenzoate are the evaluated values. The evaluated value for benzophenone is 0.68(5) eV, while the NIST value is 0.620 eV. All these are confirmed by CURES-EC. The values for the para-substituted chloro-, methoxy-, and methyl-substituted benzophenones are confirmed by reduction potentials. For compounds with two CF_3 groups on a benzene ring the available CURES-EC parameters give upper limits that are 0.3 eV higher than experiment.

Table 10.8 lists the E_a for the substituted acetophenones and benzaldehydes determined by ECD. For comparison the NIST and CURES-EC values are given. The revised ECD value for o-CF_3-acetophenone, 0.79(5) eV, is higher than the earlier 0.643 eV listed in the NIST tables for this same substance. The values for p-F acetophenone and p-F-benzaldehyde are also larger as a result of the assumption of two states. The ECD E_a for diacetyl-benzene and m- and p-CF_3-acetophenone are consistent with the TCT E_a.

Table 10.9 gives the E_a for substituted benzophenones and other carbonyl compounds determined using the ECD. Many of the E_a for the benzophenones are not

TABLE 10.9 Molecular Electron Affinities (in eV) of the Anisoles, Benzophenones, Benzonitriles, and Benzoates Using the ECD Method

Molecule	AE_a	NIST	ECD	CURES-EC
Benzophenone	0.68(5)	0.620	0.68(5)	0.68
Benzophenone-4-methoxy	0.61(5)	—	0.61(5)	0.60
Benzophenone-4-methyl	0.64(5)	—	0.64(5)	0.64
Benzophenone-4-ethyl	0.64(5)	—	0.64(5)	0.64
Benzophenone-p-F	0.70(5)	0.620	0.70(5)	0.70
Benzophenone-p-Cl	0.78(10)	—	0.78(10)	0.81
Benzophenone-p-Br	0.90(10)	—	0.90(10)	0.88
Benzophenone-p-I	1.10(20)	—	1.10(20)	1.10
Benzophenone-p-NO_2	1.50(20)	—	1.50(20)	1.50
Benzophenone-m-CF_3	1.00(10)	1.078	1.00(20)	1.20
Benzonitrile	0.26(2)	0.258	0.25(2)	0.26
Methyl benzoate	0.20(5)	0.180	0.20(5)	0.21
1-2-Dimethyl phthalate	0.60(10)	0.550	0.60(10)	0.60
1,3-Dimethyl phthalate	0.60(10)	0.550	0.60(10)	0.61
Diethyl phthalate	0.62(10)	0.540	0.62(10)	0.65
Anisole-pentafluoro	0.55(5)	0.542	0.55(5)	0.58
Anisole-tetrafluoro	0.25(5)	0.217	0.25(5)	0.28

listed in the NIST tables. The E_a of methyl benzoate and the esters of benzene di-carboxylic acids have been determined with the ECD. Other derivatives of these compounds have been determined by TCT measurements and scaled to the value for methyl benzoate. The E_a of more than 14 aromatic and aliphatic carbonyl compounds have been determined by both the ECD and TCT methods. These values and E_a for other carbonyl compounds determined by one of these methods are presented in Table 10.10. All the values agree within the stated uncertainties and are confirmed by CURES-EC [63–68].

The electron affinity of the acetate radical was determined from the –EDEA obtained from ECD data for ethyl acetate, benzyl acetate, and acetic anhydride. At the same time data were obtained for ethyl trifluoroacetate and ethyl trichlor-acetate, but were not analyzed or reported. These data are revisited with two states and new bond dissociation energies. This gives another example of how CURES-EC can be used to support experimental results [1, 69]. Figure 10.10 is a typical ECD plot of ln $KT^{3/2}$ versus 1,000/T for these compounds. In Table 10.11 the kinetic and thermodynamic parameters obtained from the data are shown. The values of $D - E_a$ (acetate) available in the literature were used in the data reduction procedure, and the rate constant and Q_{an} values are obtained from the ECD data. The A_1 values are close to the DeBroglie A_1. The Q_{an} values are less than 1.0. The E_1 values are less than 0.15 eV. The E_a are thus typical of the ECD, although in the lower range of E_a measured in the ECD for ethyl acetate, benzyl acetate, and acetic

TABLE 10.10 Evaluated Electron Affinities (in eV) of Esters and Carbonyl Compounds

Molecule	Selected	NIST	TCT	ECD	C-EC
Methyl benzoate-p-NO_2	1.48(9)	1.461	1.48(10)	—	1.53
Methyl benzoate-m-NO_2	1.25(9)	1.227	1.25(10)	—	1.19
Maleic anhydride	1.44(9)	1.440	1.44(5)	—	1.4
Phthalic anhydride	1.25(5)	1.245	1.25(5)	—	1.3
Methyl benzoate-p-CHO	1.18(9)	1.158	1.10(10)	—	1.15
Nitrobenzene-m-OCH_3	1.04(9)	1.040	1.04(10)	—	1.0
Nitrobenzene-p-OCH_3	0.91(9)	0.911	0.91(10)	—	0.89
Acetophenone-p-$COOCH_3$	0.96(9)	0.963	0.96(10)	1.0(10)	1.15
Acetophenone-p-Cl	0.64(5)	0.585	0.57(9)	0.64(5)	0.62
Acetophenone-m-Cl	0.67(5)	0.616	0.59(9)	0.67(5)	0.66
Acetophenone-o-CF_3	0.79(5)	0.642	—	0.79(5)	0.83
Acetophenone-m-OH	0.77(2)	—	—	0.77(2)	0.76
Acetophenone-m-CF_3	0.79(5)	0.768	0.77(9)	0.79(5)	0.80
Acetophenone-p-CF_3	0.89(10)	0.898	0.87(8)	0.89(10)	1.00
Methyl benzoate-p-$COOCH_3$	0.82(10)	0.824	0.82(10)	—	0.82
Methyl benzoate-m-$COOCH_3$	0.60(10)	0.550	—	0.60(10)	0.58
Ethyl trifluoroacetate	0.8(1)	—	—	0.8(1)	0.8
Biacetyl	0.70(5)	0.690	0.69(10)	0.71(3)	0.70
Dimethyl phthalate	0.60(10)	0.550	—	0.60(10)	0.61
Methyl benzoate	0.20(10)	0.180	—	0.20(10)	0.21
Acetic anhydride	0.21(5)	—	—	0.21(5)	0.20
Benzyl acetate	0.20(5)	—	—	0.20(5)	0.15
Ethyl acetate	0.20(5)	—	—	0.20(5)	0.20
1-Naphthaldehyde	0.69(5)	0.681	0.68(10)	0.69(5)	0.73
2-Naphthaldehyde	0.64(5)	0.642	0.64(10)	0.65(5)	0.68
Benzophenone	0.68(5)	0.620	0.68(5)	0.68(5)	0.66

The values in parentheses are the uncertainties in the last figure.

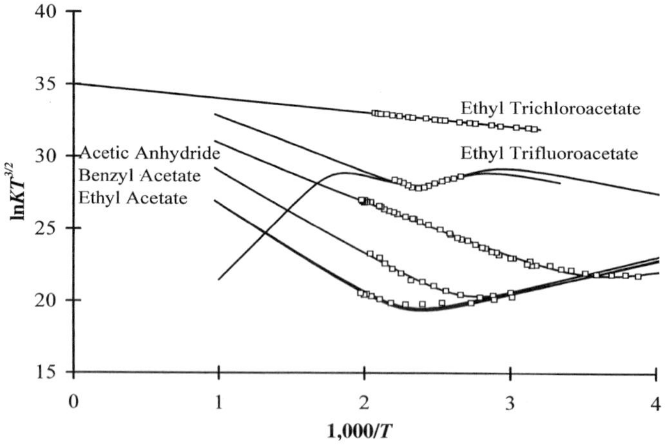

Figure 10.10 Plots of ECD data as ln $KT^{3/2}$ versus 1,000/T for acetic anhydride and acetates. From [69, 73].

TABLE 10.11 Kinetic and Thermodynamic Properties for Dissociative Electron Attachment from ECD Data for Esters

Species	$\ln(A_1)$	E_1(eV)	Q(eV)	E_a(eV)	E_2(eV)	$\ln(A_2)$
Acetic anhydride	34.2(4)	0.01	0.4	0.21(5)	0.22(5)	24.3(2)
Benzyl acetate	35.8(4)	0.04(1)	0.3	0.20(5)	0.24(5)	22.0(2)
Ethyl acetate	35.3(4)	0.04(1)	0.8	0.20(5)	0.24(5)	21.0(20)
Ethyl trifluoroacetate	35.2(4)	0.14(2)	0.5	0.51(5)	0.8(1)	24.8(5)
Ethyl trifluoroacetate	35.1(4)	0.22(5)	0.4	0.8(1)	—	—
Ethyl trifluoroacetate	35.1(4)	0.13(2)	0.4	0.6(1)	—	—
Ethyl trichloroacetate	35.0(5)	0.08(5)	—	—	—	—

The values in parentheses are the uncertainties in the last figure. Other quantities are given to the proper number of significant figures The second set of data for the trifluoroacetate assumes two states. The data for the trichloroacetate could represent C—Cl dissociation.

anhydride. The electron affinities of these molecules have been calculated using CURES-EC and support the experimental data, as shown in Table 10.10.

The data for ethyl trifluoroacetate in the low-temperature region can be definitely assigned to the formation of a stable negative ion. The increase at higher temperatures could result from activation to the ground state, as observed for tetracene. The lowest CURES-EC calculated value is 0.8 eV, while the ECD value from the lower-temperature slope is only 0.51 eV. As a result, we have calculated another curve that has two states. These results are also given in Table 10.11. The mechanism for the ECD reactions of ethyltrichloroacetate cannot be assigned because there is only one region and multiple pathways for dissociative electron capture exist. Dissociative electron capture could take place via the C—Cl bond rather than the C—O bond.

10.4.3 Electron Affinities of Organic Nitro Compounds the ECD and TCT Methods

Table 10.12 lists the E_a of aromatic nitrocompounds determined by both the TCT and ECD methods. The values are the weighted averages. The ECD and TCT values agree within the uncertainty. They are confirmed by reduction potential data and CURES-EC calculations (not listed). This is the largest number of E_a for related compounds that have been measured by any two techniques [15, 16, 63–68]. The ECD E_a for the dimethyl-nitrobenzenes, m-Cl-nitrobenzene, p-Cl-nitrobenzene, and 2-Cl-6-Me-nitrobenzene are not published elsewhere. The E_a for the dimethyl-nitrobenzenes are obtained from data in the linear region. Those for the chlorinated compounds are obtained from the fit to the data in both regions. The E_a for m-Cl-nitrobenzene is a lower limit because there are no data in the α region. These confirm the TCT values.

Figures 10.11 and 10.12 illustrate the ECD data showing structure and two states in the data for p-Cl-nitrobenzene and dimethylnitrobenzene. These data were collected along with those for other nitrobenzenes but not published because the

TABLE 10.12 Electron Affinities (in eV) for Aromatic Nitrocompounds Using the TCT and ECD Methods

Molecule	AE_a	NIST	ECD	TCT	$E_{1/2}$
Nitrobenzene (NB)	1.00(1)	1.006	1.00(2)	1.01(5)	1.00
Nitrobenzene-F_5	1.50(10)	1.450	1.50(10)	1.45(10)	1.50
Nitrobenzene-mF	1.20(5)	1.236	1.30(15)	1.18(5)	1.20
Nitrobenzene-oF	1.09(5)	1.075	1.10(15)	1.09(5)	1.11
Nitrobenzene-pF	1.10(5)	1.119	1.16(5)	1.09(5)	1.09
m-Nitrotoluene	0.98(3)	0.989	0.97(3)	0.98(5)	1.00
o-Nitrotoluene	0.90(3)	0.924	0.88(3)	0.94(5)	0.89
p-Nitrotoluene	0.95(3)	0.954	0.96(3)	0.95(5)	0.95
2,3-diMe-NB	0.85(3)	0.854	0.85(3)	0.87(5)	0.87
3,4-diMe-NB	0.92(5)	0.924	0.90(2)	0.92(5)	0.93
2,4-diMe-NB	0.88(5)	0.880	—	0.88(5)	—
2,6-diMe-NB	0.78(5)	0.811	0.77(6)	0.78(5)	0.78
2,4,6-triMe-NB	0.73(5)	0.711	—	0.73(5)	0.72
3-Cl, 6-Me-NB	1.09(5)	—	1.09(5)	—	—
o-Cl-NB	1.13(5)	1.162	—	1.13(5)	1.20
m-Cl-NB	1.27(5)	1.280	1.25(15)	1.26(5)	1.28
p-Cl-NB	1.24(4)	1.258	1.25(5)	1.23(5)	1.25

structure was not understood [15]. Shown in the plots are the resolved curves for both states: the best combined curve and the best single-state curve. Also shown is a least-squares fit through the highest-temperature points, which indicates that the Q_{an} for the ground state is low. These data support the TCT value of 1.23(5) eV for p-chloronitrobenzene. This clearly establishes the ability of both methods to measure ground-state E_a. The observation of structure in the ECD data shows that even at 1 atm pressure detachment can result from excited states. Thus, it is also certainly reasonable that excited-state anions may exist in TCT or photon methods at much lower pressures. A study of the effect of pressure on the thermal electron detachment rate of azulene demonstrated that the "high pressure limit" is not generally reached in TCT experiments [70].

Table 10.13 lists the E_a for a representative number of nitrocompounds that have been determined by only the TCT methods. These provide a range of values that were measured using this technique and examples of the substituent effects. When two determinations are made by different investigators, the weighted average is used. The most important point in this table is the CURES-EC method support of the TCT values. The E_a of the halogenated nitrobenzenes are also supported by reduction potential data. There are presently no higher estimates of the electron affinities of these compounds so these values are the current "best" adiabatic electron affinities.

Table 10.14 gives the E_a for acetophenones, benzaldehydes, benzonitriles, and benzophenones measured using only the TCT method. The values were scaled to

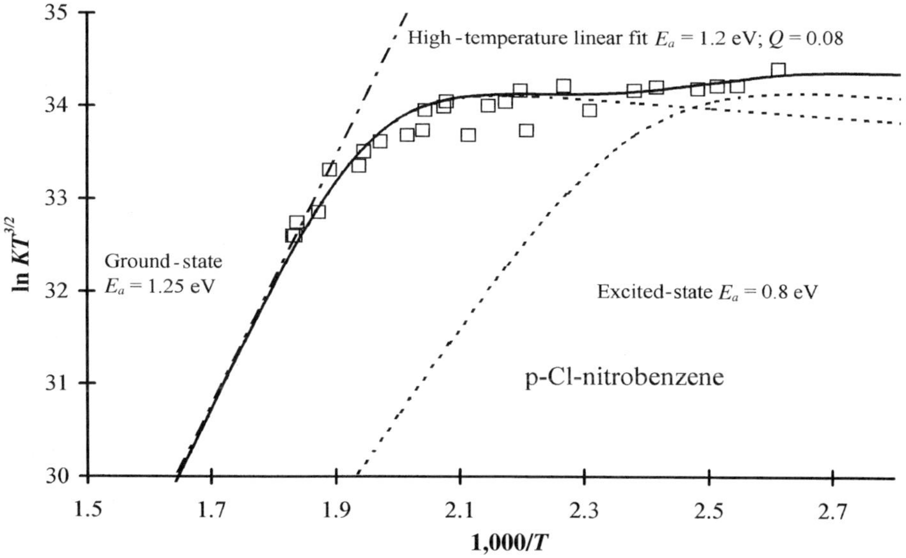

Figure 10.11 Plots of ECD data as ln $KT^{3/2}$ versus 1,000/T for p-Cl-nitrobenzene. The data were determined with a concentric electrode Ni-63 detector using dibromobenzene as an internal standard. They were also collected along with the data in [15], but electron affinities were not obtained because of the structure.

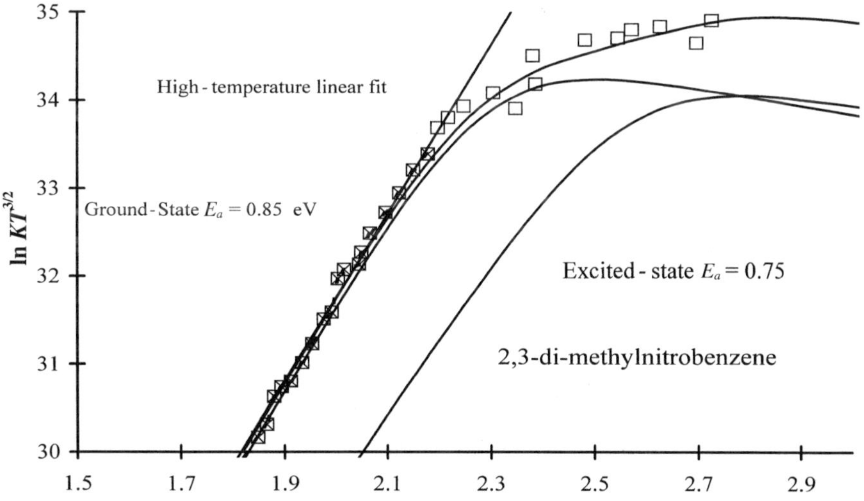

Figure 10.12 Plots of ECD data as ln $KT^{3/2}$ versus 1,000/T for dimethyl-nitrobenzene. The data were determined with a concentric electrode Ni-63 detector using dibromobenzene as an internal standard. They were also collected along with the data in [15], but electron affinities were not obtained because of the structure.

TABLE 10.13 Electron Affinities (in eV) for Aromatic Nitrocompounds Using the TCT Method

Molecule	AE_a	NIST	TCT	C-EC
Nitrobenzene-2,3-di-Cl	1.29(10)	1.292	1.30(10)	1.40
Nitrobenzene-3,4-di-Cl	1.44(10)	1.444	1.44(10)	1.50
Nitrobenzene-3,5-di-Cl	1.52(8)	1.500	1.52(8)	1.50
Nitrobenzene-o-Br	1.18(10)	1.162	1.18(10)	1.22
Nitrobenzene-m-Br	1.32(10)	1.318	1.32(10)	1.35
Nitrobenzene-p-Br	1.29(10)	1.292	1.29(10)	1.48
Nitrobenzene-o-CF_3	1.33(10)	1.331	1.33(10)	1.25
Nitrobenzene-m-CF_3	1.41(10)	1.414	1.41(10)	1.45
Nitrobenzene-p-CF_3	1.50(10)	1.500	1.50(10)	1.55
Nitrobenzene-m-OCH_3	1.04(10)	1.040	1.04(10)	1.00
Nitrobenzene-p-OCH_3	0.91(10)	0.911	0.91(10)	0.89
Nitrobenzene-p-NH_2	0.92(10)	0.915	0.92(10)	0.90
Nitrobenzene-m-NH_2	0.95(10)	0.945	0.95(10)	0.92
Nitrobenzene-m-CN	1.57(10)	1.565	1.57(10)	1.60
Dinitrobenzene-m	1.66(10)	1.657	1.66(10)	1.68
Dinitrobenzene-o	1.65(10)	1.652	1.65(10)	1.68
Dinitrobenzene-p	2.00(10)	2.003	2.00(10)	2.01
1-Nitronaphthalene	1.23(10)	1.227	1.23(10)	1.30
1,5-Dinitronaphthalene	1.77(10)	1.765	1.77(10)	1.80

TABLE 10.14 Molecular Electron Affinities (in eV) Using the TCT Method

Molecule	AE_a	NIST	C-EC
Acetophenone-m-NO_2	1.33(10)	1.33	1.40
Acetophenone-o-NO_2	1.40(10)	1.40	1.45
Acetophenone-p-NO_2	1.57(10)	1.57	1.50
Acetophenone-F_5	0.88(10)	0.876	0.88
Acetophenone-p-COOMe	0.96(10)	0.963	1.15
Benzaldehyde-F_5	1.10(10)	1.097	1.15
Benzaldehyde-m-CN	1.02(10)	0.990	1.08
Benzaldehyde-p-CN	1.28(10)	1.250	1.28
Benzaldehyde-m-CF_3	0.85(10)	0.815	0.90
Benzaldehyde-p-CF_3	0.97(10)	0.941	1.05
Benzaldehyde-3,5-DiCF_3	1.26(10)	1.232	1.54
Benzaldehyde-3,5-DiCl	1.03(10)	0.989	1.01
Benzaldehyde-p-Cl	0.68(10)	0.659	0.72
Benzaldehyde-m-Cl	0.71(10)	0.668	0.70
Benzaldehyde-p-CHO	1.30(10)	1.236	1.30
Benzaldehyde-m-CHO	1.03(10)	0.971	1.06
Benzaldehyde-p-NO_2	1.69(10)	1.691	1.66
Benzaldehyde-m-NO_2	1.43(10)	1.431	1.39
Benzaldehyde-o-NO_2	1.56(10)	1.557	1.50

the ECD values of the parents. The evaluated values reflect the increased values for acetophenone, benzaldehyde, and benzophenone. The majority of these are confirmed by CURES-EC and relative values have been supported by ab initio calculations. The LUMO are positive rather than negative, so a constant displacement of 2.2 eV is required to obtain absolute E_a. These compounds are the natural bridge between the TCT and ECD values and the ab initio and CURES-EC values.

10.5 ELECTRON AFFINITIES OF CHARGE TRANSFER COMPLEX ACCEPTORS

The electron affinities of a number of the standard π charge transfer complex acceptors have been measured using TCT. By scaling the values originally obtained for these acceptors, unmeasured E_a can be estimated. The values for the substituted benzoquinones differ from the values published in 1975, as discussed in Chapter 6 [71]. By shifting the methyl-substituted compounds by a different amount than the halogenated compounds, more accurate values were obtained. The difference is a systematic uncertainty that can be removed empirically with the larger number of calibration points. The random uncertainties will, however, remain. The values for 1,4 naphthoquinone and anthraquinone can be scaled by 0.1 eV to obtain more precise values. The E_a of s-trinitrobenzene has the largest systematic error, about 0.9 eV. The charge transfer complex maxima for other trinitrocompounds are close to the value for 1,3,5 trinitrobenzene. Comparable values for the E_a of other trinitro compounds are obtained by scaling. Table 10.15 lists the E_a that have been scaled to the selected E_a, the values determined from charge transfer complex data in 1975, and the current "best" values. This procedure introduces additional scale factors: 0.20 and 0.30 eV for the halobenzoquinones, 0.1 eV for anthraquinone and naphthoquinone, 0.55 eV for the dinitrobenzenes, and 0.9 eV for the trinitrocompounds. Figure 10.13 is a precision and accuracy plot for the E_a of the calibration points determined from the charge transfer spectra versus the evaluated E_a. If we assume a unit slope, the intercept is slightly negative. The standard deviation is ± 0.08 eV. Table 10.16 gives the E_a of other compounds scaled with these same factors.

In many cases the E_a of these acceptors have been determined from reduction potentials by using $-\Delta\Delta G$ values that vary from 2.6 eV for nitrobenzene to 2.0 for the tetrahalobenzoquinones. Figure 10.14 shows the calibration points for the reduction potentials given in Table 10.17. The standard deviation is ± 0.08 eV. Five values of $-\Delta\Delta G$ are defined, 2.00 eV and 2.05 eV, the same as for some aromatic hydrocarbons, 2.25 eV for the benzoquinones, cyanonitrobenzenes and dinitrobenzenes, 2.45 eV for halogen substituted nitro compounds, and 2.55 eV for nitrobenzenes and alkyl nitro benzenes. In Table 10.18 are E_a using these $-\Delta\Delta G$ and the average of these values with those obtained from E_{CT}.

TABLE 10.15 Calibration Data (in eV) for Charge Transfer Complex Acceptors

Compound	Evaluated	1975	Difference	Adjusted
Hexacyanobutadiene	3.29(15)	3.08	0.21	3.18
Tetracyanoethylene	2.95(10)	2.77	0.16	2.87
Tetracyanoquinodimethane	2.80(6)	2.84	−0.04	2.94
Hexacyanobenzene	2.54(15)	2.56	−0.02	2.66
Tetracyanobenzene	2.20(15)	2.00	0.20	2.10
Maleic anhydride	1.44(9)	1.33	0.11	1.43
Hexafluorobenzene	0.86(2)	0.86	0.00	0.86
9,10-Anthraquinone	1.59(6)	1.57	0.02	1.67
Nitrobenzene	1.00(1)	1.19	−0.19	1.00
p-Benzoquinone	1.860(5)	1.83	0.04	1.83
Chloranil	2.77(5)	2.44	0.33	2.74
Fluorobenzoquinone	2.20(15)	1.93	0.23	2.13
Tetrabromo-o-benzoquinone	2.44(20)	2.48	−0.04	2.78
2,5-Dichlorobenzoquinone	2.40(3)	2.30	0.13	2.50
2,6-Dichlorobenzoquinone	2.48(3)	2.23	0.25	2.53
Trichlorobenzoquinone	2.61(5)	2.41	0.20	2.71
Methyltrichlorobenzoquinone	2.54(5)	2.22	0.32	2.52
Methylbenzoquinone	1.85(5)	1.74	0.15	1.74
2,5-Dimethylbenzoquinone	1.76(6)	1.73	0.03	1.73
2,6-Dimethylbenzoquinone	1.77(5)	1.67	0.10	1.67
Tetramethylbenzoquinone	1.61(9)	1.67	−0.06	1.67
1,4-Naphthoquinone	1.80(5)	1.71	0.10	1.81
1,2-Dinitrobenzene	1.65(10)	1.07	0.58	1.57
1.3-dinitrobenzene	1.66(10)	1.26	0.40	1.77
1,3,5-Trinitrobenzene	2.63(15)	1.73	0.90	2.63

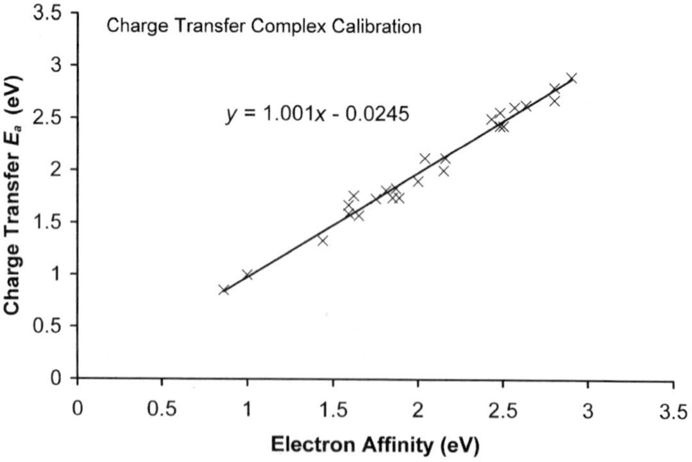

Figure 10.13 Precision and accuracy plots of the calibration data for charge transfer complex energy for the maximum absorbance. The charge transfer complex data were compiled in [71]. The data are given in Table 10.15.

TABLE 10.16 Electron Affinities (in eV) from Charge Transfer Complex Data

Compound	CT	Previous	Offset
Acetylbenzoquinone	2.29(10)	2.09	0.2
Bromobenzoquinone	2.21(10)	2.01	0.2
Di-bromobenzoquinone	2.61(10)	2.31	0.3
Di-bromo-dimethylbenzoquinone	2.35(10)	2.05	0.3
p-Tetrabromobenzoquinone	2.78(10)	2.48	0.3
o-Tetrachlorobenzoquinone	2.67(10)	2.57	0.3
Cyanobenzoquinone	2.22(10)	2.22	0.0
2,3-Dicyanobenzoquinone	2.78(10)	2.78	0.0
Tetracyanobenzoquinone	2.87(10)	2.87	0.0
Trifluoromethylbenzoquinone	2.38(10)	2.18	0.2
Iodobenzoquinone	2.20(10)	2.00	0.2
p-Tetraiodobenzoquinone	2.83(10)	2.43	0.4
Methoxybenzoquinone	1.69(10)	1.69	0.0
Dimethylaminobenzoquinone	1.47(10)	1.47	0.0
2,5-Di-t-butylbenzoquinone	1.57(10)	1.57	0.0
Nitrobenzoquinone	2.63(10)	2.63	0.2
Dibromo-1,4-naphthoquinone	2.41(10)	2.11	0.3
Bromochloro-1,4-naphthoquinone	2.30(10)	2.00	0.3
2,3-Dichloro-1,4-naphthoquinone	2.20(10)	2.00	0.2
2,3-Dicyano-1,4-naphthoquinone	2.66(10)	2.66	0.0
Tetracyanoxylene	1.65(10)	1.65	0.0
Pentacyanotoluene	2.19(10)	2.19	0.0
1,2,3-Trinitrobenzene	2.18(10)	1.28	0.9
1,2,4-Trinitrobenzene	2.59(10)	1.69	0.9
1,2,3,5-Tetranitrobenzene	2.81(10)	1.91	0.9
2,4-Dinitrotoluene	1.67(10)	1.17	0.5
2,4,6-Trinitrotoluene	2.57(10)	1.67	0.9
2,4,6-Trinitro-p-xylene	2.37(10)	1.47	0.9
2,4,6-Trinitrobenzoic acid	2.64(10)	1.74	0.9
2,4,6-Trinitroanisole	2.52(10)	1.62	0.9

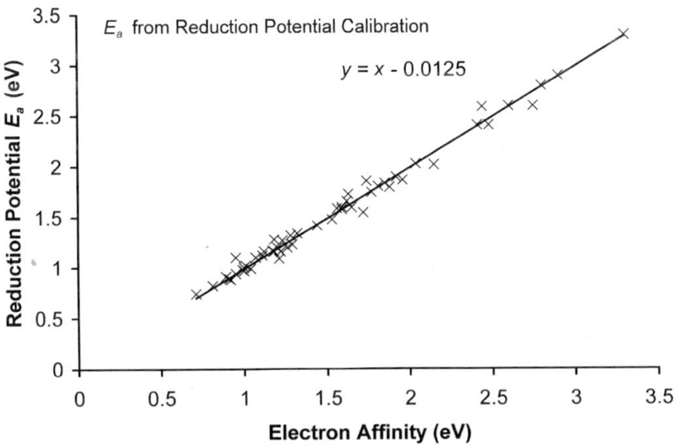

Figure 10.14 Precision and accuracy plots of the calibration data for half-wave reduction potentials. The data were compiled in [58, 71]. The data are given in Table 10.17.

TABLE 10.17 Calibration Data (in eV) for Reduction Potentials

	E_a	$-\Delta\Delta G$	E_a (reduction)
Trifluoromethylnitrobenzene	1.41(9)	2.55	1.42
Nitrobenzene	1.00(1)	2.55	1.02
m-Methoxynitrobenzene	1.04(1)	2.55	0.99
m-Nitrotoluene	0.98(3)	2.55	0.98
Dimethylaminonitrobenzene	0.98(3)	2.55	1.10
p-Nitrotoluene	0.95(3)	2.55	0.94
o-Nitrotoluene	0.90(3)	2.55	0.88
p-Methoxynitrobenzene	0.89(5)	2.55	0.89
3,4-Dimethylnitrobenzene	0.90(5)	2.55	0.91
2,3-Dimethylnitrobenzene	0.80(5)	2.55	0.82
2,6-Dimethylnitrobenzene	0.74(4)	2.55	0.74
p-Bromonitrobenzene	1.29(10)	2.45	1.34
m-Bromonitrobenzene	1.29(10)	2.45	1.23
m-Chloronitrobenzene	1.27(5)	2.45	1.32
p-Chloronitrobenzene	1.24(4)	2.45	1.20
1-Nitronaphthalene	1.23(9)	2.45	1.27
p-Nitrobenzophenone	1.28(9)	2.45	1.20
m-Fluoronitrobenzene	1.22(5)	2.45	1.16
2-Nitronaphthalene	1.16(9)	2.45	1.28
o-Fluoronitrobenzene	1.09(5)	2.45	1.17
o-Chloronitrobenzene	1.13(5)	2.45	1.16
p-Fluoronitrobenzene	1.10(5)	2.45	1.12
o-Nitrobiphenyl	1.07(10)	2.45	1.10
Maleic anhydride	1.44(9)	2.40	1.42
2,3-Di-t-butylbenzoquinone	1.92(9)	2.25	1.90
p-Benzoquinone	1.860(5)	2.25	1.80
2,6-Di-t-butylbenzoquinone	1.88(9)	2.25	1.80
Methylbenzoquinone	1.85(5)	2.25	1.84
1,4-Naphthoquinone	1.80(5)	2.25	1.81
2,5-Dimethylbenzoquinone	1.76(6)	2.25	1.75
2-Methyl-1,4-naphthoquinone	1.75(4)	2.25	1.86
p-Cyanonitrobenzene	1.71(9)	2.25	1.55
o-Dinitrobenzene	1.65(10)	2.25	1.60
Trimethylbenzoquinone	1.69(5)	2.25	1.73
m-Dinitrobenzene	1.66(10)	2.25	1.66
Tetramethylbenzoquinone	1.61(4)	2.25	1.58
9,10-Anthraquinone	1.59(6)	2.25	1.60
t-Butyl-9,10-anthraquinone	1.56(6)	2.25	1.59
Ethyl-9,10-anthraquinone	1.56(6)	2.25	1.59
m-Cyanonitrobenzene	1.57(5)	2.25	1.48
Phenylbenzoquinone	2.04(6)	2.25	2.03
p-Dinitrobenzene	2.00(9)	2.20	1.87
Phthalic anhydride	1.26(5)	2.20	1.09
Dicyanoethylene	1.00(10)	2.15	0.97
Hexacyanobutadiene	3.29(15)	2.05	3.30
Tetracyanoethylene	2.95(10)	2.05	2.90

TABLE 10.17 (*Continued*)

	E_a	$-\Delta\Delta G$	E_a (reduction)
Tetracyanoquinodimethane	2.80(6)	2.05	2.80
p-Fluoranil	2.70(10)	2.05	2.60
p-Chloranil	2.77(5)	2.00	2.65
o-Bromanil	2.44(20)	2.00	2.64
2,6-Dichlorobenzoquinone	2.48(3)	2.00	2.46
2,3-Dichlorobenzoquinone	2.40(3)	2.00	2.46
Tetracyanobenzene	2.20(15)	2.00	2.07
o-Dicyanobenzene	1.00(10)	2.00	1.00

TABLE 10.18 Electron Affinities (in eV) from Reduction Potential and Charge Transfer Complex Data

Compound	Eval	E_{CT}	$E_{1/2}$
Acetylbenzoquinone	2.21(5)	2.29	2.18
Bromobenzoquinone	2.19(5)	2.21	2.17
Di-bromo-dimethylbenzoquinone	2.25(5)	2.25	2.25
p-Tetrabromobenzoquinone	2.78(5)	2.78	2.78
Cyanobenzoquinone	2.30(5)	2.22	2.37
Trifluoromethylbenzoquinone	2.32(5)	2.38	2.26
Iodobenzoquinone	2.18(5)	2.20	2.16
Methoxybenzoquinone	1.76(5)	1.69	1.87
Dimethylaminobenzoquinone	1.55(5)	1.47	1.64
2,5-Di-t-butylbenzoquinone	1.76(8)	—	1.76
2,3-Dichloro-1,4-naphthoquinone	2.20(5)	2.20	2.20
2,3-Dicyano-1,4-naphthoquinone	2.66(5)	2.66	2.66

The evaluated values are the averages.

10.6 SUBSTITUENT EFFECT

From the changes in the electron affinities for the substituted compounds, general substitution effects can be identified. The replacement of an aromatic hydrogen atom by an F atom raises the E_a by 0.1 to 0.2 eV, while the replacement by a single Cl, Br or I increases the E_a 0.2 to 0.3 eV. The substitution of a CH_3 group can lower the E_a as in the case of nitrobenzene or can raise it as in the case of benzene. The substitution of a single CF_3 group raises the E_a by as much as 0.4 eV. The effect of substituting OH for an H atom is to increase the E_a by about 0.5 eV. The substitution of a $COOCH_3$ will increase the E_a by about 0.4 eV. The substitution of an acetyl group will increase the E_a by about 0.6 eV while the substitution of a CHO or CN group will increase the E_a by a greater amount. The substitution of a nitro group on nitrobenzene increases the E_a by about 1 eV for the para position

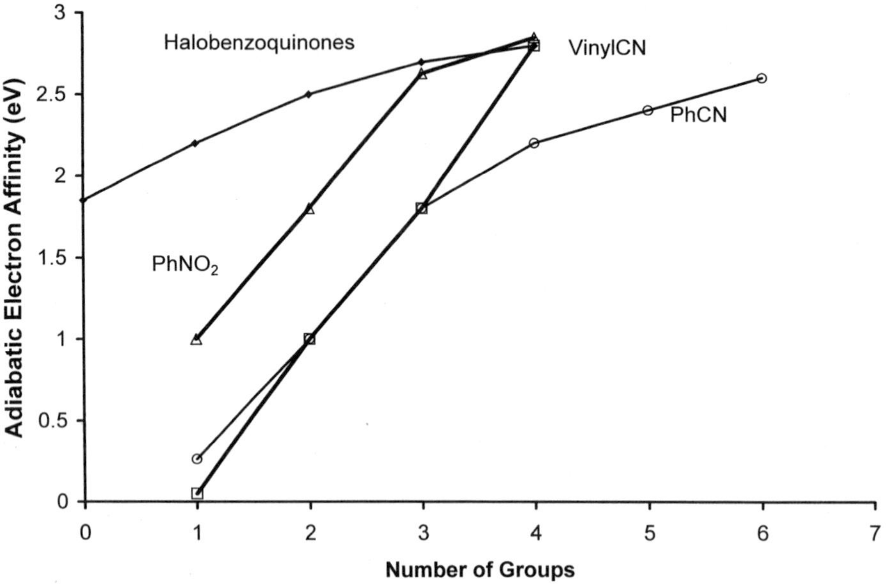

Figure 10.15 The electron affinities of halogenated benzoquinones, cyanoethylene, cyanobenzenes, and nitrobenzenes versus the number of substituents, data from [20].

but only increases the E_a by about 0.6 eV in the meta and ortho positions. This difference is supported by the CURES-EC calculations. The increase in the E_a of naphthalene is 1 eV for the first nitro group and about 0.6 eV for a second nitro group. The largest increases are observed for the substitution of the first NO_2 group, from 1 eV to 1.5 eV. The effects are larger for the molecules with the lower E_a values, such as benzonitrile and methyl benzoate, and are smallest for nitrobenzene [68]. The E_a from half-wave reduction potentials and charge transfer maxima support the substitution effects. The substitution of alkyl groups decreases the E_a by 0.05 eV to 0.1 eV per alkyl group, and the substitution of a halogen atom increases it by 0.1 eV to 0.25 eV. The substitution of four halogens in benzoquinone raises the E_a from 1.86 eV to 2.8 eV regardless of the halogen. Upon substitution of an additional NO_2 group in trinitrobenzene, the increase in the E_a is only 0.2 eV, indicating a saturation effect. Indeed, the electron affinities of organic molecules seem to plateau at 4 eV. This E_a maxima is observed in Figure 10.15. Values for radicals are observed at higher values.

With these generalizations the E_a of other compounds can be estimated and used to optimize the CURES-EC results. The E_a of phenanthraldehyde should be about 0.6 eV greater than that for phenanthrene or $0.3 + 0.6 = 0.9$ eV. The experimental value is 0.9 eV. The values for the chloroanthracenes should be approximately 0.2 eV greater than that of anthracene or $0.2 + 0.68 = 0.88$ eV. The experimental values are 0.8 to 0.9 eV. Multiple substitutions for the NO_2 and CN groups in benzene level off with three groups. The substitution of halogens in benzoquinone level off with three substitutions.

10.7 SUMMARY

Morse potential energy curves for nitromethane and nitrobenzene were calculated from experimental data. These curves are the most precisely defined curves for the organic compounds. The electron affinities measured by the magnetron, AMB, photon, collisional ionization, TCT, NIMS, and ECD methods generally agree within the uncertainties. Since each of these methods is capable of accurately and precisely measuring the E_a of a given compound, when a value deviates from the highest precise value by more than the experimental uncertainty, the possibility of an excited state should be considered. For ECD and PES data it is possible to interpret the data in terms of the simultaneous observation of two nondetaching negative-ion states. New electron affinities for substituted nitrobenzenes and esters were reported from ECD data. The experimental data for charge transfer complex acceptors from absorption maxima and half-wave reduction potentials were calibrated with experimental E_a to obtain more accurate values for the electron affinities.

REFERENCES

1. National Institute of Standards and Technology (NIST). *Chemistry WebBook*, **2003**. Available at http://webbook.nist.gov.
2. Christodoulides, A. A.; McCorkle, D. L.; and Christophorou, L. G. "Electron Affinities of Atoms, Molecules and Radicals" in *Electron-Molecule Interactions and Their Applications*. New York: Academic Press, **1984**.
3. Chen, E. C. M.; Welk, N.; Chen, E. S.; and Wentworth, W. E. *J. Phys. Chem. A* **1999**, 103, 9072.
4. Chen, E. C. M. and Wentworth, W. E. *J. Phys. Chem.* **1983**, 87, 45.
5. Jager, K. and Henglein, A. *Z. Natforschg.* **1967**, 22a, 700.
6. Stockdale, J. A.; Davis, F. J.; Compton, R. N.; and Klots, C. E. *J. Chem. Phys.* **1974**, 60, 4279.
7. DeDomenico, A. and Franklin, J. L. *Int. J. Mass. Spectrom. Ion Phys.* **1972**, 9, 171.
8. Compton, R. N.; Reinhardt, P. W.; and Cooper, C. D. *J. Chem. Phys.* **1978**, 68, 4360.
9. Grimsrud, E.; Caldwell, G.; and Kebarle, P. *J. Amer. Chem. Soc.* **1985**, 107, 4627.
10. Lobo, R. F. M.; Moutinho, A. M. C.; Lachmann, K.; and Los, J. *J. Chem. Phys.* **1991**, 95, 166.
11. Lecomte, F.; Carles, S.; Desfrancois, C.; and Johnson, M. A. *J. Chem. Phys.* **2000**, 113, 10973.
12. Compton, R. N.; Carman, H. S.; Desfrancois, C.; Abdoul-Carmine, H.; Schermann, J. P.; Hendricks, J. H.; Lyapustina, S. A.; and Bowen, K. H. *J. Chem. Phys.* **1996**, 105, 3475.
13. Gutsev, G. L. and Bartlett, R. J. *J. Chem. Phys.* **1996**, 105, 8785.
14. Chen, E. C. M.; Wiley, J. R.; Batten, C. F.; and Wentworth, W. E. *J. Phys. Chem.* **1994**, 98, 88.
15. Chen, E. C. M.; Chen, E. S.; Milligan, M. S.; Wentworth, W. E.; and Wiley, J. R. *J. Phys. Chem.* **1992**, 96, 2385.

16. Fukuda, E. K. and McIver, R. T., Jr. *J. Amer. Chem. Soc.* **1985**, 107, 2291.
17. Chowdhury, S.; Heinis, T.; Grimsrud, E. P.; and Kebarle, P. *J. Phys. Chem.* **1986**, 90, 2747.
18. Henglein, A. and Muccini, G. A. *J. Chem. Phys.* **1959**, 31, 1426.
19. Laramee, J. A.; Kocher, C. A.; and Deinzer, M. L. *Anal. Chem.* **1992**, 64, 2316.
20. Page, F. M. and Goode, G. C. *Negative Ions and the Magnetron*. New York: Wiley-Interscience, **1969**.
21. Kleyn, A. W. and Moutinho, A. M. C. *J. Phys. B* **2001**, 34, R1.
22. Dispert, H. and Lacmann, K. *Int. J. Mass Spectrom. Ion Phys.* **1978**, 28, 49.
23. Lacmann, K.; Maneira, M. J. P.; Moutinho, A. M. C.; and Weigman, U. *J. Chem. Phys.* **1983**, 78, 1767.
24. Spyrou, S. M.; Sauers, I.; and Christophorou, L. G. *J. Chem. Phys.* **1983**, 78, 7200.
25. Compton, R. N.; Reinhardt, P. W.; and Cooper, C. D. *J. Chem. Phys.* **1978**, 68, 4360.
26. Tang, S. Y.; Mathur, B. P.; Roth, E. W.; and Reck, G. P. *J. Chem. Phys.* **1976**, 64, 1270.
27. McNamee, P. E.; Lacmann, K.; and Herschbach, D. R. *Farad. Disc. Chem. Soc.* **1973**, 55, 318.
28. Klots, C. E.; Compton, R. N.; and Raaen, V. F. *J. Chem. Phys.* **1974**, 60, 1177.
29. Cooper, C. D.; Frey, W. F.; and Compton, R. N. *J. Chem. Phys.* **1978**, 69, 2367.
30. Compton, R. N.; Reinhardt, P. W.; and Cooper, C. D. *J. Chem. Phys.* **1974**, 60, 2953.
31. Cooper, C. D. and Compton, R. N. *J. Chem. Phys.* **1973**, 59, 3550.
32. Cooper, C. D.; Naff, W. T.; and Compton, R. N. *J. Chem. Phys.* **1975**, 63, 2752.
33. Nalley, S. J. and Compton, R. N. *Chem. Phys. Lett.* **1971**, 9, 529.
34. Reinstra-Kiracofe, J. C.; Tschumper, G. S.; Schaefer, H. F.; Nandi, S.; and Ellison, G. B. *Chem. Rev.* **2002**, 102, 231.
35. Mock, R. S. and Grimsrud, E. P. *J. Amer. Chem. Soc.* **1989**, 111, 2861.
36. Schiedt, J. and Weinkauf, R. *J. Chem. Phys.* **1999**, 110, 304.
37. Marks, J.; Comita, P. B.; and Brauman, J. I. *J. Amer. Chem. Soc.* **1985**, 107, 3718.
38. Nakajima, A.; Taguwa, T.; Hoshino, K.; Sugioka, T.; Naganuma, T.; Ono, F.; Watanabe, K.; and Nakao, K. *Chem. Phys. Lett.* **1993**, 214, 22.
39. Lyons, L. E. and Palmer, L. D. *Chem. Phys. Lett.* **1973**, 21, 442.
40. Datskos, P. G.; Carter, J. G.; and Christophorou, L. G. *Chem. Phys. Lett.* **1995**, 239, 38.
41. Drzaic, P. S. and Brauman, J. I. *J. Amer. Chem. Soc.* **1982**, 104, 13.
42. Duncan, M. A.; Knight, A. M.; Negishi, Y.; Nagao, S.; Nakamura, Y.; Kato, A.; Nakajima, A.; and Kaya, K. *Chem. Phys. Lett.* **1999**, 309, 49.
43. Lyapustina, S. A.; Xu, S. K.; Nilles, J. M.; and Bowen, K. H. *J. Chem. Phys.* **2000**, 112, 6643.
44. Schiedt, J.; Knott, W. J.; Le Barbu, K.; Schlag, E. W.; and Weinkauf, R. *J. Chem. Phys.* **2000**, 113, 9470.
45. Scheidt, J. and Weinkauf, R. *Chem. Phys. Lett.* **1997**, 266, 201.
46. Scheidt, J. and Weinkauf, R. *Chem. Phys. Lett.* **1997**, 274, 18.
47. Wenthold, P. G.; Hrovat, D. A.; Borden, W. T.; and Lineberger, W. C. *Science* **1996**, 272, 1456.
48. Gygax, R.; Peters, H. L.; and Brauman, J. I. *J. Amer. Chem. Soc.* **1979**, 101, 2567.
49. Wentworth, W. E. and Ristau, W. *J. Phys. Chem.* **1969**, 73, 2126.

50. Kato, S.; Lee, H. S.; Gareyev, R.; Wenthold, P. G.; Lineberger, W. C.; DePuy, C. H.; and Bierbaum, V. M. *J. Amer. Chem. Soc.* **1997**, 119, 7863.
51. Denault, J. W.; Chen, G. D.; and Cooks, R. G. *J. Amer. Soc. Mass Spectrom.* **1998**, 9, 1141.
52. Chen, G. D. and Cooks, R. G. *J. Mass Spectrom.* **1995**, 30, 1167.
53. Chen, G.; Cooks, R. G.; Corpuz, E.; and Scott, L. T. *J. Amer. Soc. Mass Spectrom.* **1996**, 7, 619.
54. Ruoff, R. S.; Kadish, K. M.; Boulas, P.; and Chen, E. C. M. *J. Phys. Chem.* **1995**, 99, 8843.
55. Crocker, L.; Wang, T. B.; and Kebarle, P. *J. Amer. Chem. Soc.* **1993**, 115, 7818.
56. Chen, E. S. D.; Chen, E. C. M.; Sane, N.; Talley, L.; Kozanecki, N.; and Shultze, S. *J. Chem. Phys.* **1999**, 110, 9319.
57. Chen, E. C. M. and Chen, E. S. D. *J. Chromatogr. A* **2002**, 952, 173.
58. Chen, E. C. M. and Wentworth, W. E. *Mol. Cryst. Liq. Cryst.* **1989**, 171, 271.
59. Wojnarvits, L. and Foldiak, G. *J. Chromatogr.* **1981**, 206, 511.
60. Dewar, M. J. S.; Hashmall, J. A.; and Trinajstic, J. *J. Amer. Chem. Soc.* **1970**, 92, 5555.
61. Bergman, I. *Trans. Farad. Soc.* **1954**, 50, 829.
62. Bergman, I. *Trans. Farad. Soc.* **1956**, 52, 690.
63. Kebarle, P. and Chowdhury, S. *Chem. Rev.* **1987**, 87, 513.
64. Rains, L. J.; Moore, H. W.; and McIver, R. T. *J. Chem. Phys.* **1978**, 68, 3309.
65. Fukuda, E. K. and McIver, R. T. In *Lecture Notes in Chemistry*, edited by H. Hartman and K. P. Wanzek. Berlin: Springer, **1982**, p. 164.
66. Fukuda, E. K. and McIver, R. T. *J. Chem. Phys.* **1982**, 77, 4942.
67. Burinsky, D. J.; Fukuda, E. K.; and Campana, J. E. *J. Amer. Chem. Soc.* **1984**, 106, 2270.
68. Huh, C.; Kang, C. H.; Lee, H. W.; Nakamura, H.; Mishima, M.; Tsuno, Y.; and Yamataka, H. *Bull. Chem. Soc. Japan* **1999**, 72, 1083.
69. Steelhammer, J. C. M. S. thesis, University of Houston, **1968**.
70. Williamson, D. H.; Knighton, W. B.; and Grimsrud, E. P. *Internat. J. Mass Spectrom.* **2000**, 195/196, 481.
71. Chen, E. C. M. and Wentworth, W. E. *J. Chem. Phys.* **1975**, 63, 3183.
72. Wentworth, W. E.; Kao, L. W.; and Becker, R. S. *J. Phys. Chem.* **1975**, 79, 1161.
73. Wentworth, W. E.; Chen, E. C. M.; and Steelhammer, J. C. *J. Phys. Chem.* **1968**, 72, 2671.

CHAPTER 11

Thermal Electrons and Environmental Pollutants

11.1 INTRODUCTION

The ECD and NIMS are important tools for the analysis of environmental pollutants. The determination of freon levels in the atmosphere and their effects on ozone depletion required the ECD. Identifying the highly toxic chlorinated dioxins was only possible with the ECD and NIMS. Despite these successes very little fundamental data for the reactions of electrons with these molecules in the gas phase have been obtained. This chapter evaluates the available data and presents new data.

The electron affinities of the alkyl halides including the freons and newer coolants are evaluated. The E_a of c-$CF_3C_6F_{11}$ and c-$CF_3C_5F_9$ and the EDEA are determined from unpublished atmospheric pressure negative-ion mass spectra. Activation energies and bond dissociation energies are determined from molar ECD responses at a single temperature. These are compared with activation energies obtained from relative rate constants determined by complementary methods. Electron affinities and activation energies for electron attachment to freon substitutes are obtained by analyzing ECD responses measured as a function of temperature; these results were published in 1997. The two-state model gives new results that are compared with the original single-state interpretation.

Few electron affinities of halogenated olefins and aromatic hydrocarbons have been determined in the gas phase. The electron affinities of the chloroethylenes and chlorobenzenes were determined using the ECD and those of C_6F_6, C_6F_5X, C_6HCl_5, and C_6Cl_6 using TCT. The only E_a for the polychlorinated naphthalenes and chlorinated biphenyls and the complete set of E_a for the chlorobenzenes were obtained from reduction potentials in aprotic solvents. The ECD data for fluorobenzenes and chlorobenzenes, some of which have not been previously published, are analyzed and evaluated. The electron affinities of the halogenated benzenes are calculated using CURES-EC. Unpublished ECD data for trifluoromethylhalobenzenes are analyzed to obtain E_a. Morse potential energy curves for CCl_4,

The Electron Capture Detector and the Study of Reactions with Thermal Electrons
by E. C. M. Chen and E. S. D. Chen
ISBN 0-471-32622-4

CH_2Cl_2, and $C_6F_5X(-)$ are calculated from electron affinities, electron impact data, and dissociation energies.

The electron affinities of chlorinated biphenyls and naphthalenes were estimated from half-wave reduction potentials by assuming that the solution energy differences were constant. Now that it is possible to estimate solution energy differences, more accurate E_a are obtained. The E_a for all the isomers of the chlornaphthalenes are calculated using CURES-EC.

The negative-ion mass spectra for more than 300 environmental pollutants have been reported at two temperatures: 373 K and 523 K. The electron affinities of aromatic hydrocarbons, phthalates, chloroethylenes, chlorobenzenes, chloronphthalenes, chlorinated biphenyls, and nitrobenzenes have been measured, but the E_a of others, such as the bromobenzenes and chlorinated dioxins, have not. If we know the E_a of the parent compounds, the electron affinities and bond dissociation energies of these compounds can be estimated and compared with the NIMS data. These values are examined using CURES-EC.

One use of the electron affinities of molecules is to predict the sensitivity and temperature dependence of the ECD to compounds that might be analyzed. Many environmental pollutants have different multiple substituents. Pesticides are highly chlorinated organic compounds. The chlorinated biphenyls, naphthalenes, and dioxanes are among the most toxic compounds. The temperature dependence of these compounds in the ECD is important, but has not been extensively studied. When the electron affinities and bond dissociation energies are known, the temperature dependence can be calculated from the kinetic model. This is done for the chlorinated biphenyls and naphthalenes, and the calculated temperature dependence is then compared with experiment. These calculations offer clues about the best conditions for analysis.

11.2 ALKYL HALIDES

11.2.1 Morse Potential Energy Curves

The temperature dependence of the alkyl halides was one of the first subjects to be studied using the ECD. These are the simplest to analyze because often there is only one temperature region when dissociative thermal electron attachment is exothermic. This means that the EDEA, the energy of dissociative electron attachment, is positive: $\text{EDEA} = E_a(X) - D(R - X)$. The alkyl bromides, iodides, and chlorides are among the few organic compounds that have positive EDEA. Like the homonuclear diatomic halogen ions, the ground-state anionic curves for these molecules are $M(3)$, with positive values for all three Herschbach metrics—EDEA, E_a, and VE_a.

Originally, the temperature dependence for these compounds was interpreted in terms of unimolecular dissociation. The activation energy was equal to the energy of the crossing of a single curve with the molecular curve on the approach side of the molecule. The negative-ion curve dissociates to the lowest limit, $R + X(-)$. The interpretation is now different since there are three potential energy curves in the

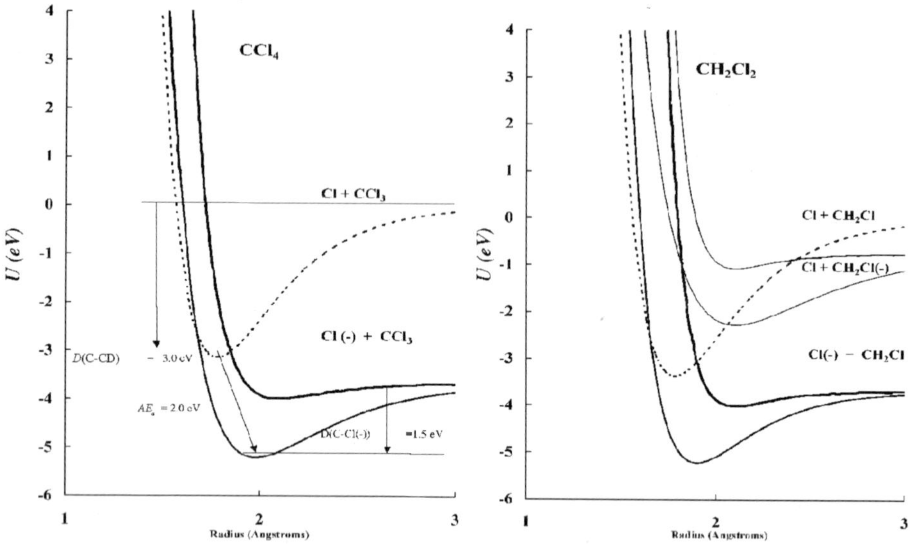

Figure 11.1 Morse potential energy curves for CCl_4 and CH_2Cl_2, DEC(1) compounds.

region of the internuclear distance of the neutral. One curve is the polarization or dipole bound curve with an E_a less than 0.1 eV. The other two are a bonding curve and an antibonding curve resulting from the addition of the electron to the C—X bond to give a three-electron bond [1–6].

This is exemplified in the data for the chloromethanes [1, 3–5]. Figure 11.1 illustrates the Morse potential energy curves for the bonding and antibonding states for CH_2Cl_2 and CCl_4. The electron affinity of the bonding curve for CCl_4 is 2.04(10) eV, as determined by MGN and AMB studies. The excited-state E_a is 0.80(34) eV by TCT [7–10]. The dissociation energy of the ground-state anion is $D(\text{C—Cl}) - E_a(\text{Cl}) + E_a(\text{CCl}_4) = 3.1 - 3.67 + 2.04 = 1.5$ eV, which gives a bond order of 0.5 from 1.5/3.1. Since the C—X bond in the neutral has two bonding electrons, whereas the bond in the anion has one net bonding electron, the simple bond order is predicted to be 0.5. The excited-state bond order is predicted to be lower and is $0.23/3.1 = 0.08$. Both curves cross the neutral curve close to zero, but the ground state is an $M(3)$ curve and the excited state is a $D(2)$ curve since the VE_a is negative and the vertical transition leads to dissociation. The ground-state curve is an $Mc(3)$ curve because it can lead to the long-range negative ion.

The activation energy from the ECD data for CH_2Cl_2 arises from the crossing of the excited-state curve and the neutral. The bond dissociation energy of the excited-state anion is calculated by assuming a bond order of 0.1. The ground-state curve for CH_2Cl_2 is defined by the ECD activation energy and the electron affinity of 1.3 ± 0.1 eV [7]. For CH_2Cl_2 two curves are drawn to the complementary dissociation limit, $CH_2Cl(-) + Cl$. The two higher curves for CH_2Cl_2 are $D(0)$ because all three metrics are negative and the vertical transition leads to dissociation. The

TABLE 11.1 Thermodynamic Properties for Dissociative Electron Attachment to Alkyl Halides (in eV)

Compound $R-X$	E_1 ECD	$E_a(X)$	$D(R-X)$ in the Literature	$D(R-X)$ ECD	Intercept
CF_3—Cl	0.25 ± 0.02	3.616	3.74	$>3.31, <3.93$	34.1 ± 0.3
CH_3—Cl	0.54 ± 0.02	3.616	3.59	3.61 ± 0.09	35.6 ± 0.5
t-C_4H_9—Cl	0.47 ± 0.02	3.616	3.58	3.54 ± 0.09	36.5 ± 0.5
1,2C_2H_4Cl—Cl	0.38 ± 0.01	3.616	3.46	3.45 ± 0.09	36.1 ± 0.1
CH_2Cl—Cl	0.33 ± 0.02	3.616	3.44	3.40 ± 0.09	33.9 ± 0.5
CF_2Cl—Cl	0.15 ± 0.02	3.616	3.30	3.22 ± 0.09	36.0 ± 0.5
$CHCl_2$—Cl	0.14 ± 0.02	3.616	3.30	3.21 ± 0.09	35.5 ± 0.2
$CFCl_2$—Cl	0.01 ± 0.02	3.616	3.17	3.08 ± 0.09	36.0 ± 0.2
CCl_3—Cl	0.00 ± 0.05	3.616	3.08	3.07 ± 0.09	33.6 ± 0.2
CH_3—Br	0.25 ± 0.02	3.365	3.02	3.07 ± 0.09	33.2 ± 0.5
n-C_3H_7—Br	0.34 ± 0.02	3.365	3.14	3.16 ± 0.09	36.3 ± 0.5
CH_2Br—Br	0.05 ± 0.02	3.365	2.97	2.87 ± 0.09	36.8 ± 0.2
$CHBr_2$—Br	0.01 ± 0.02	3.365	2.84	2.83 ± 0.09	34.4 ± 0.5
CH_3—I	0.03 ± 0.02	3.055	2.45	2.54 ± 0.09	34.3 ± 0.2
C_2H_5—I	0.05 ± 0.02	3.057	2.39	2.56 ± 0.09	35.2 ± 0.2
CHFCl—Cl	0.25 ± 0.02	3.616	—	3.32 ± 0.09	34.0 ± 0.2
1,3-C_3H_6Cl—Cl	0.34 ± 0.02	3.616	—	3.41 ± 0.09	35.2 ± 0.3
1,3-C_3H_6Br—Br	0.18 ± 0.02	3.365	—	3.00 ± 0.09	35.9 ± 0.2

bound curve is $Mc(0)$ because it crosses the long-range curve below the dissociation limit, while the higher antibonding curve is $Dc(0)$ because it will lead to dissociation when crossing the long-range curve. Previously only the $D(2)$ curve crossing on the right-hand side of the minimum was considered [1, 3, 5].

11.2.2 Experimental Activation Energies

The activation energy for thermal electron attachment to CCl_4 is 0.0 ± 0.05 eV, while that for CH_2Cl_2 is 0.33 ± 0.02 eV. The activation energies for thermal electron attachment to the mono chloroalkanes are 0.45 eV to 0.55 eV. The A_1 and E_1 for the chloromethanes and values for other alkyl chlorides, bromides, and iodides are given in Table 11.1. The A_1 values for all these compounds are within an order of magnitude of the De Broglie A_1. Examples of the temperature dependence for the chloromethanes were given in Chapter 5 and are used to verify the suitability of ECD equipment to obtain fundamental data. Since only one temperature region is observed for most of these compounds, only one slope and one intercept are obtained. The activation energies establish the energy of the crossing of the excited-state negative ion and neutral curves.

An empirical relationship $D(R-X) = E^* + E_a(X) - C$ relates the bond dissociation energy, electron affinity of the halogen, and activation energy for these types of reactions. For compounds studied in the ECD the quantity C, obtained from the

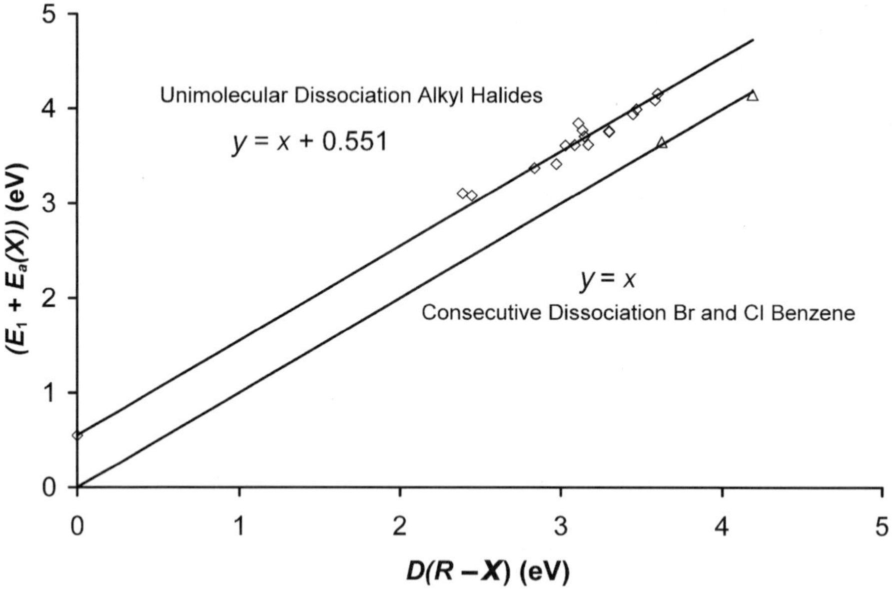

Figure 11.2 Correlation of activation energies for thermal electron attachment with bond dissociation energies for DEC(1) and DEC(2) compounds.

plot shown in Figure 11.2, is 0.55(9) eV. The bond dissociation energies for other compounds are obtained from these data using $C = 0.55(9)$ and the experimental E_1. These are compared with experiment in Table 11.1 [11, 12]. This relationship will be valid if dissociation takes place via a curve crossing on the right-hand side of the neutral, as in Figure 11.1 for CH_2Cl_2. If, however, the ground-state curve also participates in the dissociation, this gives a lower limit to the bond dissociation energy. Also shown in Figure 11.2 is a unit slope zero intercept line that is valid for compounds in which dissociation only results from a single state. In the case of CF_3Cl the $D(R - \mathrm{Cl})$ is $E^* + E_a(\mathrm{Cl}) - 0.55 = 3.31$ eV. As seen in Table 11.1, the value found in the literature is 3.73 eV. If, on the other hand, the dissociation only occurs from the ground state, then $D(R - \mathrm{Cl}) = E^* + E_a(\mathrm{Cl}) = 3.93$ eV, which is an upper limit. This interpretation suggests that two curves participate in the dissociation. In the case of the chloroalkanes there is a region at low temperatures with a low slope. These data points cannot be explained with a single state [1, 5].

From the data in Table 11.1 the consistency of A_1 is apparent. If we assume a constant A_1 about equal to the DeBroglie A_1, and with relative molar responses of similar compounds in the ECD at the same temperature, the activation energy for thermal electron attachment can be obtained from $E_{1x} - E_{\mathrm{ref}} = RT(\ln(R_{\mathrm{ref}}/R_X))$. This is only applicable to alkyl chlorides, bromides, or iodides. The activation energy of dissociative reactions for alkylfluorides is too large to be observed in the ECD.

Shown in Table 11.2 and 11.3 are the log (relative responses) and activation energies and bond dissociation energies obtained from this analysis [11–14]. The

TABLE 11.2 Thermodynamic Properties for Dissociative Electron Attachment to Alkyl Halides from ECD Data at 373 K (in eV)

Compound $R-X$	E_1 ECD	$E_a(X)$	$D(R-X)$ in the Literature	$D(R-X)$ ECD (463 K)	ln(*RR*) ECD
CHF_2–Cl	0.45 ± 0.10	3.616	—	3.52 ± 0.10	2.30
C_2F_5–Cl	0.33 ± 0.10	3.616	3.60	3.40 ± 0.10	5.13
1,1C_2F_4Cl–Cl	0.21 ± 0.10	3.616	—	3.28 ± 0.10	8.81
CF_2Cl–Cl	0.16 ± 0.10	3.616	3.31	3.23 ± 0.10	10.30
$CHCl_2$–Cl	0.16 ± 0.02	3.616	3.30	3.23 ± 0.10	10.41
CF_3–Br	0.12 ± 0.10	3.365	3.05	2.94 ± 0.10	11.37
1,2,2$C_2F_3Cl_2$–Cl	0.11 ± 0.10	3.616	—	3.18 ± 0.10	11.98
CF_3CHCl–Br	0.07 ± 0.10	3.365	2.85	2.89 ± 0.10	12.89
C_3F_7–I	0.06 ± 0.10	3.057	2.17	2.57 ± 0.10	13.30
1,2C_2F_4Br–Br	0.06 ± 0.10	3.365	—	2.88 ± 0.10	13.55
$CHBr_2$–Br	0.06 ± 0.10	3.365	2.84	2.86 ± 0.10	13.81
$CFCl_2$–Cl	0.04 ± 0.10	3.616	3.17	3.11 ± 0.10	13.99
CF_2Br–Br	0.03 ± 0.10	3.365	2.84	2.83 ± 0.10	14.34
CCl_3–Cl	0.02 ± 0.10	3.616	3.08	3.09 ± 0.10	14.60
1,2 C_2F_4I–I	0.00 ± 0.10	3.055	—	2.50 ± 0.10	15.20

uncertainties in the E_1 are nominally ± 0.10 eV. The deviation between the values determined using multiple data points given in Table 11.1 and the single-temperature data in Tables 11.2 and 11.3 is less than ± 0.05 eV. The dissociation energies are approximately the same for the primary *RX* compounds. The values for the bromo and iodo compounds are lower than for the chlorocompounds.

TABLE 11.3 Thermodynamic Properties for Dissociative Electron Attachment to Alkyl Halides from ECD Data at 463 K (in eV)

Compound $R-X$	E_1 ECD	$E_a(X)$	$D(R-X)$ in the Literature	$D(R-X)$ ECD (463 K)	ln(*RR*) ECD
C_4H_9–Cl	0.58 ± 0.10	3.616	—	3.65 ± 0.10	0.00
C_5H_9–Cl	0.58 ± 0.10	3.616	—	3.65 ± 0.10	0.00
C_6H_{13}–Cl	0.58 ± 0.10	3.616	—	3.65 ± 0.10	0.09
C_8H_{17}–Cl	0.57 ± 0.10	3.616	—	3.64 ± 0.10	0.47
2-C_4H_9–Cl	0.55 ± 0.10	3.616	—	3.62 ± 0.10	0.69
t-C_4H_9–Cl	0.48 ± 0.10	3.616	—	3.54 ± 0.10	2.48
1,4C_4H_8Cl–Cl	0.47 ± 0.10	3.616	—	3.53 ± 0.10	2.71
1,1C_4H_8Cl–Cl	0.39 ± 0.10	3.616	—	3.46 ± 0.10	4.70
1,2C_2H_4Cl–Cl	0.37 ± 0.10	3.616	—	3.44 ± 0.10	5.25
C_3H_7–Br	0.36 ± 0.10	3.365	—	3.18 ± 0.10	5.54
C_4H_9–Br	0.36 ± 0.10	3.365	—	3.18 ± 0.10	5.63
$CHCl_2$–Cl	0.14 ± 0.02	3.616	3.30	3.21 ± 0.10	11.00
C_4H_9–I	0.12 ± 0.10	3.055	2.15	2.52 ± 0.10	11.40
1,1C_2H_4Br–Br	0.12 ± 0.10	3.365	—	2.83 ± 0.10	11.61
CCl_3–Cl	0.06 ± 0.10	3.616	3.08	3.13 ± 0.10	12.90

TABLE 11.4 Thermodynamic Properties for Dissociative Electron Attachment to Alkyl Halides from Rate Constants at 298 K (in eV)

Compound $R-X$	E_1 Rates	$E_a(X)$	E_1 ECD	$D(R-X)$ Rates	ln(RR) Rates
CH_3–Cl	0.48 ± 0.10	3.616	0.54	3.55 ± 0.10	0.00
CF_3–Cl	0.39 ± 0.10	3.616	0.25	3.46 ± 0.10	3.50
CHF_2–Cl	0.38 ± 0.10	3.616	0.45	3.45 ± 0.10	4.21
CHFCl–Cl	0.37 ± 0.10	3.616	0.25	3.44 ± 0.10	6.21
CH_2Cl–Cl	0.29 ± 0.10	3.616	0.33	3.36 ± 0.10	7.36
CH_3–Br	0.29 ± 0.10	3.365	0.25	3.10 ± 0.10	7.41
1,1C_2H_4Cl–Cl	0.25 ± 0.10	3.616	—	3.32 ± 0.10	8.85
1,2C_2H_4Cl–Cl	0.24 ± 0.10	3.616	0.38	3.31 ± 0.10	9.27
CH_3CHCl–Cl	0.20 ± 0.10	3.616	—	3.27 ± 0.10	10.59
CH_3CCl_2–Cl	0.16 ± 0.10	3.616	—	3.23 ± 0.10	12.36
CF_2Cl–Cl	0.14 ± 0.10	3.616	0.15	3.21 ± 0.10	13.12
$CHCl_2$–Cl	0.12 ± 0.02	3.616	0.14	3.19 ± 0.10	13.81
1,1C_2F_4Cl–Cl	0.11 ± 0.10	3.616	—	3.18 ± 0.10	14.28
CF_3CFCl–Cl	0.09 ± 0.10	3.616	—	3.16 ± 0.10	15.11
CF2ClCFCl–Cl	0.08 ± 0.10	3.616	—	3.14 ± 0.10	15.42
CH_3–I	0.04 ± 0.02	3.055	0.03	2.54 ± 0.10	16.96
$CFCl_2$–Cl	0.01 ± 0.10	3.616	0.01	3.08 ± 0.10	18.01
$C_2F_3Cl_2$–Cl	0.01 ± 0.10	3.616	—	3.08 ± 0.10	18.35
CCl_3–Cl	0.00 ± 0.10	3.616	0.00	3.07 ± 0.10	18.57

Although dissociative thermal electron attachment has been studied by many different techniques, the ECD method is the only one that has been extensively applied to reactions at 1 atm pressure as a function of temperature. Other procedures give values for the rate constants that agree with those obtained from the ECD at a specified temperature. To predict the temperature dependence of the rate constants determined by other methods, the single-point data relationship is applied to data in the literature. The ln of the relative rate constants are given in Table 11.4 [15]. The activation energies agree well with those values obtained with the ECD, as seen in Table 11.4. The largest difference is 0.14 eV, while the standard deviation is ±0.1 eV. These represent another determination of the bond dissociation energies using the linear correlation. For reactions with responses that are close to the maximum values, the rate constants are given by the DeBroglie A_1 value and the E_1 are small.

11.2.3 Alkyl Fluorocompounds

Some of the above alkyl halides contain fluorine. In the case of CF_3Cl the activation energy indicated that two negative-ion curves contribute to the dissociative thermal electron attachment. In 1997 S. R. Sousa and S. E. Bialkowski carried out a classic study of the ECD temperature dependence of alternative fluorocarbon freon replacements [16]. The study used a commercial ECD and the fundamental concepts discussed in this book to obtain rate constants and energies for compounds. One of these, CF_2Cl_2, was used as an internal standard in all the measurements. The

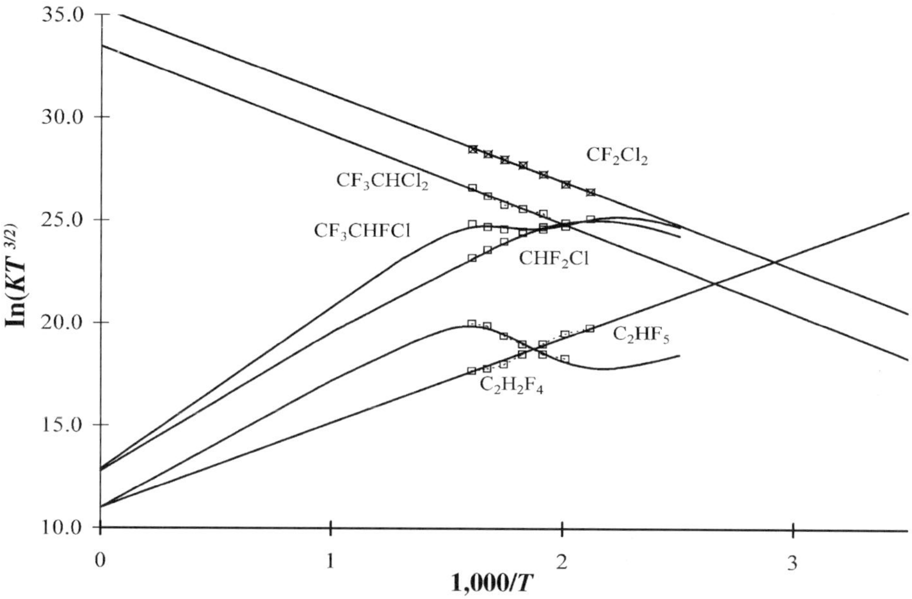

Figure 11.3 Plots of ECD data as ln $KT^{3/2}$ versus 1,000/T for freon substitutes [16]. The parameters are given in Table 11.5. The analysis appears in this work.

data were analyzed in terms of either dissociative or nondissociative capture because the multistate model had not been presented. The basic classification of the compounds from Figure 11.3 is as follows: (1) CCl_2F_2 (CFC-12), CF_3CHCl_2 (HCFC123), and CF_3CHF_2 with only a single negative slope; (2) $CHClF_2$ (HCFC-22), CF_3CH_2F (HFC134a) with only a single positive slope; and (3) CF_3CHClF with two slopes. These responses have been converted to absolute electron capture coefficients by scaling to the response for CF_2Cl_2 and adjusting for the relative molar responses reported at 573 K.

The molecules with two chlorines on a single carbon are analogous to methylene chloride and have similar E_1 and A_1 values. The reported activation energies for CCl_2F_2 and CF_3CHCl_2 are 0.28 eV and 0.32 eV, respectively. The value for CCl_2F_2 is 0.15 eV higher than the value previously determined at lower temperatures in Table 11.1. Indeed, the higher value gives a C—Cl bond dissociation energy of 3.34 ± 0.10 eV, as compared to the value of 3.30 eV to be found in the literature. This suggests that the lower-temperature data partly result from molecular ion formation with an E_a of about 0.4 eV. The negative slope of the data for CF_3CHCl_2 gives the C—Cl bond energy of 3.40 eV. It is also consistent with an E_a of approximately 0.5 eV.

The two molecules with a single positive slope ($CHClF_2$ and CF_3CH_2F) and the one with two slopes (CF_3CHF_2) indicate nondissociative thermal electron attachment. The negative slope for the latter is the activation energy for electron attachment to the ground state. There are no reasonable dissociative pathways.

TABLE 11.5 ECD and NIMS Parameters for Freon Substitutes and Fluorocarbons [16]

Compound	E_1 (eV)	ln(A_1) (eV)	E_a (eV)	Q_{an}	E_a(Lit) (eV)
CCl_2F_2	0.28	36	—	—	—
CF_3CHCl_2	0.32	34	—	—	—
$CHClF_2$	0.35	33.9 ± 0.5	0.50 ± 0.15	1	0.28
CF_3CHClF	0.37	33.9 ± 0.5	0.60 ± 0.15	1	—
CF_3CHClF^*	0.23	33.9 ± 0.5	0.50 ± 0.15	1	—
CF_3CHF_2	0.57	33.9 ± 0.5	0.45 ± 0.15	0.2	—
$CF_3CHF_2^*$	0.09	33.9 ± 0.5	0.26 ± 0.15	0.2	—
CF_3CH_2F	0.01	33.9 ± 0.5	0.38 ± 0.15	0.2	0.45
$CF_3C_6F_{11}$	0.03	34.9 ± 0.5	1.06 ± 0.05	1.1	1.07
$CF_3C_5F_9$	0.04	35.0 ± 0.5	1.02 ± 0.05	1.0	—

* Excited state.

The negative slope for CF_3CHClF may also be an activation energy for attachment to the ground state. The positive slope indicates stable negative-ion formation. The data for $CHClF_2$ were analyzed to yield an E_a of 0.28 eV. However, this is a lower limit because the intercept is high. As shown in Figure 11.3, the use of a Q value of 1 gives a higher E_a of 0.5 eV. By using two states, a curve for CF_3CHClF is constructed with an E_a of 0.5 eV and 0.55 eV. If $Q < 0.1$, the E_a of CF_3CH_2F is 0.45 eV, as reported. If $Q = 0.1$, the E_a is 0.35 eV. The curve drawn through the data uses two states, $Q = 0.5$ and $E_1 = 0.37$ eV, for CF_3CHF_2 and E_a of 0.35 eV and 0.40 eV.

Table 11.5 lists the electron affinities and activation energies obtained using two states and nominal intercepts. The larger uncertainties in the electron affinities reflect possible systematic uncertainties due to the values of Q. The E_a are confirmed by the CURES-EC calculations. These are the only measurements of a positive valence-state electron affinity of the fluoroethanes and chlorofluoroethanes. The parent negative ion of C_4F_{10} has been observed and the E_a of CF_2Cl_2 measured as 0.4 eV [8]. The values in Table 11.5 are thus reasonable.

11.2.4 Electron Affinities of the Alkyl Halides

The electron affinities of the alkyl and cycloalkyl halides reported in the NIST tables are tabulated and evaluated in Appendix II. The NIST value is returned by a sequential search, using CX, CClF, or CHX. In some cases two values that differ by more than the uncertainty have been reported so excited-state values can be assigned. Many of the E_a have only been measured by one technique; thus, the experimental values could refer to an excited state.

The electron affinities of C_2Cl_6, CBr_4, CCl_4, $CHCl_3$, and CH_2Cl_2 have been measured using the magnetron method. The value for C_6Cl_6 is lower than for $CHCl_3$, but the CURES-EC value given in Table 10.1 is the same as for CCl_4. Therefore, the ground-state value is estimated to be 1.80(15) eV. The AMB E_a for CCl_4 confirmed the magnetron value and provide estimates of the electron affinities for CF_2Cl_2, $CFCl_3$, CF_3X, and CH_3I. The AMB E_a for $CFCl_3$ was confirmed by TCT studies.

It is 1.0 ± 0.3 eV that is lower than the value obtained for $CHCl_3$ and consequently could exist for an excited state. We estimate the ground-state E_a of all the fully halogenated compounds to be larger than the experimental values obtained through AMB, except for CF_3I. Excited-state values have been measured for CCl_4 and $CHCl_3$ in TCT studies. The AE_a is approximately 2 eV for CBr_4 and CCl_4 [7–10, 17–20].

The electron affinities of the chloroethylenes and two bromoethylenes have been measured using the ECD. Both the halogen negative ions and parent negative ions are observed. The electron affinities of C_2Cl_4 and C_2HCl_3 measured in NIMS are verified using $E_{1/2}$ values. The NIMS data for these compounds are shown in Figure 5.12. The ECD values are assigned to excited states. By analogy the adiabatic electron affinities of $C_2H_2Cl_2$ are estimated to be 0.15 eV to 0.25 eV and that of vinyl chloride to be 0.1 ± 0.05 eV [21–23]. These values are confirmed by CURES-EC.

The electron affinity of perfluromethylcyclohexane, C_7F_{14}, has been measured by thermal charge transfer studies. The electron affinity of perfluorocyclobutane has been measured by the TCT, kinetic equilibrium, and endothermic charge transfer methods. Two values were reported: 1.05 ± 0.1 eV and 0.6 ± 0.02 eV. The lower value was assigned to an isomer, but could also be for an excited state. Interestingly, both these values are higher than the 0.4 ± 0.3 eV obtained from endothermic charge transfer studies, indicating multiple bound states [24–26].

The electron affinity of C_7F_{14} can be determined from unpublished API-MS data for the parent negative ion. These data are shown in a plot of ln $KT^{3/2}$ versus 1000/T for C_7F_{14} in Figure 11.4. The good fit is apparent. The ln(A_1) is 34.9(2); the Q_{an} is

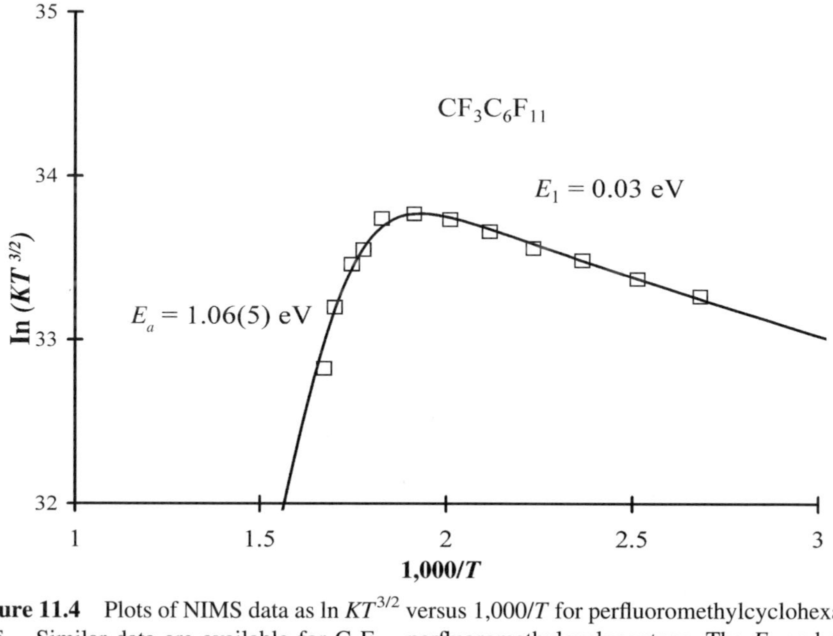

Figure 11.4 Plots of NIMS data as ln $KT^{3/2}$ versus 1,000/T for perfluoromethylcyclohexane C_7F_{14}. Similar data are available for C_6F_{12}, perfluoromethylcyclopentane. The E_a and Q_{an} values are given in Table 11.5. The analysis appears in this work.

1.1(2); the E_1 is 0.03 ± 0.01 eV, while the E_a is 1.06(5). The E_a agrees with the TCT value and the kinetic parameters support measurements of the rate constant for thermal electron attachment [27, 28]. For perfluoromethylcyclopentane, C_6F_{12} ($m/z = 300$), the parameters are $\ln(A_1) = 35.0(2)$; $Q_{an} = 1.0(2)$; $E_1 = 0.04(1)$ eV, and $E_a = 1.02(5)$. These values are given in Table 11.5. There are no other values for these parameters in the literature. The activation energy for the formation of the $[M - F](-)$ by dissociation from the ground state is 1.35 ± 0.10 eV. With the electron affinity of the radical, 3.9 eV, the C—F bond dissociation energy is 4.15(10) eV = (1.35 − 1.06 = 3.9). The comparable activation energy from perfluromethylcyclopentane is 1.35(10) eV. Assuming the same bond dissociation energy, $E_a(C_6F_{11})$ is $4.15 - 1.35 + 1.02 = 3.92(15)$ eV.

11.3 AROMATIC HALIDES

11.3.1 Electron Affinities of Fluoro- and Chlorobenzenes

The electron affinities of C_6F_6 have been determined by seven different methods. The TCT and kinetic equilibrium methods give 0.5(1) eV. The ECD, PES, NIMS, and a second TCT method yield 0.86 eV. The magnetron and endothermic charge transfer values are 1.20(15) eV and 1.8(3) eV. The ECD value of 0.86(2) eV is the most precise. Ten independent ECD determinations by different investigators using different equipment produced this value. A TCT study bracketed the E_a of C_6F_6 between that of benzophenone and nitrobenzene or 0.85(15) eV. The magnetron value overlaps the ECD value at twice the nominal uncertainty. The endothermic charge transfer value with a lower limit of 1.2 eV is a high outlier. The reduction potential and CURES-EC values support the ECD value of 0.86 ± 0.02 eV. The lower value can be assigned to an excited state. Two states are observed in the ECD data. The ECD E_a of C_6F_5H is 0.73(5) eV. A lower limit of 0.4 eV has been obtained from TCT measurements. The TCT E_a of perfluorotoluene, C_7F_8, is 0.86(10) eV. This is the same as for the ground state of C_6F_6. The substitution of a CF_3 group increases the E_a by about 0.4 eV. The ground-state E_a should be about $0.73 + 0.4 = 1.15$ eV. The CURES-EC value is 1.3(1) eV. The endothermic charge transfer E_a is 1.7(3) eV [26, 28–34]. Thus, we propose that the TCT E_a is for an excited state and the AE_a is 1.0 ± 0.2 eV.

In the 1960s ECD data for the mono, di, and tetra fluorobenzenes were obtained but not published because of the unexplained structure in the data. We have analyzed the ECD data for all the fluorobenzenes using two states and report on their electron affinities. Figure 11.5 provides the ECD data. Table 11.6 lists the ECD parameters. The $\ln(A_{1i})$, 34.5 to 36.5, are within an order of magnitude of $\ln(\text{DeBA}) = 36$. The Q values are within the observed range of 10^{-4} to 1 [35]. The CURES-EC calculations agree with the AE_a. The electron affinities range from 0.13(5) eV to 0.86(2) eV for C_6F_6. There is a regular change in the E_a values with the number of fluorine atoms. Each fluorine replacement increases the E_a by about 0.15 eV. These values are supported by experimental reduction potentials [36].

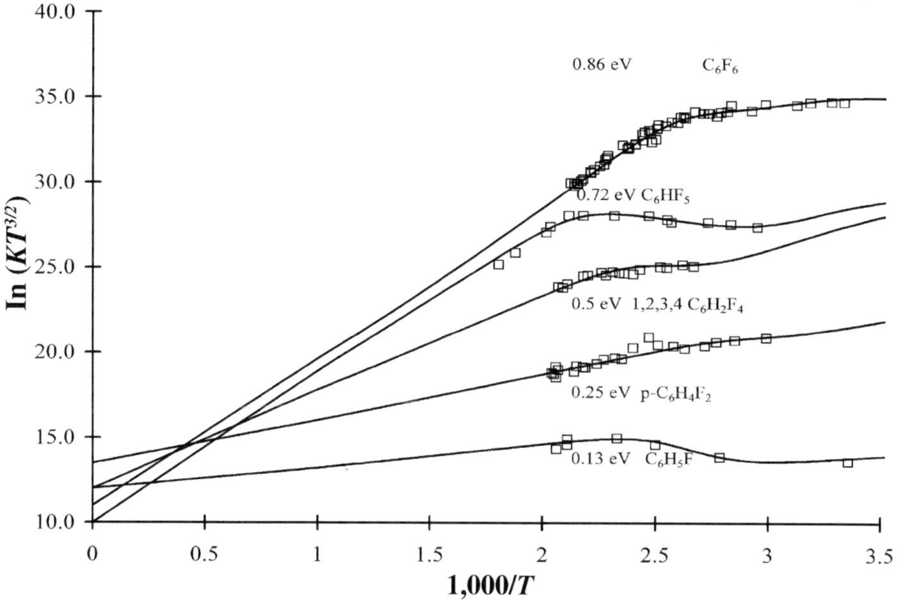

Figure 11.5 ECD data as ln $KT^{3/2}$ versus 1,000/T for fluorobenzenes illustrating multiple states and nondissociative capture, from [34, 35]. The analysis appears in this work. The parameters for the calculated curves are given in Table 11.6.

The electron affinities of several pentafluorophenyl compounds have been determined using the TCT method. Values for $C_6F_5NO_2$ and C_6F_5Cl have been measured using the ECD. The increment per fluorine addition ranges from 0.1 eV/F for the nitro compound to 0.16 eV/F for C_6F_5CN. The largest increment per fluorine atom among the substituted benzene compounds occurs for 1,4-$C_6F_4(CN)_2$, which is 0.2 eV/F atom [37].

TABLE 11.6 ECD Parameters for Fluorobenzenes

Compound	ln(A_1) (eV)	E_1 (eV)	Q_{an}	E_a (eV)
C_6H_5F	35.2	0.63	0.80	0.13 ± 0.05
$C_6H_5F^*$	35.2	0.20	0.40	0.07 ± 0.05
p-$C_6H_4F_2$	35.4	0.39	0.22	0.25 ± 0.05
p-$C_6H_4F_2$*	35.4	0.19	1.00	0.19 ± 0.05
1,2,3,4-$C_6H_2F_4$	35.5	0.28	0.48	0.52 ± 0.05
1,2,3,4-$C_6H_2F_4^*$	35.8	0.10	0.88	0.40 ± 0.05
C_6HF_5	34.9	0.17	0.31	0.72 ± 0.05
$C_6HF_5^*$	35.5	0.05	1.04	0.43 ± 0.05
C_6F_6	34.7	0.04	0.04	0.86 ± 0.02
$C_6F_6^*$	34.9	[0.01]	1.00	0.61 ± 0.05

*Excited state.

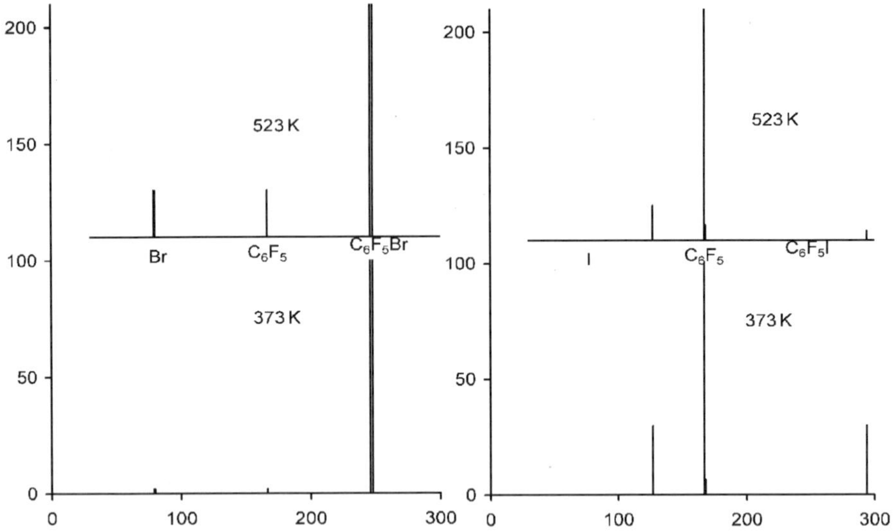

Figure 11.6 Negative-ion mass spectra for C_6F_5Br and C_6F_5I, from [39].

The electron affinities of C_6Cl_6 and C_6HCl_5 have been determined by TCT [37, 38]. The value for C_6Cl_6 has also been determined by ECD [23, 35]. The ECD value is about 0.15 eV higher than the TCT value, as is the ECD C_6F_5Cl value. In the ECD data for C_6F_5Cl an excited state with a lower electron affinity is observed. By analogy to the fluorobenzenes two states are anticipated. Dissociative electron attachment is not observed in the NIMS spectra taken at 373 K and 523 K. For the majority of these compounds the activation energy is greater than 1 eV, so the only feasible reaction is nondissociative thermal electron attachment. In the case of C_6F_5Br and C_6F_5I both dissociative and nondissociative capture is observed. The relative intensities of the $C_6F_5(-)$, Br(−), or I(−) ions give the electron affinity of C_6F_5. The relative intensity of Br(−) to the $C_6F_5(-)$ ions is about 1 while the ratio for the I(−) to the $C_6F_5(-)$ ions is 0.2. In the spectrum of $C_6F_5I(-)$ the parent negative ion is observed at 373 K, but is much smaller at 523 K. These data are illustrated in Figure 11.6. The E_a of the C_6F_5 radical falls between that of I, 3.07 eV, and Br, 3.37 eV or 3.22(10) eV [33, 37, 39, 40].

CURES-EC calculated the E_a for the C_6X_5 radicals as 3.15, 2.90, 3.05, and 2.90 eV, for X from F to I. Since the E_a of the phenyl radical is 1.10 eV the E_a increments per X are 0.4 eV/F, 0.36 eV/Cl, 0.4 eV/Br, and 0.36 eV/I. The first and second fluorine atoms have a larger increment since the experimental E_a of p-C_6H_4F is 1.7 eV, while that of $C_6H_3F_2$ is 2.1 eV. The CURES-EC values are 1.72 eV and 2.15 eV. The last three replacements contribute an average of 0.3 eV/F. The experimental E_a of the mono-chlorophenide radicals range from 1.4(2) eV to 1.7(2) eV, giving increments from 0.3 to 0.6 eV/Cl. The CURES-EC values for o-C_6H_4Cl and p-C_6H_4Cl are 1.4 eV and 1.7 eV [37].

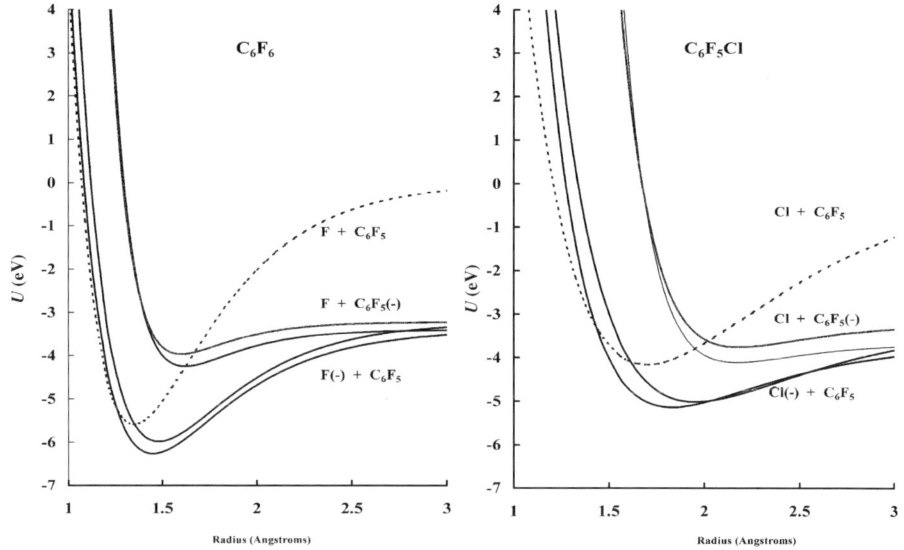

Figure 11.7 Morse potential energy curves for C_6F_5Cl and C_6F_6, Eql(2/2) compounds. Calculated from electron impact, ECD, and TCT electron affinities.

With the E_a of the molecules and radicals, bond dissociation energies, and electron impact data, the Morse potential energy curves for the C_6F_5X compounds can be calculated. They are shown in Figures 11.7 and 11.8. There are two curves dissociating to each of the complementary limits $C_6F_5 + X(-)$ and $C_6F_5(-) + X$. The

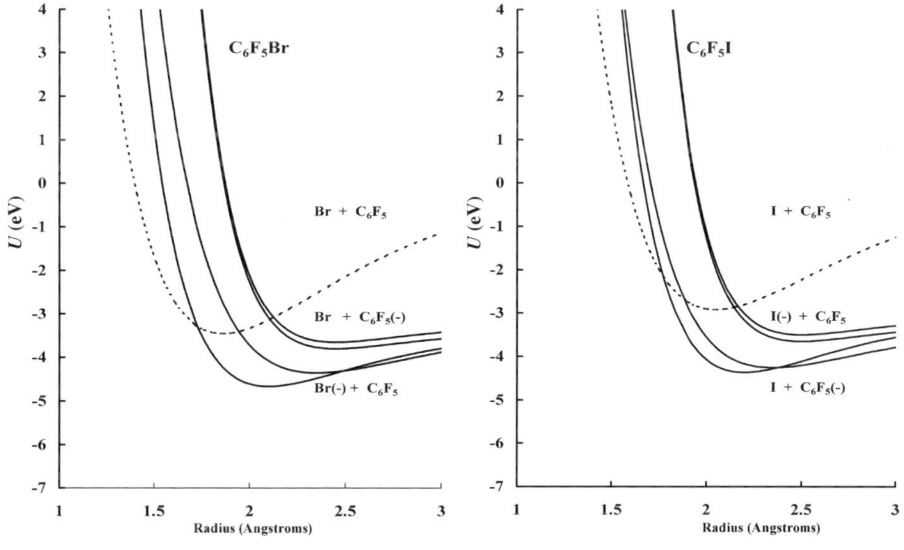

Figure 11.8 Morse potential energy curves for C_6F_5Br and C_6F_5I DEC(2) compounds. Calculated from electron impact and TCT electron affinities.

EDEA for C_6F_5I, C_6F_5Br, C_6F_5Cl, and C_6F_6 are 0.2, −0.2, −0.5, and −1.2 eV. Dissociative electron attachment occurs in C_6F_5I and C_6F_5Br. The two bonding curves are drawn with the experimental E_a for C_6F_6 and C_6F_5Cl, while the two other curves are drawn with an assumed bond energy, the VE_a, and electron impact ion distributions. The activation energy of about 0.4 eV needed for the reaction to form the ground state of C_6F_5Cl results from this "backside crossing." The activation energy to the first excited state is small, but rapid stabilization to the ground state is possible. These curves are both $M(2)$ since the E_a and VE_a are positive but the EDEA is negative. Electron attachment will primarily take place via the first excited state, but the ground state will participate at higher temperatures. The two regions are observed in the ECD data for C_6F_5Cl. The TCT value, 0.82(10) eV, is assigned to the excited state and the higher ECD value, 1.00(10) eV, to the ground state [33–35]. At sufficiently high temperatures or with hyperthermal electrons, dissociative electron attachment will take place. The curves for C_6F_6 are constructed in a similar manner.

The bonding curves for C_6F_5I and C_6F_5Br are qualitative since only one E_a has been measured. In the case of C_6F_5Br the low-temperature reaction will be dominated by nondissociative capture. At higher temperatures both immediate and sequential dissociation from the first excited state can take place. Because both reactions have the same activation energy, the intensities of the Br(−) and C_6F_5(−) ions should be the same. For C_6F_6I the molecular ion is formed at low temperatures via the first excited state. This anion state can also dissociate to yield I(−) and C_6F_5(−). The activation energy for the formation of I(−) is greater than for C_6F_5(−), as reflected in the NIMS data. In electron impact studies the parent negative ions of C_6F_6, C_6F_5Cl, and C_6F_5Br were detected at thermal energies, but the parent negative ion of C_6F_5I was not observed [39, 40].

The electron affinities of the lower chlorobenzenes and several chlorobenzenes substituted by the fluoro and trifluoromethyl groups have been measured using the ECD. These data have been revisited using two states with dissociation that give higher electron affinities [34]. The E_a for the chlorobenzenes range from 0.17 eV to 1.15 eV. This corresponds to an increment in the E_a of 0.2 eV/Cl atom. Figure 11.9 shows the ECD data for the chlorofluorobenzene. Figure 11.10 presents the data for representative mixed halogenated compounds [6, 34]. The ECD parameters given in Table 11.7 are used to calculate the curves in these figures. In these cases the energy for dissociative electron attachment and the bond dissociation energy are obtained from the ECD data. The bond dissociation energies for several aromatic C—X bonds determined in this manner are given in Table 11.8. The value for the iodobenzene is an upper limit because dissociative thermal electron attachment is exothermic. The electron affinities for the CF_3-substituted compounds calculated using CURES-EC are higher than the experimental values. This could indicate that the ECD values are for excited states, or it might simply mean that the CURES-EC calculations give upper limits. The E_a of C_6F_5X, where X is NO_2, CN, CHO, $CO(CH_3)$, and aromatic halogen compounds, are evaluated in the appendices [33, 37].

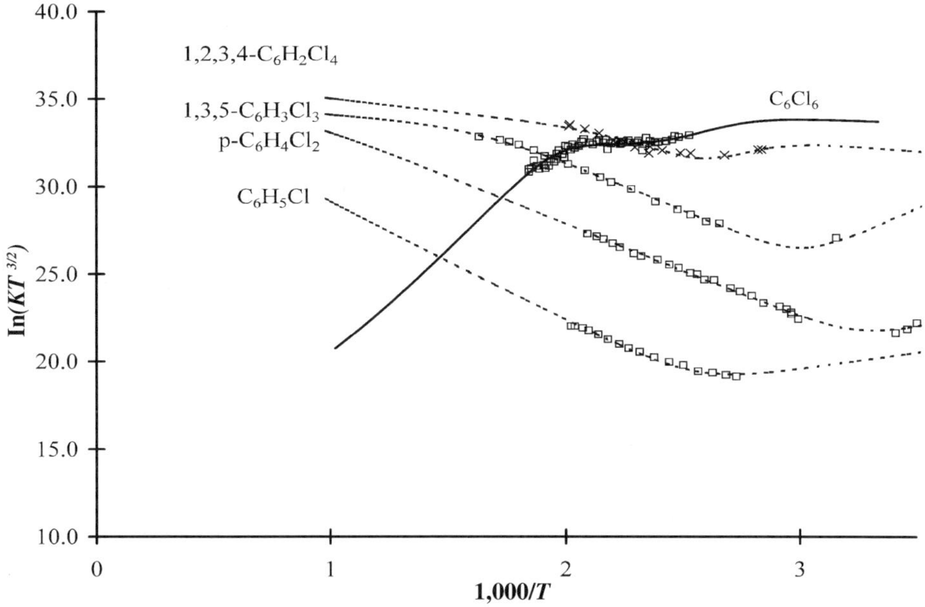

Figure 11.9 ECD data as ln $KT^{3/2}$ versus 1,000/T for chlorobenzenes illustrating multiple states and dissociative and nondissociative capture, data from [6, 34]. The analysis appears in this work. The parameters for the calculated curves are given in Table 11.7.

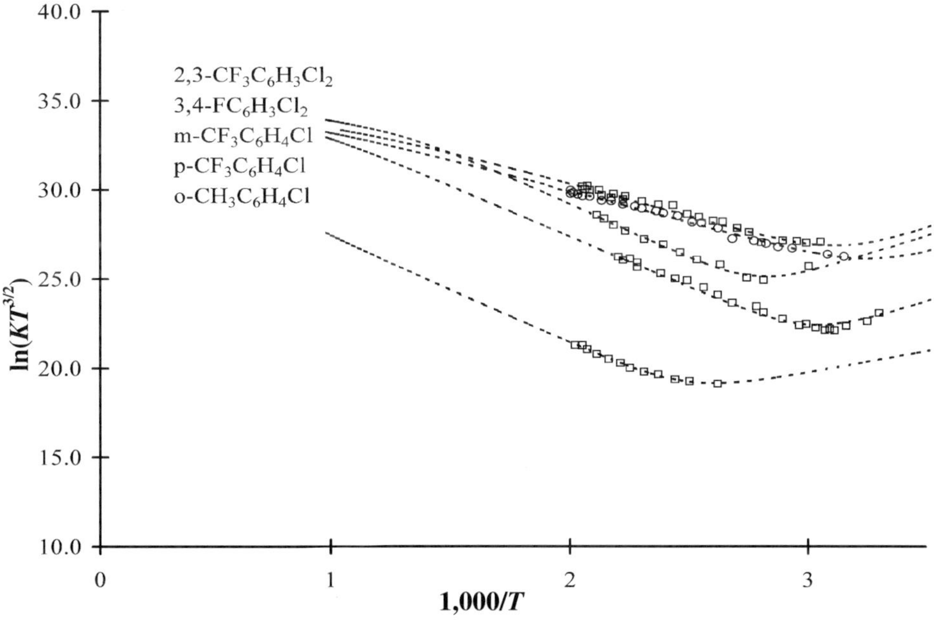

Figure 11.10 Plots of ECD data as ln $KT^{3/2}$ versus 1,000/T for mixed halogenated benzenes. The parameters for the calculated curves are given in Table 11.7.

TABLE 11.7 Kinetic and Thermodynamic Properties of Dissociative Electron Attachment

Species	ln(A_1) (eV)	E_1 (eV)	Q	E_a (eV)	E_2 (eV)	ln(A_2) (eV)	D(C–Cl)
3-Cl-acetophenone	34.9	0.03	0.99	0.67(5)	1.05(5)	28.5(3)	3.96(5)
4-Cl-acetophenone	34.8	0.00	0.60	0.64(5)	1.05(5)	29.4(4)	4.01(5)
C_6H_5Cl	35.1	0.07	0.38	0.17(10)	0.73(5)	24.2(3)	4.09(5)
p-$C_6H_4Cl_2$	35.2	0.03	0.45	0.29(5)	0.72(5)	29.0(3)	4.03(5)
m-$C_6H_4Cl_2$	35.1	0.04	0.81	0.30(5)	0.69(5)	25.7(3)	3.97(5)
o-$C_6H_4Cl_2$	35.1	0.03	0.70	0.30(5)	0.65(5)	25.7(3)	3.92(5)
s-$C_6H_3Cl_3$	35.6	0.04	0.64	0.48(5)	0.91(5)	32.9(3)	4.04(5)
s-$C_6H_2Cl_4$	35.8	0.03	0.94	0.69(5)	1.10(5)	34.2(3)	4.02(5)
1-Cl-naphthalene	35.6	0.14	1.01	0.34(5)	0.82(5)	29.7(2)	3.94(5)

The value in parentheses are uncertainties in the last figure.

The electron affinities of the three mono-chloroanthracenes measured by TCT and ECD agree within the uncertainties. The E_a of bromoanthracene and bromopyrene measured by the collisional ionization techniques could be for excited states based on CURES-EC and the substitution effects. The E_a of 9-bromoanthracene, 0.62 eV, is lower than that of anthracene and lower than the CURES-EC value of 0.95 eV. The evaluated value for 9-chloroanthracene is 0.86 eV. The value for the bromocompound should be higher. The E_a for bromopyrene, 0.75 eV, is larger than that for pyrene but is still lower than the CURES-EC value of 1.1 eV. Many of the

TABLE 11.8 Bond Dissociation Energies (in eV) from ECD Data

Species	–EDEA	$E_a(X)$	D(C–X)	D(Lit)
C_6H_5I	0.10(5)	3.055	<3.15	2.85
C_6H_5Br	0.25(3)	3.365	3.62	3.65
C_6H_5Cl	0.56(3)	3.615	4.18	4.15
3-Cl-acetophenone	0.38(3)	3.615	3.99	—
4-Cl-acetophenone	0.41(3)	3.615	4.02	—
p-$C_6H_4Cl_2$	0.43(3)	3.615	4.05	—
m-$C_6H_4Cl_2$	0.39(3)	3.615	4.02	—
o-$C_6H_4Cl_2$	0.35(3)	3.615	3.97	—
s-$C_6H_3Cl_3$	0.43(3)	3.615	4.05	—
s-$C_6H_2Cl_4$	0.41(3)	3.615	4.03	—
1,4$CF_3C_6H_4Cl$	0.41(4)	3.615	4.03	—
1,2$CF_3C_6H_4Cl$	0.34(5)	3.615	3.97	—
1,2,4$CF_3C_6H_3Cl_2$	0.28(5)	3.615	3.91	—
1,3,4-$FC_6H_3Cl_2$	0.41(5)	3.615	4.03	—
1-Cl-naphthalene	0.32(3)	3.615	3.94	—
1,3$CF_3C_6H_4Br$	0.10(3)	3.365	3.47	—
1,2FC_6H_4Br	0.20(3)	3.365	3.57	—
1-Br-naphthalene	0.06(3)	3.365	3.42	—

TABLE 11.9 Electron Affinities (in eV) of Chlorobenzenes from Reduction Potentials, ECD, and CURES-EC [22 and this work]

Species	E_a(ER)	E_a(ECD)	E_a(CURES-EC)
C_6H_5Cl	0.10	0.17(5)	0.10
p-$C_6H_4Cl_2$	0.27	0.29(5)	0.25
m-$C_6H_4Cl_2$	0.27	0.29(5)	0.25
o-$C_6H_4CCl_2$	0.25	0.30(5)	0.23
1,3,5-$C_6H_3Cl_3$	0.49	0.48(5)	0.47
1,2,3-$C_6H_3Cl_3$	0.51	—	0.45
1,2,4-$C_6H_3Cl_3$	0.47	—	0.49
1,2,4,5-$C_6H_2Cl_4$	0.66	0.65(5)	0.71
1,2,3,4-$C_6H_2Cl_4$	0.73	—	0.72
1,2,3,5-$C_6H_2Cl_4$	0.68	—	0.70
C_6Cl_5H	0.90	0.85(10)	0.90
C_6Cl_6	1.15	1.15(5)	1.15

TCT values could exist for excited states. Based on this analysis, the AE_a should be larger, as indicated in the appendices [41, 42].

11.3.2 Electron Affinities from Reduction Potentials and CURES-EC

The electron affinities of the chlorobenzene isomers have been determined by scaling half-wave reduction potentials [22]. With higher gas phase values higher values are obtained from reduction potentials. These are compared to the ECD and CURES-EC values in Table 11.9. The CURES-EC-calculated values for the above compounds support experimental quantities and suggest that the E_a of all halogenated benzenes can be calculated. The CURES-EC values are listed in Table 11.10. The E_a

TABLE 11.10 Calculated Valence-State Electron Affinities (in eV) of Halobenzenes [this work]

Species	X=F	X=Cl	X=Br	X=I
C_6H_6	−0.756	−0.756	−0.756	−0.756
C_6H_5X	0.10	0.10	0.30	0.50
p-$C_6H_4X_2$	0.19	0.25	0.53	0.76
m-$C_6H_4X_2$	0.21	0.25	0.54	0.79
o-$C_6H_4CX_2$	0.16	0.23	0.50	0.74
1,3,5-$C_6H_3X_3$	0.45	0.47	0.72	1.01
1,2,3-$C_6H_3X_3$	0.39	0.45	0.73	0.92
1,2,4-$C_6H_3X_3$	0.35	0.49	0.78	0.92
1,2,4,5-$C_6H_2X_4$	0.50	0.71	0.95	1.17
1,2,3,4-$C_6H_2X_4$	0.55	0.72	0.95	1.20
1,2,3,5-$C_6H_2X_4$	0.54	0.70	0.98	1.16
C_6X_5H	0.72	0.90	1.24	1.50
C_6X_6	0.85	1.15	1.55	1.77

for the bromobenzenes range from 0.3 eV to 1.55 eV; those for the iodobenzenes range from 0.5 eV to 1.77 eV. For each substitution the E_a increases by about 0.25 eV. The E_a for the C_6X_6 compounds are larger than the values for the C_6F_5X compounds. The differences increase in the order F = 0 to I = 0.36 eV. By comparison with the Morse potential energy curves for C_6F_5X, the electron impact and NIMS data for these compounds should be similar.

The electron affinities of several chlorinated biphenyls and chlorinated naphthalenes have been determined from half-wave reduction potentials [22]. The electron affinity of 1-Cl naphthalene measured in the ECD has been used to scale the values for the other chloronaphthalens. The solution energy differences were set to the same value, 2.05(5) eV, for compounds with the same number of chlorine atoms up to three chlorines. For four or more chlorines up to the fully chlorinated naphthalene the mddG is 1.95(5) eV. For the fully chlorinated compound the mddG is 1.85(5) eV. This gives for E_a a range of 0.3 to 1.57 eV or 0.2 eV/Cl atom from naphthalene to the fully chlorinated naphthalene. The CURES-EC calculations support these values. They are given in Table 11.11.

The electron affinities of the chlorinated biphenyls are lower than those of the chlorinated napthalenes. The electron affinity of phenylpentachlorobenzene, $C_{12}H_5Cl_5$, is about 1.0 eV, while that of $C_{12}Cl_{10}$ is only 0.3 eV higher. The first five chlorine atoms raise the electron affinity of biphenyl by about 0.8 eV, while the second five only increase the electron affinity by approximately 0.2 eV. The CURES-EC calculations agree with the $E_{1/2}$ values for substitution on a given ring, but are higher for the substitution on two rings. For example, the decachlorobiphenyl CURES-EC value is 1.5 eV versus the reduction potential value of 1.3 eV. These differences can be seen in the values listed in Table 11.12. The CURES-EC method gives upper limits for these E_a.

11.3.3 Negative-Ion Mass Spectra and Electron Affinities

The negative-ion mass spectra of halogenated ethylenes, benzenes, biphenyls, and naphthalenes can be related to experimental electron affinities [39]. The ground-state E_a for C_2Cl_4 and C_2HCl_3 are obtained from the temperature dependence of the parent negative-ion intensities. The NIMS data at 373 K and 523 K support these values. The parent negative ions are observed at 373 K, but the relative intensity of the parent negative ion at 523 K is much smaller for C_2HCl_3 than C_2Cl_4. Dissociative thermal electron attachment is observed at both temperatures [39].

The mono and dichlorobenzenes do not show parent negative ions at 373 K or 523 K. This does not imply that the electron affinity is negative, but rather that other processes occur. Both dissociative and nondissociative capture are observed for the tri and tetrachlorobenzenes. For the penta and hexachlorobenzenes no dissociative attachment is observed. For the biphenyls with two chlorines ion molecule reactions predominate. With three chlorines, two on one ring and the third on the other ring, molecular ions are observed at 373 K. With four or more chlorines molecular ions are observed at both high and low temperatures, and dissociative electron

TABLE 11.11 Electron Affinities of Chloronaphthalenes from Reduction Potentials and CURES-EC [22 and this work]

Species	E_a(ER)	E_a(C-EC)	mddG
1-$C_{10}H_7Cl$	0.37	0.36	2.05
2-$C_{10}H_7Cl$	0.33	0.35	2.05
1,2-$C_{10}H_6Cl_2$	0.58	0.64	2.05
1,3-$C_{10}H_6Cl_2$	0.56	0.52	2.05
1,4-$C_{10}H_6Cl_2$	0.56	0.51	2.05
1,5-$C_{10}H_6Cl_2$	0.55	0.51	2.05
1,6-$C_{10}H_6Cl_2$	0.51	0.50	2.05
1,7-$C_{10}H_6Cl_2$	0.52	0.49	2.05
1,8-$C_{10}H_6Cl_2$	0.61	0.65	2.05
2,3-$C_{10}H_6Cl_2$	0.54	0.50	2.05
2,6-$C_{10}H_6Cl_2$	0.58	0.64	2.05
2,7-$C_{10}H_6Cl_2$	0.58	0.64	2.05
1,2,3-$C_{10}H_5Cl_3$	0.81	0.74	2.00
1,2,4-$C_{10}H_5Cl_3$	0.80	0.75	2.00
1,2,6-$C_{10}H_5Cl_3$	0.78	0.75	2.00
1,2,6-$C_{10}H_5Cl_3$	0.74	0.78	2.00
1,2,7-$C_{10}H_5Cl_3$	0.75	0.74	2.00
1,2,8-$C_{10}H_5Cl_3$	0.85	0.74	2.00
1,3,5-$C_{10}H_5Cl_3$	0.78	0.77	2.00
1,3,6-$C_{10}H_5Cl_3$	0.73	0.79	2.00
1,3,7-$C_{10}H_5Cl_3$	0.73	0.78	2.00
1,3,8-$C_{10}H_5Cl_3$	0.82	0.77	2.00
1,4,5-$C_{10}H_5Cl_3$	0.82	0.75	2.00
1,4,6-$C_{10}H_5Cl_3$	0.74	0.77	2.00
1,6,7-$C_{10}H_5Cl_3$	0.76	0.76	2.00
1,2,3,4-$C_{10}H_4Cl_4$	1.02	0.98	1.95
1,2,3,5-$C_{10}H_4Cl_4$	1.00	1.00	1.95
1,2,3,7-$C_{10}H_4Cl_4$	0.97	0.92	1.95
1,2,4,6-$C_{10}H_4Cl_4$	0.97	1.02	1.95
1,3,5,7-$C_{10}H_4Cl_4$	0.97	1.03	1.95
1,3,5,8-$C_{10}H_4Cl_4$	1.04	1.02	1.95
1,4,5,8-$C_{10}H_4Cl_4$	1.07	1.00	1.95
1,4,6,7-$C_{10}H_4Cl_4$	0.99	1.02	1.95
1,2,3,5,7-$C_{10}H_3Cl_5$	1.07	1.29	1.85
$C_{10}Cl_8$	1.57	1.63	1.85

attachment takes place for molecules with up to five or six chlorines. With additional chlorine substitution only the parent negative ion is observed. In the chlorinated biphenyls containing a hydroxyl group, the loss of HCl and Cl_2 is observed. In the case of tri and tetra-chlorinated p-terphenyls, parent negative ions are observed at 373 K and 523 K, but the intensity at 523 K is much lower

TABLE 11.12 Electron Affinities (in eV) of Chlorinated Biphenyls from Reduction Potentials and CURES-EC [22 and this work]

Species	$E_a(E_{1/2})$	E_a(CEC)	Difference
4-$C_{12}H_9Cl$	0.44	0.40	−0.04
3-$C_{12}H_9Cl$	0.39	0.41	0.02
2-$C_{12}H_9Cl$	0.40	0.38	−0.02
2,3-$C_{12}H_8Cl_2$	0.54	0.61	0.07
2,4-$C_{12}H_8Cl_2$	0.52	0.66	0.14
2,5-$C_{12}H_8Cl_2$	0.56	0.66	0.10
3,4-$C_{12}H_8Cl_2$	0.63	0.66	0.13
3,5-$C_{12}H_8Cl_2$	0.60	0.65	0.05
2,3-$C_{12}H_8Cl_2$	0.54	0.61	0.07
2,4′-$C_{12}H_8Cl_2$	0.46	0.48	0.02
2,2′-$C_{12}H_8Cl_2$	0.37	0.60	0.23
3,3′-$C_{12}H_8Cl_2$	0.47	0.66	0.19
4,4′-$C_{12}H_8Cl_2$	0.50	0.66	0.16
2,3,5-$C_{12}H_7Cl_3$	0.82	0.83	0.01
2,3,6-$C_{12}H_7Cl_3$	0.66	0.70	0.04
2,4,5-$C_{12}H_7Cl_3$	0.76	0.89	0.13
3,4,5-$C_{12}H_7Cl_3$	0.90	0.86	−0.04
2,3,4,5-$C_{12}H_6Cl_4$	1.02	1.06	0.04
2,3,4,6-$C_{12}H_6Cl_4$	0.92	0.88	−0.04
2,3,5,6-$C_{12}H_6Cl_4$	0.91	0.91	0.00
3,4,3′,4′-$C_{12}H_6Cl_4$	0.94	1.05	0.11
2,5,2′,5′-$C_{12}H_6Cl_4$	0.80	0.80	0.00
3,5,3′,5′-$C_{12}H_6Cl_4$	0.98	1.02	0.04
2,6,2′,6′-$C_{12}H_6Cl_4$	0.58	0.61	0.03
2,3,4,5,7-$C_{12}H_5Cl_5$	1.03	1.03	0.00
2′,5′,2,4,5-$C_{12}H_5Cl_5$	0.83	0.78	−0.05
2′,4′,5′,2,4,5-$C_{12}H_4Cl_6$	1.04	1.16	0.12
2′,4′6′,2,4,6-$C_{12}H_4Cl_6$	0.89	0.96	0.07
$C_{12}Cl_{10}$	1.39	1.57	0.18

than at 373 K. At 523 K dissociative electron attachment predominates. The CURES-EC calculated E_a for p-terphenyl is 0.4 eV so the electron affinities for trichlorophenyl should be about 1 eV.

Chloronaphthalene undergoes ion molecule reactions with methane to give the M—H ion at 373 K. At 523 K the major ion is chloride, in agreement with the ECD data and a −EDEA of 0.5 eV. The parent negative ion for the dichloronaphthalenes is observed at 373 K, indicating an electron affinity greater than 0.4 eV. Dissociative thermal electron attachment is also observed, indicating that the –EDEA is about 0.5 eV. With three or more chlorines we observe only the parent negative ion, showing stabilization to a ground state with a higher E_a.

The parent negative ion is not observed for monobromobenzene, but is observed for the tribromobenzenes. For all the higher substituted bromobenzenes dissociative thermal electron attachment takes place along with molecular ion formation. A small amount of the parent negative ion of bromonaphthalene is observed at 373 K, while the formation of the bromide ion is the predominant process. At 523 K only the bromide ion is formed, indicating an increase in the rate of dissociation and a decrease in the rate of stabilization of the intermediate negative ion to the ground state. For brominated biphenyls molecular ion formation is observed at 373 K and dissociation predominates at 523 K. The initial electron attachment occurs via an excited state that is stabilized to the ground state at 373 K. At 523 K dissociation competes with stabilization because of the smaller EDEA of the bromo compound relative to the chloro compound. The parent negative ion is observed for decafluorobiphenyl at 373 K and 523 K. The parent negative ions of bezo[a]pyrene, fluoranthene, benzo[g,h,i]perylene, diphenylfulvene, tetrachlorodimethylphthalate, 2,3dichloronaphthoquinone, 2,4-dichloronaphthol, nitronaphthalene, cyanonaphthalene, and naphthaldehyde are observed in NIMS. All these molecules have a E_a greater than 0.7 eV.

On the basis of these observations, for this NIMS equipment, an electron affinity of approximately 0.4 to 0.5 eV is required to observe the parent negative ion at 373 K. If the E_a is greater than 0.7 eV, and there are no thermally accessible dissociative pathways, the molecular ion will dominate at both 373 K and 523 K. In the case of dissociative electron attachment the EDEA will determine the abundance of the parent negative ion. The possibility of ion molecule and elimination reactions could also reduce the parent negative-ion intensity for molecules with a large electron affinity.

11.4 NEGATIVE-ION MASS SPECTROMETRY

The pesticides are highly chlorinated organic compounds, while some herbicides contain nitro groups, halogens, and alkylamino groups. Chlorinated dioxanes and phenols are among the most toxic compounds. The electron affinities of these compounds have not been measured, but the E_a of related compounds can be used to predict their electron affinities using substitution effects. These predictions can be compared with the NIMS data at 373 K and 523 K. Nitrobenzene has an established E_a of 1.0 eV and only undergoes nondissociative reactions in NIMS, while iodobenzene only shows dissociative reactions because the EDEA is positive (exothermic).

The electron affinities of phenol and the chlorinated phenols have not been measured. Since the parent negative ion for phenol is not observed in NIMS at 373 K, the E_a should be less than 0.5 eV. Asssuming a value of 0.3 eV and noting that the effect of the substitution of a hydroxyl group in aromatic compounds increases the E_a more than the substitution of a chlorine atom, the E_a of chlorinated phenols should be larger than that of the chlorobenzenes. The E_a of the trichlorophenols

TABLE 11.13 Electron Affinities (in eV) of Chlorinated Anthracenes, Xanthene, and Dibenzofuran from ECD Data [43 and this work]

Species	K (523)	$RT\ \ln(K/K_R)$	E_a	E_a(Lit)
1-Chloroanthracene	2,900	0.19	0.87	0.83
2-Chloroanthracene	3,500	0.20	0.88	0.80
9-Chloroanthracene	2,600	0.19	0.87	0.86
Anthracene(ref)	40	0.00	0.68	0.68
Dibenzofuran	1.6	0.02	0.32	—
Xanthene	14	0.12	0.42	—
Phenanthrene(ref)	1	0.00	0.30	0.30

should be greater than that for the tetrachlorobenzenes, or about 0.8 eV. All trichlorophenols show parent negative ions at 373 K and 523 K. The E_a of pentachlorophenol with two additional chlorine substitutions is predicted to be 1.3 eV. The E_a of the tribromophenols should be higher than that for the trichlorocompounds or approximately 0.9 eV, while that for the pentabromophenol should be about 1.5 eV. The E_a for pentachlorphenol obtained via CURES-EC is 1.25 eV, while that for pentabromophenol is 1.45 eV.

The electron affinities of the anilines can be compared to those of the phenols and halobenzenes. The E_a of the trichloranilines should be less than that for the trichlorphenols or about 0.6 eV. The NIMS for the anilines are very similar to that of the phenols, except that less loss of HX occurs. The E_a for the tribromoanilines should be about 0.75 eV, while that for pentachloroaniline and pentabromoaniline should be 1.1 eV and 1.4 eV. The CURES-EC values are 1.2 eV and 1.45 eV.

The E_a of the parents of the chlorinated dioxins and dibenzofuran have not been reported. A value for dibenzofuran can be obtained from a single-point ECD determination, as shown in Table 11.13, from $\ln(K_x/K_{\text{ref}}) = E_a(X) - E_a(\text{ref})$. Also shown are data for dibenzofuran, xanthene, and the chloroanthracenes [43]. The E_a is determined by referring to compounds with a measured E_a. The uncertainties in these values are ± 0.15 eV since they are determined from a single data point. The chloroanthracene values agree with the TCT values within the error. The CURES-EC and $E_{1/2}$ values support the gas phase measurements. The CURES-EC values for dibenzofuran and xanthene, are 0.3 eV and 0.4 eV respectively, in agreement with the gas phase values.

With an estimate for the E_a of dibenzofuran, 0.42 ± 0.15 eV, the E_a of dioxin will be larger and is estimated at 0.6 ± 0.15 eV. This agrees with the CURES-EC value. The electron affinities of the chlorinated dibenzofurans are predicted to range from 0.42 eV to 2.0 eV based on nominal increments. The CURES-EC value for octachlorodibenzofuran is 2.0 eV. The parent negative ions are observed at both temperatures in NIMS for two and four chlorines on each ring. No dissociative reactions are observed. The E_a of the dioxins should vary from 0.6 eV to 2.2 eV. The CURES-EC value for the perchlorinated compound is 2.1 eV. The E_a of 1,2,3,4-tetrachlorodibenzo-p-dioxin is predicted to be $0.6 + 0.8 = 1.4$ eV. The CURES-EC value is 1.35 eV. For the dioxins in NIMS the parent negative ions are formed at both temperatures, but dissociative reactions are also observed.

The NIMS of all the substituted nitrobenzenes and m-dinitrobenzenes for which electron affinities have been determined are consistent with the measured E_a values. In addition, those compounds for which substituent effects have been established are consistent with predicted E_a. All the nitrobenzenes substituted with groups other than the alkyl or amino groups should have an E_a greater than 1 eV. Thus, the parent negative ion should dominate in the absence of dissociative reactions. For molecules with E_a between 0.5 eV and 1.5 eV, the E_a could be obtained from the measurement of the molar parent negative-ion intensities as a function of temperature. With higher temperatures a higher E_a can be measured. If dissociative reactions take place, the NIMS data can give estimates of bond dissociation energies and radical electron affinities.

The electron affinities of the multihalogenated nitrobenzenes can be estimated from the measured substitution effects for nitrobenzene. The electron affinities are increased by about 0.1 eV/F for one to five substitutions. By evenly spacing the values between the extreme values 1.1 eV and 1.5 eV, the E_a for nitrobenzene substituted with two, three, or four fluorines are 1.2, 1.3, and 1.4 eV. For the chloronitrobenzenes the first two chlorines increase the E_a by 0.2 eV/Cl so the E_a for pentachloronitrobenzene is predicted to be 2.0 eV. The CURES-EC value is 1.9 eV. The parent negative ions for the tri and tetrachloronitrobenzenes are formed at 373 K and 523 K. The loss of NO_2 and NO also occurs for compounds with three or more chlorines. Since $NO_2(-)$ is not observed with the loss of NO_2, the electron affinity of the halogenated aromatic radical must be larger than that of $NO_2(-)$. The E_a of the trichlorophenide radical formed from the dissociation of the trichloronitrobenzene(−) ion is greater than 2.3 eV. The predicted value based on the phenyl radical is $1.1 + 3 \times 0.4 = 2.3$ eV. Similarly, the experimental data for tetrachloronitrobenzene and pentachloronitrobenzene show the loss of NO_2, consistent with an E_a greater than 2.3 eV. The linearly extrapolated E_a for tetrachlorophenide and pentachlorophenide radicals are 2.6 eV and 3.1 eV. The experimental value for trichlorophenide is 2.4(1) eV from gas phase acidities and using the magnetron method. That of the pentafluorophenide radical determined from NIMS is 3.22(10) eV.

Atmospheric pressure negative-ion mass spectrometry (API-NIMS) has been used to determine the electron affinities of fluorocarbons, as described in Section 11.2.4. Another important application of API-NIMS to atmospheric and environmental problems is determining the hydration energies of anions. The determination of hydration energies supports the determination of molecular electron affinities. In Chapter 4, illustrations of data were given. We have determined the hydration energies of the Cl(−), I(−), and Br(−), $O_2(-)$ for $n = 1$ to 5 or 7. These are the only such values measured at atmospheric pressure. The entropy changes of the $O_2(-)$ complexes up to $n = 5$ are 23.1(1), 23.3(6), 21.8(6), 22.9(9), and 23.3(6) entropy units. The weighted average of these is 22.7(6). This value is used for the $n = 6$

TABLE 11.14 Thermodynamic Properties for Stepwise Enthalpy, Entropy, and Gibb's Free Energy Changes for the Hydration of $O_2(-)$ [23]

$i-1, i$	$-\Delta H^o_{i-1,i}$ (kcal/mole)	$-\Delta S^o_{i-1,i}$ (e.u.)	$-\Delta G^o_{i-1,i}$ (kcal/mole)	$-\Delta H^o_{i-1,i}$ [a] (kcal/mole)	$-\Delta S^o_{i-1,i}$ [a] (e.u.)	$-\Delta G^o_{i-1,i}$ [a] (kcal/mole)
0,1	20.6(16)	23.1(10)	13.7(7)	18.4	20.1	12.5
1,2	16.4(4)	23.3(6)	9.4(2)	17.2	25.1	9.71
2,3	13.2(3)	21.8(6)	6.5(1)	15.4	28.2	7.02
3,4	12.1(4)	22.9(6)	5.6(1)	—	—	—
4,5	11.9(4)	23.3(9)	5.0(1)	—	—	—
5,6	12.1(4)	22.7(6)	5.2(1)	—	—	—
6,7	12.0(4)	22.7(6)	5.2(1)	—	—	—

[a]Data from the literature in [42].
Standard state is 1 atm.

and $n = 7$ hydrates. The thermodynamic properties for the $O_2(-)$ hydrates are given in Table 11.14 and plotted in Figure 11.11. These are compared to experiment for the first three hydrates in Table 11.14 [44]. The general trends are the same. More definitive comparisons are not possible since the uncertainties in the values found in the literature are not given. The APIMS data indicate the filling of a shell at four waters or two waters bonded to each oxygen atom.

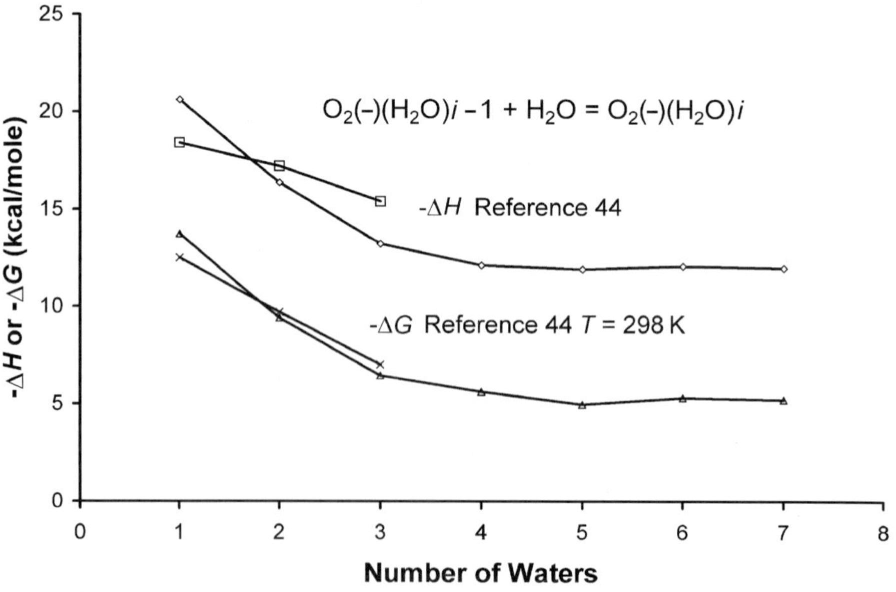

Figure 11.11 Energies for the sequential hydration of $O_2(-)$ determined using API mass spectrometry, from [23].

11.5 CALCULATION OF THE ECD AND NIMS TEMPERATURE DEPENDENCE

One objective of measuring the electron affinities of molecules is to predict the sensitivity and temperature dependence of the ECD and NIMS to compounds that might be analyzed. The chlorinated pesticides and dinitrobenzene herbicides give large responses in the ECD. Thus, their temperature dependence will be very small. Figure 11.12 gives the ECD data for several pesticides. High sensitivity and low temperature dependence are ideal for the analysis of these compounds. If NIMS is to be used, there could be problems with monitoring a specific ion, for example, the molecular ion. The dissociative reactions will deplete the intensity of the parent negative ion at higher temperatures and lower the overall sensitivity to the compound. If NIMS is to be used in the total ion mode, then it could be replaced with an ECD.

For the compounds with electron affinities less than 1.5 eV, such as the chlorinated dibenzodioxins, benzofurans, biphenyls, and naphthalenes, the temperature dependence of the ECD will be important but has not been measured. With estimates of the electron affinities, kinetic parameters, and bond dissociation energies, temperature dependence can be predicted from the kinetic model. This is illustrated for the chlorinated naphthalenes and biphenyls. These calculations may suggest the best conditions for analysis.

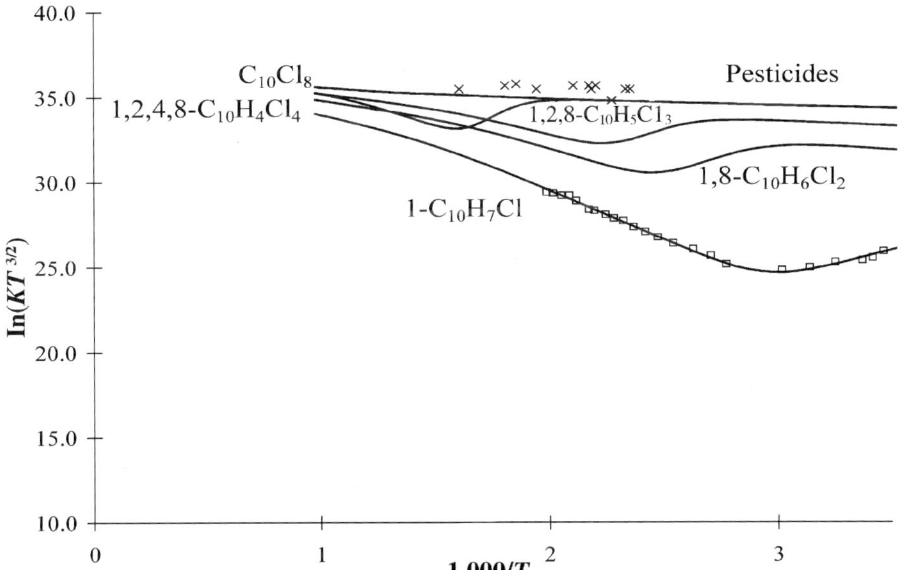

Figure 11.12 Calculated ln $KT^{3/2}$ versus 1,000/T for chlorinated naphthalenes and pesticides, from [6]. The pesticide data appear in [45].

From Table 11.11 we see that the electron affinity of 1-chloronaphthalene is 0.30 eV. The E_a of octachlornaphthalene is 1.55 ± 0.15 eV. The electron affinity of the dichlorocompound is nominally 0.6 eV, the trichlorocompound, 0.75 eV, and the tetrachlorocompound, 0.95 eV. The ECD curves were drawn with the parameters for 1-Cl-naphthalene. The bond dissociation energies and kinetic parameters were kept the same. The calculated curves for the chloronaphthalenes are shown in Figure 11.12. The experimental data for p,p′-DDT, heptachlor, dieldrin, and lindane are slightly higher, but have the same temperature dependence as perchloronaphthlene [45]. The negative-ion mass spectrometry of the chloronapthalenes shows dissociative electron capture for the compounds with one to three chlorines at both 373 K and 523 K. However, those with four to eight chlorines only form the parent negative ion. The calculated ECD curves agree with these data. At still higher temperatures the compounds with four to eight chlorines will dissociate with a lowering of the parent negative-ion intensity. The use of selected ion monitoring of the parent negative ion for analysis should be done at a lower temperature.

The electron affinities of the chlorinated biphenyls are lower than those of the chlorinated napthalenes. Nominal values are taken from Table 11.12. The temperature dependence of the three isomers of the monchlorobiphenyl will be similar to meta, ortho, and para dichlorobenzene data. Likewise, the temperature dependence of the compounds with two chlorines on the same ring will be similar to that of the trichlorobenzenes. The response of the fully chlorinated compound will be similar to that of hexachlorobenzene. The isomers with eight and nine chlorines only show nondissociative capture. Approximate curves for the chlorinated biphenyls are illustrated in Figure 11.13 and compared with experimental data obtained using the PDECD [45].

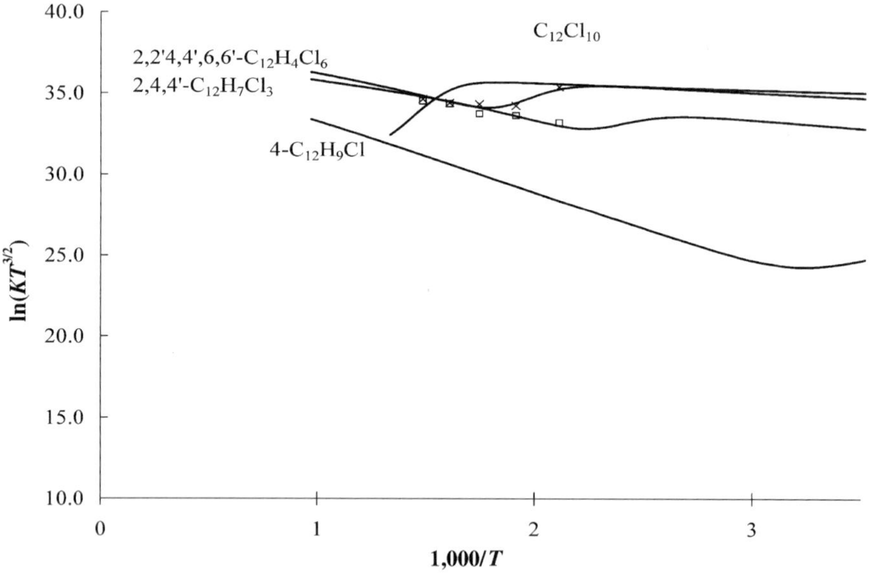

Figure 11.13 Calculated ln $KT^{3/2}$ versus 1,000/T for chlorinated biphenyls, from [45].

11.6 SUMMARY

The electron affinities of halogenated aromatic and aliphatic compounds and nitro compounds have been evaluated. Additional electron affinities for halogenated benzene, freons, heterocyclic compounds, dibenzofuran, and the chloro- and fluorobenzenes are reported from ECD data. The first positive E_a for the fluorochloroethanes were obtained from published ECD data. The E_a of halogenated aromatic radicals have been estimated from NIMS data. The AE_a of all the halobenzenes have been calculated using CURES-EC. The E_a of chlorinated biphenyls and chlorinated napthalenes obtained from reduction potentials have been revised based on variable solution energy differences.

The TCT E_a of perfluoromethylcyclohexane is confirmed with NIMS data and the E_a of perfluoromethylcyclopentane is reported. The EDEA for the loss of the tertiary fluorine atom is determined for both these compounds. In many cases the NIST values for fluorinated hydrocarbons could be for excited states. In all but a few of the cases, there is only one experimental determination of the E_a so confirmation using the CURES-EC procedure is important.

The activation energies for dissociative thermal electron reactions have been correlated with bond dissociation energies when available. Estimates of activation energies for dissociative reactions have been made by assuming a fixed value for A_1 and using relative molar responses or relative rate constants to calculate activation energy differences. Additional bond dissociation energies are obtained from these data.

The temperature dependence of negative-ion mass spectra of environmental compounds has been analyzed in terms of the electron affinities and EDEA values. Morse potential energy curves have been constructed using the Herschbach parameters to illustrate the different types of thermal electron reactions. The temperature dependence of the ECD response for chlorinated naphthalenes and biphenyls has been predicted and is compared with experimental data for several pesticides and chlorinated biphenyls.

REFERENCES

1. Wentworth, W. E.; Becker, R. S.; and Tung, R. *J. Phys. Chem.* **1967**, 71, 1652.
2. Wentworth, W. E. and Chen, E. C. M. *J. Gas Chromatogr.* **1967**, 5, 170.
3. Wentworth, W. E; George, R.; and Keith, H. *J. Chem. Phys.* **1969**, 51, 1791.
4. Wentworth, W. E. and Steelhammer, J. C. *J. Chem. Phys.* **1969**, 51, 1802.
5. Wentworth, W. E. and Steelhammer, J. C. *Advs. Chem. Sers.* **1968**, 82, 75.
6. Steelhammer, J. C. M. S. thesis, University of Houston, **1968**.
7. Page, F. M. and Goode, G. C. *Negative Ions and the Magnetron*. New York: Wiley-Interscience, **1969**.
8. Dispert, H. and Lacmann, K. *Int. J. Mass Spectrom. Ion Phys.* **1978**, 28, 49.
9. Lacmann, K.; Maneira, M. J. P.; Moutinho, A. M. C.; and Weigman, U. *J. Chem. Phys.* **1983**, 78, 1767.

10. Staneke, P. O.; Groothuis, G.; Ingemann, S.; and Nibbering, N. N. M. *Int. J. Mass Spectrom. Ion Proc.* **1995**, 142, 83.
11. McMillen, J. F. and Golden, D. M. *Ann. Revs. Phys. Chem.* **1982**, 33, 493.
12. Chen, E. C. M.; Albyns, K.; Dussak, L.; and Wentworth, W. E. *J. Phys. Chem.* **1989**, 93, 6827.
13. Clemons, C. A. and Altshuler, A. P. *Anal. Chem.* **1966**, 38, 133.
14. Devaux, P. and Guichon, G. *J. Gas. Chromatog.* **1967**, 5, 341.
15. Christodoulides, A. A.; McCorkle, D. L.; and Christophorou, L. G. "Electron Attachment Processes," in *Electron-Molecule Interactions and Their Applications.* New York: Academic Press, **1984**.
16. Sousa, S. R. and Biakowski, S. E. *Anal. Chem.* **1997**, 69, 3871.
17. Spyrou, S. M.; Sauers, I.; and Christophorou, L. G. *J. Chem. Phys.* **1983**, 78, 7200.
18. Compton, R. N.; Reinhardt, P. W.; and Cooper, C. D. *J. Chem. Phys.* **1978**, 68, 4360.
19. Tang, S. Y.; Mathur, B. P.; Roth, E. W.; and Reck, G. P. *J. Chem. Phys.* **1976**, 64, 1270.
20. McNamee, P. E.; Lacmann, K.; and Herschbach, D. R. *Farad. Disc. Chem. Soc.* **1973**, 55, 318.
21. Chen, E. C. M.; Wiley, J. R.; Batten, C. F.; and Wentworth, W. E. *J. Phys. Chem.* **1994**, 98, 88.
22. Wiley, J. R.; Chen, E. C. M.; Chen, E. S. D.; Richardson, P.; Reed, W. R.; and Wentworth, W. E. *J. Electroanal. Chem. Interfac.* **1991**, 307, 169.
23. D'sa, E. D. Ph.D. dissertation, University of Houston, **1987**.
24. Hiraoka, K.; Mizuno, T.; Eguchi, D.; Takao, K.; Iino, T.; and Yamabe, S. *J. Chem. Phys.* **2002**, 116, 7574.
25. Miller, T. M.; Morris, R. A.; Miller, A. E. S.; Viggiano, A. A.; and Paulson, J. F. *Int. J. Mass Spectrom. Ion Proc.* **1994**, 135, 195.
26. Lifshitz, C.; Tiernan, T. O.; and Hughes, B. M. *J. Chem. Phys.* **1973**, 59, 3182.
27. Mahan B. H. and Young, C. E. *J. Chem. Phys.* **1966**, 44, 2192.
28. Grimsrud, E. P.; Chowdhury, S.; and Kebarle, P. *J. Chem. Phys.* **1985**, 83, 1059.
29. Wentworth, W. E.; Limero, T.; and Chen, E. C. M. *J. Phys. Chem.* **1987**, 91, 241.
30. Christophorou, L. G. and Datkos, P. G. *Int. J. Mass Spectrom. Ion Proc.* **1995**, 149/150, 59.
31. Nakajima, A.; Taguwa, T.; Hoshino, K.; Sugioka, T.; Naganuma, T.; Ono, F.; Watanabe, K.; and Nakao, K. *Chem. Phys. Lett.* **1993**, 214, 22.
32. Rains, L. J.; Moore, H. W.; and McIver, R. T. *J. Chem. Phys.* **1978**, 68, 3309.
33. Dillow, G. W. and Kebarle, P. J. *Am. Chem. Soc.* **1989**, 111, 5592.
34. Lin, S. N. Ph.D. dissertation, University of Houston, **1969**.
35. Chen, E. C. M.; Carr, S. D.; Wentworth, W. E.; and Chen, E. S. D. *J. Chromatogr. A* **1998**, 827, 91.
36. Campbell, B. H. *Anal. Chem.* **1972**, 44, 1659.
37. National Institute of Standards and Technology (NIST). *Chemistry WebBook*, **2003**. Available at http://webbook.nist.gov.
38. Knighton, W. B.; Bognar, J. A.; and Grimsrud, E. P. *J. Mass Spectrom.* 30, 557.

39. Stemmler, E. A. and Hites, R. A. *Electron Capture Negative Ion Mass Spectra*. New York: VCH, **1988**.

40. Naff, W. T.; Compton, R. N.; and Cooper, J. D. *J. Chem. Phys.* **1971**, 59, 212.

41. Heinis, T.; Chowdhury, S.; and Kebarle, P. *Org. Mass Spectrom.* **1993**, 28, 358.

42. Chen, G. D. and Cooks, R. G. *J. Mass Spectrom.* **1995**, 30, 1167.

43. Grimsrud, E. P. In The Electron Capture Detector, edited by A. Zlatkis, and C. Poole. New York: Elsevier, **1981**, p. 91.

44. Arshardi, M. and Kebarle, P. *J. Phys. Chem.* **1970**, 74, 1483.

45. Wentworth, W. E.; Huang, J.; Sun, K.; Zhang, Y.; Rao, H.; Cai, H.; and Stearns, S. D. *J. Chromatogr. A* **1999**, 842, 229.

46. Shuie, L. R. Ph.D. dissertation, University of Houston, **1984**.

CHAPTER 12

Biologically Significant Molecules

12.1 INTRODUCTION

The interaction of thermal electrons with biologically significant molecules is essential to all living organisms. It has long been known that energetic electrons and radiation can damage and destroy DNA. Until recently little accurate fundamental thermodynamic or kinetic information on the reaction of thermal electrons with molecules of biological significance has been available. Since 1990 the techniques presented in this book have been used to obtain additional data and develop an interpretation. The negative-ion states of NO and O_2 were assigned. The first experimental electron affinities and gas phase acidities of the purines and pyrimidines were measured using half-wave reduction potentials and negative-ion mass spectrometry. These were confirmed by CURES-EC calculations. Morse potential energy curves for cytosine and thymine were calculated. Electron affinities of the Watson Crick hydrogen-bonded base pairs were predicted using CURES-EC. A model for electron transport was proposed. These results will be integrated with other experimental and theoretical studies to provide a summary of the field's current status.

Before 2002 the experimental electron affinities of the simple molecules O_2 and NO had not been assigned to electronic states. The role of O_2 and its anion in biochemical processes is well known, but only in the past 20 years have we begun to understand the role and mechanism of NO in living organisms. However, more than a century ago nitroglycerin, discovered by Alfred Nobel, was found to act biologically as an NO donor, resulting in headaches among and affecting the blood pressure of those manufacturing the explosive. Indeed, one of first drugs to treat angina is another NO donor, amyl nitrate. In 1992 the journal *Science* named NO the "molecule of the year." In 1998 Furchgott, Ignarro, and Murad won the Nobel Prize in chemistry for "describing the role of nitric oxide as a signaling molecule in the cardiovascular system," returning NO full circle to the work of Alfred Nobel [1, 2].

The Electron Capture Detector and the Study of Reactions with Thermal Electrons
by E. C. M. Chen and E. S. D. Chen
ISBN 0-471-32622-4

In Chapter 9 the ECD data for NO and O_2 were analyzed in terms of multiple negative ion states. Because the $A\,E_a$ of NO was considered to be less than 0.1 eV, no stable bound excited states of NO(–) have been previously considered. The isoelectronic principle and the measurement of higher E_a by three independent techniques support the postulate of stable bound excited states of NO(–). Figure 12.1 presents a comparison of the states for the isoelectronic NO(–) and O_2. By analogy at least three bound states of NO(–) are predicted [3].

R. R. Freeman collected the original ECD data for O_2 and NO in the 1970s. The ECD results were not published because it was believed that there were no stable excited anion states for these molecules and because the change in slope was not understood [4]. Later the low-temperature data were published as an example of the use of the ECD to determine E_a [5]. The high-temperature data were not published until 2002 when multiple states for anions in the ECD were considered. Other data showing the excited states of O_2(–) and the recognition that 12 of the 24 predicted states of O_2(–) were bound and stable led to the assignments previously given in Chapter 9 [3, 6, 7].

The ECD data for NO shown in Chapter 9 were resolved into four curves, three of which are predicted by analogy to the three states of O_2 shown in Figure 12.1. The largest value of the E_a, 0.85 ± 0.1 eV, has been measured by the MGN, EI, and ECD techniques and can be accurately assigned to the ground state. The remaining

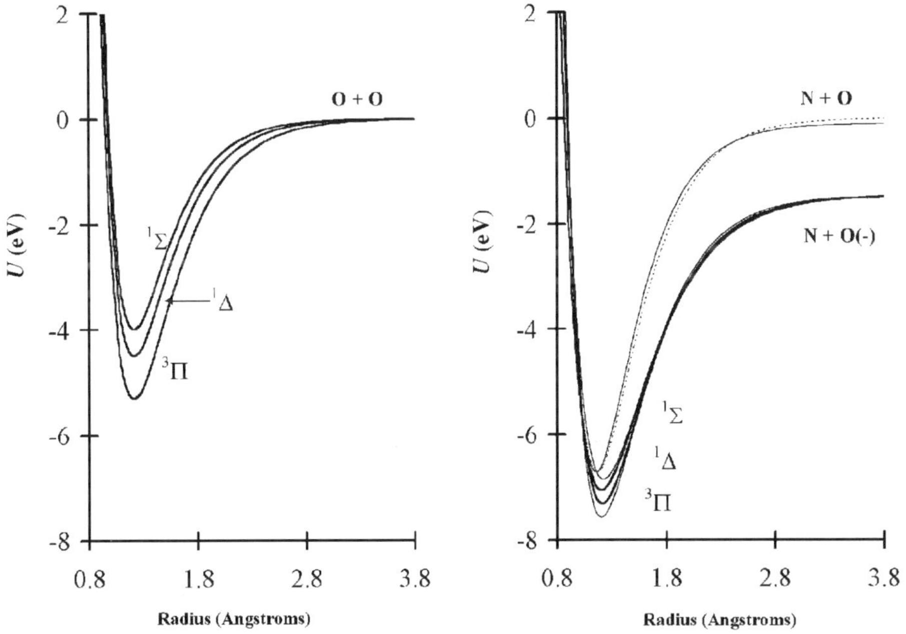

Figure 12.1 Morse potential energy curves for NO(–) and O_2 illustrating the similar electronic structure of the isoelectronic species.

ECD values and the E_a of less than 0.1 eV determined by PES and other experimental techniques require further theoretical predictions to be accurately assigned. It is clearly important to consider all the experimental E_a for both O_2 and NO so that values differing by more than the uncertainty are assigned to different states.

The present situation for the electron affinities of adenine, guanine, cytosine, uracil, and thymine (AGCUT) is similar to that for NO. Prior to 1990 the electron affinities had not been measured. Indeed, theoretical calculations of the valence-state E_a alternated between positive and negative [8, 9]. Presently the E_a of the purines and pyrimidines are measured using reduction potentials and scaling to gas phase values. These E_a are confirmed by the CURES-EC calculations [10–20]. These are assigned to the AE_a, but the existence of excited valence states is not generally recognized. The values for C, U, and T have been confirmed by PES of the respective anion hydrates, Ah-PES [21–24]. Scaled ab initio and density functional calculations give values that confirm the E_a for the pyrimidines [25, 26]. The vertical E_a for C and T and AGCUT have been measured [27, 28]. In the NIST tables only the dipole bound E_a of AGCUT are given. The energies for the hydrated excited-state anions listed in NIST appear in the appendix [21–24, 29].

The general method for estimating the electron affinities of biologically significant molecules from solution energy differences and half-wave reduction potentials, the use of CURES-EC, and substitution and replacement effects are reviewed and applied to other molecules. The electron affinities of heterocyclic aromatic molecules listed in the NIST tables, including acridine, have been determined by TCT measurements and are evaluated [16, 29]. The E_a for halogenated uracils have been calculated by density functional methods and are calculated using CURES-EC [30, 31].

The gas phase acidities of AGCUT have been estimated from negative-ion mass spectra and that of U has been measured by TCT experiments [32, 33]. By consolidating electron impact, electron transmission, and NIMS data and by calculating bond dissociation energies, Morse potential energy curves have been constructed in the N—H and C—NH_2 dimensions for C and only the N—H dimension for T. We review these curves [7, 26]. The gas phase acidities and electron affinities of AGCUT have been calculated by CURES-EC and standard AM1 and PM3 procedures. The experimental and CURES-EC GPA are revised based on the determinations of the GPA of uracil [32–36]. The gas phase acidities of halogenated uracils are calculated by CURES-EC.

At present there are no experimental values for the electron affinities of the amino acids. The gas phase acidities of alanine, glycine, and histidine have been measured and the E_a of the corresponding radical determined. This follows the procedure first used to determine the E_a of the acetate radical [37–42]. These values are verified using CURES-EC. Predictions of the GPA for representative amino acids are made.

The values of the electron affinities and rate constants for thermal electron attachment to the purines, pyrimidines, and heterocyclic compounds can be used to predict the temperature dependence of the ECD and NIMS response. These are similar to those made for the chlorinated biphenyls and naphthalenes in

Chapter 11. By analogy to the ECD data for the esters of simple acids, the ECD temperature dependence for derivatives of the amino acids is predicted. The temperature dependence of derivatives used to analyze nucleic acids and amino acids will be estimated based on the calculated data for the parent compounds and reported minimum detectable amounts [43–48].

Once the E_a of the purines and pyrimidines were measured and confirmed by the CURES-EC method, other fundamental questions involving the purines and pyrimidines and DNA were investigated. These are the electron affinities of the hydrogen-bonded base pairs; the conduction of electrons by DNA; the role of electron bound complexes in radiation damage in DNA; the change in geometry upon electron addition to DNA, and the difference between the electron affinity of guanine and adenine in the hydrogen-bonded form and the isolated global minimum [49–55]. These calculations were made using the commercial software and PCs available in 1998 [49, 50]. The electron affinities of GC and AT were calculated by density functional procedures in 2002 and 2003 [51–54]. These values are compared with those predicted by CURES-EC.

12.2 ELECTRON AFFINITIES OF PURINES AND PYRIMIDINES

12.2.1 Predictions of Electron Affinities

Before 1990 there were no accurate experimental or theoretical values for the E_a of AGCUT. It was recognized that ionization potentials and electron affinities were important to charge transfer in biological processes, but there were three potential measures of the E_a. These were donor acceptor complex data, reduction potentials, and theoretical calculations, each of which resulted in different measures of the E_a. Much of our work during the past decade has attempted to reconcile these differences.

In 1975 the anion of T was observed in a mass spectrometer, indicating a positive valence-state E_a for T. In 1990 the E_a of AGCUT were predicted using substitution, replacement, and conjugation effects [10–14]. In order to estimate the E_a of substituted compounds, that of the parent compounds is required. In 1974 I. Nenner and G. J. Schulz estimated the $A\,E_a$ of quinoline (0.36 eV), pyradazine (0.40 eV), pyrimidine (0.00 eV), pyrazine (0.40 eV), and s-triazine (0.45 eV) from electron transmission spectra and half-wave reduction potentials [15]. No adiabatic electron affinities of aromatic nitrogen heterocyclic compounds were measured in the gas phase before 1989 [16].

In 1990 the $A\,E_a$ for C, U, and T were predicted to range from 0.6 eV to 0.75 eV, while the $A\,E_a$ of A and G were predicted to be higher. These estimates were based on an E_a of 0.3 eV for pyrimidine and the observation that C, U, and T are hydroxyl-, methyl-, and amino-substituted pyrimidines. From the data for pairs of molecules shown in Table 12.1, quantitative measures of the substitution, conjugation, and replacement effects were postulated. The replacement of a CH by N in an aromatic system increases the E_a by 0.2 eV to 0.8 eV. The substitution of a hydroxyl

TABLE 12.1 Substitution and Replacement Effects on Electron Affinities [10–12, 15–20]

Molecule	E_a (eV)	Method	Effect
Benzene C_6H_6	−0.73(10)	$E_{1/2}$	Parent
Pyridine C_5H_5N	0.00(20)	$E_{1/2}$, ECD	CH to N
Nitrobenzene $C_6H_5NO_2$	1.00(1)	Many	Parent
Nitroaniline $C_6H_6N_2O_2$	0.95(10)	TCT	NH_2 sub
Benzonitrile C_7H_5N	0.28(5)	ECD	Parent
Cyanopyridine $C_6H_5N_2$	1.05(5)	$E_{1/2}$	CH to N
Naphthalene C_8H_{10}	0.16(2)	ECD	Parent
Quinoline C_7H_9N	0.50(5)	$E_{1/2}$	CH to N
Hydroxyquinoline C_7H_8NO	0.85(5)	$E_{1/2}$	OH sub
Phenanthrene $C_{14}H_{10}$	0.30(2)	ECD	Parent
Benzoquinoline $C_{13}H_9N$	0.50(15)	ECD	CH to N
Anthracene $C_{14}H_{10}$	0.68(2)	ECD	Parent
Acridine $C_{13}H_9N$	0.91(10)	TCT	CH to N
Pyradazine $C_4H_4N_2$	0.31(5)	$E_{1/2}$	CH to N
Pyrimidine $C_4H_4N_2$	0.20(5)	$E_{1/2}$	Parent
Pyrazine $C_4H_4N_2$	0.36(5)	$E_{1/2}$	CH to N
as-Triazine $C_3H_3N_3$	0.90(5)	$E_{1/2}$	CH to N
Acetophenone C_8H_8O	0.338(2)	ECD	Parent
Hydroxyacetophenone	0.77(5)	ECD	OH sub
Purine	1.04(5)	$E_{1/2}$	Parent

group increases the E_a by about 0.3 eV to 0.5 eV, while the methyl and amino groups have a small effect on the E_a. Since G and A are substituted purines and related to C and T by extension of the aromatic system, their E_a should be larger than those for C, U, and T. The E_a of purine was later estimated to be 1 eV so the E_a of adenine and guanine are predicted to be 1.0 eV and 1.3 eV [10].

12.2.2 Electron Affinities from Reduction Potentials

Once the predictions were in place, the next step was to measure the E_a. They were obtained by measuring the half-wave reduction potentials of the bases. These relative values were scaled to measured gas phase values for acridine and anthracene [11]. To validate the experimental procedure, the E_a for molecules with different solution energy differences and experimental gas phase electron affinities were determined. These values all agreed with values found in the literature within the experimental uncertainty of ± 0.03 V. Recall that the general classes of charge densities, q and $-\Delta\Delta G$, are [E_L, 0,1.7]; [F, 0.05, 1.8]; [A, 0.2, 2.0]; [B, 0.4, 2.2]; [C, 0.6, 2.4]; [D, 0.8, 2.6]; [E_H, 1, 2.7]; and $E_a = E_{\text{ref}} - 2.20 - n(0.20) + E_{1/2}$, where $-2.20 - n(0.20)$ is $-\Delta\Delta G$ (see Chapter 7). Thus, the $-\Delta\Delta G$ for anthracene and acridine are different from those of the nitrocompounds by about 0.4 eV, the difference between groups $A(n = -1)$ and $C(n = 1)$. The localized charge densities in anions of A, G, C, U, and T are about the same as for the acridine anion so the

TABLE 12.2 Reduction Potentials and Electron Affinities [1–12, 15–20]

Compound	E_a	$-E_{1/2}$ (V vs. Ag/AgCl)	$-E_{1/2}$ (V vs. Ag/AgCl)	E_a (eV)	E_a (eV)	Constant[a] (eV)
Anthracene	0.68(2)	1.93	1.96	0.69	0.66	2.62
Acridine	0.91(10)	1.57	1.57	0.91	0.91	2.48
Nitrobenzene	1.00(2)	1.11	1.10	1.00	1.01	2.11
m-Nitrotoluene	0.98(5)	1.13	1.12	0.98	0.99	2.11
o-Nitrotoluene	0.90(5)	1.17	1.16	0.89	0.90	2.06
p-Nitrotoluene	0.95(5)	1.23	1.22	0.94	0.95	2.17
Pyridazine	0.31(5)	2.17	2.16	0.31	0.32	2.48
Pyrazine	0.36(5)	2.12	2.09	0.36	0.39	2.48
s-Triazine	0.47(5)	2.01	2.02	0.47	0.46	2.48
as-Triazine	0.91(5)	1.57	—	—	0.91	2.48
s-Tetrazine	1.67(5)	0.81	—	—	1.67	2.48
Pyrimidine	0.19(5)	2.29	2.32	0.19	0.16	2.48
Cytosine	0.56(5)	1.92	—	0.56	—	2.48
Uracil	0.80(5)	1.68	—	0.80	—	2.48
Thymine	0.79(5)	1.69	—	0.79	—	2.48
Purine	1.02(5)	1.46	1.44	1.02	1.04	2.48
Adenine	0.95(5)	1.53	1.59	0.95	0.89	2.48
Guanine	1.51(5)	0.97	—	1.51	—	2.48

[a] $E_a = E_{\text{ref}} - \text{mddG} + E_{1/2} = \text{constant} + E_{1/2}, E_{\text{ref}} = 4.58$, and mddG = 2.1 for the purines and pyrimidines. The constant is 4.58 − 2.1 = 2.48.

$-\Delta\Delta G$ was set to 2.1 eV. The specific value of $E_{\text{ref}} - 2.20 - n(0.20)$ for the purines and pyrimidines is $E_a = 4.58 - 2.10 + E_{1/2} = 2.48 + E_{1/2}$.

The electron affinities of AGCUT were obtained by scaling to all the reference compounds since the $-\Delta\Delta G$ values for the nitro compounds and anthracene agreed with those that had been determined previously. The experimental data defining these E_a are shown in Table 12.2. The first column contains the evaluated values of the E_a for the reference compounds. The second column gives the experimental reduction potentials used to determine the E_a of AGCUT. In the third column the values of the reduction potentials scaled to the Ag/AgCl electrode value for acridine in DMSO, those available in the literature, are given. The specific electrode supporting the electrolyte and solvent is important for obtaining accurate E_a values. The next two columns contain the E_a values obtained by using $E_a = E_{\text{ref}} - -\Delta\Delta G + E_{1/2}$. The values in the last column are $E_{\text{ref}} - -\Delta\Delta G$. For example, for the purines and pyrimidines the $E_a = 2.48 + E_{1/2} = 2.48 - 0.97 = 1.51$ eV for guanine. The agreement between the E_a obtained from the reduction potentials found in the literature and the values determined in our study is good. The $E_{1/2}$ for adenine (−1.59 V versus Ag/AgCl) and purine (−1.44 V versus Ag/AgCl) give independent E_a for adenine: $E_a(\text{adenine}) = 2.48 - 1.59 = 0.89(10)$ eV and $E_a(\text{purine}) = 2.48 - 1.44 = 1.04(10)$ eV. These E_a are the largest precise values and are selected as the AE_a.

Subsequently published $E_{1/2}$ (V versus NHE, in dimethylformamide) for the ribonucleosides are A, −2.57; C, −2.26 ; U, −2.04; T, −2.15, and purine, −1.84 (±0.05 V) [17]. If we assume the E_a of rpurine is 1.04 eV, these give rC, 0.62; rU, 0.84, and rT, 0.73 (±0.1 eV), the same as for C, U, and T. The (rA) is 0.31 ± 0.1 eV, supporting an excited state for rA(−). These are the only experimental E_a for the ribonucleosides [17]. It should be noted that the E_a of ribonucleoside is primarily determined by the base.

12.2.3 Gas Phase Measurements of Electron Affinities

The PES of dipole bound U and valence-state anion complexes with Xe, Kr, and one water molecule were reported in one study. The PES of dipole bound C, T, and U and their anion hydrate complexes for $n = 1$ to 5 were observed in another [23, 24]. In the bare anion the major peak occurs for the dipole bound anion, but other small peaks are observed at higher energies. In the U•Xe(−) PES "valence-state" peaks are observed at 0.35, 0.5, 0.65, 0.8, and 0.9 eV. In the spectra of the hydrates valence states are also observed but with greater intensities. The onsets and peaks for the U•H_2O(−) are different in the two studies. In the first one the peak occurs at 1.0 eV and the onset at 0.5 eV, while in the second the peak is at 0.8 eV and the onset at 0.4 eV [21–24]. By assigning the peak to the E_a of the monohydrate and subtracting the hydration energy of 0.2 eV, the E_a of U are 0.8 eV and 0.6 eV, respectively. In the series of multihydrated anions the onsets and peaks increase with the number of water molecules, as shown in Figure 12.2. The peaks for cytosine occur initially at higher energies than the peaks for U and T

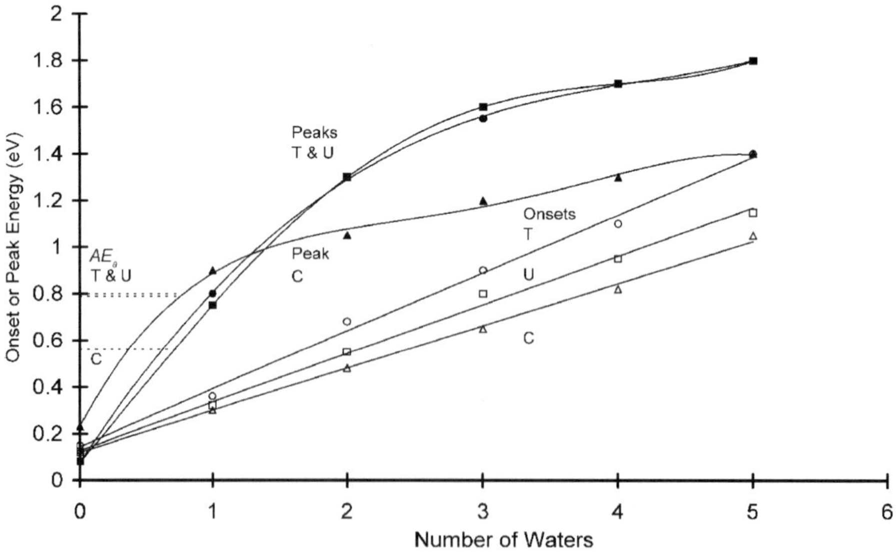

Figure 12.2 Onsets and peaks for the photoelectron spectra of the anion hydrates of cytosine, thymine, and uracil. From [24].

but drop off sooner. This indicates the establishment of a shell at two or three water molecules for cytosine but four or five for U and T [24]. The energies are given in the appendix.

For cytosine two DBEA are reported: 0.230 eV and 0.085 eV. Interestingly, the dipole bound states are observed in the PES of the dihydrates due to two-photon absorption. The first photon removes the waters and then a second photon detaches the electron from the dipole bound state. These spectra for cytosine are shown in Figure 12.3. The spectrum of the bare cytosine is shown in the inset offset by 1.5 eV. The peak at 0.230 eV is assigned to the dipole bound anion of keto-cytosine, whereas the peak at 0.085 eV is assigned to the enol form of the anion. The dipole moment of the enol form is about 4 Debye, while that for the keto form is about 6.5 Debye. The two peak intensities are quite different in the spectrum for the monomer. In the spectrum for the dihydrate the intensities are about the same. The double-photon process explains this difference. The absorption of the initial photon by the dihydrate leaves equal concentrations of two forms that are then photodetached. In the spectrum of the bare anion the distribution reflects the equilibrium concentrations. The extra structure in the spectrum of the dihydrate can be attributed to excited states of the anion and offers a different interpretation of the onset.

Since the dipole bound anions are formed by double-photon absorption, the valence states could also be formed, in which case the onsets and initial peaks results from the photodetachment of the isolated ground-state or excited-state anions. These can be a measure of the E_a. The onsets that are lower than the

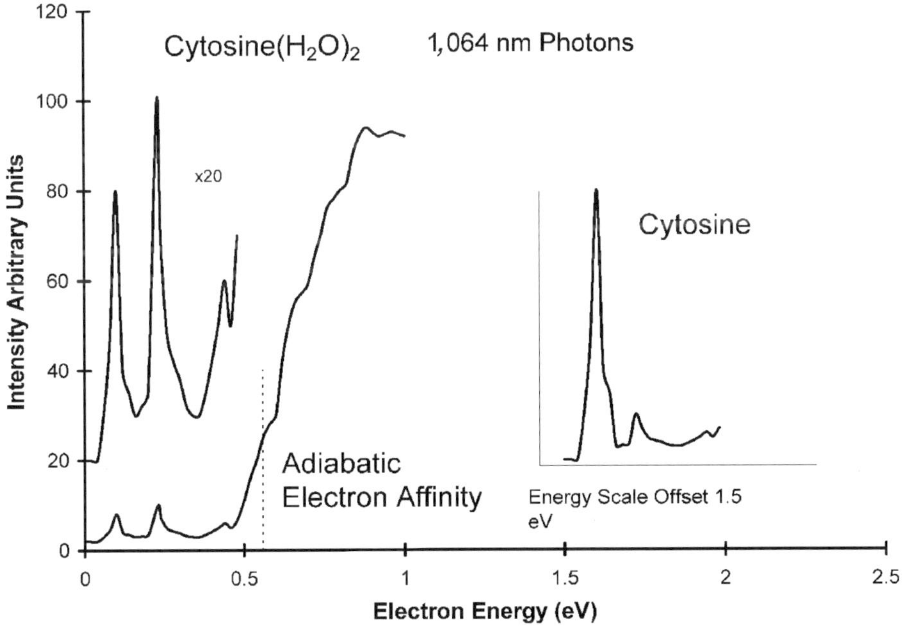

Figure 12.3 Photoelectron spectra of the anion hydrates of cytosine [24].

TABLE 12.3 Electron Affinities and Excited-State Electron Affinities (in eV)

Molecule	$E_{1/2}$	PES	NIST
Uracil	0.80(5)	0.8, 0.7(1)	—
Uracil*	—	(0.3, 0.5)[a]	—
Uracil*	—	0.15(10)	—
Uracil-DBEA	—	—	0.09(1)
Thymine	0.79(5)	0.7(1)	—
Thymine*	—	(0.3, 0.5)[a]	—
Thymine*	—	0.12(10)	—
Thymine DBEA	—	—	0.06(1)
Cytosine	0.56(5)	0.6(1)	—
Cytosine*	—	(0.3, 0.45)[a]	—
Cytosine*	—	0.13(10)	—
Cytosine-DBEA	—	—	0.09(1)
Cytosine-DBEA	—	—	0.23(1)
Adenine	0.95(5)	—	—
Adenine-DBEA	—	—	0.01(1)
Guanine	1.51(5)	—	(0.3)[b]

[a]Peaks observed in PES but not originally assigned.
[b]Predicted based on the value for cytosine and the similarity of the dipole moments of cytosine and guanine.
* Excited states.

$A\,E_a$ are assigned to excited states. The weighted average DBEA for U, T, and A are 0.065(7), 0.090(7), and 0.012(5) eV, respectively [21–24, 29]. The only valence-state E_a reported for the parent was obtained by extrapolation of onsets to $n = 0$ to give E_a^*, and 0 eV to 0.2 eV for C, U, and T, as shown in Figure 12.2 [23]. The other onsets have been assigned to the E_a of the hydrates and are listed in the NIST tables and the appendix. The peak energies could also be attributed to the photodetachment of the hydrates and give a higher E_a. Based on the observation of excited anion states in other molecules and the peaks observed in the spectra of the isolated anions, it was postulated that excited-state E_a of 0.3 eV and 0.5 eV are observed for C, U, and T. These values are given in Table 12.3.

The vertical electron affinities of AGCUT have been measured by electron transmission spectroscopy and those of C and T by electron impact spectra [25, 26]. The electron transmission spectra give VE_a of −0.32, −0.29, −0.54, −0.46, and −0.22 eV for C, T, A, G, and U, respectively. In the same order, the CURES-EC values are 0.2, −0.15, −0.5, −0.3 and −0.15 eV. It must be recalled that electron transmission and electron impact spectra can only give negative values of VE_a. From the lowest energy peak in the electron impact distribution of the parent negative ions, the VE_a are 0, or −1.40 eV for C and −0.18 eV for T. The observation of the parent negative ions in mass spectrometry at 0 eV to 0.2 eV electron energies for both C and T indicates a positive electron affinity greater than about 0.5 eV. Surprisingly, the electron impact spectra of C and T also yield parent negative-ion peaks at about 3.4 eV and 5.4 eV. These states must be valence states because they exist long enough to be recorded in the mass spectrometer. In addition to the parent negative ions fragment

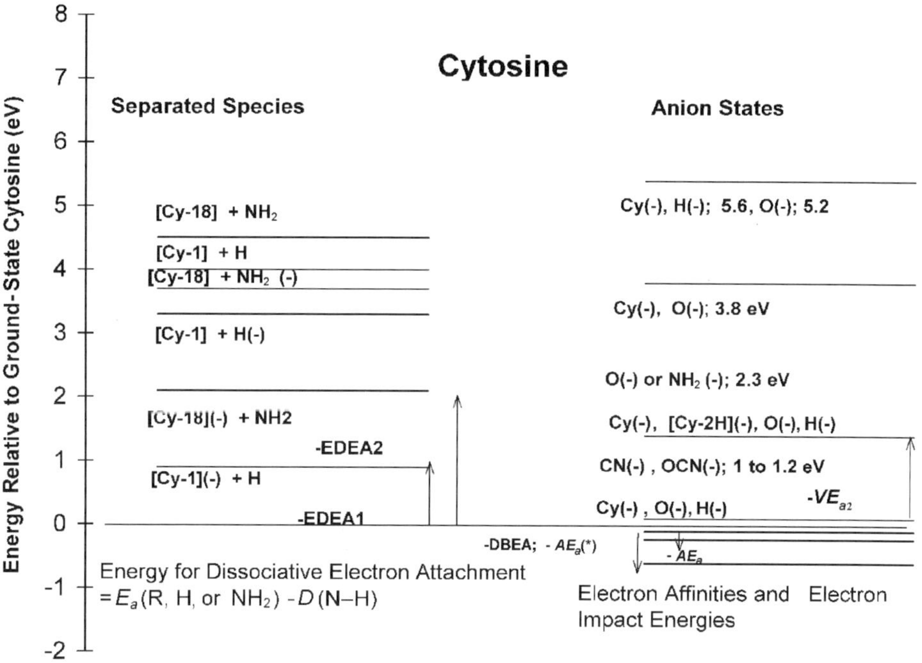

Figure 12.4 Energy-level diagrams constructed for cytosine from electron affinities, electron impact and transmission data, and CURES-EC calculations. From [7, 27, 28].

ions are also observed. Energy-level diagrams can be obtained from the electron impact data, as shown in Figures 12.4 and 12.5. For example, limits to the bond dissociation energies are obtained from the onsets of various ions. These data are used to construct Morse potential energy curves for C and T to explain the parent negative-ion peaks at different energies in the electron impact spectra.

12.2.4 Theoretical Electron Affinities

In 1966 a semi-empirical SCF calculation of the valence-state E_a of A, G, C, and T reported that the energies of the LUMO were positive, indicative of negative valence-state E_a in contrast to earlier calculations that indicated positive E_a [8, 9]. Indeed, up to the early 1990s and even today many theoretical calculations still predict negative valence-state E_a for G and A [26, 27]. All the calculations anticipate a small positive polarization E_a for adenine and larger positive dipole bound E_a for guanine. A comparison of the theoretical estimates for the electron affinities of AGCUT is given in Table 12.4. The only theoretical values that agree with all the $E_{1/2}$ values are obtained by CURES-EC [13]. By scaling ab initio $631 + G(d)//6 - 31\,G^*$ LUMO to the AE_a of naphthalene using $AE_a = VE_a +$ 1.05, the E_a are found to be as follows: U, 0.9; T, 0.8; C, 0.7; G, 0.9, and A, 0.5 eV [26]. The largest unscaled density functional values are obtained with the B3P86 density functional and are the following E_a: U, 0.75; T, 0.71; C, 0.54;

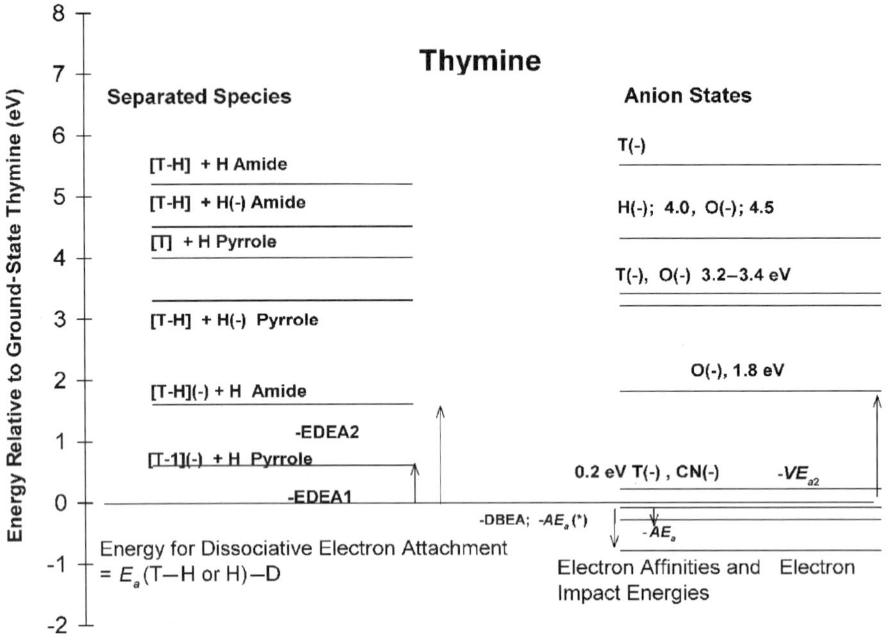

Figure 12.5 Energy-level diagrams constructed for thymine from electron affinities, electron impact and transmission data, and CURES-EC calculations. From [7, 27, 28].

G, 0.36, and A, 0.01 eV. By using different functionals, lower E_a are calculated [27]. The highest values for C, U, and T agree with experimental values, but those of G and A are systematically lower than experiment by 1.1 eV and 0.5 eV. The density functional calculations all result in a change in the geometry of the anion relative to that of the neutral. However, the changes are not as large as those obtained in the CURES-EC optimization, as illustrated for the keto and enol forms of the C anions in Figure 12.6.

TABLE 12.4 Experimental and Theoretical Electron Affinities (in eV) of Purines and Pyrimidines and Halogenated Uracils

Molecule	Experimental	CEC	ab initio	Density Functional
Uracil	0.80(5)	0.80	0.9	0.06 to 0.75
Thymine	0.79(5)	0.78	0.8	0.0 to 0.71
Cytosine	0.56(5)	0.58	0.7	−0.14 to 0.54
Adenine	0.95(5)	0.94	0.5	−0.66 to −0.08
Guanine	1.51(5)	1.51	0.9	−0.36 to 0.36
5-F-uracil	—	1.00	—	0.45, 0.48
5-Cl-uracil	—	1.14	—	0.58, 0.60
5-Br-uracil	—	1.20	—	0.63, 0.64

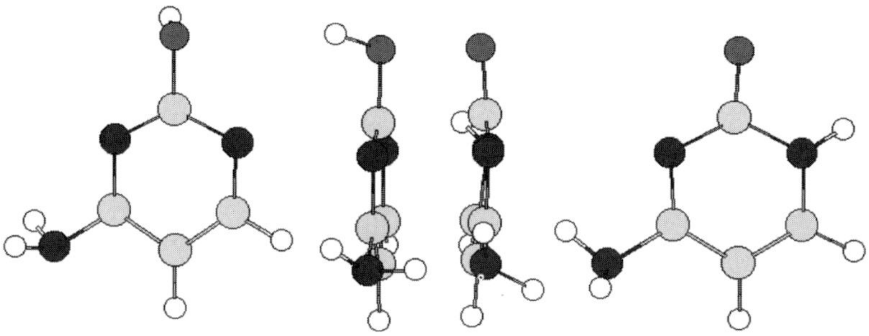

Figure 12.6 Geometry-optimized forms of cytosine anion in the keto and enol forms, AM1 calculations. Note the twisted amino group and bent N—H bond in the keto anion and the bent O—H bond in the enol form.

The use of 5-fluorouracil in the treatment of cancer is well known, giving added significance to the electron affinities of the substituted uracils. The electron affinity of the 5-fluoro-, chloro-, and bromo-uracils have been calculated by two groups using density functional theory [30, 31]. We calculated these quantities utilizing CURES-EC and obtain systematically higher values. In all cases the E_a of the halogenated uracil is larger than that of uracil. The density functional value for U is 0.18 eV, while that for 5-FU is 0.45 or 0.48, an increase of 0.3 eV upon substitution of a fluorine atom. There is a systematic increase in the E_a for the other halogenated uracils in both the density functional and CURES-EC calculated values. The increments agree with the substitution effects discussed in Chapter 11.

12.3 ELECTRON AFFINITIES OF BIOLOGICAL MOLECULES FROM REDUCTION POTENTIALS

Absolute electron affinities can be obtained by classification of biological molecules to establish different values of $-\Delta\Delta G$. This is illustrated for riboflavin, vitamin K, vitamin A, polyazines, and hydroxyprimidines. The E_a of these compounds are also predicted by substitution and replacement rules. Those for the diazines range from 0.2 eV to 0.4 eV. The values for 1,2,4 triazine and 1,2,4,5 tetrazine are 0.9 eV and 1.7 eV. The replacement of an additional CH by N increases the E_a by 0.6 eV. Therefore, the predicted values for pentazine and hexazine are 2.2 eV and 2.9 eV. The CURES-EC method gives better approximations to these E_a and can differentiate between isomers.

The prototype molecule for vitamin A, a conjugated olefin, is decapentaene. The half-wave reduction potentials of the linear conjugated olefins can be converted to absolute E_a. The olefins are classified as A molecules, with $-\Delta\Delta G = 2.0$ eV. The values calculated from the reduction potentials are the only absolute E_a of the

TABLE 12.5 Electron Affinities of Biologically Related Compounds (in eV)

	Experimental		Predicted		
Compounds	Gas	$E_{1/2}$	S&R	CEC	MCCI
Pyridine	0.0	—	—	0.00	AM1(0011)
Pyrimidine	>0	0.2	0.3	0.22	AM1(2222)
Pyrazine	>0	0.36	0.3	0.38	AM1(0023)
Pyradazine	>0	0.31	0.3	0.33	AM1(0023)
1,3,5-Triazine	>0	0.5	0.9	0.50	PM3(0032)
1,2,3-Triazine	—	—	0.9	1.08	AM1(0000)
1,2,4-Triazine	—	0.91	0.9	0.89	AM1(1100)
1,2,3,4-Tetrazine	—	—	1.6	1.62	AM1(0000)
1,2,3,5-Tetrazine	—	—	1.6	1.50	AM1(0000)
1,2,4,5-Tetrazine	—	1.67	1.6	1.67	PM3(0033)
Pentazine	—	—	2.2	2.17	AM1(3300)
Hexazine	—	—	2.8	2.88	AM1(3300)
1-OH-pyrimidine-ol	—	—	0.65	0.71	AM1(3300)
3-OH-pyrimidine-ol	—	—	0.65	0.61	AM1(3300)
6-OH-pyrimidine-ol	—	—	0.65	0.58	AM1(3300)
3-OH-pyrimidine-one	—	—	0.65	0.77	AM1(3300)
Ethylene	—	—	—	−1.04	AM1(3300)
Butadiene	—	0.0	0.0	0.00	AM1(2200)
Hexatriene	—	0.44	0.4	0.41	AM1(0021)
Octatetrene	—	0.66	0.6	0.64	M/3(2100)
Decapentaene	—	0.85	0.8	0.80	M/3(2100)
Cubene	0.50	0.5	—	0.50	M/3(0021)
Vitamin A alcohol	—	0.75	0.8	0.76	M/3(0033)
Vitamin A acetate	—	0.9	0.8	0.91	M/3(0023)
1,4-Naphthoquinone	1.81	—	1.88	1.89	AM1(0033)
Vitamin K	—	1.75	1.8	1.81	AM1(0033)
Guanine	—	1.51	1.3	1.51	AM1(3300)
Riboflavin	—	1.65	1.3	1.66	M/3(0000)

conjugated olefins. The prototype for vitamin K is naphthoquinone with an $-\Delta\Delta G$ of 2.2 eV and the prototype for riboflavin is guanine. These initial assignments are based on the similarity of the molecules to the prototype molecules. The calculated charge densities agree with the intuitive assignment. The E_a determined from the $E_{1/2}$ values, the predicted E_a, CURES-EC E_a for these compounds, and prototype molecules are given in Table 12.5. The agreement between the predicted and CURES-EC values is good. The predicted and calculated values of the E_a for the hydroxy-pyrimidines support the value for cytosine since they are all larger than for cytosine that has an additional NH_2 group.

The electron affinities of aromatic heterocyclic compounds have been measured by TCT and are included in the NIST tables [21]. At the same time the electron affinities of pyridine, the diazines, and aza-substituted phenanthrenes were

TABLE 12.6 Selected Adiabatic Electron Affinities (in eV) for AGCUT, Related Compounds, and Nitrogen Heterocyclic Molecules

Molecule	Selected	NIST	TCT/ET	$E_{1/2}$	C-EC
Vitamin K	1.81(10)	—	—	1.81(10)	1.83
1,2,4,5-Tetrazine	1.67(10)	—	—	1.67(10)	1.67
Riboflavin	1.60(10)	—	—	1.60(10)	1.66
Guanine	1.51(5)	—	—	1.51(5)	1.51
9,10-Diazaanthracene	1.31(10)	1.31	1.31(10)	1.00(10)	1.35
1,6,7-Triazanaphthalene	1.00(10)	1.00(10)	1.00(10)	—	1.01
Adenine	0.95(5)	0.01	—	0.95(5)	0.95
1,2-Diazaphenanthrene	0.92(10)	0.92	0.92(10)	0.99(10)	0.95
9-Azaanthracene	0.91(10)	0.91	0.9(10)	0.91(10)	0.91
1,2,4-Triazine	0.91(10)	—	—	0.91(10)	0.89
Vitamin A acetate	0.90(10)	—	—	0.90(10)	0.91
Uracil	0.80(5)	—	0.8(1)	0.80(5)	0.80
Thymine	0.79(5)	—	0.7(1)	0.79(5)	0.77
Vitamin A alcohol	0.75(10)	—	—	0.75(10)	0.76
1,3,5-Triazine	0.50(10)	0.45	0.45(10)	0.50(10)	0.50
1,2-Diazanaphthalene	0.72(10)	0.72	0.72(10)	0.82(10)	0.75
1,4-Diazanaphthalene	0.71(10)	0.71	0.71(10)	0.80(10)	0.73
1,3-Diazanaphthalene	0.58(10)	0.58	0.58(10)	0.71(10)	0.60
Cytosine	0.56(5)	—	0.6(1)	0.56(5)	0.58
Quinoline	0.50(15)	—	—	0.40(15)	0.45
Pyrazine	0.36(10)	—	—	0.36(10)	0.38
Pyradazine	0.31(10)	—	—	0.31(10)	0.33
Pyrimidine	0.20(10)	—	—	0.20(10)	0.22

The values in parentheses are the uncertainties in the last figure.

estimated from half-wave reduction potentials using two parameters. If a unit slope is assumed, the equation is the same as for the purines and pyrimidines: $E_a = 2.48(5) + E_{1/2}$. Table 12.6 presents the E_a measured in the gas phase and with $E_{1/2}$ values. One of the molecules is acridine, originally used as a standard for the calibration of the E_a for AGCUT. Another is 9,10-diazaanthracene, which is analogous to the structure in riboflavin. Riboflavin with additional nitrogen atoms and hydroxyl groups should have an E_a greater than 9,10-diazaanthracene. The reduction potential data for riboflavin give an E_a of 1.65 eV, as compared to the TCT value of 1.31 eV for 9,10-diazaanthracene. The CURES-EC calculations support the TCT values. The substitution of a single N for a CH in naphthalene increased the E_a by about 0.3 eV. The increase from anthracene to acridine is 0.2 eV. With the data in Table 12.6 we can determine the effect of multiple replacements from gas phase data. The E_a for the diazanaphthalenes are all roughly 0.7 eV or 0.5 eV greater than naphthalene, or 0.25 eV per nitrogen replacement. Triazanaphthalene is 1.0 eV or another 0.3 eV higher. The monocyclic N_{10} has a predicted electron affinity of 4.0 eV or 0.4 eV/N. This is compared to the cyclic N_6 with a predicted E_a of 2.9 eV. The E_a of phenazine, 9,10-diazaanthracene, is

0.4 eV higher than that for acridine. The increase in the E_a for the replacement of a CH by an N is about 0.3 eV. Based on these values and an E_a of 0.30 eV for phenanthrene, the E_a for 9- or 10-azaphenanthrene should be about 0.5 ± 0.15 eV, as obtained from the single-point ECD data given in Table 12.4. Likewise, the value for the 9,10-diazaphenanthrene measured by TCT is yet another 0.3 eV higher or 0.9 eV.

12.4 GAS PHASE ACIDITIES OF NUCLEIC ACIDS

Along with the determinations of the reduction potentials for the purines and pyrimidines, the negative-ion mass spectra of AGCUT and purine were obtained. Experiments on SF_6, C_6F_5Cl, and C_6F_6 (see Chapter 5) demonstrated that thermal electron capture took place in the ion source. It was anticipated that the parent negative ions would be formed. However, instead the parent minus a hydrogen atom $P_{minH}(-)$ was observed. Thus, dissociative thermal electron attachment occurs and the energy for dissociative electron attachment was small, between 0.5 eV and 1.0 eV. The electron affinities of the radicals and bond dissociation energies were unknown at that time, so the data were set aside.

When it was demonstrated that the semi-empirical calculations could reproduce the experimental E_a of AGCUT, the gas phase acidities were calculated. Similar calculations had been made earlier using the standard AM1 and PM3 parameters without MCCI [32]. The calculated N–H bond dissociation energies are 3.95 eV for G, A, and T and 4.15 eV for C and U. These results are shown in Table 12.7 with the -EDEA that range from 0.5 eV to 0.9 eV. Also shown are density functional calculated values. Based on current experimental data for the gas phase acidity of U, 14.4 ± 0.2 eV determined by TCT, the earlier values are systematically low but within the uncertainty of ± 0.2 eV [32–36]. In order to obtain the highest possible GPA values (weaker acids), the minimum electron affinities of the P_{minH} have been recalculated with AM1(0033). The minimum E_a is 3.25 eV for cytosine-H and 3.48 eV for guanine-H. This range of 0.3 eV is considerably smaller than that for the molecules, 1 eV. The E_a are comparable to the values for the halogens, 3.07 eV

TABLE 12.7 Gas Phase Acidities and Electron Affinities (in eV) of Radicals and Dissociation Energies

Molecule	Exp-GPA	D(N–H)	$E_a(R)$	−EDEA	GPACEC	GPAPM3	GPADF
Uracil	14.4(2)	4.15	3.38	0.77	14.4(2)	—	14.4(1)
Thymine	14.2(3)	3.95	3.36	0.59	14.2(2)	14.0	14.5(1)
Cytosine	14.6(3)	4.17	3.25	0.92	14.5(2)	14.2	15.0(1)
Adenine	14.2(3)	3.96	3.40	0.56	14.2(2)	14.2	—
Guanine	14.2(3)	4.10	3.48	0.59	14.2(2)	14.2	—
5-F-uracil	—	4.10	3.48	0.52	14.2(2)	—	—
5-Cl-uracil	—	4.14	3.50	0.64	14.2(2)	—	—
5-Br-uracil	—	4.12	3.66	0.46	14.1(2)	—	—

to 3.61 eV. When combined with the bond dissociation energies, AGCUT are among the strongest organic acids. The gas phase acidities range from 14.2 eV to 14.5 eV for adenine to cytosine. All the theoretical values agree within the uncertainties. The GPA are remarkably similar for A, G, and T. The density functional GPA for U is less than that for T, as opposed to the CURES-EC values. The GPA for C is the largest (weakest acid), which agrees with the order of the collisional dissociation of deprotonated dinucleotides, which is A > T > G > C or G > A > T > C depending on the nature of the terminus base [36]. Based on these calculations, the NIMS values can be scaled to the experimental GPA for U and the relative values of -EDEA to obtain the EDEA given in Table 12.7. The uncertainty is taken as 0.3 eV based on the range of observed values of the -EDEA in ECD studies, which is 0.5 eV to 1.0 eV.

The GPA, calculated bond dissociation energies, and EDEA of the halogenated uracils are reasonable. These are also strong organic acids. The major difference in these molecules and uracil is the strength of the C—X bond. The -EDEA in the C—F dimension in fluorouracil is greater than for the N—H dissociation of the anion. In the case of the chloro and bromo compounds the -EDEA is comparable to that in the N—H dimension. There will be two dissociation limits in the C—X dimension, one leading to the halogen anion plus the U-yl radical and the other leading to the U-yl radical anion plus a halogen. The E_a of the U-yl radical was calculated to be 2.30 eV using density functional theory and the AM1(0000) CURES-EC. The formation of the halogen anions has been observed experimentally [33]. Potential energy curves illustrating these dissociations were calculated in 2002 [32, 33]. The E_a for the molecules were calculated and it was concluded that all the halouracil anions in the gas phase have multiple bound electronic states. This supports the calculations of Herschbach negative-ion Morse potential energy curves in multiple dimensions from experimental data for C and T [7].

12.5 MORSE POTENTIAL ENERGY CURVES FOR THYMINE AND CYTOSINE

With gas phase acidities and electron impact data Morse potential energy curves can be calculated to explain the differences in the electron impact data for C and T. Recall that the VE_a for C is 0.0 eV and −1.4 eV, while that for T is −0.2 eV based on the electron impact spectra. The obvious difference is that C has an NH_2 group attached to the ring, whereas T does not so C can dissociate in the C—NH_2 dimension. The C—NH_2 bond energy is calculated to be 4.42 eV, while the E_a of the radical formed by the loss of the NH_2 group is 2.35 eV. This radical is analogous to that formed in the halouracils by the loss of the halide ion. In the ground state of the anions of G, A, and C, the NH_2 group is twisted relative to the plane, N—H is bent, and the rings are nonplanar, as shown in Figure 12.6. In the case of T the negative ion is nonplanar and the pyrrole N—H bent.

From the electron impact data four valence-state curves can be drawn. In Figure 12.4 the energy-level diagrams obtained from experimental data were shown.

Table 12.8 lists the Morse parameters and experimental data used to construct these curves. These curves could not be constructed without the gas phase acidities and bond dissociation energies obtained from the electron impact and NIMS experiments. The curves in the C—NH_2 dimension for cytosine are shown in Figure 12.7, those for T in the N—H dimension are illustrated in Figure 12.9. The curves for C and T in this dimension are similar. The ion distributions for the parent negative ions are shown in Figures 12.8 and 12.10. From these curves the vertical transitions lead to a temporary anion that can dissociate or be stabilized to the stable ground-state parent negative ion. Thus, the parent negative ion for C observed at 1.4 eV results from the curve leading to dissociation in the C—NH_2 dimension. The low-energy dissociative curve is also evident in Figure 12.9 [7].

The formation of the parent negative ions at 3.4 eV and 4.5 eV proceeds via vertical processes that form a temporary negative ion which can then be stabilized to

TABLE 12.8 Morse Parameters and Dimensionless Constants

Molecule	k_A	k_B	k_R	D_e (eV)	r_e (pm)	ν (cm^{-1})	AE_a	$-VE_a$
Cytosine N—H	1.000	1.000	1.000	4.27	99.4	3,470	—	—
Dipole bound	0.895	1.064	0.908	3.76	100	3,470	0.23	−0.23
X	0.357	1.419	0.367	1.48	100	2,900	0.56	−0.45
A	0.485	0.867	1.015	1.05	124	1,590	0.15	0.32
B	0.239	0.761	1.022	0.24	186	625	—	3.30
C	0.485	0.867	1.015	0.99	138	1,450	—	3.40
D	0.232	0.677	0.977	0.23	196	550	—	5.40
Cytosine C—NH_2	1.000	1.000	1.000	4.51	137	1,720	—	—
Dipole bound	0.841	1.090	0.841	3.90	137	1,700	0.08	−0.08
X′	0.621	1.254	0.621	2.91	137	1,700	0.56	−0.56
A′	0.610	1.096	0.814	2.14	146	1,275	−0.1	0.33
B′	0.369	1.012	0.559	1.15	151	860	—	1.4
C′	0.699	1.000	0.909	2.50	146	1,260	—	1.4
D′	0.207	0.878	0.484	0.43	170	450	—	5.0
Thymine N—H pyrrole	1.000	1.000	1.000	4.17	99.4	3,470	—	—
Dipole bound	0.844	1.089	0.844	3.52	99.4	3,470	0.06	−0.06
X	0.364	1.447	0.371	1.49	100	3,000	0.79	−0.79
A	0.405	0.934	0.764	0.89	129	1,500	0.15	0.30
B	0.256	0.752	1.145	0.24	186	625	—	3.40
C	0.547	0.871	1.139	1.10	136	1,550	—	3.40
D	0.225	0.725	0.892	0.24	182	600	—	5.40
Thymine N—H (amide)	1.000	1.000	1.000	5.21	99.4	3,470	—	—
Dipole bound	0.875	1.069	0.875	4.57	99.4	3,470	0.06	−0.06
X′	0.422	1.342	0.429	2.16	100	3,000	0.79	−0.79
A′	0.515	0.897	0.888	0.88	129	1,700	0.15	0.30
B′	0.201	0.880	0.802	0.26	176	685	—	3.40
C′	0.616	0.911	1.071	1.85	129	1,880	—	3.40
D′	0.187	0.821	0.704	0.26	178	635	—	5.40

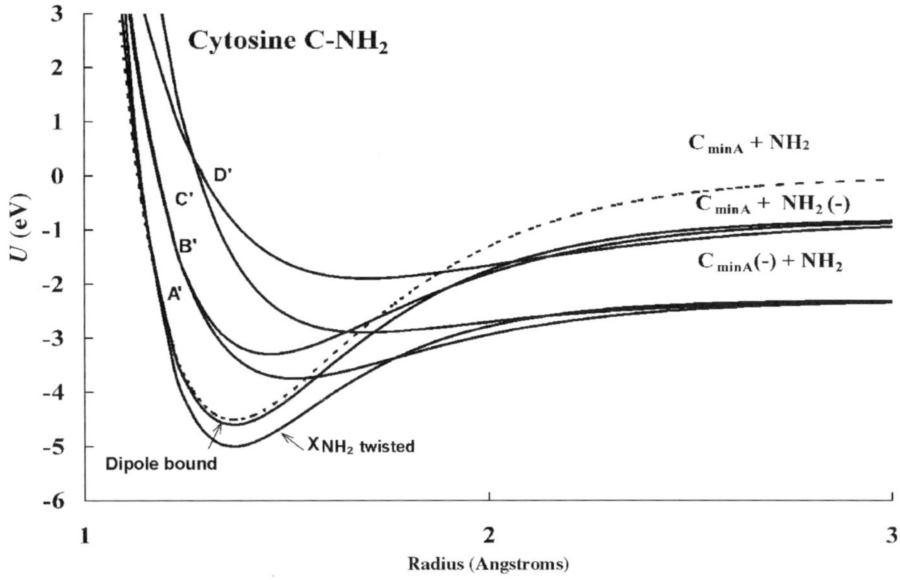

Figure 12.7 Morse potential energy curves for cytosine in the C—NH_2 dimension calculated from E_a and data in Figure 12.4 [7, 28]. The molecular anion formed at 1.5 eV comes from the vertical transition to the A' and B' curves. The parameters are given in Table 12.8.

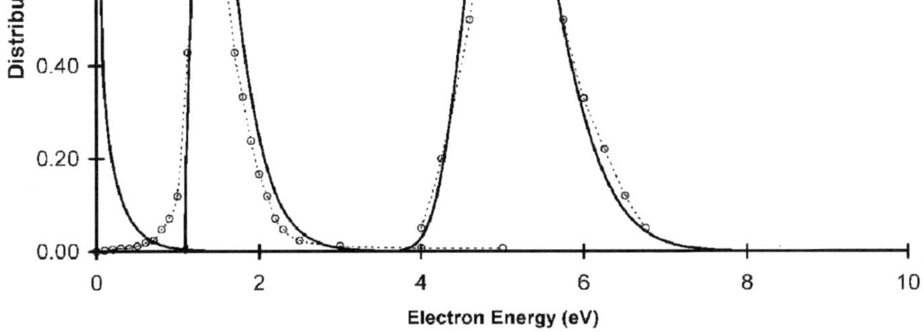

Figure 12.8 The calculated and experimental molecular ion distributions for electron impact on cytosine [7, 28].

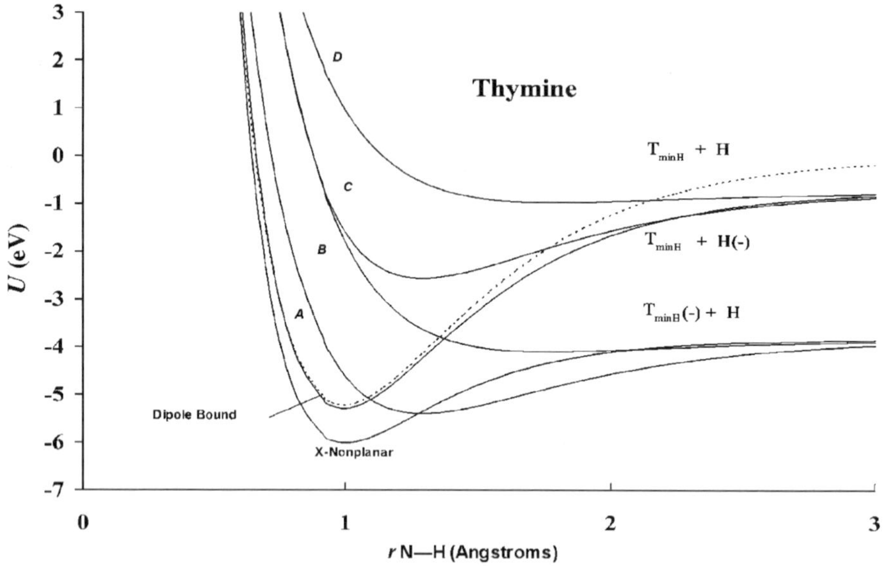

Figure 12.9 Morse potential energy curves for thymine in the N—H dimension calculated from E_a and data in Figure 12.5 [7, 28]. The molecular anion formed at 0.2 eV comes from the vertical transition to the *A* curve, while those at 0.8 eV come from the *B* and *C* curves and the one at 6 eV comes from the *D* curve. There is no peak at 1.5 eV, as occurs in cytosine. The parameters are given in Table 12.8.

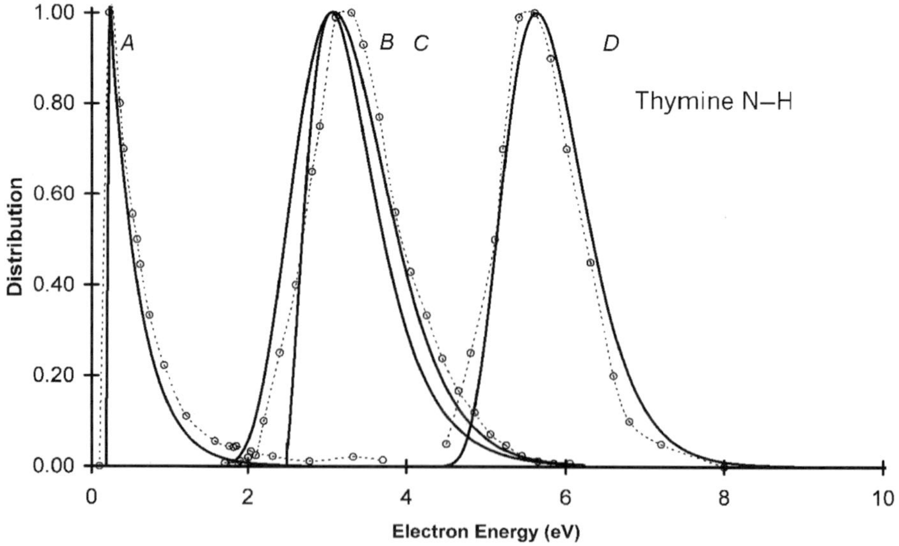

Figure 12.10 The calculated and experimental molecular ion distributions for electron impact on thymine [7, 28].

the ground state or dissociate via an exothermic process. Since the VE_a and E_a are positive but the EDEA is negative, the ground-state curve is an $M(2)$ curve and the excited-state curves for the states with positive E_a and negative VE_a are $M(1)$ curves. The higher-energy curves are $D(0)$ curves but also $Mc(0)$ because they can be stabilized to the ground-state anion via the dipole bound state. In these curves the anions are formed with a diminished attractive force, as indicated by the values of k_A in Table 12.8 that are lower than 1. This is different from the diatomic molecules in the ground state. For these molecules both the attractive and repulsive portions of the curve are increased.

12.6 GAS PHASE ACIDITIES AND ELECTRON AFFINITIES OF THE AMINO ACIDS

The activation energies for dissociative electron capture of the esters and acetic anhydride give $D(\text{C—O}) - E_a(\text{acetate})$. By incorporating the data for bond dissociation energies found in the literature, the E_a of the acetate radical can be obtained. In the sequential NIST table the E_a of the acetate radical is 3.40 ± 0.30 eV by laser photodetachment, but the value from the ECD data using more recent bond dissociation energies is 3.18 ± 0.05 eV, while an earlier electron impact value is 3.30 ± 0.20 eV. The weighted average of these values is 3.19 ± 0.05 eV, the current best value. The electron affinities of many radicals of saturated organic acids vary from 3.14 eV to 3.24 eV. This provides another example of an evaluation of the E_a that requires searching the complete NIST tables rather than only taking the most recent value. The MNDO(0031) CURES-EC value for the electron affinity of the acetate radical is 3.16 eV [37, 38].

The gas phase acidities of only three amino acids—glycine, alanine, and L-histidine—have been measured [39–42]. These are higher than the GPA for the purines and pyrimidines because the O–H bond in the amino acids is stronger than the N–H bond in the purines and pyrimidines, while the E_a of the radicals are only slightly larger than those of the P–H radicals. We have calculated the electron affinities of the molecules and radicals for these amino acids using CURES-EC. The energies of the negative ions of the free acids are about equal to the energies of the neutral molecule for acetic acid, glycine, and alanine. Thus, the current best estimate of the $A\,E_a$ for acetic acid, glycine, and alanine is approximately zero. The values for tyrosine and phenylalanine are about 0 eV to 0.3 eV, while that for histidine is higher.

The experimental and CURES-EC electron affinities of the radicals are glycine-H: 3.37(10), 3.38 eV; alanine-H: 3.42(13), 3.40 eV; and histidine-H: 3.84(19), 3.90 eV. The value for histidine-H is one of the largest experimental or CURES-EC E_a for an amino acid radical. Its GPA is also among the lowest of the amino acids and is comparable to that of uracil, 14.4(2) eV. Glycine with a GPA of 14.7(2) eV is a weaker gas phase acid than histidine or cytosine. The E_a of the radical from aspartic acid and tyrosine are about 3.8 eV for AM1(0000), giving them a GPA comparable to that of cytosine. A most remarkable observation is that very little experimental or theoretical E_a or GPA are available for the amino acids.

12.7 THE CALCULATION OF THE ECD AND NIMS TEMPERATURE DEPENDENCE

Very little experimental work using ECD has been done on heterocyclic molecules. Some data have been obtained at a single temperature and were used to obtain E_a. Some E_a have been measured with TCT and/or $E_{1/2}$ techniques and can be used to predict the ECD or NIMS temperature dependence. The temperature dependence of the compounds with similar E_a can be predicted by assuming a single state and value of Q_{an}. The ECD response for pyridine has been determined to be low. The response for substituted pyridines should be similar to those for substituted benzenes. The compounds with E_a less than about 0.6 eV should exhibit only a single region. The ECD temperature dependence of quinoline should be like that of benzaldehyde, while that for acridine should be like the temperature dependence of benz[a]pyrene.

The ECD response for the esters of the amino acids should be similar to that of the acetates. The CURES-EC calculated value for the methyl ester of glycine is 0.2 eV, approximately the same as that for ethyl acetate. By forming the trifluoromethyl ester of glycine, the electron affinity is increased to over 1.5 eV, making the compound a maximum capturing analyte. The amino acids have been analyzed in the ECD by making such electronegative derivatives. Very low limits of detection can be achieved, similar to those of some pesticides. For these cases the temperature dependence should be low [43–46].

The electron attachment cross-section for thymine is reported to be as large as that for SF_6 [27]. Thus, with a value of $Q = 1$, $A_1 = A_{max}, E_1 < 0.2$ eV and the assumption of stable negative-ion formation of an excited state at about 0.5 eV, the curves shown in Figure 12.11 are calculated using the standard ECD equation for two states. These curves can also be calculated by assuming dissociative capture. With the N—H bond dissociation energy the quantity $D - E_a$ is about 0.5 eV to 0.9 eV (see Table 12.7). With this value the temperature dependence (if we assume dissociative electron attachment) is shown in Figure 12.12. This indicates a higher ECD response than if nondissociative capture is assumed. For both cases the response for cytosine is much smaller (three orders of magnitude at 500 K) than for the other compounds. The optimum temperatures for the analyses would be the highest temperatures. The response factors for adenine and guanine reach saturation. The response for thymine does not occur at saturation, but will be large. Perhaps the most important part of the calculation is the lower response predicted for cytosine. To our knowledge there are no routine analyses of these compounds using the ECD. Techniques have been developed to determine concentrations of the nucleic acids by forming derivatives. It was reported that as few as 0.3 pg to 5.0 pg of thymine or 5-fluorouracil could be detected. With this large a response and the predicted temperature dependence of the parent compounds, these derivatives should have very little temperature dependence [47, 48].

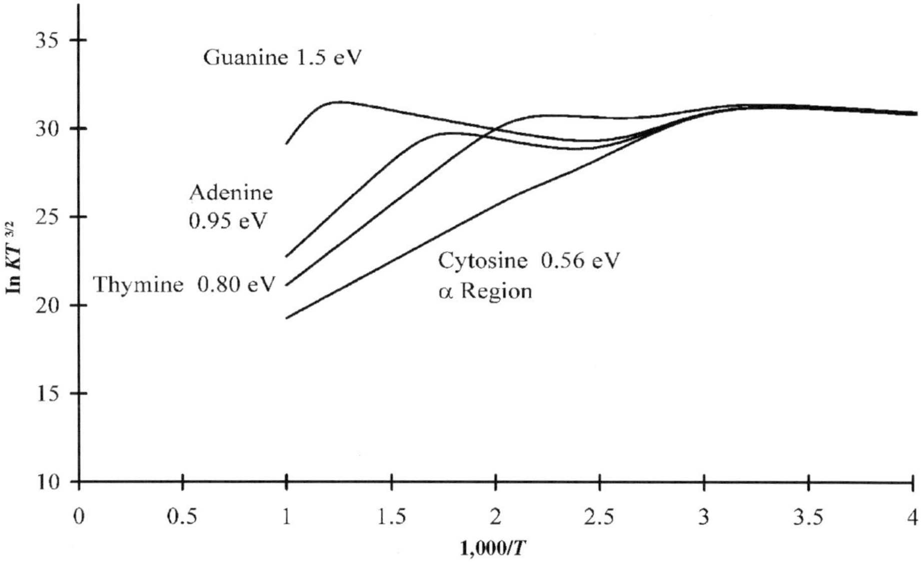

Figure 12.11 The calculated ECD temperature dependence for AGCUT if we assume nondissociative capture.

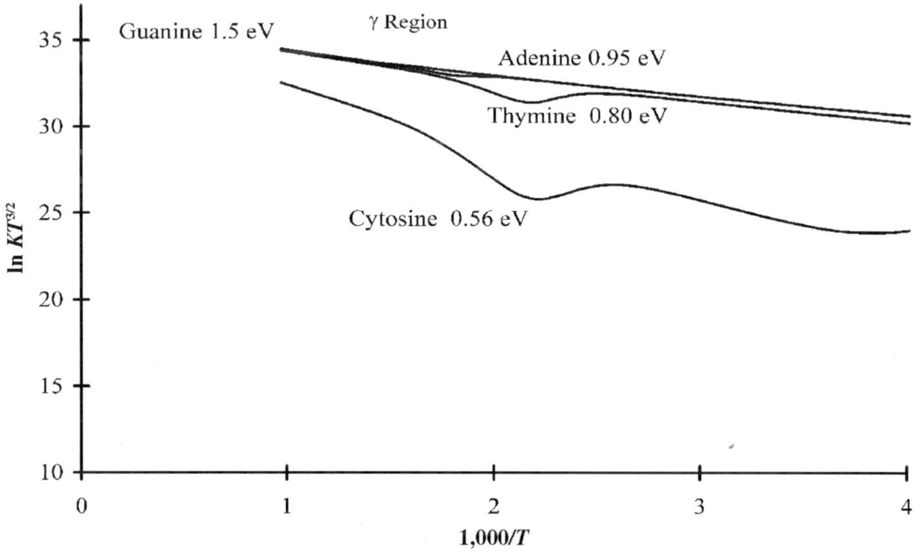

Figure 12.12 The calculated ECD temperature dependence for AGCUT if we assume dissociative capture.

12.8 ELECTRON AFFINITIES OF AT AU AND GC

In 1998 a unique model for electron transport in DNA was proposed based on the role of anionic donor acceptor complexes [49, 50]. DNA has a covalent backbone and an inner anhydrous π stacked core. This core consists of hydrogen-bonded base pairs. Conduction can occur along the backbone or through space via the inner core. The π system can be modeled as a series of donor acceptor complexes. In the neutral state there will be little charge transfer. Once an electron is added to a base pair, rapid, length-independent electron transfer will take place through the π way. This will occur since the electron affinities of the base pairs are within the range of thermal activation. The changes in geometry and hydrogen bonding will make up for the fundamental differences in the E_a.

In order to establish a donor acceptor mechanism for electron transport in DNA, values for the electron affinities and ionization potentials of the hydrogen-bonded base pairs are required. As early as 1962 the importance of such interactions was considered:

> From molecular orbital calculations on the adenosine-thymidine and guanosine-cytidine base pairs of DNA.... the second pair is both a better electron acceptor and electron donor than the first.... There is considerable overlap between the electron systems resulting in the conjugated electrons extending over both bases in the pair. This point needs to be investigated further as it is of considerable significance. [51]

In 1998 the E_a of AT, AU, and GC were calculated using the CURES-EC method. These were the first to give positive values for the electron affinities. Subsequently, all the calculations emphasize the positive nature of the E_a. The CURES-EC E_a are given in Table 12.9 and compared with the experimental values for the individual bases and values calculated using the density functional theory available in 2002 [52–54]. Although the density functional values are lower than the CURES-EC values, the difference between the E_a of GC and AT is about the same, 0.20 eV. This activation energy is very small for an electron transfer reaction. This difference supports the possibility of thermal activation of electron conduction.

TABLE 12.9 Electron Affinities (in eV) of Purines and Pyrimidines and Watson Crick Hydrogen-Bonded Base Pairs

Molecule	Exp	CEC	B3P86	BP86	B3LYP
Adenine	0.95(5)	0.96	0.01	−0.05	−0.28
Guanine	1.51(5)	1.50	0.36	0.11	−0.10
Cytosine	0.56(5)	0.58	0.54	0.13	0.03
Uracil	0.80(5)	0.80	0.75	0.31	0.24
Thymine	0.79(5)	0.76	0.71	0.28	0.20
G-C	—	1.35	1.15	0.71	0.60
A-T	—	1.15	0.88	0.58	0.30
A-U	—	1.05	—	—	0.32

The density functional values are from [40–42].

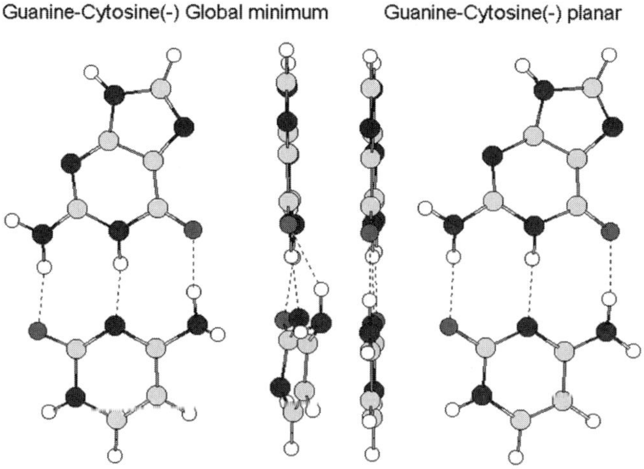

Figure 12.13 Planar and global minimum anions of GC calculated using AM1. Note the twisted NH_2 group and the differences in the lengths of the hydrogen bonds in the global minimum.

The geometry of the anions of GC and AT is significantly different from that of the neutral. The anions and neutrals are shown in the side and top view for AT and GC in Figures 12.13 and 12.14. The change for GC is larger than for AT. Both are distorted from planarity and the lengths of hydrogen bonds change. The changes for AT are smaller than those for GC. The energetics and geometry changes show that the hydrogen-bonded base pairs can act as both donors and acceptors of electrons. Pairwise electron transfer down the π stack via an anion complex is possible. When

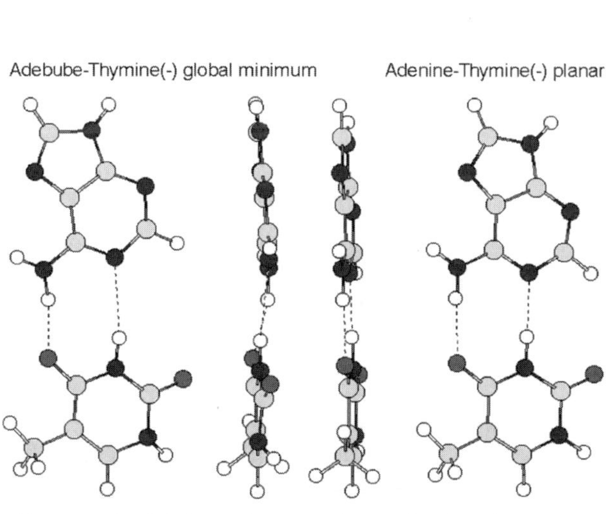

Figure 12.14 Planar and global minimum anions of AT calculated using AM1. The distortions in the base pair are still present, but are smaller for AT than for GC.

the electron is added to one base pair, the anion becomes the donor and the adjacent base pair the acceptor. The charge transfer energy is related to the quantity IP $- E_a$. For anion complexes the IP of the donor is the E_a of the first base pair, while the E_a of the acceptor is that of the adjacent base pair. The anion stabilization energy is related to the differences in the E_a. This is only about 0.2 eV, as shown in Table 12.9. It is much smaller than the energy required for many biological processes such as hydrogen bonding. The net effect is rapid charge transfer through the π stack after an electron is added to the system. This model merely shows that conduction is thermodynamically and geometrically feasible, but it requires that the bases be aligned so the proper overlap is achieved. These interactions are noncovalent and occur through space. The model requires that hydrogen-bonded base pairs be present. In the cases where rapid electron transfer does not take place, the breakage of the hydrogen bonds or an irreversible geometry change could slow down the electrons and lead to DNA damage.

The electron affinities and charge distributions in stacked hydrogen-bonded base pairs in the B form of DNA were calculated and used to test the model. Not surprisingly, for a stack of the same base pairs, GC/GC or AT/AT, READS-TCT = q(GC1)/q(GC2) = q(AT1)/q(AT2) = 1. A greater proportion of the charge resides on the GC in the stacked GC/AT, or AT/GC: READS (GC/AT) = 1.5; and READS (AT/GC) = 0.65. However, it was surprising that the charge resides primarily on the pyrimidine and not the purine, if we consider the larger adiabatic electron affinity of the purines. In the isolated Watson Crick hydrogen-bonded base pairs the READS(G/C) = 0.05 and READS(A/U) = READS(A/T) = 0.2 at the global minimum. Density functional calculations give similar results [52–54]. Based on this observation, it was postulated that the geometry of G and A in the hydrogen-bonded base pair exists for an excited anion state. The inverted charge distribution for the Watson Crick hydrogen-bonded base pairs is responsible for the pyrimidine anions observed in radiation-damaged DNA [55].

12.9 RADIATION DAMAGE IN DNA

When DNA is irradiated, the resulting anions are primarily C(−) and T(−) [49]. However, the adiabatic electron affinities AE_a are Pu > Pyr. The abundance of the Pyr(−) could imply inaccurate values of the AE_a or lower-energy excited states of A(−) and G(−) in DNA. The CURES-EC AE_a of G, at 1.5 eV, agrees with the experimental value, but an E_a^* of 0.3 ± 0.05 eV is obtained for the local minimum dipolar planar anion. The dipole moment of G is similar to that of C and thus the DBEA should be about the same, 0.25 eV. The E_a^* and the predicted DBEA are approximately the same.

In Figure 12.15 the planar local minimum form of G(−) is shown with a three-dimensional spin density plot. The spin density is localized on the hydrogen bonding sites and is similar to a dipole bound anion. Figure 12.16 presents a similar plot for the global minimum form of G(−). The NH_2 group is twisted out of the plane,

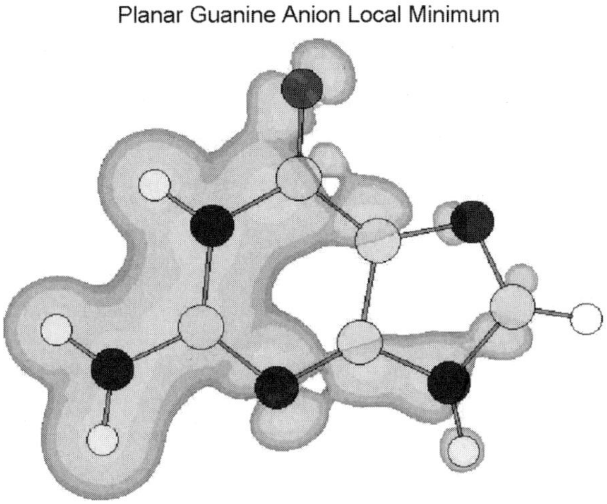

Figure 12.15 Three-dimensional spin densities for the planar guanine anion calculated using AM1. The spin density is localized on the hydrogen bonding sites and is similar to a dipole bound state.

the N—H bond is bent, and the spin distribution is distributed over the complete molecule as expected for a valence-state anion. The twisting of the NH_2 group requires the addition of energy. The local minimum planar form is formed initially, and when the bond twists, the global minimum is formed.

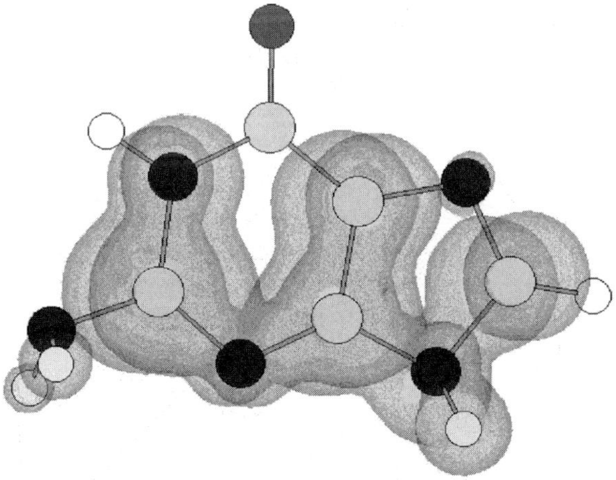

Figure 12.16 Three-dimensional spin densities for the global minimum guanine anion calculated using AM1. The spin density is distributed over the complete molecule, as expected for a valence state.

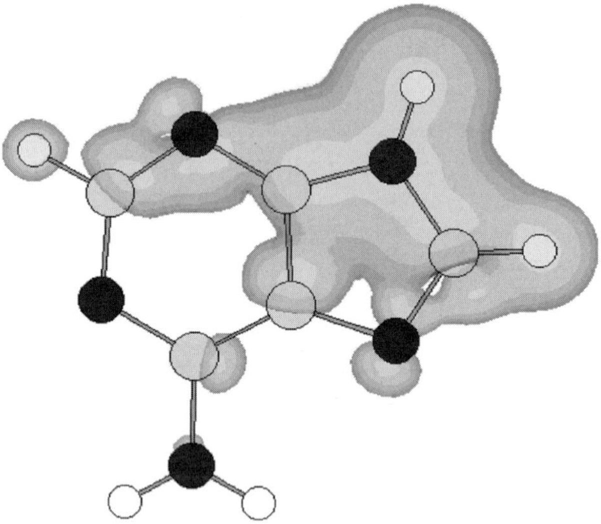

Figure 12.17 Three-dimensional spin densities for the planar adenine anion calculated using AM1. The spin density is localized away from the major structure and is similar to a dipole bound state.

The CURES-EC AE_a of A, at 0.95 eV, agrees with experimental value, but an E_a^* of 0.0 eV is obtained for the local minimum planar anion. This is close to the DBEA measured for A. The lower DBEA is expected because of the lower dipole moment for A, 2.18 Debye. In Figures 12.17 and 12.18 the planar local minimum

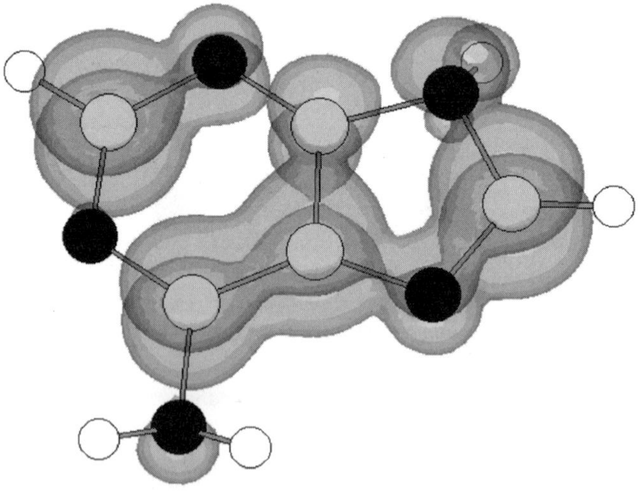

Figure 12.18 Three-dimensional spin densities for the global minimum adenine anion calculated using AM1. The spin density is distributed over the complete molecule, as expected for a valence state.

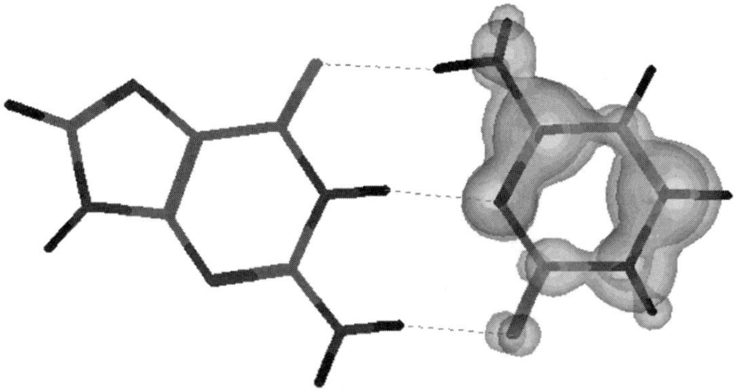

Figure 12.19 Three-dimensional spin densities for the GC anion keto cytosine calculated using AM1. READS(G/C) = 0.05. In this structure the guanine is in an excited state with an E_a of 0.3 eV, as compared to the global minimum with an E_a of 1.51 eV.

and global minimum forms of A(−) are shown with a three-dimensional spin density plot. The ground-state anion of A, like that of G and C, is nonplanar with a twisted NH_2 group, as seen in Figure 12.18. Unlike G(−) the spin density is localized away from the hydrogen bonding sites for A(−) in the planar state, as shown in Figure 12.17. This spin density is also similar to a dipole bound anion. In the global minimum for A(−) the spin distribution involves the complete molecule.

In Figure 12.19 the three-dimensional spin densities of the anion with a GC keto form are shown. This is compared to a similar plot in Figure 12.20 for the anion of the GC enol base pair. In the GC keto(−) form of cytosine READS(G/C) = 0.05, or 95% of the spin occurs in the cytosine. For the GC enol(−) READS(G/C) = 3.0, or

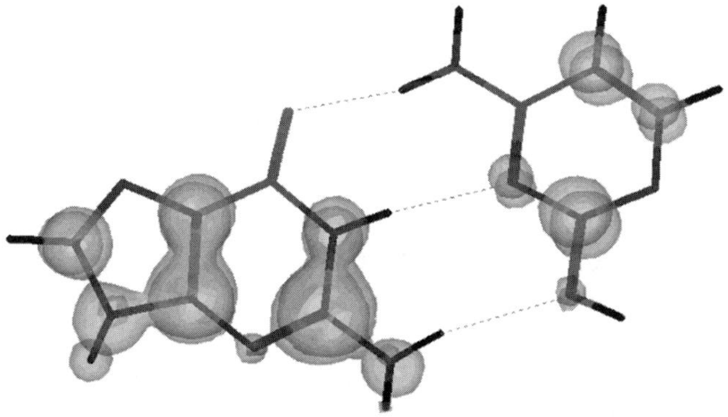

Figure 12.20 Three-dimensional spin densities for the GC anion enol cytosine anion calculated using AM1. READS(G/C) = 3.0. The E_a of the cytosine in the enol form is lower and therefore the spin density is distributed to the planar guanine.

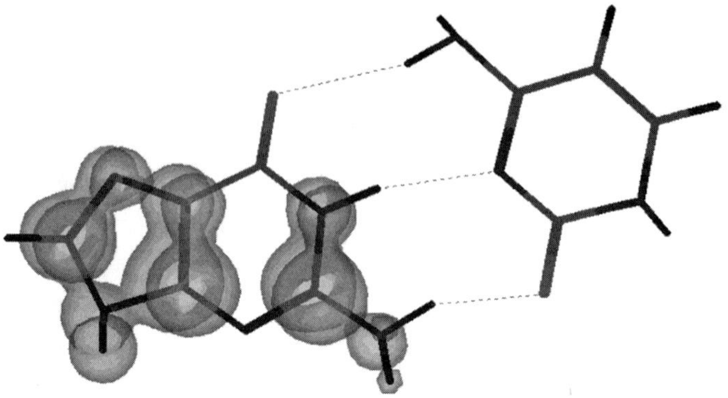

Figure 12.21 Three-dimensional spin densities for the GC(−3) anion keto cytosine anion calculated using AM1. As more electrons are added to the hydrogen-bonded base pair, the spin distribution is distributed to the guanine.

67% of the spin resides on G. The spin density on the cytosine in the GC keto(−) form is similar to the spin density of the isolated anion of cytosine in that it is distributed over the complete molecule. The spin density in the GC enol(−) form is different in that the spin density is localized on a few atoms.

The spin density of the isolated GC(−3) is shown in Figure 12.21. The charge is distributed to the G across the hydrogen bonds and the net spin density is localized on the G(−) rather than distributed to both the G(−) and C(−). The NH_2 bond is twisted to a greater degree than in the singly charged species. The three structures shown in Figures 12.19 through 12.21 were obtained by starting with the global minimum of the normal keto hydrogen-bonded GC(−), (see Figure 12.19) changing the keto structure to the enol structure and geometry optimizing to generate Figure 12.20. The GC(-3) was generated by starting with the GC(−), changing the charge to −3, and then geometry optimizing to obtain Figure 12.21.

This suggested that similar calculations should be made for stacked [GC:GC:GC](−1) in the B DNA form. These structures are generated by first using the HYPERCHEM database to build a DNA segment of three GCs in the B form. Then the sugars and phosphates are removed and the bases terminated with hydrogens at the standard N—H distance. Then the stack is geometry-optimized in the neutral form. The last step is to geometry-optimize the stack with a single negative charge. Starting with this optimized form, additional electrons are added and the geometry optimized.

The calculations of the charge and spin densities are a part of the HYPERCHEM program. The following data are obtained: READS $(G_i/C_i) = 0.05$; READS(GC1/GC2) = 1.05, and READS(GC2/GC3) = 1.25. Since the base pairs are the same, these latter ratios should be 1. In Figure 12.22 the three-dimensional spin densities of GC1:GC2:GC3(−1) are shown. In GC1:GC2:GC3(−3) the charge moves down the stack to the first and third base pairs. The larger charge density and spin density

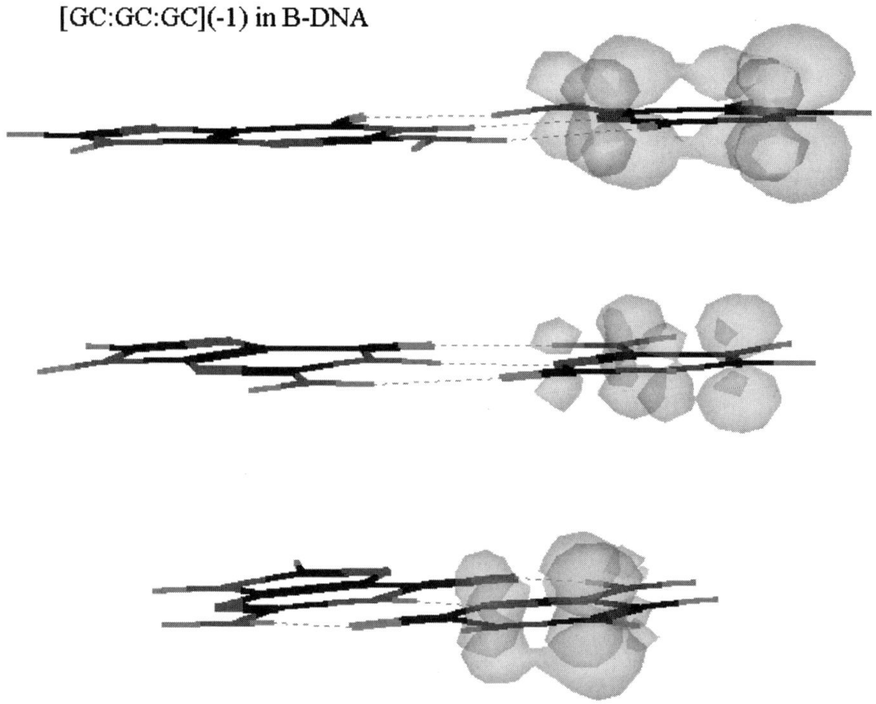

Figure 12.22 Three-dimensional spin densities for the GC:GC:GC(−1) anion stacked in the B DNA form calculated using AM1. The spin distribution in the individual base pairs is inverted, but the spin density down the stack is distributed equally.

on the first and third GC are indicated in Figure 12.23. In addition, the spin distribution is moved from the C(−) in GC(−1) to both the G and C in GC(−3). The spin density on the middle GC is much lower than that on the first and third GC. A similar behavior is obtained with AT:AT:AT. Only two electrons are required to begin moving spin to the A side, and with three electrons the spin density resides more on the A side than on the T side.

Although these data were generated to explain the nonadiabatic spin distribution in radiation-damaged DNA, the effect of adding multiple electrons to a stacked sequence can be extended to electron transport. One electron is delocalized down the stack, but when multiple electrons are added, the charge moves to the purines. Figures 12.22 and 12.23 clearly show the movement of the electrons down the stack without a bridge or bond. This demonstrates that there is no thermodynamic barrier when the base pairs are the same. In addition, even with a small barrier the movement can occur by geometrical changes and thermal activation. The projection of these movements to longer pieces of DNA supports the proposed mechanism of electron transport through the π electron system in DNA without a bridge or bonded backbone.

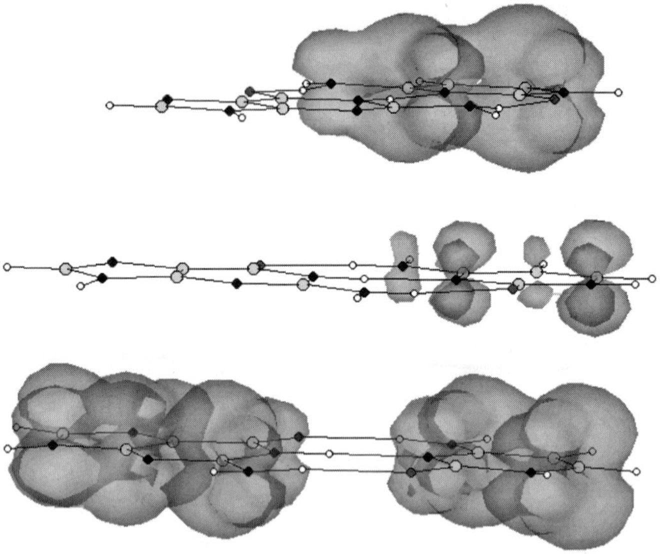

Figure 12.23 Three-dimensional spin densities for the GC:GC:GC(−3) anion stacked in the B DNA form calculated using AM1. When additional electrons are added to the stack, the spin density moves down the stack and across the hydrogen bonds.

12.10 SUMMARY

The electron affinities of biologically significant molecules—NO, O_2, the purines and pyrimidines, and heterocyclic aromatic molecules—have been systematically evaluated. The highest precise value is assigned to the adiabatic electron affinity. Significantly lower values are assigned to excited states. The E_a of AGCUT were obtained by scaling half-wave reduction potentials. Interpreting the PES of isolated anions and anion hydrates gives the E_a of C, U, and T. The E_a of heterocyclic aromatic molecules and the uracil radical have been determined by TCT. The NIMS method has been applied to the estimation of the gas phase acidities of AGCUT. Vertical electron affinities of AGCUT have been determined by electron transmission spectra and for C and T from electron impact spectra. Morse potential energy curves in two dimensions have been constructed to explain differences in the electron impact spectra of C and T. All these have been confirmed by CURES-EC calculations. The temperature dependence of ECD and NIMS spectra has been predicted. A proposed electron conduction model based on pairwise electron bound complexes was reviewed and supported by the CURES-EC calculated electron affinities of Watson Crick GC, AT, and AU. The CURES-EC values were compared with the density functional values.

One of the most important conclusions of this chapter is that there remains much to be done to characterize the thermal electron reactions of even the nucleic acids and amino acids. With these data it might be possible to realize the importance of electrons to bioenergetics, as proposed by Szent-Gyorgi as early as 1957 [56]. The use of quantum mechanical calculations and any experimental techniques available to study these reactions with biological molecules will be important in the future.

REFERENCES

1. Koshland, D. E. *Science*, **1992**, 258, 1861.
2. Available at http://www.nobel.se/medicine/educational/poster/1998.
3. Chen, E. S. D.; Wentworth, W. E.; and Chen, E. C. M. *J. Mol. Struct.* **2002**, 606, 1.
4. Freeman, R. R. Ph.D. dissertation, University of Houston, **1971**.
5. Chen, E. C. M. and Wentworth, W. E. *J. Phys. Chem.* **1983**, 87, 45.
6. Chen, E. S. D. and Chen, E. C. M. *J. Phys. Chem. A* **2003**, 107, 169.
7. Chen, E. C. M. and Chen, E. S. D. *J. Phys. Chem. B* **2000**, 104, 7835.
8. Berthold, H.; Giessner-Prettre, G.; and Pullman, A. *Theoret. Chem. Acta* **1966**, 5, 53.
9. Younkin, J. M.; Smith, L. J.; and Compton, R. N. *Theoret. Chem. Acta* **1976**, 41, 157.
10. Chen, E. S. D.; Chen, E. C. M.; and Wentworth, W. E. *Biochem. Biophys. Res. Comm.* **1990**,171, 97.
11. Wiley, J. R.; Robinson, J. M.; Ehdaie, S.; Chen, E. S. D.; Chen, E. C. M.; and Wentworth, W. E. *Biochem. Biophys. Res. Comm.* **1991**, 180, 841.
12. Chen, E. S. D.; Chen, E. C. M.; Sane, N.; and Schultz, S. *Bioelectrochem. Bioenerget.* **1999**, 48, 69.
13. Zhang, O. and Chen, E. C. M. *Biochem. Biophys. Res. Comm.* **1995**, 217, 255.
14. Anbar, M. and St. John, G. A. *Science* **1975**, 190, 781.
15. Nenner, I. and Schultz, G. J. *J. Chem. Phys.* **1975**, 62, 1747.
16. Dillow, G. W. and Kebarle, P. *Can. J. Chem.* **1989,** 67, 1628.
17. Seidel, C. A. M.; Schulz, A.; and Sauer, M. H. M. *J. Phys. Chem.* **1996**, 100, 5541.
18. Miller, D. A.; Skogerboe, K.; and Grimsrud, E. P. *Anal. Chem.* **1981**, 53, 464.
19. Keiichi, T. and Elving, P. J. *Anal. Chem.* **1969**, 41, 286.
20. Wiberg, W. B. and Lewis, T. P. *J. Amer. Chem. Soc.* **1969**, 92, 7155.
21. Desfrancois, C.; Aboul-Carmine, H.; and Schermann, J. P. *J. Chem. Phys.* **1996**, 104, 7792.
22. Hendricks, J. H.; Lyapustina, S. A.; De-Clercq, H. L.; Snodgrass, J. T.; and Bowen, K. H. *J. Chem. Phys.* **1996**, 104, 7788.
23. Hendricks, J. H.; Lyapustina, S. A.; De-Clercq, H. L.; Snodgrass, J. T.; and Bowen, K. H. *J. Chem. Phys.* **1998**, 108, 8.
24. Scheidt, J.; Weinkauf, R.; Neumark, D. M.; and Schlag, E. W. *Chem. Phys.* **1998**, 239, 511.
25. Sevilla, M. D.; Besler, B.; and Colson, A.-O. *J. Phys. Chem.* **1995**, 99, 1060.
26. Wesolowski, S. S.; Leininger, M. L.; Pentchev, P. N.; and Schaefer, H. F. *J. Amer. Chem. Soc.* **2001**, 123, 4023.
27. Huels, M. A.; Hahndorf, I.; Illenberger, E.; and Sanche, L. *J. Chem. Phys.* **1998**, 108, 1309.
28. Aflatooni, K.; Gallup, G. A.; and Burrow, P. D. *J. Phys. Chem. A* **1998**, 102, 6205.

29. National Institute of Standards and Technology (NIST). *Chemistry WebBook*, **2003**. Available at http://webbook.nist.gov.
30. Aboul-Carime, H.; Huels, M. A.; Illenberger, E.; and Sanche, L. *J. Amer. Chem. Soc.* **2001**, 123, 5354.
31. Li, X.; Sanche, L.; and Sevilla, M. D. *J. Phys. Chem. A* **2002**, 106, 11248.
32. Rodgers, M. T.; Campbell, S.; Marzluff, R. M.; and Beauchamp, J. L. *Int. J. Mass Spectrom. Ion Proc.* **1994**, 137, 121.
33. Chandra, A. K.; Nguyen, M. T.; Uchimaru, T., and Zeegers-Huyskens, T. *J. Phys. Chem. A* **1999**, 103, 8853.
34. Kurinovich, M. A. and Lee, J. K. *J. Amer. Chem. Soc.* **2000**,122, 6258.
35. Feng, M. J.; Austin, T. J.; Chew, F.; Gronert, S.; and Wu, W. *Biochemistry* **2000**, 39,1778.
36. Huang, Y. and Kenttamaa, H. *J. Phys. Chem. A* **2003**, 107, 4893.
37. Wentworth, W. E.; Chen, E. C. M.; and Steelhammer, J. C. *J. Phys. Chem.* **1968**, 72, 2671.
38. Tsuda, S. and Hamill, W. H. *Adv. Mass Spectrom.* **1964**, 3, 249.
39. Ding, C. F.; Wang, X. B.; and Wang, L. S. *J. Phys. Chem. A* **1998**, 102, 8633.
40. Caldwell, G.; Renneboog, R.; and Kebarle, P. *Can. J. Chem.* **1989**, 67, 661.
41. Locke, M. J. and McIver, R. T. *J. Amer. Chem. Soc.* **1983**, 105, 4226.
42. O'Hair, R. J.; Bowie, J. H.; and Gronert, S. *Int. J. Mass Spectrom. Ion Proc.* **1992**, 117, 23.
43. Landowne, R. A. *Chim. Anal.(Paris)* **1965**, 47, 589.
44. Zumwaldt, R. W.; Kuo, K.; and Gehrk, C. W. *J. Chromatogr.* **1971**, 57, 193.
45. Bengtsson, G., Oldham, G.; and Westerdahl, G. *Anal. Biochem.* **1981**, 111, 163.
46. Robertson, B. H.; Wolcott, M. H.; Bennett, J. C. *Proc. Soc. Exp. Biol. Med.* **1977**, 155, 287.
47. Stadler, J. *Anal. Biochem.* **1978**, 86, 477.
48. Van den Berg, H. W.; Murphy, R. F.; Hunter, R.; and Elmore, D. T. *J. Chromatogr.* **1978**, 145, 311.
49. Chen, E. S. D. and Chen, E. C. M. *Bioelectrochem. Bioenerget.* **1998**, 46, 15.
50. Chen, E. S. D. and Chen, E. C. M.; Wentworth, W. E. *Biochem. Biophys. Res. Comm.* **2001**, 289, 421.
51. Zimm, B. H. and Kallenbach, N. R. *Annu. Rev. Phys. Chem.* **1962**, 13, 176.
52. Richardson, N. A.; Wesolowski, S. S.; and Schaefer, H. F. *J. Amer. Chem. Soc.* **2002**, 124, 10163.
53. Richardson, N. A.; Wesolowski, S. S.; and Schaefer, H. F. *J. Phys. Chem. B* **2003**, 107, 848.
54. Li, X.; Cai, Z.; and Sevilla, M. D. *J. Phys. Chem. A* **2002**, 106, 9345.
55. Steenken, S. *Chem. Rev.* **1989**, 89, 503.
56. Szent-Gyorgi, A. *Bioenergetics*: New York: Academic Press, **1957**.

APPENDICES

Appendix I is a glossary of terms, acronyms, and symbols.

Appendix II presents the structures of organic compounds. Structure 1 provides the number, names, and adiabatic electron affinities of the Bergman Dewar set. Structure 2 gives the adiabatic electron affinities, gas phase acidities, and names of the DNA and RNA bases. Structure 3 shows the charge transfer complex acceptors. Structure 4 gives the numbering system of naphthalene and biphenyl and compares the structures of acenaphthylene and biphenylene.

Appendix III examines the general least-squares procedure. The normal least-squares solution is compared to the general least-squares solution that allows multiple variables, variable weights in all data, and the use of data determined from other experiments to be combined with that from a specific experiment. The data reduction for a linear plot of $\ln(KT^{3/2})$ versus 1,000/T for acetophenone as a function of reaction time is illustrated. The intercept can be improved by using the weighted average value of the E_a of acetophenone. This is an example of combining data and their uncertainties in the general least-squares solution.

The tables in Appendix IV summarize the evaluated values of the electron affinities given in this book. The electron affinities of the atoms and homonuclear diatomic molecules are given in two tables, A1.1 and A1.2. The references for both tables are combined. The electron affinities of the hydrocarbons are given in Tables A2.1 and A2.2. Tables A2.3 and A2.4 provide the electron affinities of the halogenated hydrocarbons. The odd-numbered tables are ordered by value and the even-numbered tables are ordered by molecular weight. The references for the hydrocarbons are given separately from those of the CHX compounds. Tables A3.1 and A3.2 list the values for the CHNX molecules. These were combined because there are so few halogenated compounds. Tables A4.1 and A4.2 contain the electron affinities of the CHO and CHOX compounds, while Tables A5.1 and A5.2 contain those of the CHON and CHONX compounds.

Searching the NIST tables by combination of elements, for example, CHO, generated Tables A2 through A5. The list contains both radicals and molecules.

The Electron Capture Detector and the Study of Reactions with Thermal Electrons
by E. C. M. Chen and E. S. D. Chen
ISBN 0-471-32622-4

Approximately 170 entries were returned. These were saved in a text file. Less than half of these are for molecules. After eliminating the radicals, the text file was loaded into a spreadsheet, the molecular weights calculated, and the electron affinities in the NIST tables evaluated by taking the weighted average of the values for the same state of a molecule. Values not listed in the NIST table were also included in the weighted average. The revised ECD values were used to adjust the TCT values scaled to the ECD values for benzaldehyde, acetophenone, and benzophenone. For example, the weighted average of the ECD E_a for benzaldehyde is 0.457(5), about 0.03 eV higher than the NIST value. The TCT values scaled to the NIST value have been revised upward by this amount.

In Table A6.1 the names and electron affinities of the Bergman Dewar hydrocarbon set are given. The structures for these compounds are shown in Appendix III along with the electron affinities and Bergman Dewar number. The gas phase electron affinities that are significantly different from the NIST values are tabulated in Table A6.2. This is simply a compilation of the values in the earlier appendices. Table A6.3 lists the gas phase values determined primarily by ECD that are not listed in the NIST tables. Included are some values that could apply to excited states. The excited-state values for cytosine, thymine, and uracil obtained by interpreting hydrated PES spectra are given. Table A6.4 presents the data for hydrated purines. Table A6.5 contains the electron affinities of charge transfer complex acceptors not in the NIST tables. Tables A6.6 and A6.7 list the electron affinities obtained from half-wave reduction potentials also not contained in the NIST tables.

APPENDIX I

Glossary of Terms, Acronyms, and Symbols

Accuracy The agreement between the measured quantity and the "true" value. The difference is due to systematic uncertainties, as opposed to random uncertainties that define precision.

A_i Pre-exponential term for rate constants. Subscript 1 stands for attachment, −1 detachment, 2 dissociation, D recombination of electrons, N recombination of anions, ET electron transfer of ions, $-ET$ reverse of electron transfer.

AB General molecule.

$AB(-)$ Anion of AB.

AE_a Adiabatic electron affinity, energy difference between the ground state of the anion and the most stable state of the neutral molecule.

AM1 A particular semi-empirical self-consistent field calculation. It stands for Austin Model-1.

AMB Alkali metal beam formation of ion pairs.

β Exponential constant in the Morse potential function. With μ as the reduced mass $\beta = \nu_e(2\pi^2\mu/D_e[X_2])^{1/2}$.

C_1 Constant relating the energy for a charge transfer absorption maximum to the electron affinity of the acceptor and the ionization potential of the donor. See equation 2.23.

C_2 Second constant relating the energy for a charge transfer absorption maximum to the electron affinity of the acceptor and the ionization potential of the donor. See equation 2.23.

CEC Abbreviation for CURES-EC in the tables.

CURES-EC The use of semi-empirical multiconfiguration configuration interaction quantum mechanical procedures to estimate electron

The Electron Capture Detector and the Study of Reactions with Thermal Electrons
by E. C. M. Chen and E. S. D. Chen
ISBN 0-471-32622-4

	correlation in the calculation of electron affinities, gas phase acidities, ionization potentials, and bond dissociation energies.
DEC (1 or 2)	Compounds under dissociative electron capture in the ECD. DEC(1) refers to molecules that can dissociate unimolecularly via a single potential energy curve. DEC(2) refers to molecules that can dissociate via a negative-ion intermediate.
$D(m)$	Designation of HIMPEC with $m = 0$ to 3. This indicates dissociation in the Franck Condon region and the number of values of E_a, VE_a, and EDEA that are positive. This stands in contrast with $M(m)$.
$Dc(m)$	Designation of HIMPEC with $m = 0$ to 3. This indicates dissociation in crossing the long-range curve and the number of values of E_a, VE_a, and EDEA that are positive. This stands in contrast with $Mc(m)$.
D_{AB}	Bond dissociation energy of AB.
DBE_a	Electron affinity due to the attraction of the permanent dipole moment.
D_e	Minimum energy in a potential energy curve.
DeBA	Maximum value of A_1 calculated from the DeBroglie wavelength of the electron. The value of ln(DeBA) is about 36 at 400 K.
$e(-)$	Electron.
E_i	Activation energy for kinetic rate constant. Subscript 1 stands for attachment, -1 detachment, 2 dissociation, D recombination of electrons, N recombination of anions, ET electron transfer of ions, $-ET$ reverse of electron transfer.
$E_{1/2}$	Polarographic half-wave reduction potential measured in aprotic solvents.
E_a	Electron affinity, energy between the most stable state of an anion in an electronic state and the most stable state of the neutra molecule.
E_{abs}	The maximum in the absorption spectra of negative ions.
EB	Formation of an anion by the impact of energetic electron beams.
ECD	Electron capture detector.
E_{CT}	Absorption maximum for charge transfer complexes.
EDEA	The electron affinity of dissociating species minus the bond dissociation energy.
Eql(a/b)	Classification of compounds that form stable negative ions and have one or two temperature regions.
E_{ref}	Reference potential for a specific reference electrode in polarographic reduction potential determinations. The value for the saturated calomel electrode is 4.71 V.

EnCT	Endothermic charge transfer. Also called energetic ion beam electron transfer.
E_{pd}	Photodetachment energy, the energy difference between the anion and the neutral in the geometry of the anion.
E_{rr}	Rearrangement energy, the $E_a - VE_a$.
ES	Electron swarm.
ET	Electron transmission, the electron current transmitted through a gas.
EvV	The value selected as the "best" current value. For measurements of the same quantity with different methods, the least-squares best value is the weighted average.
GPA	Gas phase acidity, also called the deprotonation energy of a molecule, *AH*. It is the energy for the reaction $AH = H(+) = A(-)$.
g( )	Partition functions of $(A(-))$, (A), and $e(-)$; $g(e(-)) = S(e-)$ $(2\pi m_e kT)^{3/2}/h^3$. S is the spin multiplicity of the electron, the other quantities are the fundamental constants, and T is the temperature.
HIMPEC	Herschbach ionic Morse potential energy curves. A classification of negative-ion potential energy curves originally proposed by Herschbach and recently modified.
IP	Ionization potential.
I_b	Electron concentration in the absence of *AB*.
I_e	Electron concentration in the presence of *AB*.
k_A	Dimensionless constant that modifies the attraction of the Morse potential of anions.
k_B	Dimensionless constant that modifies the exponent of the Morse potential of anions.
k_R	Dimensionless constant that modifies the repulsion of the Morse potential of anions.
k_1	Rate constant for thermal electron reaction, $k_1 = A_1 T^{-1/2} \exp(-E_1/RT)$.
k_{-1}	Rate constant for thermal electron detachment, $k_{-1} = A_{-1} T \exp(-E_{-1}/RT)$.
k_2	Rate constant for molecular ion dissociation, $k_2 = A_2 T \exp(-E_2/RT)$.
k_{ET}	Rate constant for electron transfer.
k_{-ET}	Rate constant for reverse electron transfer.
k'_D	Rate constant for electron recombination.
k_D	Rate constant for electron recombination multiplied by positive-ion concentration.

k'_N	Rate constant for ion recombination.
k_N	Rate constant for ion recombination multiplied by positive-ion concentration.
K_{ECD}	ECD molar response. Subsequently just K.
K_{eq}	Equilibrium constant for thermal electron reactions, $K_{eq} = k_1/k_{-1}$.
Le	Leaving group in dissociative electron attachment.
LUMO	Lowest unoccupied molecular orbital.
$M(m)$	Designation of HIMPEC with $m = 0$ to 3. This indicates formation of a molecular ion in the Franck Condon region and the number of values of E_a, VE_a, and EDEA that are positive. This contrasts with $D(m)$.
$Mc(m)$	Designation of HIMPEC with $m = 0$ to 3. This indicates formation of a molecular ion in crossing the long-range curve and the number of values of E_a, VE_a, and EDEA that are positive. This contrasts with $Dc(m)$.
MCCI	Multiconfiguration configuration interaction. The modification of the wave functions used to calculate the energies utilizing semi-empirical calculations.
mddG	The solution energy difference for a reaction in polarographic determinations is $-\Delta\Delta G$ and depends on the solvent and specific reaction.
MGN	Magnetron method for measuring electron affinities.
ν	Vibrational frequency in a Morse potential energy curve.
NIMS	Negative-ion mass spectrometry.
P	Positive ion.
PE_a	Polarization electron affinity. The attraction is due to the polarizability of the molecule.
PD	Photodetachment, the removal of an electron from an ion by photons.
PES	Photoelectron spectroscopy, the measurement of the intensity and energy of electrons photodetached from an ion by a fixed-energy photon beam.
Precision	The reproducibility of a measurement. This is determined by random uncertainties, as opposed to systematic uncertainties.
P and A	A graph of two sets of values for the same quantity measured or calculated by two different methods. The deviations from a zero intercept unit slope line will identify systematic (inaccurate) and random (imprecise) uncertainties.
Q_{an}	Ratio of the partition function of the anion to that of the neutral without the spin multiplicity term for the anion.
r_e, r	Internuclear distance r at the minimum of a potential energy curve.

SCF — Self-consistent field quantum mechanical procedure.

Term symbol — The standard term symbol gives a pre-superscript of $2S+1$, where S is the total spin. The major symbol is the total angular momentum. The post-superscript and subscript are a symmetry term and a spin orbital coupling term. The electronic configuration determines the term symbol.

Timeline — A chronological plot of values of a quantity measured with different techniques. The deviations from a constant value can be identified as random or systematic uncertainties to establish accuracy and precision.

URX3O — Acronym for the optimization procedure of CURES-EC UHF, RHF, extremes, three(3), optimization. The UHF, RHF(3300), and RHF(0033) energies are calculated and compared to the experimental values. If the experimental value fits between the maximum and minimum values, then the agreement can be optimized.

$U(AB)$ — Morse potential energy curve for a molecule.

$U(AB(-))$ — Morse potential energy curve for the anion of AB.

VE_a — Vertical electron affinity, energy difference between an anion in the geometry of the neutral molecule and the most stable state of the neutral.

Temperature Regions

From low (298 K) to high temperatures (600 K) these regions are as follows:

1. The β region, where $(K_N \gg (k_{-1} + k_2))$ and $K = k_1/2k_D$
2. The α region, where $(k_{-1} \gg (k_N + k_2))$ and $K = [K_N/2k_D][k_1/k_{-1}]$
3. The γ region, where $(k_2 \gg k_N)$ and $(k_{-1} \gg k_2)$, and $K = [k_1k_2/2k_Dk_{-1}]$
4. The δ region, where $(k_2 \gg (k_{-1} + k_N))$ and $K = k_1/2k_D$

APPENDIX II

Structures of Organic Molecules

Names, Numbers, and Adiabatic Electron Affinities of Bergman Dewar Set

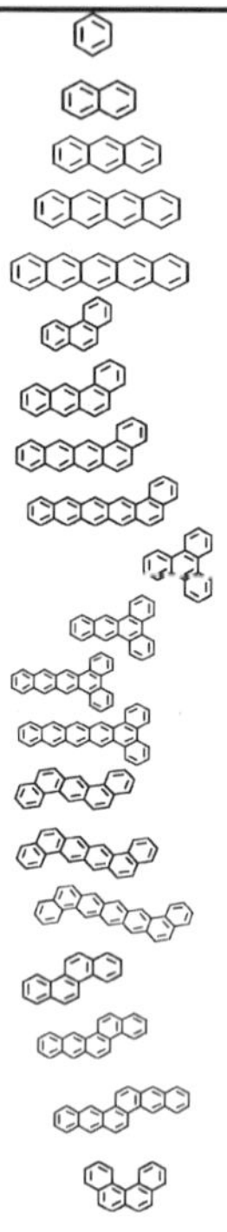
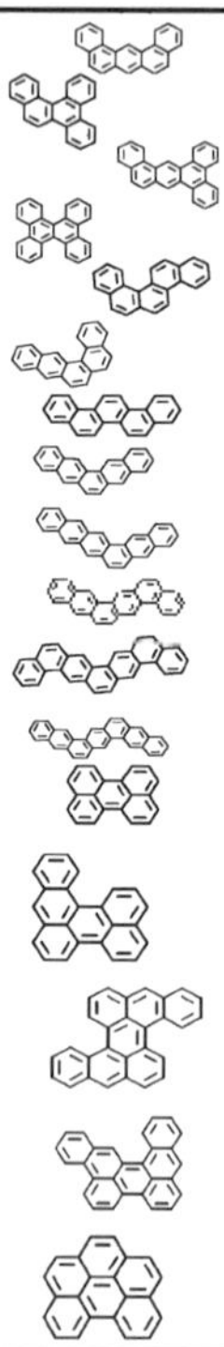
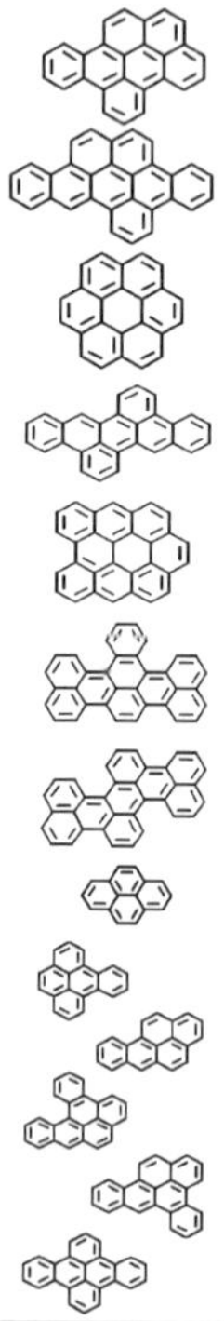

No	AEa (eV)
0	0+
1	0.16
2	0.68
3	1.08
4	1.39
5	0.30
6	0.72
7	1.04
8	1.38
9	0.29
10	0.69
11	1.02
12	1.30
13	0.70
14	1.03
15	1.41
16	0.42
17	0.89
18	1.05
19	0.58
20	0.66
21	0.58
22	0.61
23	0.79
24	0.60
25	0.78
26	0.54
27	0.75
28	1.06
29	0.78
30	0.76
31	0.91
32	0.98
33	1.24
34	1.45
35	1.28
36	0.89
37	0.86
38	1.03
39	0.80
40	1.10
41	1.41
42	1.70
43	1.68
44	0.61
45	0.56
46	0.82
47	1.10
48	0.88
49	0.59

Structure 1

The Electron Capture Detector and the Study of Reactions with Thermal Electrons
by E. C. M. Chen and E. S. D. Chen
ISBN 0-471-32622-4

No	Structure	AEa (eV)	No	Structure	AEa (eV)	No	Structure	AEa (eV)
50	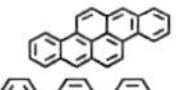	1.16	59		1.14	72	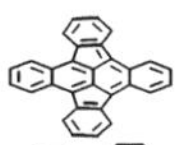	2.04
51	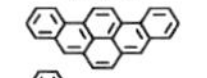	1.06	60		1.32	73	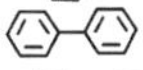	0.13
52		0.67	61	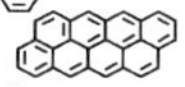	1.65	74	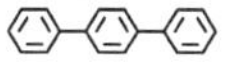	0.27
53	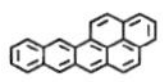	1.23	62	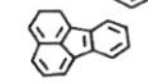	0.82	75	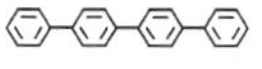	0.41
54	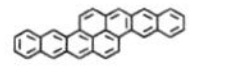	1.55	63	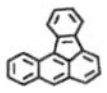	1.25	76	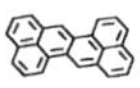	1.44
55		1.28	64	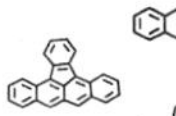	1.58	77	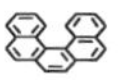	0.58
56	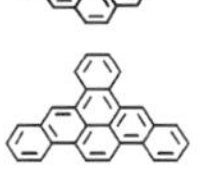	1.03	65	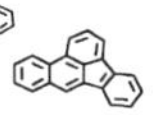	0.95	78	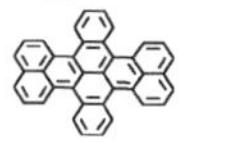	1.90
57	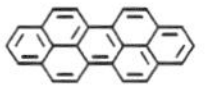	1.25	66	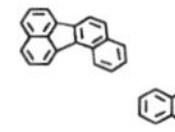	1.01	79	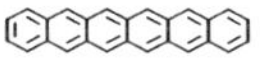	1.58
58	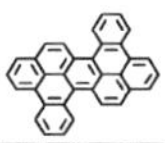	1.18	67	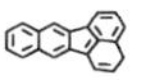	0.89	80	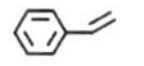	0.10
			68	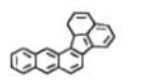	1.17	81	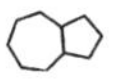	0.84
			69	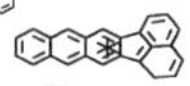	0.90			
			70	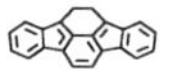	1.33			
			71	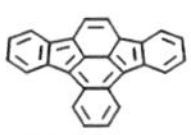	1.71			

Structure 1 (*Continued*)

Adiabatic Electron Affinities, Gas Phase Acidities, and Names of the DNA and RNA Bases

Adenine E_a = 0.95(5) eV
GPA = 14.2(3) eV

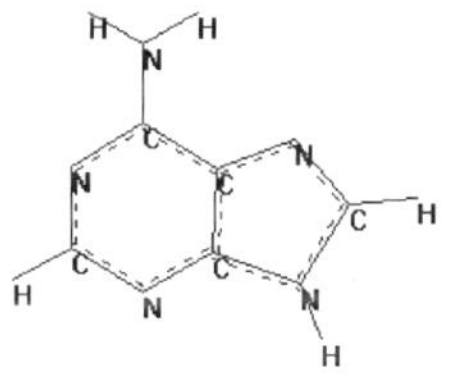

Uracil E_a = 0.80(5) eV
GPA = 14.4(2)

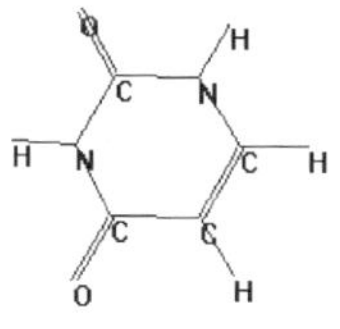

Thymine E_a = 0.79(5) eV
GPA = 14.2(3) eV

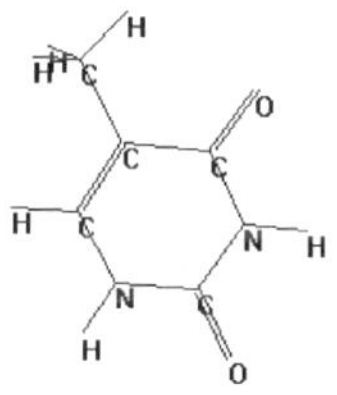

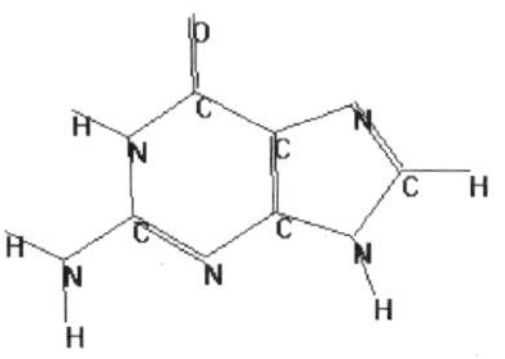

Guanine E_a = 1.51(5) eV
GPA = 14.2(3) eV

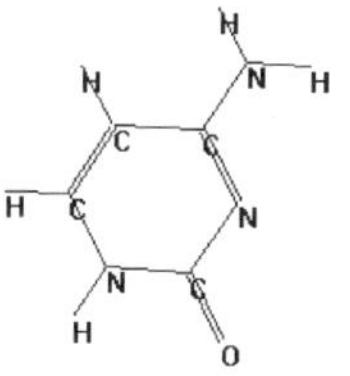

Cytosine E_a = 0.56(5) eV
GPA = 14.6(3) eV

Structure 2

Charge Transfer Complex Acceptors

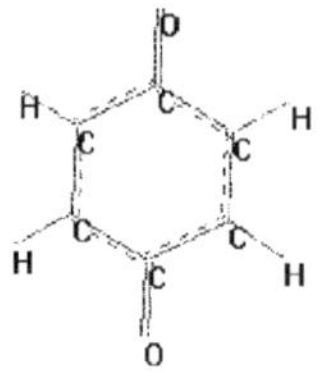

1,4-Benzoquinone

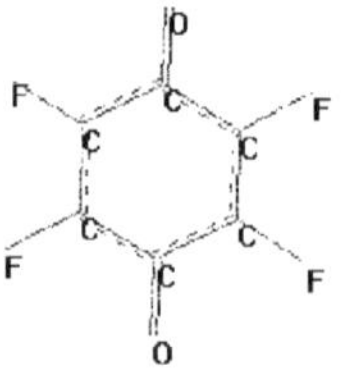

Fluoranil

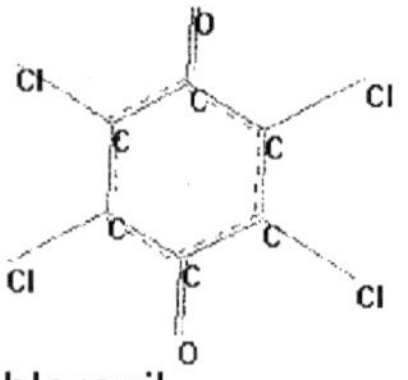

Chloranil

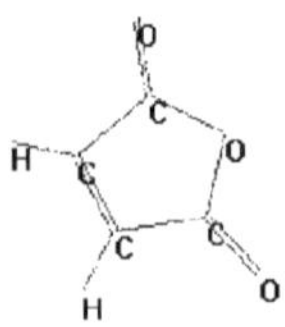

Maleic anhydride

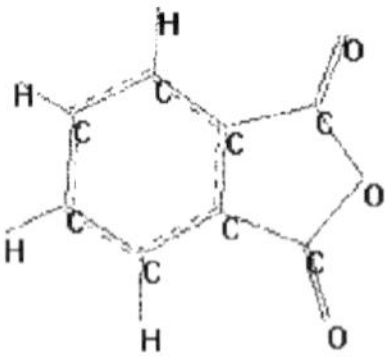

Phthalic anhydride

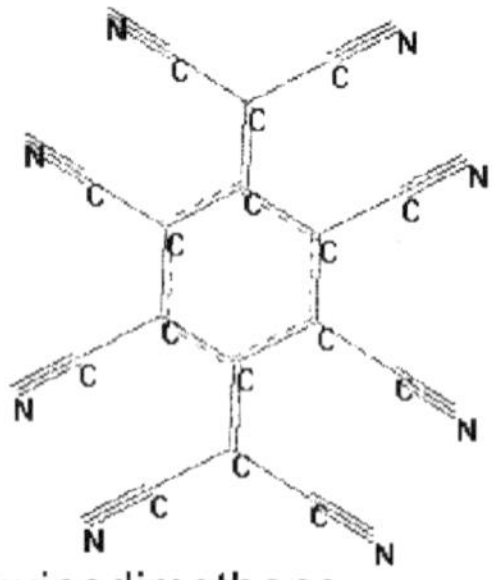

TCNQ
Tetracyanoquinodimethane

Structure 3

Aromatic Hydrocarbons and Numbering System

Naphthalene

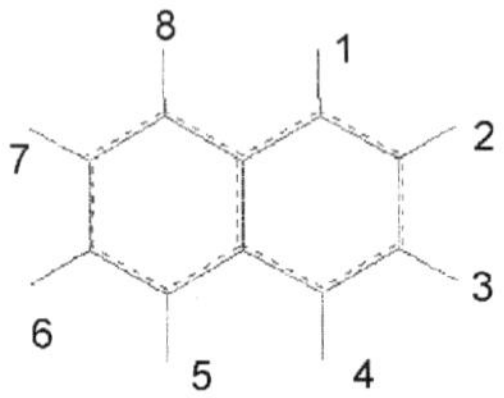

Acenaphthylene

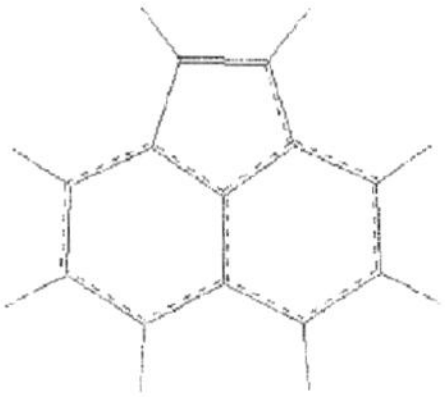

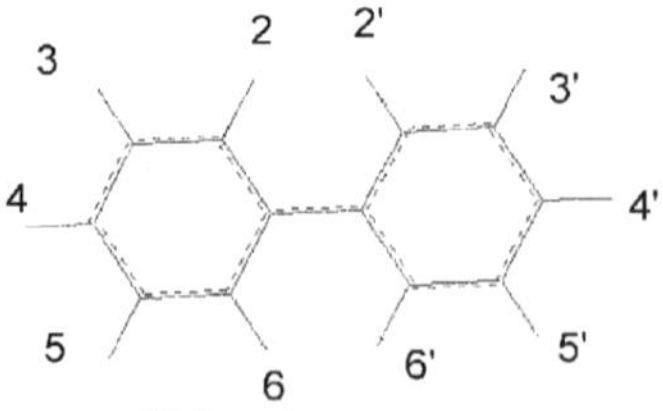

Biphenyl

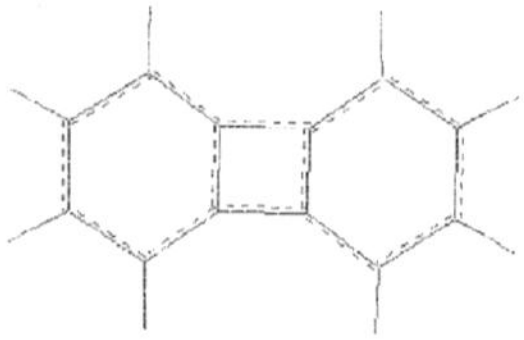

Biphenylene

Structure 4

APPENDIX III

General Least Squares

> Any hesitancy to use statistical analysis has probably stemmed from the time-consuming and tedious calculations. This is especially true in least squares where the function relating the variables is complicated and may involve multiple parameters to be determined from variables which have comparable errors. . . . With the advent of high speed computers, the difficulty in this task can be eliminated.
>
> — W. E. Wentworth
> *Journal of Chemical Education*

Here examples of least-squares adjustments of experimental data will be presented. For measurements with equal uncertainties the average is the least-squares "best" estimate. For a series of measurements with unequal uncertainties the values should be weighted according to their uncertainty. This was discussed for the determination of the current "best" value for the E_a of bromine. The adjustment for a linear two-parameter equation with equal errors in only one direction, as is generally carried out in trendline programs, is compared to the adjustment of data with unequal errors in both the X and Y. The determination of experimental electron affinities from ln $KT^{3/2}$ versus 1,000/T for acetophenone is an example of this type of adjustment. The unequal weights result because the original data, K and T, are transformed into a linear equation.

Next we consider the general problem, where the variables x, y, z, . . . and parameters a, b, c,. . . are related by a mathematical relationship obtained from fundamental principles. That is, there is a fundamental function $F(x, y, z,\ldots; a, b, c,\ldots) = 0$. An example is the extended equations for ln $KT^{3/2}$ versus 1,000/T, where the variables are K and T, and the parameters are the pre-exponential and energy quantities for the rate constants of attachment, detachment, and dissociation for the ground and excited states. There can be as many as eight parameters and two variables. The experimental variables and errors in each of the s_x and s_y variables must be estimated. The parameters a, b, . . . and errors in parameters, s_a, s_b, . . . , result from the adjustment procedure. In addition, the covariance terms s_{ab} and s_{ac} are obtained. These are used in the propagation of error in other quantities.

The Electron Capture Detector and the Study of Reactions with Thermal Electrons
by E. C. M. Chen and E. S. D. Chen
ISBN 0-471-32622-4

First, it is necessary to describe the general principle of least squares and to define the symbols that will be used in the examples. For this, we will return to Deming:

> In all adjustments of observations, simple or complicated, the principle of least squares requires the minimization of the sum of the weighted squares of the residuals. This sum may be written
>
> $$S = \mathrm{Sum} = \Sigma\, w(\mathrm{res})^2 = \Sigma\{w_x V_x^2 + w_y {V_y}^2\} \qquad \text{AIII.0}$$
>
> S is called the sum of the weighted squares.
>
> The weights are proportional to the reciprocal of the errors squared in the experimental values. The quantity V_x and V_y are the x and y residuals. The principle of least squares is the minimization of S. The method of least squares is a rule or set of rules for proceeding with the actual computation. [Chap 4, 36, p. ?]

If the errors in the experimental parameters can be estimated or are known for a given experimental measurement, then the errors obtained in parameters from a specific experiment can be compared with these values and the quality of the results from a specific experiment evaluated. The "known" errors are designated σ(o). In the statistical treatment of data this is designated as the "true" standard deviation, while the standard deviation from a given experiment is designated with the symbol *s* or σ(ext).

The standard formulation of least-squares adjustments of a linear equation can be solved in a closed form and the errors in the parameters calculated directly. This formulation is often given as an example of the use of partial differential calculus in a practical situation. If the errors are equal and exist primarily in the *y* variable, the mathematical function to be minimized is

$$[F^2(x, y, a, b)] = \Sigma(y_i - a - bx_i)^2 \qquad \text{(AIII.1)}$$

The derivative of F with respect to a and b (F_a and F_b) is then taken and set equal to zero, where $F(x, y, a, b) = y - a - bx$, to give

$$F_a = 2\Sigma(y_i - a - bx_i)(-1) = 0 \qquad \text{(AIII.2)}$$

$$F_b = 2\Sigma(y_i - a - bx_i)(-x_i) = 0 \qquad \text{(AIII.3)}$$

These can be solved directly for a and b and their errors. This is the procedure that is used in standard trendlines. The problem with this procedure is that the random errors only occur in the *y* variable and the errors are assumed to be equal. In many cases there are unequal errors in both variables.

We define the summations as $[y] = \Sigma y_i [x] = \Sigma x_i$; $[xy] = \Sigma x y_i$; $[xx] = \Sigma x_i^2$ $n = \Sigma 1$. Then

$$na + b[x] = [y] \quad \text{and} \quad a[x] + b[xx] = [xy] \qquad \text{(AIII.4)}$$

and using Cramer's rule, we obtain

$$a = ([y][xx] - [x][xy]/D;\ b = n[xy] - [x][y]/D;\ D = n[xx] - [x][x] \qquad \text{(AIII.5)}$$

$$s_a^2 = \{[F^2(x, y, a, b)]/(n-2)\}[xx]/D;\ s_b^2 = \{[F^2(x, y, a, b)]/(n-2)\}n/D \qquad \text{(AIII.6)}$$

Thus, by simply calculating $[x]$, $[y]$, $[xx]$, $[xy]$, and $[F^2(x, y, a, b)]$, the quantities in equations AIII.5 and AIII.6 can be obtained.

Figure AIII.1 is an example of the fit for ln $KT^{3/2} = Y = \text{aest} + \text{best}\ X$. The errors are all assumed to fall in the Y direction. The values of the parameters are aest = 12.37(7) and best = 3,572(138). This gives an E_a of 0.310(14) eV and a Q_{an}

X	Y	XX	XY	y-ycalc	d2		
0.00330	24.08	1.09E-05	7.95E-02	-1.14E-01	1.30E-02	D=	3.00E-05
0.00304	23.79	9.24E-06	7.23E-02	5.37E-01	2.89E-01	n=	14.00
0.00296	23.03	8.76E-06	6.82E-02	6.69E-02	4.48E-03	a=	12.25
0.00290	22.62	8.41E-06	6.56E-02	-1.26E-01	1.58E-02	b=	3620.60
0.00285	22.52	8.12E-06	6.42E-02	-4.48E-02	2.01E-03	sb=	133.64
0.00261	21.51	6.81E-06	5.61E-02	-1.86E-01	3.46E-02	sa=	0.06
0.00273	21.98	7.45E-06	6.00E-02	-1.50E-01	2.26E-02		
0.00261	21.51	6.81E-06	5.61E-02	-1.86E-01	3.46E-02		
0.00251	21.25	6.30E-06	5.33E-02	-8.38E-02	7.03E-03		
0.00238	20.91	5.66E-06	4.98E-02	4.69E-02	2.20E-03		
0.00226	20.53	5.11E-06	4.64E-02	1.01E-01	1.03E-02		
0.00207	19.89	4.28E-06	4.12E-02	1.49E-01	2.23E-02		
0.00203	19.62	4.12E-06	3.98E-02	2.41E-02	5.79E-04		
0.00198	19.38	3.92E-06	3.84E-02	-3.49E-02	1.22E-03	s2=	0.46
3.62E-02	3.03E+02	9.59E-05	7.91E-01	-1.32E-12	4.59E-01		

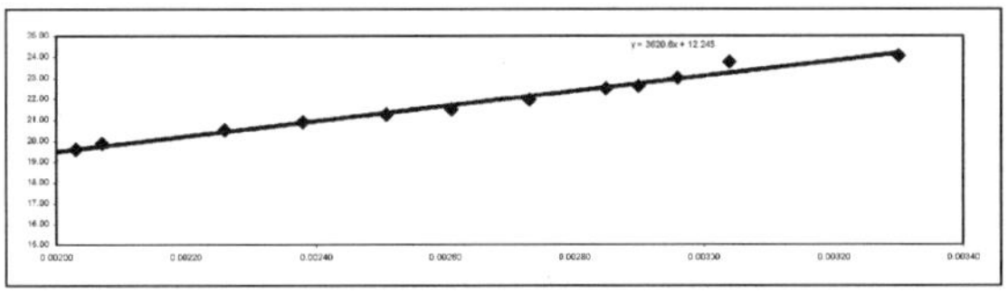

Ea= 0.314

X	Y	XX	XY	y-ycalc	d2		
0.0033	24.08	=A2*A2	=A2*B2	=B2-b*A2-a	=E2^2	D=	=n*sxx-sx*sx
0.00304	23.79	=A3*A3	=A3*B3	=B3-b*A3-a	=E3^2	n=	14
0.00296	23.03	=A4*A4	=A4*B4	=B4-b*A4-a	=E4^2	a=	=(sy*sxx-sx*sxy)/D
0.0029	22.62	=A5*A5	=A5*B5	=B5-b*A5-a	=E5^2	b=	=(n*sxy-sx*sy)/D
0.00285	22.52	=A6*A6	=A6*B6	=B6-b*A6-a	=E6^2	sb=	=((H15/(n-2)*n/D)^0.5)
0.00261	21.51	=A7*A7	=A7*B7	=B7-b*A7-a	=E7^2	sa=	=(H15/(n-2)*sxx/D)*0.5
0.00273	21.98	=A8*A8	=A8*B8	=B8-b*A8-a	=E8^2		
0.00261	21.51	=A9*A9	=A9*B9	=B9-b*A9-a	=E9^2		
0.00251	21.25	=A10*A10	=A10*B10	=B10-b*A10-a	=E10^2		
0.00238	20.91	=A11*A11	=A11*B11	=B11-b*A11-a	=E11^2		
0.00226	20.53	=A12*A12	=A12*B12	=B12-b*A12-a	=E12^2		
0.00207	19.89	=A13*A13	=A13*B13	=B13-b*A13-a	=E13^2		
0.00203	19.62	=A14*A14	=A14*B14	=B14-b*A14-a	=E14^2		
0.00198	19.38	=A15*A15	=A15*B15	=B15-b*A15-a	=E15^2	s2=	=F16
0.03625	302.66	=SUM(C2	=SUM(D2	=SUM(E2:E15)	=SUM(F		

Ea= b*2/23060 eV

Figure AIII.1 A linear least-squares fit to the function $Y = \text{aest} + \text{best}X$, where Y is $\ln(KT^{3/2})$ and X is $1/T$ for acetophenone at a reaction time of 50 μs. The plot is the trendline that shows the equivalence of the trendline and the normal least squares.

of 2.00(10). Both of these are different from the "best" values of 0.338(2) eV and 1.00(5).

Other problems with this simple approach are that estimates of the parameters obtained from other experiments or which are indicated by theory cannot be included in the adjustment. For example, if the intercept has been determined by a series of other experiments to be $a \pm s_a$, it is appropriate to include that value in the adjustment of new data to estimate the slope b. Alternatively, if the slope b is 1.0 ± 0.05, then this should be included in the adjustment to obtain a more precise value of a. Also, the quality of the fit of the data to the function is often not realistic because of the assumptions concerning the random errors. Finally, it is difficult to consider additional variables and/or parameters or nonlinear functions. Thus, the use of a general least-squares operation is possibly more valuable than this simple approach and with modern spreadsheets is no more difficult.

The general treatment of least squares presented by Deming eliminates these problems and is not much more complicated than the closed form of the equations given above. In addition, the transformation of a nonlinear equation to a linear one can be accomplished quite simply with the proper weights. We will not present a derivation of the procedure since it has been described previously. We will simply present examples of the steps in the procedure for a two-parameter case. The extension to more than two variables is simple since there is no increase in the size of the matrix to be inverted. The extension to more than two parameters does involve an increase in the size of the matrix but is apparent.

The exact solution of the minimization equations obtained from AIII.2 is not usually attempted. Rather, an "iterative" method is used. Reasonable approximations or "guesses" of the parameters are made and improvements to the guesses are calculated. As long as the guesses are relatively good, the procedure will "converge" so that the changes in the guesses are small and the value of S is a minimum. With a macro and a modern spreadsheet such as EXCEL, the repeated application of this procedure will result in convergence within a short period of time. The eventual result is the "best" values of the parameters consistent with the experimental data and, more significantly, estimates of the errors as given by the variances and covariances of the parameters so that other quantities can be calculated from "new" data.

Wentworth summarized these advantages as follows:

> Of primary importance are the estimates of the errors of the parameters which may be used directly in interpreting or evaluating the parameters or in the propagation of errors upon the calculation of subsequent quantities. The actual parameters should be the most probable values if the errors are truly random, following a Gaussian distribution. From the experimental data and the associated errors, an experimental apparatus (procedure) can be pre-evaluated in regards to the desired accuracy of the results. If the errors of the observations are well characterized, a statistical test can be applied to evaluate how well the data fit a given function. This can aid the investigator in deciding whether a theoretical or empirical function of the variables is satisfactory and whether a further critical evaluation of the theoretical function is justified.

Other advantages are that it is possible to include estimates of parameters and their errors in the adjustment procedure to obtain more precise values of unmeasured parameters. The procedures can be carried out rapidly using modern desktop computers. The calculation procedure is no more complicated than for the closed solution form.

The first step in the procedure is to assemble the data x_i, y_i, sx_i, and sy_i. These are entered in columns on the spreadsheet. Then the expression for the function is written in terms of the variables and parameters, for example, $F1$, $F1 = y - a - bx$, or $F2 = \ln(KT^{3/2}) - a - b/T$. The next step is to find the derivatives $F'x$, $F'y$, $F'a$, and $F'b$:

$$F1'x = -b; \quad F1'y = 1; \quad F1'a = -1; \quad F1'b = x$$

$$F2'T = \{1/(KT^{3/2})\}\{(3/2)T^{1/2}\} + b/T^2; \quad F2'K = 1/(KT^{3/2});$$

$$F2'a = -1; \quad F2'b = -1/T$$

The analytical expressions for these derivatives are then entered into the spreadsheet. Subsequently, the term $\text{Li} = (F'x)^2(sx)^2 + (F'y^2)(sy)^2$ is calculated in a separate column. Next a and b are estimated. These are defined as aest and best. Then the function is calculated in a separate column using these estimates; it is designated F_0. Next the quantities $F'aF'a/L$, $F'bF'b/L$; $F'aF'b/L$; F_0F_0/L; $F'aF_0/L$; and $F'bF_0/L$ are calculated and the sums over the data points taken. These sums are used in a matrix to calculate the corrections to the estimated parameters. New estimates are then calculated and the procedure carried out until the changes in the parameters are small.

The equations to be solved can be written as

$$[a'a']\Delta a + [a'b']\Delta b = [F_0a'] \quad \text{and} \quad [a'b']\Delta a + [b'b]\Delta b = [F_0b']$$

where the brackets indicate sums and the term $a'a'$ is $Fa'Fa'/L$, etc. In matrix form this is

$$\begin{vmatrix} [a'a'] & [a'b]\,i \\ [a'b'] & [b'b'] \end{vmatrix} \begin{vmatrix} \Delta a \\ \Delta b \end{vmatrix} = \begin{vmatrix} [F'_{0a}] \\ [F'_{0b}] \end{vmatrix}$$

This equation can be solved by Cramers' rule as above or by finding the inverse matrix.

In many spreadsheets there is a matrix solution routine that will carry out this procedure. Such is convenient for larger matrices. For example, for the least-squares solutions for two negative-ion states for ECD data, there are eight parameters to be determined. The elements of the inverse matrix are designated as d's.

The corresponding table is

X	sX	Y	sY	x'	y'	L	a'	b'	F_0	F_0F_0	$a'a'$	$a'b'$	$b'b'$	F'_{0a}	F'_{0b}
–	–	–	–	–	–	–	–	–	–	–	–	–	–	–	–
										–	–	–	–	–	–
										—	—	—	—	—	—
										$[F_0F_0]$	$[a'a']$	$[a'b']$	$[b'b']$	$[F'_{0a}]$	$[F'_{0b}]$
										$g0$	$g1$	$g2$	$g3$	$g4$	$g5$

These calculations can be completed using the spreadsheet functions. The iteration is controlled using a Visual Basic MACRO. The inverse matrix gives the errors and covariances. The errors in the parameters will be obtained from the inverse matrix. They are $(sa)^2 = daa^*S$, where S is $[F_0F_0]/(n-2)$ for the two parameters that are determined from the data. Likewise, $(sb)^2 = dbb^*S$ and sab, the covariance term, is dab^*S.

With more than two variables the L terms are simply expanded to include more terms. For a two-parameter equation the size of the matrix remains 2. For a six-parameter equation with two variables, the size of the matrix is a symmetrical 6×6. Thus, only 27 sums need be calculated. The 6×6 square matrix is inverted and multiplied by the 1×6 matrix to obtain corrections in the six parameters. These are then adjusted and the process iterated to convergence. The iteration is controlled with a Visual Basic Macro. The rigorous inclusion of estimates for parameters from other experiments is easily incorporated into this procedure. The parameters and errors must be input. Next the program simply adds terms to the appropriate sums. For example, if the value of a has been determined to be ax with an uncertainty of sa, then the quantity $1/(sa^*sa)$ is added to $[a'a']$ and this quantity is multiplied by (aest $-ax$) and added to $[F_0a']$. The adjustment is made as before, as are the parameters and uncertainties obtained. This has been demonstrated by Wentworth, Hirsch, and Chen [Chapt. 5, 37].

The calculation of another quantity from the parameters and their errors can then be carried out. If a property $p = a + (a + b)/b^3$ has some physical significance, then the error in P may be calculated using the propagation of error as

$$
\begin{aligned}
S_p^2 &= p_a^2 s_a^2 + p_b^2 s_p^2 + 2p_a p_b s_{ab} \\
&= (1 + 1/b^3)^2 s_a^2 + \{(a/b^3) - 3(a+b)/b^2\}^2 s_b^2 \\
&\quad + 2(1 + 1/b^3)\{(a/b^3) - 3(a+b)/b^2\} s_{ab}
\end{aligned}
$$

In the majority of presentations the last term, the covariance term, is not included. However, it can be a very important portion of the error in the calculated quantity. The general least-squares procedure calculates this quantity as indicated above.

An example of the calculation for $F2 = \ln KT^{3/2} -$ aest $-$ best(1,000/T) is shown in Figure AIII.2. The derivatives are given above. Only five sums should be taken, but the equations must be iterated to minimize the function. In Figure AIII.2 the parameters from the program with equal errors in T and a constant fractional error of 10% in K give essentially the same results as a linear least squares. The values of the parameters are aest = 12.37(29) and best = 3,572(83).

If the "known" E_a of 0.338(2) eV is used in the data analysis by adding 1/(sa^*sa) to the $[a'a']$ term and this quantity is multiplied by (aest $-$ax) and applied to $[F_0a']$, a rigorous least-squares solution can be obtained, as shown in Figure AIII.3. The added terms were originally zero. The intercept is 11.71(10) and the slope 3,847(26). This gives an E_a of 0.334(3) eV and a Q_{an} of 1.00(1).

Y	sY	X	Sx	$LnYX^{3/2}$	F0	Fa	Fb	FX	FY	L	F0F0/L	FaFa/L	FaFb/L	FbFb/L	F0Fa/L	F0Fb/L
5.54E+06	5.54E+05	300	5	24.08	-0.22	-1	-0.00333	0.0448	1.81E-07	0.0601	8.01E-01	16.63	5.54E-02	1.85E-04	3.65	1.22E-02
3.58E+06	3.58E+05	329	5	23.79	0.54	-1	-0.00304	0.0376	2.79E-07	0.0454	6.31E+00	22.02	6.69E-02	2.03E-04	-11.79	-3.58E-02
1.61E+06	1.61E+05	338	5	23.03	0.07	-1	-0.00296	0.0358	6.20E-07	0.0420	1.06E-01	23.81	7.05E-02	2.08E-04	-1.59	-4.70E-03
1.04E+06	1.04E+05	345	5	22.62	-0.12	-1	-0.00290	0.0344	9.59E-07	0.0396	3.85E-01	25.24	7.31E-02	2.12E-04	3.12	9.04E-03
9.18E+05	9.18E+04	351	5	22.52	-0.05	-1	-0.00285	0.0333	1.09E-06	0.0378	5.95E-02	26.47	7.54E-02	2.15E-04	1.26	3.58E-03
8.03E+05	8.03E+04	360	5	22.43	0.11	-1	-0.00278	0.0318	1.25E-06	0.0353	3.54E-01	28.36	7.88E-02	2.19E-04	-3.17	-8.81E-03
5.07E+05	5.07E+04	366	5	21.99	-0.16	-1	-0.00273	0.0308	1.97E-06	0.0338	7.66E-01	29.63	8.10E-02	2.21E-04	4.76	1.30E-02
2.95E+05	2.95E+04	383	5	21.52	-0.20	-1	-0.00261	0.0283	3.39E-06	0.0301	1.34E+00	33.28	8.69E-02	2.27E-04	6.69	1.75E-02
2.14E+05	2.14E+04	398	5	21.25	-0.11	-1	-0.00251	0.0264	4.67E-06	0.0274	4.36E-01	36.52	9.18E-02	2.31E-04	3.99	1.00E-02
1.40E+05	1.40E+04	421	5	20.91	0.04	-1	-0.00238	0.0238	7.15E-06	0.0241	6.60E-02	41.47	9.85E-02	2.34E-04	-1.65	-3.93E-03
8.86E+04	8.86E+03	443	5	20.53	0.08	-1	-0.00226	0.0216	1.13E-05	0.0217	3.08E-01	46.10	1.04E-01	2.35E-04	-3.77	-8.51E-03
4.10E+04	4.10E+03	484	5	19.89	0.13	-1	-0.00207	0.0184	2.44E-05	0.0184	8.80E-01	54.21	1.12E-01	2.31E-04	-6.91	-1.43E-02
3.03E+04	3.03E+03	493	5	19.62	-0.01	-1	-0.00203	0.0178	3.30E-05	0.0179	5.54E-03	55.88	1.13E-01	2.30E-04	0.56	1.13E-03
2.33E+04	2.33E+03	504	5	19.39	-0.08	-1	-0.00198	0.0171	4.30E-05	0.0173	4.08E-01	57.86	1.15E-01	2.28E-04	4.86	9.64E-03
														0.00E+00		0.00E+00
											1.22E+01	497.47	1.22E+00	3.08E-03	0.00	-2.26E-16
											g0	g1	g2	g3	g4	g5

ea =	0.310	a=	12.37	a=	12.37
		b=	3579.88	b=	3579.88
		sa=	0.29	sa=	0.29
		sb=	82.50	sb=	82.50
		sab=	-33.45	sab=	-33.45

a=	12.24612
b=	3620.598
sb=	133.6365
sa=	0.061167
Ea=	0.314
	Linear ls

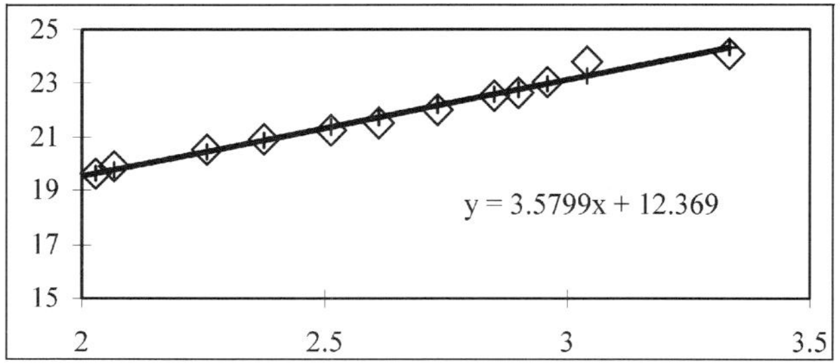

Figure AIII.2 Nonlinear least squares with equal weights in the temperature and equal fractional errors in K. The parameters are slightly different from the simple linear least-squares solution and the variances and covariances are calculated.

Y	sY	X	Sx	LnYX$^{3/2}$	F0	Fa	Fb	FX	FY	L	F0F0/L	FaFa/L	FaFb/L	FbFb/L	F0Fa/L	F0Fb/L
5.54E+06	5.54E+05	300	5	24.08	-0.46	-1	-0.00333	0.0477	1.81E-07	0.0670	3.10E+00	14.93	4.98E-02	1.66E-04	6.80	2.27E-02
3.58E+06	3.58E+05	329	5	23.79	0.38	-1	-0.00304	0.0401	2.79E-07	0.0502	2.84E+00	19.92	6.05E-02	1.84E-04	-7.53	-2.29E-02
1.61E+06	1.61E+05	338	5	23.03	-0.07	-1	-0.00296	0.0381	6.20E-07	0.0463	1.03E-01	21.59	6.39E-02	1.89E-04	1.49	4.42E-03
1.04E+06	1.04E+05	345	5	22.62	-0.24	-1	-0.00290	0.0367	9.59E-07	0.0436	1.36E+00	22.93	6.64E-02	1.93E-04	5.58	1.62E-02
9.18E+05	9.18E+04	351	5	22.52	-0.15	-1	-0.00285	0.0355	1.09E-06	0.0415	5.72E-01	24.09	6.86E-02	1.96E-04	3.71	1.06E-02
8.03E+05	8.03E+04	360	5	22.43	0.02	-1	-0.00278	0.0339	1.25E-06	0.0387	1.52E-02	25.87	7.19E-02	2.00E-04	-0.63	-1.74E-03
5.07E+05	5.07E+04	366	5	21.99	-0.24	-1	-0.00273	0.0328	1.97E-06	0.0369	1.51E+00	27.08	7.40E-02	2.02E-04	6.40	1.75E-02
2.95E+05	2.95E+04	383	5	21.52	-0.24	-1	-0.00261	0.0301	3.39E-06	0.0327	1.82E+00	30.57	7.98E-02	2.08E-04	7.45	1.95E-02
2.14E+05	2.14E+04	398	5	21.25	-0.13	-1	-0.00251	0.0281	4.67E-06	0.0297	5.34E-01	33.69	8.47E-02	2.13E-04	4.24	1.07E-02
1.40E+05	1.40E+04	421	5	20.91	0.06	-1	-0.00238	0.0253	7.15E-06	0.0260	1.39E-01	38.51	9.15E-02	2.17E-04	-2.31	-5.49E-03
8.86E+04	8.86E+03	443	5	20.53	0.13	-1	-0.00226	0.0230	1.13E-05	0.0232	7.67E-01	43.08	9.72E-02	2.20E-04	-5.75	-1.30E-02
4.10E+04	4.10E+03	484	5	19.89	0.23	-1	-0.00207	0.0195	2.44E-05	0.0195	2.71E+00	51.21	1.06E-01	2.19E-04	-11.79	-2.44E-02
3.03E+04	3.03E+03	493	5	19.62	0.10	-1	-0.00203	0.0189	3.30E-05	0.0189	5.60E-01	52.90	1.07E-01	2.18E-04	-5.44	-1.10E-02
2.33E+04	2.33E+03	504	5	19.39	0.04	-1	-0.00198	0.0181	4.30E-05	0.0182	9.11E-02	54.91	1.09E-01	2.16E-04	-2.24	-4.44E-03
														9.61E-04		-1.85E-02
											1.61E+01	461.27	1.13E+00	3.80E-03	0.00	1.03E-15
											g0	g1	g2	g3	g4	g5

ea =	0.334	a=	11.71	a=	12.37
		b=	3847.34	b=	3579.88
		sa=	0.10	sa=	0.29
		sb=	25.54	sb=	82.50
		sab=	-3.20	sab=	-33.45

a=	12.37
b=	3572.01
sb=	137.85
sa=	0.07
linear	

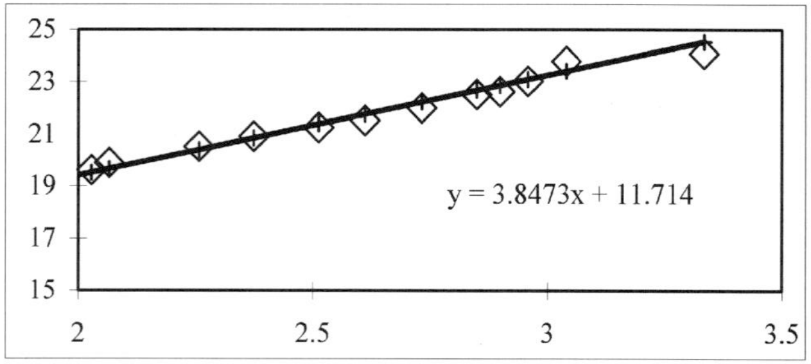

Figure AIII.3 The weighted average value for the slope is included in this data reduction with its uncertainties. This simply requires the addition of two terms to the sums, *g*4 and *g*5. If estimates exist for both the slope and intercept, these can be added to the data to obtain the rigorous solution and uncertainties.

Y	sY	X	Sx	$LnYX^{3/2}$	F0	Fa	Fb	FX	FY	L	F0F0/L	FaFa/L	FaFb/L	FbFb/L	F0Fa/L	F0Fb/L
5.54E+06	5.54E+05	300	5	24.08	-0.45	-1	-0.00333	0.0470	1.81E-07	0.0653	3.08E+00	15.33	5.11E-02	1.70E-04	6.87	2.29E-02
3.58E+06	3.58E+05	329	1	23.79	0.37	-1	-0.00304	0.0395	2.79E-07	0.0116	1.16E+01	86.51	2.63E-01	7.99E-04	-31.65	-9.62E-02
1.61E+06	1.61E+05	338	5	23.03	-0.09	-1	-0.00296	0.0375	6.20E-07	0.0452	1.66E-01	22.11	6.54E-02	1.94E-04	1.91	5.66E-03
1.04E+06	1.04E+05	345	5	22.62	-0.26	-1	-0.00290	0.0361	9.59E-07	0.0426	1.65E+00	23.47	6.80E-02	1.97E-04	6.21	1.80E-02
9.18E+05	9.18E+04	351	5	22.52	-0.18	-1	-0.00285	0.0350	1.09E-06	0.0406	7.87E-01	24.65	7.02E-02	2.00E-04	4.41	1.26E-02
8.03E+05	8.03E+04	360	5	22.43	-0.01	-1	-0.00278	0.0333	1.25E-06	0.0378	7.06E-04	26.46	7.35E-02	2.04E-04	0.14	3.80E-04
5.07E+05	5.07E+04	366	5	21.99	-0.27	-1	-0.00273	0.0323	1.97E-06	0.0361	2.00E+00	27.69	7.56E-02	2.07E-04	7.44	2.03E-02
2.95E+05	2.95E+04	383	5	21.52	-0.28	-1	-0.00261	0.0297	3.39E-06	0.0320	2.52E+00	31.21	8.15E-02	2.13E-04	8.87	2.32E-02
2.14E+05	2.14E+04	398	5	21.25	-0.17	-1	-0.00251	0.0276	4.67E-06	0.0291	1.03E+00	34.37	8.64E-02	2.17E-04	5.94	1.49E-02
1.40E+05	1.40E+04	421	5	20.91	0.00	-1	-0.00238	0.0249	7.15E-06	0.0255	5.90E-04	39.22	9.32E-02	2.21E-04	-0.15	-3.61E-04
8.86E+04	8.86E+03	443	5	20.53	0.07	-1	-0.00226	0.0227	1.13E-05	0.0228	2.11E-01	43.81	9.89E-02	2.23E-04	-3.04	-6.87E-03
4.10E+04	4.10E+03	484	5	19.89	0.15	-1	-0.00207	0.0192	2.44E-05	0.0193	1.22E+00	51.94	1.07E-01	2.22E-04	-7.98	-1.65E-02
3.03E+04	3.03E+03	493	5	19.62	0.02	-1	-0.00203	0.0186	3.30E-05	0.0186	3.03E-02	53.62	1.09E-01	2.21E-04	-1.28	-2.59E-03
2.33E+04	2.33E+03	504	5	19.39	-0.04	-1	-0.00198	0.0179	4.30E-05	0.0180	9.51E-02	55.63	1.10E-01	2.19E-04	2.30	4.56E-03
														0.00E+00		0.00E+00
											2.44E+01	536.03	1.35E+00	3.51E-03	0.00	-1.58E-15
											g0	g1	g2	g3	g4	g5

ea =	0.328	a=	11.93	a=	12.37
		b=	3780.96	b=	3579.88
		sa=	0.38	sa=	0.29
		sb=	105.91	sb=	82.50
		sab=	-56.64	sab=	-33.45
		a=	12.37		
		b=	3572.01		
		sb=	137.85		
		sa=	0.07		
		linear			

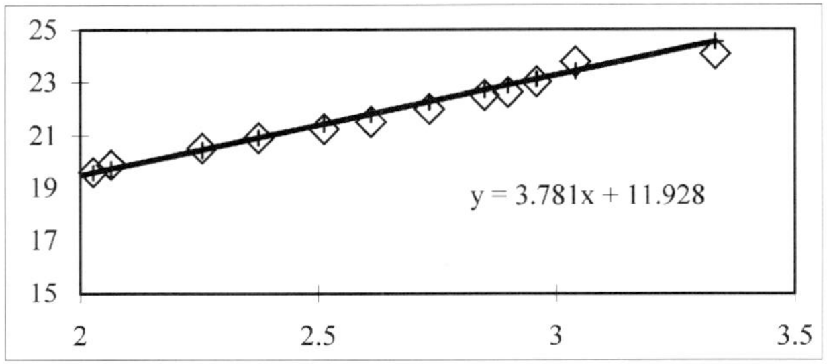

Figure AIII.4 The error for one temperature is reduced to 1 K, as opposed to 5 K for the other data point. This illustrates the importance of proper weighting.

To illustrate the effect of weighting the values, a smaller error in T, 1° instead of 5°, will be assumed for one point. If the second data point had been part of a concentration study where the temperature was stabilized and multiple samples injected, this might be possible. As shown in Figure AIII.4, the values of the slope and intercept are 11.93(18) and 3,781(106), giving an E_a of 0.328(10) and Q_{an} of 1.00(2). This is an example of how important weighting can be and how other data and their uncertainties may be used to improve the value from the data treatment.

APPENDIX IV

Tables of Evaluated Electron Affinities

TABLE A1.1 Atoms (in eV)

N	Atom	AE_a	Uncertainty	Method	Reference	Year
1	H	0.754209	3.0E-06	calc	[9]	1962
2	He	0+	—	—	this work	2003
3	Li	0.618069	4.4E-05	THD	[10]	1996
4	Be	0+	—	—	this work	2003
5	B	0.279723	4.4E-05	THD	[11]	1998
6	C	1.262119	4.4E-05	THD	[12]	1998
7	N	0+	—	—	this work	2003
8	O	1.461110	7.0E-08	THD	[13]	1999
9	F	3.401290	3.0E-06	THD	[14]	2001
10	Ne	0+	—	—	this work	2003
11	Na	0.547930	2.5E-05	THD	[2]	1985
12	Mg	0+	—	—	this work	2003
13	Al	0.432830	5.0E-05	THD	[15]	1998
14	Si	1.389521	2.0E-05	THD	[12]	1998
15	P	0.7464	4.0E-04	THD	[2]	1985
16	S	2.077103	3.0E-06	THD	[16]	1995
17	Cl	3.612740	3.0E-05	THD	[17]	1987
18	Ar	0+	—	—	this work	2003
19	K	0.50147	1.2E-04	THD	[18]	2000
20	Ca	0.024546	8.7E-05	THD	[19]	1996
21	Sc	0.19	2.0E-02	PES	[20]	1981
22	Ti	0.08	1.0E-02	PES	[21]	1987
23	V	0.53	1.0E-02	PES	[22]	1998
24	Cr	0.676	1.2E-04	PES	[22]	1998
25	Mn	0+	—	—	this work	2003
26	Fe	0.151	3.0E-03	PES	[23]	1986
27	Co	0.6633	6.1E-04	THD	[24]	1998
28	Ni	1.15716	1.3E-04	THD	[24]	1998
29	Cu	1.235792	4.4E-04	THD	[25]	1992
30	Zn	0+	—	—	this work	2003

The Electron Capture Detector and the Study of Reactions with Thermal Electrons
by E. C. M. Chen and E. S. D. Chen
ISBN 0-471-32622-4

TABLE A1.1 (*Continued*)

N	Atom	AE_a	Uncertainty	Method	Reference	Year
31	Ga	0.41	4.0E-02	THD	[26]	1998
32	Ge	1.232712	1.5E-04	THD	[12]	1998
33	As	0.814	8.0E-03	PES	[27]	1998
34	Se	2.020682	4.4E-05	THD	[28]	1988
35	Br	3.363583	4.4E-05	THD	[29]	1989
36	Kr	0+	—	—	this work	2003
37	Rb	0.485920	2.0E-05	THD	[30]	1978
38	Sr	0.05206	4.4E-05	THD	[31]	1997
39	Y	0.308	1.2E-02	PES	[20]	1981
40	Zr	0.427	1.4E-02	PES	[2]	1985
41	Nb	0.894	2.5E-02	PES	[2]	1985
42	Mo	0.7472	2.0E-04	PES	[22]	1998
43	Tc	0.55	1.5E-01	EST	[2]	1985
44	Ru	1.04638	2.0E-04	THD	[32]	1999
45	Rh	1.142890	2.0E-04	THD	[24]	1998
46	Pd	0.56214	1.2E-04	THD	[24]	1998
47	Ag	1.30447	2.0E-05	THD	[22]	1998
48	Cd	0+	—	—	this work	2003
49	In	0.404	9.0E-03	THD	[33]	1998
50	Sn	1.112067	1.5E-04	THD	[12]	1998
51	Sb	1.474020	2.0E-05	THD	[34]	1997
52	Te	1.970876	7.0E-06	THD	[35]	1996
53	I	3.059000	1.0E-05	THD	[36]	1992
54	Xe	0+	—	—	this work	2003
55	Cs	0.471640	6.0E-05	THD	[37]	1998
56	Ba	0.144620	6.0E-05	THD	[38]	1995
57	La	0.470	2.6E-02	THD	[39]	1998
58	Ce	0.955	2.6E-02	THD	[40]	2002
59	Pr	0.962	2.6E-02	THD	[41]	2002
60	Nd	0.050	—	EST	[4]	1997
61	Pm	0+	—	EST	[4]	1997
62	Sm	0.05	—	EST	[4]	1997
63	Eu	0.05	—	EST	[4]	1997
64	Gd	0.10	—	EST	[4]	1997
65	Tb	0.10	—	EST	[4]	1997
66	Dy	0.15	—	EST	[4]	1997
67	Ho	0+	—	EST	[4]	1997
68	Er	0+	—	EST	[4]	1997
69	Tm	1.029	2.2E-02	THD	[42]	2002
70	Yb	0.01	—	EST	[4]	1997
71	Lu	0.34	—	THD	[43]	2001
72	Hf	0.10	—	EST	[4]	1997
73	Ta	0.323	1.2E-02	PES	[2]	1985
74	W	0.815	4.0E-03	PES	[44]	1992
75	Re	0.150	1.0E-01	SI	[45]	1970
76	Os	1.0778	1.5E-04	THD	[46]	2000

TABLE A1.1 (*Continued*)

N	Atom	AE_a	Uncertainty	Method	Reference	Year
77	Ir	1.5644	1.5E-04	THD	[47]	1999
78	Pt	2.12510	5.0E-05	THD	[47]	1999
79	Au	2.30863	3.0E-05	THD	[2]	1985
80	Hg	0+	—	EST	this work	2003
81	Tl	0.377	1.3E-02	PD	[48]	2000
82	Pb	1.10	5.0E-02	PD	[49]	1973
83	Bi	0.942362	1.3E-05	PES	[50]	2001
84	Po	1.9	3.0E-01	EST	this work	2003
85	At	2.8	2.0E-01	EST	this work	2003
86	Rn	0+	—	EST	this work	2003
87	Fr	0.460	—	EST	this work	2003
88	Ra	0.170	—	EST	this work	2003
89	Ac	0+	—	EST	this work	2003
90	Th	0.05	—	EST	this work	2003
91	Pa	0.05	—	EST	this work	2003
92	U	0.05	—	EST	this work	2003
93	Np	0+	—	EST	this work	2003
94	Pu	0.05	—	EST	this work	2003

TABLE A1.2 Main Group Homonuclear Diatomic Molecules (in eV)

AN	Mol	AE_a	Uncertainty	Method	Reference	Year
3	Li_2	0.509	0.009	PES	[51]	1994
5	B_2	1.300	0.400	PES	[52]	1993
6	C_2	3.269	0.006	M	[6]	2003
8	O_2	1.070	0.100	M	this work	2003
9	F_2	3.080	0.050	M	[6]	2003
11	Na_2	0.430	0.015	PES	[53]	1989
13	Al_2	1.460	0.060	PES	[54]	1998
14	Si_2	2.200	0.010	PES	[55]	1993
15	P_2	0.610	0.025	PES	[56]	1985
16	S_2	1.690	0.015	M	[6]	2003
17	Cl_2	2.450	0.020	M	Table 9.1	2003
19	K_2	0.497	0.015	PES	[53]	1989
20	Ca_2	0.025	—	EST	this work	2003
29	Cu_2	0.840	0.010	PES	[57]	1990
31	Ga_2	1.600	0.100	PES	[58]	1994
32	Ge_2	2.074	0.001	PES	[59]	1995
33	As_2	0.739	0.001	PES	[60]	1998
34	Se_2	1.940	0.070	PES	[61]	1989
35	Br_2	2.560	0.020	M	Table 9.1	2003
37	Rb_2	0.498	0.015	LPD	[53]	1989
38	Sr_2	0.052	—	EST	this work	2003
47	Ag_2	1.100	0.008	PES	[57]	1990
49	In_2	1.270	0.100	PES	[62]	1990

TABLE A1.2 (*Continued*)

AN	Mol	AE_a	Uncertainty	Method	Reference	Year
50	Sn_2	1.965	0.010	PES	[63]	1999
51	Sb_2	1.282	0.008	PES	[64]	1992
52	Te_2	1.920	0.070	PES	[65]	1992
53	I_2	2.524	0.005	PES	[66]	1997
55	Cs_2	0.469	0.015	PES	[53]	1989
56	Ba_2	0.145	—	EST	this work	2003
79	Au_2	1.940	0.001	PES	[57]	1990
81	Tl_2	0.950	0.100	PES	[62]	1990
82	Pb_2	1.366	0.008	PES	[65]	1992
83	Bi_2	1.271	0.008	PES	[67]	1991

The electron affinities of the Group II homonuclear diatomic molecules should be greater than the electron affinity of the atom.

REFERENCES

1. Chen, E. C. M. and Wentworth, W. E. *J. Chem. Educ*. **1975**, 52 486.
2. Hotop, H. and Lineberger, W. C. *J. Phys. Chem. Ref. Data* **1985**, 14, 731.
3. Wheeler, J. C. *J. Chem. Educ*. **1997**, 74, 123.
4. Nadeau, M-J.; Garwan, M. A.; Zhao, X-L.; and Litherland, A. E. *Nuclear Instr. Meth. Phys. Res. B* **1997** 123, 521.
5. Andersen, T.; Haugen, H. K.; and Hotop, H. *J. Phys. Chem. Ref. Data* **1999**, 28, 1511.
6. National Institute of Standards and Technology (NIST). *Chemistry WebBook*, **2003**. Available at http://webbook.nist.gov.
7. Rienstra-Kiracofe, J. C.; Tschumper, G. S.; Schaefer, H. F.; Nandi, S.; and Ellison, G. B. *Chem. Rev.* **2002**, 102, 231.
8. Sheer, M. D. *J. Res. Nat. Bur. Stds*. **1970**, A74, 37.
9. Pekeris, C. L. *Phys. Rev*. **1962**, 126, 1470.
10. Haeffler, G.; Hanstorp, G.; Kiyan, I.; Klinkmiller, A. E.; Ljungblad, U.; and Pegg, D. J. *Phys. Rev. A* **1996**, 53, 4127.
11. Scheer, M.; Bilodeau, R. C.; and Haugen, H. K. *Phys. Rev. Lett.* **1998**, 80, 2562–2565.
12. Scheer, M.; Bilodeau, R. C.; Brodie, C. A.; and Haugen, H. K. *Phys. Rev. A* **1998**, 58, 2844.
13. Valli, C.; Blondel, C.; and Delsart, C. *Phys. Rev. A* **1999**, 59, 3809.
14. Blondel, C.; Delsart, C.; and Goldfarb, F. *J. Phys. B—Atom. Molec. Opt. Phys.*, **2001**, 34, L281.
15. 15 Scheer, M.; Bilodeau, R. C.; Thogresen, J.; and Haugen, H. K. *Phys. Rev. A* **1998**, 57, R1493.
16. Blondel, C. *Phys. Scr.* **1995**, T58, 31.
17. Trainham, R.; Fletcher, G. D.; and Larson, D. J. *J. Phys. B* **1987**, 20, L777.

18. Andersson, K. T.; Sandstrom, J.; Kiyan, I. Y.; Hanstorp, D.; and Pegg, D. J. *Phys. Rev. A* **2000**, 62, 22503.
19. Petrunin, V. V.; Andersen, H. H.; Balling, P.; and Andersen, T. *Phys. Rev. Lett.* **1996**, 76, 744.
20. Feigerle, C. S.; Herman, Z.; and Lineberger, C. W. *J. Electron. Spectrosc. Relat. Phenom.* **1981**, 23, 441.
21. Ilin, R. N.; Sakharov, V. I.; and Serenkov, I. T. *Opt. Spectros. (USSR),* **1987**, 62, 578.
22. Bilodeau, R. C.; Scheer, M.; and Haugen, H. K. *J. Phys. B* **1998**, 31, 3885.
23. Leopold, D. G. and Lineberger, W. C. *J. Chem. Phys.* **1986**, 85, 51.
24. Scheer, M.; Brodie, C. A.; Bilodeau, R. C.; and Haugen, H. K. *Phys. Rev. A* **1998**, 58, 2051.
25. Taylor, K. J.; Pettiettehall, C. L.; Cheshnovsky, O.; and Smalley, R. E. *J. Chem. Phys.* **1992**, 96, 3319.
26. Williams, W. W.; Carpenter, D. L.; Covington, A. M.; Koepnick, M. C.; Calabrese, D.; and Thompson, J. S. *J. Phys. B* **1998**, 31, L341–L345.
27. Lippa, T. P.; Xu, S. J.; Lyapustina, S. A.; Nilles, J. M.; and Bowen, K. H. *J. Chem. Phys.* **1998**, 109, 10727.
28. Mansour, N. B.; Edge, C. J.; and Larson, D. J. *Nucl. Instrum. Meth. Phys. Res. B* **1988**, 31, 313.
29. Blondel, C.; Cacciani, P.; Delsart, C.; and Trainham, R. *Phys. Rev. A* **1989**, 40, 3698.
30. Frey, P.; Breyer, F.; and Hotop, H. *J. Phys. B: At. Mol. Opt. Phys.* **1978**, 11, L589.
31. Andersen, H. H.; Petrunin, V. V.; Kristensen, P.; and Andersen, T. *Phys. Rev. A* **1997**, 55, 3247.
32. Norquist, P. L.; Beck, D. R.; Bilodeau, R. C.; Scheer, M.; Srawley, R. A.; and Haugen, H. K. *Phys. Rev. A* **1999**, 59, 1896–1902.
33. Williams, W. W.; Carpenter, D. L.; Covington, A. M.; Thompson, J. S.; Kvale, T. J.; and Seely, D. G. *Phys. Rev. A* **1998**, 58, 3582.
34. Scheer, M.; Haugen, H. K.; and Beck, D. R. *Phys. Rev. Lett.* **1997**, 79, 4104.
35. Haeffler, G.; Klinkmüller, A. E.; Rangell, J.; Berzinsh, U.; and Hanstorp, D. *Z. Phys. D* **1996**, 38, 211.
36. Hanstorp, D. and Gustafsson, M. *J. Phys. B* **1992**, 25, 1773.
37. Scheer, M.; Thogersen, J.; Bilodeau, R. C.; Brodie, C. A.; and Haugen, H. K. *Phys. Rev. Lett.* **1998**, 80, 684.
38. Petrunin, V. V.; Volstad, J. D.; Balling, P.; Kristensen, K.; and Andersen, T. *Phys. Rev. Lett.* **1995**, 75, 1911.
39. Covington, A. M.; Calabrese, D.; Thompson, J. S.; and Kvale, T. J. *J. Phys. B* **1998**, 31, L855.
40. Davis, V. T. and Thompson, J. S. *Phys. Rev. Lett.* **2002**, 88, 073003.
41. Davis, V. T. and Thompson, J. S. *J. Phys. B* **2002**, 35, L11.
42. Davis, V. T. and Thompson, J. S. *Phys. Rev. A* **2002**, 6501, 0501.

43. Davis, V. T. and Thompson, J. S. *J. Phys. B* **2001**, 354, L433.

44. Bengali, A. A.; Casey, S. M.; Cheng, C-L.; Dick, J. P.; Fenn, T.; Villaalta, P. W.; and Leopold, D. G. *J. Amer. Chem. Soc.* **1992**, 114, 5257.

45. Sheer, M. D. *J. Res. Nat. Bur. Stds.* **1970**, 74A, 37.

46. Bilodeau, R. C.; and Haugen, H. K. *Phys. Rev. Lett.* **2000**, 85, 534–537.

47. Bilodeau, R. C.; Scheer, M.; Haugen, H. K.; and Brooks, R. L. *Phys. Rev. A* **1999**, 61, 12505.

48. Carpenter, D. L.; Covington, A. M.; and Thompson J. S. *Phys. Rev. A* **2000**, 61, 42501.

49. Feldman, D. In *Proc. 8th Intern. Conf. on Electr. Atom. Collis*, Belgrade, **1973**.

50. Bilodeau, R. C. and Haugen, H. K. *Phys. Rev. A* **2001**, 6402, 4501.

51. Sarkas, H. W.; Arnold, S. T.; Hendricks, J. H.; Slager, V. L.; and Bowen, K. H. *Z. Phys. D* **1994**, 29, 209.

52. Reid, C. J. *Int. J. Mass Spectrom. Ion Proc.* **1993**, 127, 147.

53. McHugh, K. M.; Eaton, J. G.; Lee, G. H.; Sarkas, H. W.; Kidder, L. H.; Snodgrass, J. T.; Manaa, M. R.; and Bowen, K. H. *J. Chem. Phys.* **1989**, 91, 3792.

54. Li, X.; Wu, H.; Wang, X.; and Wang, L. *Phys. Rev. Lett.* **1998**, 81, 1909.

55. Arnold, C. C.; Kitsopoulos, T. N.; and Neumark, D. M. *J. Chem. Phys.* **1993**, 99, 766.

56. Snodgrass, J. T.; Coe, J. V.; Freidhoff, C. B.; McHugh, K. M.; and Bowen, K. H. *Chem. Phys. Lett.* **1985**, 122, 352.

57. Ho, J.; Ervin, K. M.; and Lineberger, W. C. *J. Chem. Phys.* **1990**, 93, 6987.

58. Cha, C.-Y.; Ganteför, G.; and Eberhardt, W. *J. Chem. Phys.* **1994**, 100, 995.

59. Arnold, C. C.; Xu, C. S.; Burton, G. R.; and Neumark, D. M. *J. Chem. Phys.* **1995**, 102, 6982.

60. Lippa, T. P.; Xu, S. J.; Lyapustina, S. A.; Nilles, J. M.; and Bowen, K. H. *J. Chem. Phys.* **1998**, 109, 10727.

61. Snodgrass, J. T.; Coe, J. V.; McHugh, K. M.; Freidhoff, C. B.; and Bowen, K. H. *J. Phys. Chem.* **1989**, 93, 1249.

62. Gausa, M.; Ganteför, G.; Lutz, H. O.; and Meiwes-Broer, K-H. *Int. J. Mass Spectrom. Ion Proc.* **1990**, 102, 227.

63. Moravec, V. D.; Klopcic, S. A.; and Jarrold, C. C. *J. Chem. Phys.* **1999**, 110, 5079–5088.

64. Polak, M. L.; Gerber, G.; Ho, J.; and Lineberger, W. C. *J. Chem. Phys.* **1992**, 97, 8990.

65. Ho, J.; Polak, M. L.; and Lineberger, W. C. *J. Chem. Phys.* **1992**, 96, 144.

66. Zanni, M. T.; Taylor, T. R.; Greenblatt, B. J.; Miller, W. H.; and Neumark, D. M. *J. Chem. Phys.* **1997**, 107, 7613.

67. Polak, M. L.; Ho, J.; Gerber, G.; Lineberger, W. C. *J. Chem. Phys.* **1991**, 95, 3053.

TABLE A2.1 CH Molecules by Value (in eV)

EVAL	NIST	Name/Formula	MW	Mtd.	Reference
0.07(2)	0.048	Benzene, 1,2,4,5-tetramethyl- (C10H14)	134	E	[2]
0.10(5)	—	Styrene (C8H8)	104	E	[2]
0.10(5)	0.108	Benzene, 1,2,3,5-tetramethyl- (C10H14)	134	E	[2]
0.12(5)	0.121	Benzene, hexamethyl- (C12H18)	162	E	[2]
0.13(5)	0.130	Biphenyl (C12H10)	154	E	[2]
0.14(5)	0.143	Naphthalene, 2-methyl- (C11H10)	142	E	[2]
0.16(2)	−0.200	Naphthalene (C10H8)	128	E	[1]
0.16(5)	0.160	Naphthalene, 1-methyl- (C11H10)	142	E	[2]
0.16(5)	0.147	Naphthalene, 1-ethyl- (C12H12)	156	E	[2]
0.16(5)	0.160	Naphthalene, 2,6-dimethyl- (C12H12)	156	E	[2]
0.16(5)	0.156	Diphenylmethane (C13H12)	168	E	[2]
0.17(5)	0.173	Indene (C9H8)	116	E	[2]
0.17(5)	0.173	Naphthalene, 2,3-dimethyl- (C12H12)	156	E	[2]
0.18(5)	0.182	Benzene, pentamethyl- (C11H16)	148	E	[2]
0.19(5)	0.195	Naphthalene, 2-ethyl- (C12H12)	156	E	[2]
0.21(5)	0.247	Naphthalene, 1,4-dimethyl- (C12H12)	156	E	[2]
0.28(5)	0.278	Fluorene (C13H10)	166	E	[2]
0.29(2)	0.285	Triphenylene (C18H12)	228	E	[1]
0.30(2)	0.307	Phenanthrene (C14H10)	178	E	[1]
0.32(5)	0.321	Diphenylethyne (C14H10)	178	E	[2]
0.39(5)	0.390	(E)-stilbene (C14H12)	180	E	[2]
0.39(5)	0.390	Ethylene, 1,1-diphenyl- (C14H12)	180	E	[2]
0.40(5)	0.890	Biphenylene (C12H8)	152	CI	[3]
0.42(4)	0.397	Chrysene (C18H12)	228	E	[1]
0.54(3)	0.542	Picene (C22H14)	278	E	[1]
0.55(3)	0.534	Benzo[e]pyrene (C20H12)	252	E	[1]
0.58(1)	0.545	Benzo[c]phenanthrene (C18H12)	228	E	[1]
0.60(5)	0.550	Anthracene, 1-methyl- (C15H12)	192	CI	[3]
0.61(2)	0.500	Pyrene (C16H10)	202	E	[1]
0.67(3)	0.591	Dibenz[a,j]anthracene (C22H14)	278	E	[1]
0.68(2)	0.530	Anthracene (C14H10)	178	E, P, T	[1, 4, 5]
0.69(3)	0.595	Dibenz[a,h]anthracene (C22H14)	278	E	[1]
0.69(3)	—	Dibenz[a,c]anthracene (C22H14)	278	E	[1]
0.72(1)	0.390	Benz[a]anthracene (C18H12)	228	E	[1]
0.80(10)	0.550	1,3,5,7-Cyclooctatetrene (C8H8)	104	E, P, T	[1, 6–8]
0.80(5)	0.403	Acenaphthylene (C12H8)	152	E	[1]
0.80(5)	0.470	Coronene (C24H12)	300	P, CI	[9, 10]
0.82(4)	0.815	Benzo[a]pyrene (C20H12)	252	E, T	[1]
0.82(5)	0.630	Fluoranthene (C16H10)	202	E	[1]
0.84(5)	0.694	Azulene (C10H8)	128	E, P, T	[1, 11, 12]
0.89(5)	0.420	Benzo[ghi]perylene (C22H12)	276	CI	[9]
0.97(1)	0.973	Perylene (C20H12)	252	E, P, T	[1, 5, 13]
1.00(20)	1.000	Dibenzo[a,g]corannulene radical (C28H14)	350	CI	[9]
1.08(4)	1.067	Naphthacene (C18H12)	228	E, P, T	[1, 5, 14]
1.16(20)	1.160	Diindenochrysene radical (C26H12)	324	CI	[9]
1.39(5)	1.392	Pentacene (C22H14)	278	T	[5]

TABLE A2.2 CH Molecules by Molecular Weight (in eV)

EVAL	NIST	Name/Formula	MW	Mtd.	Reference
0.10(5)	—	Styrene(C8H8)	104	E	[2]
0.80(10)	0.550	1,3,5,7-Cyclooctatetrene (C8H8)	104	E, P, T	[1, 6–8]
0.17(5)	0.173	Indene (C9H8)	116	E	[2]
0.16(2)	−0.200	Naphthalene (C10H8)	128	E	[1]
0.84(5)	0.694	Azulene (C10H8)	128	E, P, T	[1, 11, 12]
0.07(2)	0.048	Benzene, 1,2,4,5-tetramethyl- (C10H14)	134	E	[2]
0.10(5)	0.108	Benzene, 1,2,3,5-tetramethyl- (C10H14)	134	E	[2]
0.14(5)	0.143	Naphthalene, 2-methyl- (C11H10)	142	E	[2]
0.16(5)	0.160	Naphthalene, 1-methyl- (C11H10)	142	E	[2]
0.18(5)	0.182	Benzene, pentamethyl- (C11H16)	148	E	[2]
0.40(5)	0.890	Biphenylene (C12H8)	152	CI	[3]
0.80(5)	0.403	Acenaphthylene (C12H8)	152	E	[1]
0.13(5)	0.130	Biphenyl (C12H10)	154	E	[2]
0.16(5)	0.147	Naphthalene, 1-ethyl- (C12H12)	156	E	[2]
0.16(5)	0.160	Naphthalene, 2,6-dimethyl- (C12H12)	156	E	[2]
0.17(5)	0.173	Naphthalene, 2,3-dimethyl- (C12H12)	156	E	[2]
0.19(5)	0.195	Naphthalene, 2-ethyl- (C12H12)	156	E	[2]
0.21(5)	0.247	Naphthalene, 1,4-dimethyl- (C12H12)	156	E	[2]
0.12(5)	0.121	Benzene, hexamethyl- (C12H18)	162	E	[2]
0.28(5)	0.278	Fluorene (C13H10)	166	E	[2]
0.16(5)	0.156	Diphenylmethane (C13H12)	168	E	[2]
0.30(2)	0.307	Phenanthrene (C14H10)	178	E	[1]
0.32(5)	0.321	Diphenylethyne (C14H10)	178	E	[2]
0.68(2)	0.530	Anthracene (C14H10)	178	E, P, T	[1, 4, 5]
0.39(5)	0.390	(E)-stilbene (C14H12)	180	E	[2]
0.39(5)	0.390	Ethylene, 1,1-diphenyl- (C14H12)	180	E	[2]
0.60(5)	0.550	Anthracene, 1-methyl- (C15H12)	192	CI	[3]
0.61(2)	0.500	Pyrene (C16H10)	202	E	[1]
0.82(5)	0.630	Fluoranthene (C16H10)	202	E	[1]
0.29(2)	0.285	Triphenylene (C18H12)	228	E	[1]
0.42(4)	0.397	Chrysene (C18H12)	228	E	[1]
0.58(1)	0.545	Benzo[c]phenanthrene (C18H12)	228	E	[1]
0.72(1)	0.390	Benz[a]anthracene (C18H12)	228	E	[1]
1.08(4)	1.067	Naphthacene (C18H12)	228	E, P, T	[1, 5, 14]
0.55(3)	0.534	Benzo[e]pyrene (C20H12)	252	E	[1]
0.82(4)	0.815	Benzo[a]pyrene (C20H12)	252	E, T	[1]
0.97(1)	0.973	Perylene (C20H12)	252	E, P, T	[1, 5, 13]
0.89(5)	0.420	Benzo[ghi]perylene (C22H12)	276	CI	[9]
0.54(3)	0.542	Picene (C22H14)	278	E	[1]
0.67(3)	0.591	Dibenz[a,j]anthracene (C22H14)	278	E	[1]
0.69(3)	0.595	Dibenz[a,h]anthracene (C22H14)	278	E	[1]
0.69(3)	—	Dibenz[a,c]anthracene (C22H14)	278	E	[1]
1.39(5)	1.392	Pentacene (C22H14)	278	T	[5]
0.80(5)	0.470	Coronene (C24H12)	300	P, CI	[9, 10]
1.00(20)	1.000	Dibenzo[a,g]corannulene (C28H14)	350	CI	[9]
1.16(20)	1.160	Diindenochrysene radical (C26H12)	324	CI	[9]

REFERENCES

1. Chen, E. C. M. and Chen, E. S. D. *J. Chromatogr A* **2002**, 952, 173.
2. Wojnarvits, L. and Foldiak, G. *J. Chromatogr.* **1981**, 206, 511.
3. Chen, G. D. and Cooks, R. G. *J. Mass Spectrom.* **1995** 30, 1167.
4. Scheidt, J. and Weinkauf, R. *Chem. Phys. Lett.* **1997**, 266, 201.
5. Crocker, L.; Wang, T. B.; and Kebarle, P. *J. Amer. Chem. Soc.* **1993**, 115, 7818.
6. Gygax, R.; Peters, H. L.; and Brauman, J. I. *J. Amer. Chem. Soc* **1979**, 101, 2567.
7. Kato, S.; Lee, H. S.; Gareyev, R.; Wenthold, P. G.; Lineberger, W. C.; DePuy, C. H.; and Bierbaum, V. M. *J. Amer. Chem. Soc.* **1997**, 119, 7863.
8. Denault, J. W.; Chen, G. D.; and Cooks, R. G. *J. Amer. Soc. Mass Spectrom.* **1998**, 9, 1141.
9. Chen, G.; Cooks, R. G.; Corpuz, E.; and Scott, L. T. *J. Amer. Soc. Mass Spectrom.* **1996**, 7, 619.
10. Duncan, M. A.; Knight, A. M.; Negishi, Y.; Nagao, S.; Nakamura, Y.; Kato, A.; Nakajima, A.; and Kaya, K. *Chem. Phys. Lett.* **1999**, 309, 49.
11. Schiedt, J.; Knott, W. J.; Le Barbu, K.; Schlag, E. W.; and Weinkauf, R. *J. Chem. Phys.* **2000**, 113, 9470.
12. Kebarle, P. and Chowhury, S. *Chem. Revs.* **1987**, 87, 513.
13. Scheidt, J. and Weinkauf, R. *Chem. Phys. Lett.* **1997**, 274, 18.
14. Reinstra-Kiracofe, J. C.; Tschumper, G. S.; Schaefer, H. F. ; Nandi, S.; and Ellison, G. B. *Chem. Rev.* **2002**, 102, 231.
15. National Institute of Standards and Technology (NIST). *Chemistry WebBook*, **2003**. Available at http://webbook.nist.gov.

TABLE A2.3 CHX Molecules by Value (in eV)

EVAL	NIST	Name/Formula	MW	Mtd.	Reference
0.2(1)	0.100	Ethene, 1,1-dichloro- (C2H2Cl2)	97	E	[1], this work
0.24(10)	0.237	Cyclobutene, hexafluoro- (C4F6)	162	Kine	[2]
0.30(5)	0.094	Benzene, 1,2-dichloro- (C6H4Cl2)	147	E	[3], this work
0.30(5)	0.277	Naphthalene, 1-chloro- (C10H7Cl)	162.5	E	[3]
0.46(5)	0.400	Trichloroethylene (C2HCl3)	131.5	E	[1, 4], this work
0.49(3)	0.340	Benzene, 1,3,5-trichloro- (C6H3Cl3)	181.5	E	[3], this work
0.66(3)	0.450	Benzene, 1,2,4,5-tetrachloro- (C6H2Cl4)	216	E	[1, 3], this work
0.70(5)	0.640	Tetrachloroethylene (C2Cl4)	166	E	[1, 4], this work
0.70(10)	0.700	Perfluoro-2-butene, (E)- (C4F8)	200	En	[5]
0.73(5)	0.434	Benzene, pentafluoro- (C6HF5)	168	E, T	[6, 7], this work
0.80(10)	0.802	2-Chloroanthracene (C14H9Cl)	212.5	E, T	[8, 9]
0.80(10)	0.720	1-Bromopyrene (C16H9Br)	281	E, T	[10]
0.83(10)	0.828	Anthracene, 1-chloro- (C14H9Cl)	212.5	E, T	[8, 9]
0.86(10)	0.859	9-Chloroanthracene (C14H9Cl)	212.5	E, T	[8, 9]

TABLE A2.3 (*Continued*)

EVAL	NIST	Name/Formula	MW	Mtd.	Reference
0.86(2)	0.520	Benzene, hexafluoro- (C6F6)	186	E, T, P	[6, 7, 9, 11]
0.90(10)	0.610	9-Bromoanthracene (C14H9Br)	257	CI	[10]
0.90(5)	0.729	Benzene, pentachloro- (C6HCl5)	250.5	E, T	[12]
1.0(2)	0.824	1,1′-Biphenyl-decafluoro- (C12F10)	334	T	[6]
1.0(2)	0.859	Benzene, pentafluoro(CF3)- (C7F8)	236	T	[6]
1.00(5)	0.815	Benzene, chloropentafluoro- (C6ClF5)	202.5	E, T	[6]
1.05(10)	1.049	Octafluorocyclobutane (C4F8)	200	T, En	[5, 13]
1.07(5)	1.062	Perfluoromethylcyclohexane (C7F14)	350	N, T	[14]
1.15(5)	0.915	Benzene, hexachloro- (C6Cl6)	285	E, T	[1, 12], this work
1.2(1)	0.110	Methyl iodide (CH3I)	142	A	[15, 16]
1.3(1)	1.149	Benzene, bromopentafluoro- (C6BrF5)	247	E	[6]
1.4(2)	0.399	Dichlorodifluoromethane (CCl2F2)	121	A	[17]
1.4(3)	1.400	Decafluorocyclohexene (C6F10)	262	En	[5]
1.5(1)	1.414	Benzene, pentafluoroiodo- (C6F5I)	294	T	[6]
1.65(10)	0.622	Chloroform (CHCl3)	119.5	M, T	[11, 18]
1.8(2)	1.479	Ethane, hexachloro- (C2Cl6)	237	M	[11]
1.80(15)	0.911	Methane, bromotrifluoro- (CBrF3)	149	A	[16]
1.80(15)	1.101	Trichloromonofluoromethane (CCl3F)	137.5	A	[17, 18]
2.04(10)	0.805	Carbon tetrachloride (CCl4)	154	M, T	[11, 18]
2.20(20)	1.570	Methane, trifluoroiodo- (CF3I)	196	A	[19–21]

TABLE A2.4 Electron Affinities of CHX Molecules by Molecular Weight (in eV)

EVAL	NIST	Name/Formula	MW	Mtd.	Reference
0.2(1)	0.100	Ethene, 1,1-dichloro- (C2H2Cl2)	97	E	[1, this work]
1.65(10)	0.622	Chloroform (CHCl3)	119.5	M,T	[11, 18]
1.4(.15)	0.399	Dichlorodifluoromethane (CCl2F2)	121	A	[17]
0.46(5)	0.400	Trichloroethylene (C2HCl3)	131.5	E	[1,4, this work]
1.80(15)	1.101	Trichloromonofluoromethane (CCl3F)	137.5	A	[17, 18]
1.20(10)	0.110	Methyl iodide (CH3I)	142	A	[15, 16]
0.30(5)	0.094	Benzene, 1,2-dichloro- (C6H4Cl2)	147	E	[3, this work]
1.80(15)	0.911	Methane, bromotrifluoro- (CBrF3)	149	A	[16]
2.04(10)	0.805	Carbon tetrachloride (CCl4)	154	M,T	[11, 18]
0.24(10)	0.237	Cyclobutene, hexafluoro- (C4F6)	162	Kine	[2]
0.30(5)	0.277	Naphthalene, 1-chloro- (C10H7Cl)	162.5	E	[3]
0.70(5)	0.640	Tetrachloroethylene (C2Cl4)	166	E	[1, 4, this work]

TABLE A2.4 (*Continued*)

EVAL	NIST	Name/Formula	MW	Mtd.	Reference
0.73(5)	0.434	Benzene, pentafluoro- (C6HF5)	168	E,T	[6, 7 this work]
0.49(3)	0.340	Benzene, 1,3,5-trichloro- (C6H3Cl3)	181.5	E	[3, this work]
0.86(2)	0.530	Benzene, hexafluoro- (C6F6)	186	E,T,P	[6, 7, 9, 11]
2.20(20)	1.570	Methane, trifluoroiodo- (CF3I)	196	A	[19–21]
0.70(10)	0.700	Perfluoro-2-butene, (E)- (C4F8)	200	En	[5]
1.05(10)	1.049	Octafluorocyclobutane (C4F8)	200	T,En	[5, 13]
1.00(5)	0.815	Benzene, chloropentafluoro- (C6ClF5)	202.5	E,T	[6]
0.80(10)	0.802	2-Chloroanthracene (C14H9Cl)	212.5	E,T	[8, 9]
0.83(10)	0.828	Anthracene, 1-chloro- (C14H9Cl)	212.5	E,T	[8, 9]
0.86(10)	0.859	9-Chloroanthracene (C14H9Cl)	212.5	E,T	[8, 9]
0.66(3)	0.450	Benzene, 1,2,4,5-tetrachloro- (C6H2Cl4)	216	E	[1, 3, this work]
1.0(2)	0.859	Perfluorotoulene- (C7F8)	236	T	[6]
1.8(2)	1.479	Ethane, hexachloro- (C2Cl6)	237	M	[11]
1.3(1)	1.149	Benzene, bromopentafluoro- (C6BrF5)	247	E	[6]
0.90(5)	0.729	Benzene, pentachloro- (C6HCl5)	250.5	E,T	[12]
0.90(10)	0.610	9-Bromoanthracene (C14H9Br)	257	CI	[10]
1.4(3)	1.400	Decafluorocyclohexene (C6F10)	262	En	[5]
0.80(10)	0.720	1-Bromopyrene (C16H9Br)	281	E,T	[10]
1.15(5)	0.915	Benzene, hexachloro- (C6Cl6)	285	E,T	[1, 12, this work]
1.5(1)	1.414	Benzene, pentafluoroiodo- (C6F5I)	294	T	[6]
1.0(2)	0.824	1,1'-Biphenyl-decafluoro- (C12F10)	334	T	[6]
1.07(5)	1.062	Perfluoromethylcyclohexane (C7F14)	350	N,T	[14–21]

REFERENCES

1. Wiley, J. R.; Chen, E. C. M.; Chen, E. S. D.; Richardson, P.; Reed, W. R.; and Wentworth, W. E. J. *Electroanal. Chem. Interfac.* **1991**, 307, 169.
2. Christophorou, L. G. and Datkos, P. G. *Int. J. Mass Spectrom. Ion Proc.* **1995**, 149/150, 59.
3. Steelhammer, J. C. and Wentworth, W. E. *J. Chem. Phys.* **1969**, 51, 1802.
4. Chen, E. C. M.; Wiley, J. R.; Batten, C. F.; and Wentworth, W. E. *J. Phys. Chem.* **1994**, 98, 88.
5. Lifshitz, C.; Tiernan, T. O.; and Hughes, B. M. *J. Chem. Phys.* **1973**, 59, 3182.
6. Dillow, G. W. and Kebarle, P. *J. Amer. Chem. Soc.* **1989**, 111, 5592.
7. Wentworth, W. E.; Limero, T.; and Chen, E. C. M. *J. Phys. Chem.* **1987**, 91, 241.
8. Heinis, T.; Chowdhury, S.; and Kebarle, P. *Org. Mass Spectrom.* **1993**, 28, 358.
9. Chen, E. C. M. and Wentworth, W. E. *Mol. Cryst. Liq. Cryst.* **1989**, 171, 271.
10. Chen, G. D. and Cooks, R. G. *J. Mass Spectrom.* **1995**, 30, 1167.
11. Page, F. M. and Goode, G. C. *Negative Ions and the Magnetron*. New York: Wiley-Interscience **1969**.
12. Knighton, W. B.; Bognar, J. A.; and Grimsrud, E. P. *J. Mass Spectrom.* **1995**, 30, 557.
13. Hiraoka, K.; Mizuno, T.; Eguchi, D.; Takao, K.; Iino, T.; and Yamabe, S. *J. Chem. Phys.* **2002**, 116, 7574–7582.

14. Grimsrud, E. P.; Chowdhury, S.; and Kebarle, P. *J. Chem. Phys.* **1985**, 83, 1059.
15. Kim, J.; Kelley, J. A.; Ayotte, P.; Nielsen, S. B.; Weddle, G. H.; and Johnson, M. A. *J. Amer. Soc. Mass Spectrom.* **1999**, 10, 810–814.
16. Moutinho, A. M. C.; Aten, J. A.; and Los, J. *Chem. Phys.* **1974**, 5, 84.
17. Dispert, H. and Lacmann, K. *Int. J. Mass Spectrom. Ion Phys.* **1978**, 28, 49.
18. Staneke, P. O.; Groothuis, G.; Ingemann, S.; and Nibbering, N. M. M. *Int. J. Mass Spectrom. Ion Proc.* **1995**, 142, 83.
19. Compton, R. N.; Reinhardt, P. W.; and Cooper, C. D. *J. Chem. Phys.* **1978**, 68, 4360.
20. Tang, S. Y.; Mathur, B. P.; Roth, E. W.; and Reck, G. P. *J. Chem. Phys.* **1976**, 64, 1270.
21. McNamee, P. E.; Lacmann, K.; and Herschbach, D. R. *Faraday Disc. Chem. Soc.* **1973**, 55, 318.

TABLE A3.1 CHNX Molecules by Value (in eV)

EVAL	NIST	Name/Formula	MW	Mtd.	Reference
0.26(2)	0.256	Benzonitrile (C7H5N)	103	E	[5, 6]
0.50(5)	0.450	1,3,5-Triazine (C3H3N3)	144	Ered	[7]
0.58(9)	0.577	Quinazoline (C8H6N2)	130	T	[8]
0.65(9)	0.646	2-Naphthalenecarbonitrile (C11H7N)	153	T	[9]
0.67(9)	0.672	3-Trifluoromethylbenzonitrile (C8H4F3N)	171	T	[9]
0.68(9)	0.676	1-Naphthalenecarbonitrile (C11H7N)	153	T	[9]
0.70(9)	0.698	2-Trifluoromethylbenzonitrile (C8H4F3N)	171	T	[10]
0.70(9)	0.698	2,6-Dichlorobenzonitrile (C7H3Cl2N)	172	T	[10]
0.71(9)	0.711	Quinoxaline (C8H6N2)	130	T	[8]
0.72(9)	0.720	Cinnoline (C8H6N2)	130	T	[8]
0.76(9)	0.759	4-Trifluoromethylbenzonitrile (C8H4F3N)	171	T	[10]
0.82(7)	0.798	3,5-Dichlorobenzonitrile (C7H3Cl2N)	172	T	[11, 12]
0.91(9)	0.906	Acridine (C13H9N)	179	T	[7]
0.91(9)	0.911	1,3-Benzenedicarbonitrile (C8H4N2)	128	T	[10]
0.92(9)	0.919	Benzo[c]cinnoline (C12H8N2)	180	T	[7]
0.95(5)	0.012	Adenine (C5H5N5)	135	DB	[1–4]
0.95(9)	0.954	Phthalonitrile (C8H4N2)	128	T	[10]
1.00(10)	1.249	Fumaronitrile (C4H2N2)	78	T, M	[10, 13]
1.00(8)	1.002	Pyrido[2,3-d]pyridazine (C7H5N3)	131	T	[7]
1.09(9)	1.093	1,4-Benzenedicarbonitrile (C8H4N2)	128	T	[10]
1.14(9)	1.140	3,5-Bis(CF3)benzonitrile (C9H3F6N)	239	T	[10]
1.28(10)	1.271	9-Anthracenecarbonitrile (C15H9N)	203	T	[9]
1.31(10)	1.305	Phenazine (C12H8N2)	180	T	[7]
1.84(2)	1.840	1,3,5-Tricyanobenzene (C9H3N3)	153	P	[14]
2.17(15)	2.173	2,3,5,6-Pyridinetetracarbonitrile (C9HN5)	179	M	[13]
2.20(15)	2.203	1,2,4,5-Tetracyanobenzene (C10H2N4)	178	M	[13]
2.54(15)	2.541	Hexacyanobenzene (C12N6)	228	M	[13]
2.80(6)	2.800	Tetracyanoquinodimethane (C12H4N4)	204	M, A	[20]
2.95(10)	3.166	Tetracyanoethylene (C6N4)	128	M, T	[13, 19]
3.29(15)	3.291	Hexacyanobutadiene (C10N6)	204	M	[13]

TABLE A3.2 CHNX Molecules by Molecular Weight (in eV)

EVAL	NIST	Name/Formula	MW	Mtd.	Reference
1.00(10)	1.249	Fumaronitrile (C4H2N2)	78	T, M	[10, 13]
0.26(2)	0.256	Benzonitrile (C7H5N)	103	E	[5, 6]
0.91(9)	0.911	1,3-Benzenedicarbonitrile (C8H4N2)	128	T	[10]
1.00(9)	0.954	Phthalonitrile (C8H4N2)	128	T	[10]
1.09(9)	1.093	1,4-Benzenedicarbonitrile (C8H4N2)	128	T	[10]
2.95(10)	3.166	Tetracyanoethylene (C6N4)	128	M, T	[13, 19]
0.58(9)	0.577	Quinazoline (C8H6N2)	130	T	[8]
0.71(9)	0.711	Quinoxaline (C8H6N2)	130	T	[8]
0.72(9)	0.72	Cinnoline (C8H6N2)	130	T	[8]
1.00(8)	1.002	Pyrido[2,3-d]pyridazine (C7H5N3)	131	T	[7]
0.95(5)	0.012	Adenine (C5H5N5)	135	DB/VS	[1–4]
0.50(5)	0.45	1,3,5-Triazine (C3H3N3)	144	Ered, ET	[7]
0.65(9)	0.646	2-Naphthalenecarbonitrile (C11H7N)	153	T	[9]
0.68(9)	0.676	1-Naphthalenecarbonitrile (C11H7N)	153	T	[9]
1.84(2)	1.84	1,3,5-Tricyanobenzene (C9H3N3)	153	P	[14]
0.67(9)	0.672	3-Trifluoromethylbenzonitrile (C8H4F3N)	171	T	[9]
0.70(9)	0.698	2-Trifluoromethylbenzonitrile (C8H4F3N)	171	T	[10]
0.76(9)	0.759	4-Trifluoromethylbenzonitrile (C8H4F3N)	171	T	[10]
0.70(9)	0.698	2,6-Dichlorobenzonitrile (C7H3Cl2N)	172	T	[10]
0.82(7)	0.798	3,5-Dichlorobenzonitrile (C7H3Cl2N)	172	T	[11, 12]
2.20(15)	2.203	1,2,4,5-Tetracyanobenzene (C10H2N4)	178	M	[13]
0.91(9)	0.906	Acridine (C13H9N)	179	T	[7]
2.17(15)	2.173	2,3,5,6-Pyridinetetracarbonitrile (C9HN5)	179	M	[13]
0.92(9)	0.919	Benzo[c]cinnoline (C12H8N2)	180	T	[7]
1.31(10)	1.305	Phenazine (C12H8N2)	180	T	[7]
1.28(10)	1.271	9-Anthracenecarbonitrile (C15H9N)	203	T	[9]
2.80(6)	2.8	Tetracyanoquinodimethane (C12H4N4)	204	M, A	[13, 15–18]
3.29(15)	3.291	Hexacyanobutadiene (C10N6)	204	M	[13]
2.54(15)	2.541	Hexacyanobenzene (C12N6)	228	M	[13]
1.14(9)	1.14	3,5-Bis(CF3)benzonitrile (C9H3F6N)	239	T	

REFERENCES

1. Chen, E. C. M. and Chen, E. S. D. *J. Phys. Chem. B* **2000**, 104, 7835.
2. Wiley, J. R.; Robinson, J. M.; Ehdaie, S.; Chen, E. S. D.; Chen, E. C. M.; and Wentworth, W. E. *Biochem. Biophys. Res. Comm.* **1991**, 180, 841.
3. Desfrancois, C.; Aboul-Carmine, H.; and Schermann, J. P. *J. Chem. Phys.* **1996**, 104, 7792.
4. Chen, E. S. D.; Chen, E. C. M.; Sane, N.; and Schultz, S. *Bioelectrochem. Bioenerget.* **1999**, 48, 69.
5. Wentworth, W. E.; Kao, L. W.; and Becker, R. S. *J. Phys. Chem.* **1975**, 79, 1161.
6. Zlatkis, A.; Lee, C. K.; Wentworth, W. E.; and Chen, E. C. M. *Anal. Chem.* **1983**, 55, 1596.

7. Nenner, I. and Schultz, G. J. *J. Chem. Phys.* **1975**, 62, 1747.
8. Dillow, G. W. and Kebarle, P. *Can. J. Chem.* **1989**, 67, 1628.
9. Heinis, T.; Chowdhury, S.; and Kebarle, P. *Org. Mass Spectrom.* **1993**, 28, 358.
10. Kebarle, P. and Chowdhury, S. *Chem. Rev.* **1987**, 87, 513.
11. Mishima, M.; Huh, C.; Nakamura, H.; Fujio, M.; and Tsuno, Y. *Tetrahed. Lett.* **1993**, 34, 4223.
12. Huh, C.; Kang, C. H.; Lee, H. W.; Nakamura, H.; Mishima, M.; Tsuno, Y.; and Yamataka, H. *Bull. Chem. Soc. Japan* **1999**, 72, 1083.
13. Page, F. M. and Goode, G. C. *Negative Ions and the Magnetron.* New York: Wiley-Interscience, **1969**.
14. Tsukuda, T.; Kondow, T.; Dessent, C. E. H.; Bailey, C. G.; Johnson, M. A.; Hendricks, J. H.; and Lyapustina, S. S. *Chem. Phys. Lett.* **1997**, 269, 17.
15. Klots, C. E.; Compton, R. N.; and Raaen, V. F. *J. Chem. Phys.* **1974**, 60, 1177.
16. Brinkman, E. A.; Gunther, E.; Schafer, O.; and Brauman, J. I. *J. Chem. Phys.* **1994**, 100, 1840.
17. Compton, R. N. and Cooper, C. D. *J. Chem. Phys.* **1977**, 66, 4325.
18. Nazarenko, V. A. and Pokhodenko, V. D. *Int. J. Mass Spectrom. Ion Phys.* **1979**, 31, 381.
19. Chowdhury, S. and Kebarle, P. *J. Amer. Chem. Soc.* **1986**, 108, 5453.
20. National Institute of Standards and Technology (NIST). *Chemistry WebBook*, **2003**. Available at http://webbook.nist.gov.

TABLE A4.1 CHO Molecules by Value (in eV)

EVAL	NIST	Name/Formula	MW	Mtd.	Reference
0.15(3)	0.15	Benzyl ethanoate (C9H10O2)	150	1	[1, 2]
0.18(5)	0.18	Benzoic acid, methyl ester (C8H8O2)	136	E	[1, 3]
0.338(2)	0.334	Acetophenone (C8H8O)	120	E	[4]
0.36(2)	0.351	1-Propanone, 1-phenyl- (C9H10O)	134	E	[4]
0.39(2)	0.373	Benzaldehyde, 4-methyl- (C8H8O)	120	E	[4]
0.41(2)	0.408	3-CH3C6H4CHO (C8H8O)	120	E	[4]
0.44(4)	0.442	Benzaldehyde, 2,4,6-trimethyl- (C10H12O)	148	E	[5]
0.457(5)	0.429	Benzaldehyde (C7H6O)	106	E	[4]
0.48(4)	0.429	Benzaldehyde, 3-methoxy- (C8H8O2)	136	E	[4]
0.49(4)	0.490	Ethanone, 1-(2,4,6-trimethylphenyl)- (C11H14O)	162	E	[5]
0.54(10)	0.540	Diethyl phthalate (C12H14O4)	222	E	[3]
0.55(10)	0.550	Dimethyl phthalate (C10H10O4)	194	E	[3]
0.55(10)	0.550	1,3-DiCOOCH3-benzene (C10H10O4)	194	E	[3]
0.6(3)	0.624	Ethanedial (C2H2O2)	58	Misc.	[6]
0.60(3)	0.598	Ethanone, 1-(1-naphthalenyl)- (C12H10O)	170	E	[3]
0.64(5)	0.642	2-Naphthalenecarboxaldehyde (C11H8O)	156	E, T	[1, 3, 7, 8]
0.68(5)	0.620	Benzophenone (C13H10O)	182	E, T	[1, 7–12]
0.69(3)	0.690	2,3-Butanedione (C4H6O2)	86	E, T	[1, 7, 8]
0.69(5)	0.681	1-Naphthalenecarboxaldehyde (C11H8O)	156	E, T	[1, 3, 7, 8]

TABLE A4.1 (*Continued*)

EVAL	NIST	Name/Formula	MW	Mtd.	Reference
0.75(5)	0.724	Phenanthrene-9-carboxaldehyde (C15H10O)	206	E	[3]
0.82(9)	0.824	1,4-DiCOOCH3-benzene (C10H10O4)	194	E, T	[2, 9, 10]
0.85(5)	0.850	Ethanone, 1-(2-hydroxyphenyl)- (C8H8O2)	136	T	[11]
0.9(1)	0.823	Cinnamylaldehyde (C9H8O)	132	E	[1]
0.96(9)	0.963	Benzoic acid, 4-acetyl-, methyl ester (C10H10O3)	178	T	[9, 10]
1.02(10)	1.015	Ethanone, 1-(9-anthracenyl)- (C16H12O)	234	T	[13]
1.03(9)	0.971	M-phthalaldehyde (C8H6O2)	134	T	[9, 10]
1.06(9)	1.062	Ethanone, 1,1′-(1,4-phenylene)bis- (C10H10O2)	162	T	[9, 10]
1.07(9)	1.158	4-HC(O)-C6H4-COOCH3 (C9H8O3)	164	T	[9, 10]
1.11(9)	1.106	Benzoic acid, 4-benzoyl-, methyl ester (C15H12O3)	240	T	[9]
1.14(9)	1.136	4-Cyclopentene-1,3-dione (C5H4O2)	132	T	[13]
1.16(9)	1.162	2,5-Furandione, 3,4-dimethyl- (C6H6O3)	126	T	[13]
1.19(9)	1.188	4-Methylphthalic anhydride (C9H6O3)	162	T	[13]
1.22(9)	1.219	1,3-Isobenzofurandione, 4-methyl- (C9H6O3)	162	T	[13]
1.26(5)	1.245	Phthalic anhydride (C8H4O3)	148	T	[11, 13]
1.30(9)	1.236	1,4-Benzenedicarboxaldehyde (C8H6O2)	134	T	[9, 10]
1.30(9)	1.297	2,5-Furandione, 3-methyl- (C5H4O3)	112	T	[13]
1.31(9)	1.310	9-Anthracenecarboxaldehyde (C15H10O)	206	E, T	[1, 13]
1.34(9)	1.344	Ethanedione, di-2-furanyl- (C10H6O4)	190	T	[13]
1.44(9)	1.440	2,5-Furandione (C4H2O3)	98	T, A	[11, 13, 14]
1.56(6)	1.557	Anthraquinone, 2-tert-butyl (C18H16O2)	264	T	[7]
1.56(6)	1.557	9,10-Anthracenedione, 2-ethyl- (C16H12O2)	236	T	[7]
1.59(6)	1.591	9,10-Anthraquinone (C14H8O2)	208	M, T	[7, 10, 15]
1.61(4)	1.622	p-Benzoquinone-tetramethyl- (C10H12O2)	164	T	[7]
1.62(5)	1.620	o-Benzoquinone (C6H4O2)	108	P	[11]
1.620(6)	1.617	MeCH=CHO anion (C3H5O-)	57	P	[17, 18]
1.69(5)	1.687	Trimethyl-p-benzoquinone	150	T	[11]
1.72(6)	1.717	2,6-Dimethoxy-p-benzoquinone (C8H8O4)	168	T	[19]
1.74(4)	1.743	Phenylmaleic anhydride (C10H6O3)	174	T	[13]
1.75(4)	1.765	Methylnaphthoquinone(C11H8O2)	172	T	[19]
1.76(6)	1.756	2-me-5-(1-me-et)-1,4-Benzoquinone (C10H12O2)	164	T	[19]
1.76(6)	1.761	2,5-Dimethyl-1-4benzoquinone (C8H8O2)	136	T	[19]
1.77(5)	1.765	2,6-Dimethyl-1-4 benzoquinone (C8H8O2)	136	T	[19]
1.80(5)	1.804	2-tBu-p-benzoquinone (C10H12O2)	172	T	[11]
1.80(5)	1.813	1,4-Naphthalenedione (C10H6O2)	158	T	[8, 11]
1.80(9)	1.804	3,5-bis(1,1-dime-et)-o-Benzoquinone (C14H20O2)	220	T	[19]
1.85(5)	1.852	p-Benzoquinone, 2-me- (C7H6O2)	122	T	[19]
1.86(6)	1.856	2,3-Dimethoxy-5-me-1,4-benzoquinone (C9H10O4)	182	T	[19]

TABLE A4.1 (*Continued*)

EVAL	NIST	Name/Formula	MW	Mtd.	Reference
1.860(5)	1.860	p-Benzoquinone (C6H4O2)	108	A, T, P	[20, 27]
1.860(5)	1.860	2,3-Dehydrobenzoquinone anion (C6H2O2-)	106	Calc.	[21]
1.87(10)	1.865	2,6-bis(1,1-dime-et)-1,4-Benzoquinone (C14H20O_2)	220	T	[19]
2.04(6)	2.042	Phenyl-p-benzoquinone (C12H8O2)	184	T	[19]

TABLE A4.2 CHO Molecules by Molecular Weight (in eV)

EVAL	NIST	Name/Formula	MW	Mtd.	Reference
1.620(6)	1.617	MeCH=CHO anion (C3H5O-)	57	P	[17, 18]
0.6(3)	0.624	Ethanedial (C2H2O2)	58	Misc.	[6]
0.69(3)	0.69	2,3-Butanedione (C4H6O2)	86	E, T	[1, 7, 8]
1.44(9)	1.44	2,5-Furandione (C4H2O3)	98	T, A	[11, 13, 14]
0.457(5)	0.429	Benzaldehyde (C7H6O)	106	E	[4]
1.860(5)	1.86	2,3-Dehydrobenzoquinone anion (C6H2O2-)	106	Calc.	[21]
1.62(5)	1.62	o-Benzoquinone (C6H4O2)	108	P	[11]
1.860(5)	1.86	p-Benzoquinone (C6H4O2)	108	A, T, P	[20, 27]
1.30(9)	1.297	2,5-Furandione, 3-methyl- (C5H4O3)	112	T	[13]
0.338(2)	0.334	Acetophenone (C8H8O)	120	E	[4]
0.39(2)	0.373	Benzaldehyde, 4-methyl- (C8H8O)	120	E	[4]
0.41(2)	0.408	3-CH3C6H4CHO (C8H8O)	120	E	[4]
1.85(5)	1.852	p-Benzoquinone, 2-me- (C7H6O2)	122	T	[19]
1.16(9)	1.162	2,5-Furandione, 3,4-dimethyl- (C6H6O3)	126	T	[13]
0.9(1)	0.823	Cinnamylaldehyde (C9H8O)	132	E	[1]
1.14(9)	1.136	4-Cyclopentene-1,3-dione (C5H4O2)	132	T	[13]
0.36(2)	0.351	1-Propanone, 1-phenyl- (C9H10O)	134	E	[4]
1.03(9)	0.971	M-phthalaldehyde (C8H6O2)	134	T	[9, 10]
1.30(9)	1.236	1,4-Benzenedicarboxaldehyde (C8H6O2)	134	T	[9, 10]
0.18(5)	0.18	Benzoic acid, methyl ester (C8H8O2)	136	E	[1, 3]
0.48(4)	0.429	Benzaldehyde, 3-methoxy- (C8H8O2)	136	E	[4]
0.85(5)	0.85	Ethanone, 1-(2-hydroxyphenyl)- (C8H8O2)	136	T	[11]
1.76(6)	1.761	2,5-Dimethyl-1-4benzoquinone (C8H8O2)	136	T	[19]
1.77(5)	1.765	2,6-Dimethyl-1-4 benzoquinone (C8H8O2)	136	T	[19]
0.44(4)	0.442	Benzaldehyde, 2,4,6-trimethyl- (C10H12O)	148	E	[5]
1.26(5)	1.245	Phthalic anhydride (C8H4O3)	148	T	[11, 13]
0.15(3)	0.15	Benzyl ethanoate (C9H10O2)	150	1	[1, 2]
1.69(5)	1.687	Trimethyl-p-benzoquinone	150	T	[11]
0.64(5)	0.642	2-Naphthalenecarboxaldehyde (C11H8O)	156	E, T	[1, 3, 7, 8]
0.69(5)	0.681	1-Naphthalenecarboxaldehyde (C11H8O)	156	E, T	[1, 3, 7, 8]

TABLE A4.2 (*Continued*)

EVAL	NIST	Name/Formula	MW	Mtd.	Reference
1.80(5)	1.813	1,4-Naphthalenedione (C10H6O2)	158	T	[8, 11]
0.49(4)	0.49	Ethanone, 1-(2,4,6-trimethylphenyl)- (C11H14O)	162	E	[5]
1.06(9)	1.062	Ethanone, 1,1′-(1,4-phenylene)bis- (C10H10O2)	162	T	[9, 10]
1.19(9)	1.188	4-Methylphthalic anhydride (C9H6O3)	162	T	[13]
1.22(9)	1.219	1,3-Isobenzofurandione, 4-methyl- (C9H6O3)	162	T	[13]
1.07(9)	1.158	4-HC(O)-C6H4-COOCH3 (C9H8O3)	164	T	[9, 10]
1.61(4)	1.622	p-Benzoquinone-tetramethyl- (C10H12O2)	164	T	[7]
1.76(6)	1.756	2-me-5-(1-me-et)-1,4-Benzoquinone (C10H12O2)	164	T	[19]
1.72(6)	1.717	2,6-Dimethoxy-p-benzoquinone (C8H8O4)	168	T	[19]
0.60(3)	0.598	Ethanone, 1-(1-naphthalenyl)- (C12H10O)	170	E	[3]
1.75(4)	1.765	Methylnaphthoquinone(C11H8O2)	172	T	[19]
1.80(5)	1.804	2-tBu-p-benzoquinone (C10H12O2)	172	T	[11]
1.74(4)	1.743	Phenylmaleic anhydride (C10H6O3)	174	T	[13]
0.96(9)	0.963	Benzoic acid, 4-acetyl-, methyl ester (C10H10O3)	178	T	[9,10]
0.68(5)	0.62	Benzophenone (C13H10O)	182	E, T	[1, 7–12]
1.86(6)	1.856	2,3-Dimethoxy-5-me-1,4-benzoquinone (C9H10O4)	182	T	[19]
2.04(6)	2.042	Phenyl-p-benzoquinone (C12H8O2)	184	T	[19]
1.34(9)	1.344	Ethanedione, di-2-furanyl- (C10H6O4)	190	T	[13]
0.55(10)	0.55	Dimethyl phthalate (C10H10O4)	194	E	[3]
0.55(10)	0.55	1,3-DiCOOCH3-benzene (C10H10O4)	194	E	[3]
0.82(9)	0.824	1,4-DiCOOCH3-benzene (C10H10O4)	194	E, T	[2, 9, 10]
0.75(5)	0.724	Phenanthrene-9-carboxaldehyde (C15H10O)	206	E	[3]
1.31(9)	1.31	9-Anthracenecarboxaldehyde (C15H10O)	206	E,T	[1, 13]
1.59(6)	1.591	9,10-Anthraquinone (C14H8O2)	208	M,T	[7, 10, 15]
1.80(9)	1.804	3,5-bis(1,1-dime-et)-o-Benzoquinone (C14H20O2)	220	T	[19]
1.87(10)	1.865	2,6-bis(1,1-dime-et)-1,4-Benzoquinone (C14H20O2)	220	T	[19]
0.54(10)	0.54	Diethyl phthalate (C12H14O4)	222	E	[3]
1.02(10)	1.015	Ethanone, 1-(9-anthracenyl)- (C16H12O)	234	T	[13]
1.56(6)	1.557	9,10-Anthracenedione, 2-ethyl- (C16H12O2)	236	T	[7]
1.11(9)	1.106	Benzoic acid, 4-benzoyl-, methyl ester (C15H12O3)	240	T	[9]
1.56(6)	1.557	Anthraquinone, 2-tert-butyl (C18H16O2)	264	T	[7]

TABLE A4.3 CHOX Molecules by Value (in eV)

EVAL	NIST	Name/Formula	MW	Mtd.	Reference
0.22(5)	0.217	2,3,5,6-Tetrafluoroanisole (C7H4F4O)	180	E	[22]
0.28(5)	0.282	mF-C6H4CH2OPh (C13H11FO)	202	E	[22]
0.52(5)	0.395	Ethanone, 1-(4-fluorophenyl)- (C8H7FO)	138	E	[4]
0.44(9)	0.442	2-Propanone, 1,1,1,3,3,3-hexafluoro- (C3F6O)	166	E	[23]
0.49(3)	0.442	o-Fluoroacetophenone (C8H7FO)	138	E	[4]
0.57(5)	0.486	Benzaldehyde, 4-fluoro- (C7H5FO)	124	E	[4]
0.5(2)	0.500	Tetrafluorosuccinic anhydride (C4F4O3)	172	A	[24]
0.55(5)	0.542	Benzene, pentafluoromethoxy- (C7H3F5O)	198	E	[22]
0.58(3)	0.577	Ethanone, 1-(3-fluorophenyl)- (C8H7FO)	138	E	[4]
0.64(5)	0.585	Ethanone, 1-(4-chlorophenyl)- (C8H7ClO)	155	E, T	[1, 9–11]
0.67(5)	0.616	Acetophenone, 3′-chloro- (C8H7ClO)	155	E	[4]
0.70(5)	0.620	P-fluorobenzophenone (C13H9FO)	200	T	[9, 10, 12]
0.61(10)	0.624	4-CF3-C6H4-COOCH3 (C9H7F3O2)	204	T	[9, 10]
0.66(4)	0.637	Benzaldehyde, 2-fluoro- (C7H5FO)	124	E	[4]
0.79(5)	0.642	2′-(Trifluoromethyl)acetophenone (C9H7F3O)	188	E	[4]
0.68(9)	0.659	Benzaldehyde, 4-chloro- (C7H5ClO)	141	T	[9, 10]
0.67(3)	0.668	3-FC6H4CHO (C7H5FO)	124	E	[4]
0.71(9)	0.668	Benzaldehyde, 3-chloro- (C7H5ClO)	141	T	[9, 10]
0.74(9)	0.746	3-CF3-C6H4-COOCH3 (C9H7F3O2)	204	T	[9, 10]
0.75(9)	0.750	Methyl 3,5-diCl-benzoate (C8H6Cl2O2)	205	T	[9, 10]
0.79(5)	0.768	3-CF3-C6H4-COCH3 (C9H7F3O)	188	T, E	[1, 4, 9, 10]
0.80(10)	0.776	4,4′-Difluorobenzophenone (C13H8F2O)	218	T	[9, 10]
0.85(10)	0.815	Benzaldehyde, 3-(trifluoromethyl)- (C8H5F3O)	174	T	[9, 10]
0.83(9)	0.833	Methanone, (4-chlorophenyl)phenyl- (C13H9ClO)	217	T	[9, 10]
0.86(9)	0.872	Methanone, (3-chlorophenyl)phenyl- (C13H9ClO)	217	T	[9, 10]
0.88(9)	0.876	2′,3′,4′,5′,6′-Pentafluoroacetophenone (C8H3F5O)	210	T	[25]
0.90(10)	0.898	4-CF3-C6H4-COCH3 (C9H7F3O)	188	T	[10]
0.97(10)	0.941	p-CF3C6H4CHO (C8H5F3O)	174	T	[10]
0.96(10)	0.958	3-(Trifluoromethyl)benzophenone (C14H9F3O)	250	T	[10]
0.98(10)	0.984	Ethanone, 2,2,2-trifluoro-1-phenyl- (C8H5F3O)	174	T	[10]
1.03(9)	0.989	3,5-Dichlorobenzaldehyde (C7H4Cl2O)	175	T	[10]
1.00(9)	1.006	Methyl 3,5-diCF3-benzoate (C10H6F6O2)	272	T	[10]
1.08(9)	1.075	4-Trifluoromethylbenzophenone (C14H9F3O)	250	T	[10]
1.10(10)	1.097	Pentafluorobenzaldehyde (C7HF5O)	196	T	[25]
1.15(9)	1.106	3,5-diCl-Benzophenone (C13H8Cl2O)	251	T	[10]

TABLE A4.3 (*Continued*)

EVAL	NIST	Name/Formula	MW	Mtd.	Reference
1.15(9)	1.149	3′,5′-Bis(trifluoromethyl)acetophenone (C10H6F6O)	256	T	[10]
1.29(9)	1.232	3,5-Bis(trifluoromethyl)benzaldehyde (C9H4F6O)	242	T	[10]
1.3(1)	0.77	Di-MeTerephthalate tetraCl- (C10H6Cl4O4)	332	E	this work
2.1(15)	1.461	Fluoro-p-benzoquinone (C6H3FO2)	126	M	[15]
1.5(2)	1.5	Hexafluoroglutaric acid anhydride (C5F6O3)	222	A	[24]
1.52(10)	1.522	Methanone, bis(pentafluorophenyl)- (C13F10O)	362	T	[25]
1.67(9)	1.67	3,6-Dichloro-phthalic anhydride (C8H2Cl2O3)	217	T	[13]
1.71(6)	1.709	9,10-Anthracenedione, 1-chloro- (C14H7ClO2)	243	T	[26]
1.90(9)	1.904	Dichloromaleic anhydride (C4Cl2O3)	167	T	[13]
1.95(5)	1.951	Cl-triMe-p-benzoquinone (C9H9ClO2)	185	T	[11]
1.96(9)	1.956	1,3-Isobenzofurandione, 4,5,6,7-tetraCl- (C8Cl4O3)	286	T	[11]
2.02(5)	2.016	2-Cl-3,6-diMe-p-benzoquinone (C8H7ClO2)	171	T	[11]
2.11(5)	2.112	2-Chloro-5-methyl-p-benzoquinone (C7H5ClO2)	157	T	[11]
2.15(5)	2.147	2-Cl-5-tBu-p-benzoquinone (C10H11ClO2)	198	T	[11]
2.21(9)	2.207	1,4-Naphthalenedione, 2,3-dichloro- (C10H4Cl2O2)	227	T	[26]
2.23(5)	2.229	2,5-diCl-3,6-diMe-p-benzoquinone (C8H6Cl2O2)	205	T	[11]
2.33(5)	2.333	2,3-diCl-tBu-p-benzoquinone (C10H10Cl2O2)	233	T	[11]
2.40(3)	2.437	2,5-Dichloro-1,4-benzoquinone (C6H2Cl2O2)	177	T	[11, 26]
2.48(3)	2.48	2,6-Dichloro-1,4-benzoquinone (C6H2Cl2O2)	177	T	[11, 26]
2.54(5)	2.545	Me-triCl-p-benzoquinone (C7H3Cl3O2)	226	T	[11]
2.61(5)	2.611	Trichlorobenzoquinone (C6HCl3O2)	212	T	[11]
2.70(10)	2.702	p-Benzoquinone, 2,3,5,6-tetrafluoro- (C6F4O2)	180	A, T	[27]
2.77(5)	2.775	p-Benzoquinone, 2,3,5,6-tetrachloro- (C6Cl4O2)	246	M, A, T	[27]

TABLE A4.4 CHOX Molecules by Molecular Weight (in eV)

EVAL	NIST	Name/Formula	MW	Mtd.	Reference
0.57(5)	0.486	Benzaldehyde, 4-fluoro- (C7H5FO)	124	E	[4]
0.66(4)	0.637	Benzaldehyde, 2-fluoro- (C7H5FO)	124	E	[4]
0.67(3)	0.668	3-FC6H4CHO (C7H5FO)	124	E	[4]
2.1(15)	1.461	Fluoro-p-benzoquinone (C6H3FO2)	126	M	[15]
0.52(5)	0.395	Ethanone, 1-(4-fluorophenyl)- (C8H7FO)	138	E	[4]
0.49(3)	0.442	o-Fluoroacetophenone (C8H7FO)	138	E	[4]
0.58(3)	0.577	Ethanone, 1-(3-fluorophenyl)- (C8H7FO)	138	E	[4]
0.68(9)	0.659	Benzaldehyde, 4-chloro- (C7H5ClO)	141	T	[9, 10]
0.71(9)	0.668	Benzaldehyde, 3-chloro- (C7H5ClO)	141	T	[9, 10]
0.64(5)	0.585	Ethanone, 1-(4-chlorophenyl)- (C8H7ClO)	155	E, T	[1, 9–11]
0.67(5)	0.616	Acetophenone, 3′-chloro- (C8H7ClO)	155	E	[4]
2.11(5)	2.112	2-Chloro-5-methyl-p-benzoquinone (C7H5ClO2)	157	T	[11]
0.44(9)	0.442	2-Propanone, 1,1,1,3,3,3-hexafluoro- (C3F6O)	166	E	[23]
1.90(9)	1.904	Dichloromaleic anhydride (C4Cl2O3)	167	T	[13]
2.02(5)	2.016	2-Cl-3,6-diMe-p-benzoquinone (C8H7ClO2)	171	T	[11]
0.5(2)	0.5	Tetrafluorosuccinic anhydride (C4F4O3)	172	A	[24]
0.85(10)	0.815	Benzaldehyde, 3-(trifluoromethyl)- (C8H5F3O)	174	T	[9, 10]
0.97(10)	0.941	p-CF3C6H4CHO (C8H5F3O)	174	T	[10]
0.98(10)	0.984	Ethanone, 2,2,2-trifluoro-1-phenyl- (C8H5F3O)	174	T	[10]
1.03(9)	0.989	3,5-Dichlorobenzaldehyde (C7H4Cl2O)	175	T	[10]
2.40(3)	2.437	2,5-Dichloro-1,4-benzoquinone (C6H2Cl2O2)	177	T	[11, 26]
2.48(3)	2.48	2,6-Dichloro-1,4-benzoquinone (C6H2Cl2O2)	177	T	[11, 26]
0.22(5)	0.217	2,3,5,6-Tetrafluoroanisole (C7H4F4O)	180	E	[22]
2.60(10)	2.702	p-Benzoquinone, 2,3,5,6-tetrafluoro- (C6F4O2)	180	A, T	[27]
1.95(5)	1.951	Cl-triMe-p-benzoquinone (C9H9ClO2)	185	T	[11]
0.79(5)	0.642	2′-(Trifluoromethyl)acetophenone (C9H7F3O)	188	E	[4]
0.79(5)	0.768	3-CF3-C6H4-COCH3 (C9H7F3O)	188	T, E	[1, 4, 9, 10]
0.90(10)	0.898	4-CF3-C6H4-COCH3 (C9H7F3O)	188	T	[10]
1.10(10)	1.097	Pentafluorobenzaldehyde (C7HF5O)	196	T	[25]
0.55(5)	0.542	Benzene, pentafluoromethoxy- (C7H3F5O)	198	E	[22]
2.15(5)	2.147	2-Cl-5-tBu-p-benzoquinone (C10H11ClO2)	198	T	[11]
0.70(5)	0.62	P-fluorobenzophenone (C13H9FO)	200	T	[9, 10, 12]
0.28(5)	0.282	mF-C6H4CH2OPh (C13H11FO)	202	E	[22]
0.61(10)	0.624	4-CF3-C6H4-COOCH3 (C9H7F3O2)	204	T	[9, 10]
0.74(9)	0.746	3-CF3-C6H4-COOCH3 (C9H7F3O2)	204	T	[9, 10]
0.75(9)	0.75	Methyl 3,5-diCl-benzoate (C8H6Cl2O2)	205	T	[9, 10]

TABLE A4.4 (*Continued*)

EVAL	NIST	Name/Formula	MW	Mtd.	Reference
2.23(5)	2.229	2,5-diCl-3,6-diMe-p-benzoquinone (C8H6Cl2O2)	205	T	[11]
0.88(9)	0.876	2′,3′,4′,5′,6′-Pentafluoroacetophenone (C8H3F5O)	210	T	[25]
2.61(5)	2.611	Trichlorobenzoquinone (C6HCl3O2)	212	T	[11]
0.83(9)	0.833	Methanone, (4-chlorophenyl)phenyl- (C13H9ClO)	217	T	[9, 10]
0.86(9)	0.872	Methanone, (3-chlorophenyl)phenyl- (C13H9ClO)	217	T	[9, 10]
1.67(9)	1.67	3,6-Dichloro-phthalic anhydride (C8H2Cl2O3)	217	T	[13]
0.80(10)	0.776	4,4′-Difluorobenzophenone (C13H8F2O)	218	T	[9, 10]
1.5(2)	1.5	Hexafluoroglutaric acid anhydride (C5F6O3)	222	A	[24]
2.54(5)	2.545	Me-triCl-p-benzoquinone (C7H3Cl3O2)	226	T	[11]
2.21(9)	2.207	1,4-Naphthalenedione, 2,3-dichloro- (C10H4Cl2O2)	227	T	[26]
2.33(5)	2.333	2,3-diCl-tBu-p-benzoquinone (C10H10Cl2O2)	233	T	[11]
1.29(9)	1.232	3,5-Bis(trifluoromethyl)benzaldehyde (C9H4F6O)	242	T	[10]
1.71(6)	1.709	9,10-Anthracenedione, 1-chloro- (C14H7ClO2)	243	T	[26]
2.77(5)	2.775	p-Benzoquinone, 2,3,5,6-tetrachloro- (C6Cl4O2)	246	M, A, T	[27]
0.96(10)	0.958	3-(Trifluoromethyl)benzophenone (C14H9F3O)	250	T	[10]
1.08(9)	1.075	4-Trifluoromethylbenzophenone (C14H9F3O)	250	T	[10]
1.15(9)	1.106	3,5-diCl-benzophenone (C13H8Cl2O)	251	T	[10]
1.15(9)	1.149	3′,5′-Bis(trifluoromethyl)acetophenone (C10H6F6O)	256	T	[10]
1.00(9)	1.006	Methyl 3,5-diCF3-benzoate (C10H6F6O2)	272	T	[10]
1.96(9)	1.956	1,3-Isobenzofurandione, 4,5,6,7-tetraCl- (C8Cl4O3)	286	T	[11]
1.3(1)	0.77	Di-Meterephthalate tetraCl-, (C10H6Cl4O4)	332	E	this work
1.52(10)	1.522	Methanone, bis(pentafluorophenyl)- (C13F10O)	362	T	[25]

REFERENCES

1. Chen, E. C. M.; and Wentworth, W. E. *Mol. Cryst. Liq. Cryst.* **1989**, 171, 271.
2. Zlatkis, A.; Lee, C. K.; Wentworth, W. E.; and Chen, E. C. M. *Anal. Chem.* **1983**, 55, 1596.
3. Kuhn, W. F.; Levins, R. J.; and Lilly, A. C. Jr. *J. Chem. Phys.* **1968**, 49, 5550

4. Wentworth, W. E.; Kao, L. W.; and Becker, R. S. *J. Phys. Chem.* **1975**, 79, 1161.
5. Wentworth, W. E.; Ristau, W. *J. Phys. Chem.* **1969**, 73, 2126.
6. Compton, R. N.; Reinhardt, P. W.; and Schweinler, H. C. *Int. J. Mass Spectrom. Ion Phys.* **1983**, 49, 113.
7. Kebarle, P. and Chowdhury, S. *Chem. Rev.* **1987**, 87, 513.
8. Heinis, T.; Chowdhury, S.; and Kebarle, P. *Org. Mass Spectrom.* **1993**, 28, 358.
9. Huh, C.; Kang, C. H.; Lee, H. W.; Nakamura, H.; Mishima, M.; Tsuno, Y.; and Yamataka, H. *Bull. Chem. Soc. Japan* **1999**, 72, 1083–1091.
10. Mishima, M.; Huh, C.; Lee, H. W.; Nakamura, H.; Fujio, M.; and Tsuno, Y. *Tetrahed. Lett.* **1995**, 36, 2265.
11. Fukuda, E. K. and McIver, R. T. Jr. *J. Amer. Chem. Soc.* **1985**, 107, 2291.
12. Chen, E. C. M. and Wentworth, W. E. *J. Phys. Chem.* **1983**, 87, 45.
13. Paul, G. and Kebarle, P. *J. Amer. Chem. Soc.* **1989**, 111, 464.
14. Compton, R. N.; Reinhardt, P. W.; and Cooper, C. D. *J. Chem. Phys.* **1974**, 60, 2953.
15. Page, F. M. and Goode, G. C. *Negative Ions and the Magnetron*. New York: Wiley, 1969.
16. Marks, J.; Comita, P. B.; and Brauman, J. I. *J. Amer. Chem. Soc.* **1985**, 107, 3718.
17. Römer, B. C. and Brauman, J. I. *J. Amer. Chem. Soc.* **1997**, 119, 2054.
18. Brinkman, E. A.; Berger, S.; Marks, J.; and Brauman, J. I. *J. Chem. Phys.* **1993**, 99, 7586.
19. Heinis, T.; Chowdhury, S.; Scott, S. L.; and Kebarle, P. *J. Amer. Chem. Soc.* **1988**, 110, 400.
20. Schiedt, J. and Weinkauf, R. *J. Chem. Phys.* **1999**, 110, 304.
21. Davico, G. E.; Schwartz, R. L.; Ramond, T. M.; and Lineberger, W. C. *J. Amer. Chem. Soc.* **1999**, 121, 6047.
22. Hernandez-Gill, N.; Wentworth, W. E.; and Chen, E. C. M. *J. Phys. Chem.* **1984**, 88, 6181.
23. McDonald, R. N. and Chowdhury, A. K. *J. Amer. Chem. Soc.* **1985**, 107, 4123.
24. Cooper, C. D. and Compton, R. N. *J. Chem. Phys.* **1974**, 60, 2424.
25. Dillow, G. W. and Kebarle, P. *J. Amer. Chem. Soc.* **1989**, 111, 5592.
26. Heinis, T.; Chowdhury, S.; Scott, S. L.; and Kebarle, P. *J. Amer. Chem. Soc.* **1988**, 110, 400.
27. National Institute of Standards and Technology (NIST). *Chemistry WebBook*, **2003**. Available at http://webbook.nist.gov.

TABLE A5.1 CHON Molecules by Value (in eV)

EVAL	NIST	Name/Formula	MW	Mtd.	Reference
0.50(2)	0.486	Methane, nitro- (CH3NO2)	61	A, E, P, T	[1, 2, 3]
0.56(5)	0.086	Cytosine (C4H5N3O)	111	text	this work
0.56(5)	0.230	Cytosine (C4H5N3O)	111	text	this work
0.65(1)	0.650	Benzene, (nitromethyl)- (C7H7NO2)	137	E	[1]
0.66(10)	0.663	Anisole, 3,5-dimethyl-4-nitro- (C9H11NO3)	181	E, CI	[4, 5]
0.74(4)	0.711	Benzene, 1,3,5-trimethyl-2-nitro- (C9H11NO2)	165	T	[5–8]

TABLE A5.1 (*Continued*)

EVAL	NIST	Name/Formula	MW	Mtd.	Reference
0.75(9)	0.750	Benzenamine, 3,5-dimethyl-4-nitro- (C8H10N2O2)	166	T	[6]
0.79(5)	0.069	Thymine (C5H6N2O2)	126	text	this work
0.80(5)	0.086	Uracil (C4H4N2O2)	112	text	this work
0.80(5)	0.811	Benzene, 1,3-dimethyl-2-nitro- (C8H9NO2)	151	E, T	[6, 7], this work
0.85(9)	0.854	Ethyldiazoacetate (C4H6N2O2)	114		[9]
0.88(5)	0.880	Benzene, 2,4-dimethyl-1-nitro- (C8H9NO2)	151	E, T	[7], this work
0.89(5)	0.911	Benzene, 1-methoxy-4-nitro- (C7H7NO3)	153	T	[6, 8]
0.90(3)	0.924	Benzene, 1-methyl-2-nitro- (C7H7NO2)	137	E, T	[7, 8, 10]
0.90(5)	0.854	Benzene, 1,2-dimethyl-3-nitro- (C8H9NO2)	151	E, T	[7, 8], this work
0.92(3)	0.924	Benzene, 1,2-dimethyl-4-nitro- (C8H9NO2)	151	E, T	[7, 10]
0.92(7)	0.850	4-Cyanobenzoic acid methyl ester (C9H7NO2)	161		[5, 6]
0.92(9)	0.915	p-Nitroaniline (C6H6N2O2)	138	T	[6]
0.95(3)	0.954	Benzene, 1-methyl-4-nitro- (C7H7NO2)	137	E, T	[6–9, 10]
0.95(9)	0.945	m-Nitroaniline (C6H6N2O2)	138	T	[11]
0.98(3)	0.989	Benzene, 1-methyl-3-nitro- (C7H7NO2)	137	T, E	[6–9, 10]
0.98(6)	0.984	Benzenamine, N,N-dimethyl-3-nitro- (C8H10N2O2)	166	E, T	[7]
0.98(8)	1.006	3-Cyanobenzaldehyde (C8H5NO)	131	T	[5, 8]
1.00(1)	1.006	Benzene, nitro- (C6H5NO2)	123	E, P, T, N	[10, 12–14]
1.02(9)	1.015	Phthalimide (C8H5NO2)	147	T	[15]
1.03(10)	1.028	Naphthalene, 2-methyl-1-nitro- (C11H9NO2)	187	T	[16]
1.04(10)	1.040	Benzene, 1-methoxy-3-nitro- (C7H7NO3)	153	T	[14]
1.07(10)	1.067	1,1′-Biphenyl, 2-nitro- (C12H9NO2)	199	T	[11]
1.1(1)	2.168	p-t-Amyl-nitrobenzene radical (C11H15NO2)	193	CI	[21]
1.10(9)	1.101	Naphthalene, 1-methoxy-4-nitro- (C11H9NO3)	203	T	[14]
1.12(10)	1.123	1,1′-Biphenyl, 3-nitro- (C12H9NO2)	199	T	[11]
1.12(9)	1.114	N-methylmaleimide (C5H5NO2)	111	T	[15]
1.12(9)	1.123	1H-pyrrole-2,5-dione, 1-ethyl- (C6H7NO2)	125	T	[15]
1.13(9)	1.127	Benzeneacetonitrile, a-oxo- (C8H5NO)	131	T	[6]
1.13(9)	1.132	Benzonitrile, 4-acetyl- (C9H7NO)	145	T	[17]
1.15(9)	1.153	N-phenylphthalic acid imide (C14H9NO2)	223	T	[15]
1.15(9)	1.153	1H-pyrrole-2,5-dione (C4H3NO2)	97	T	[15]
1.16(9)	1.184	Naphthalene, 2-nitro- (C10H7NO2)	173	T	[16]
1.19(9)	1.201	2,6-diMe-4-CO2Me-nitrobenzene (C10H11NO4)	209	T	[5, 6]
1.20(10)	1.201	1,1′-Biphenyl, 4-nitro- (C12H9NO2)	199	T	[11]

TABLE A5.1 (*Continued*)

EVAL	NIST	Name/Formula	MW	Mtd.	Reference
1.20(5)	1.210	5-Nitro-m-xylene (C8H9NO2)	151	P	[18]
1.20(8)	1.219	Benzonitrile, 4-formyl- (C8H5NO)	131	T	[5, 8]
1.23(9)	1.227	3-O2N-C6H4-COOCH3 (C8H7NO4)	181	T	[6]
1.23(9)	1.227	Naphthalene, 1-nitro- (C10H7NO2)	173	T	[16]
1.26(9)	1.258	4-Cyanobenzophenone (C14H9NO)	207	T	[6]
1.28(9)	1.284	3-Nitrobenzophenone (C13H9NO3)	227	T	[6]
1.30(9)	1.309	2,6-diMe-4-COMe-nitrobenzene (C10H11NO3)	193	T	[5, 6]
1.33(9)	1.327	3-Nitroacetophenone (C8H7NO3)	165	T	[11]
1.36(9)	1.362	N-Phenylmaleimide (C10H7NO2)	173	T	[15]
1.38(9)	1.391	2,6-diMe-4-CHO-nitrobenzene (C9H9NO3)	179	T	[5, 6]
1.39(9)	1.388	2,3-Pyridinedicarboxylic anhydride (C7H3NO3)	149	T	[15]
1.39(9)	1.431	Benzaldehyde, 3-nitro- (C7H5NO3)	151	T	[5, 11]
1.40(9)	1.396	Ethanone, 1-(2-nitrophenyl)- (C8H7NO3)	165	T	[11]
1.41(9)	1.430	2,6-diMe-4-CN-nitrobenzene (C9H8N2O2)	176	T	[5, 6]
1.43(10)	1.431	Anthracene, 9-nitro- (C14H9NO2)	223	T	[16]
1.47(5)	1.470	Benzene, 2-methyl-1,3-dinitro- (C7H6N2O4)	182	T	[7]
1.48(9)	1.461	4-Nitrobenzoic acid methyl ester (C8H7NO4)	181	T	[6, 19]
1.51(5)	—	Guanine (C5H3N5O)	149	text	this work
1.56(9)	1.557	Benzaldehyde, 2-nitro- (C7H5NO3)	151	T	[11]
1.57(5)	1.565	Benzonitrile, 3-nitro- (C7H4N2O2)	148	T	[7]
1.57(9)	1.565	Acetophenone, 4′-nitro- (C8H7NO3)	165	T	[6, 11]
1.57(9)	1.570	4-Nitrobenzophenone (C13H9NO3)	227	T	[6]
1.60(5)	1.600	Benzene, 1-methyl-2,4-dinitro- (C7H6N2O4)	182	P	[18]
1.61(9)	1.609	2-Nitrobenzoic acid nitrile (C7H4N2O2)	148	T	[17]
1.65(10)	1.648	Nitroethylene (C2H3NO2)	73	Misc.	[20]
1.65(10)	1.652	Benzene, 1,2-dinitro- (C6H4N2O4)	168	T	[7, 14]
1.66(10)	1.657	Benzene, 1,3-dinitro- (C6H4N2O4)	168	T	[7, 14]
1.66(9)	1.660	Benzene, 1,3-dimethyl-2,5-dinitro- (C8H8N2O4)	196	T	[5, 6]
1.68(8)	1.691	Benzaldehyde, 4-nitro- (C7H5NO3)	151	T	[5, 6, 11]
1.71(9)	1.726	Benzonitrile, 4-nitro- (C7H4N2O2)	148	T	[5, 6, 8]
1.77(10)	1.765	Naphthalene, 1,5-dinitro- (C10H6N2O4)	218	T	[16]
1.77(5)	1.770	Benzene, 1-methyl-2,3-dinitro- (C7H6N2O4)	182	T	[18]
1.77(5)	1.770	Benzene, 4-methyl-1,2-dinitro- (C7H6N2O4)	182	T	[18]
1.78(8)	1.782	Naphthalene, 1,3-dinitro- (C10H6N2O4)	218	T	[16]
2.00(9)	2.003	Benzene, 1,4-dinitro- (C6H4N2O4)	168	T	[7, 14]
2.05(9)	2.051	3-Nitrophthalic anhydride (C8H3NO5)	193	T	[15]
2.13(9)	2.129	4-Nitrophthalic anhydride (C8H3NO5)	193	T	[15]
2.16(10)	2.160	Benzonitrile, 3,5-dinitro- (C7H3N3O4)	193	T	[8]
2.63(3)	2.628	1,3,5-Trinitrobenzene (C6H3N3O6)	213	M	[22]

TABLE A5.2 CHON Molecules by Molecular Weight (in eV)

EVAL	NIST	Name/Formula	MW	Mtd.	Reference
0.50(2)	0.486	Methane, nitro- (CH3NO2)	61	A, E, P, T	[1, 2, 3]
1.65(10)	1.648	Nitroethylene (C2H3NO2)	73	Misc.	[20]
1.15(9)	1.153	1H-pyrrole-2,5-dione (C4H3NO2)	97	T	[15]
0.56(5)	0.086	Cytosine (C4H5N3O)	111	text	this work
0.56(5)	0.230	Cytosine (C4H5N3O)	111	text	this work
1.12(9)	1.114	N-methylmaleimide (C5H5NO2)	111	T	[15]
0.80(5)	0.086	Uracil (C4H4N2O2)	112	text	this work
0.85(9)	0.854	Ethyldiazoacetate (C4H6N2O2)	114	Misc.	[9]
1.00(1)	1.006	Benzene, nitro- (C6H5NO2)	123	E, P, T, N	[10, 12–14]
1.12(9)	1.123	1H-pyrrole-2,5-dione, 1-ethyl- (C6H7NO2)	125	T	[15]
0.79(5)	0.069	Thymine (C5H6N2O2)	126	text	this work
0.98(8)	1.006	3-Cyanobenzaldehyde (C8H5NO)	131	T	[5, 8]
1.13(9)	1.127	Benzeneacetonitrile, a-oxo- (C8H5NO)	131	T	[6]
1.20(8)	1.219	Benzonitrile, 4-formyl- (C8H5NO)	131	T	[5, 8]
0.65(1)	0.650	Benzene, (nitromethyl)- (C7H7NO2)	137	E	[1]
0.90(3)	0.924	Benzene, 1-methyl-2-nitro- (C7H7NO2)	137	E, T	[7, 8, 10]
0.95(3)	0.954	Benzene, 1-methyl-4-nitro- (C7H7NO2)	137	E, T	[6–10]
0.98(3)	0.989	Benzene, 1-methyl-3-nitro- (C7H7NO2)	137	T, E	[6–10]
0.92(9)	0.915	p-Nitroaniline (C6H6N2O2)	138	T	[6]
0.95(9)	0.945	m-Nitroaniline (C6H6N2O2)	138	T	[11]
1.13(9)	1.132	Benzonitrile, 4-acetyl- (C9H7NO)	145	T	[17]
1.51(5)	—	Guanine(C5H3N5O)	149	text	this work
1.02(9)	1.015	Phthalimide (C8H5NO2)	147	T	[15]
1.57(5)	1.565	Benzonitrile, 3-nitro- (C7H4N2O2)	148	T	[7]
1.61(9)	1.609	2-Nitrobenzoic acid nitrile (C7H4N2O2)	148	T	[17]
1.71(9)	1.726	Benzonitrile, 4-nitro- (C7H4N2O2)	148	T	[5, 6, 8]
1.39(9)	1.388	2,3-Pyridinedicarboxylic anhydride (C7H3NO3)	149	T	[15]
0.80(5)	0.811	Benzene, 1,3-dimethyl-2-nitro- (C8H9NO2)	151	E, T	[6, 7], this work
0.88(5)	0.880	Benzene, 2,4-dimethyl-1-nitro- (C8H9NO2)	151	E, T	[7], this work
0.90(5)	0.854	Benzene, 1,2-dimethyl-3-nitro- (C8H9NO2)	151	E, T	[7, 8], this work
0.92(3)	0.924	Benzene, 1,2-dimethyl-4-nitro- (C8H9NO2)	151	E, T	[7, 10]
1.20(5)	1.210	5-Nitro-m-xylene (C8H9NO2)	151	P	[18]
1.39(9)	1.431	Benzaldehyde, 3-nitro- (C7H5NO3)	151	T	[5, 11]
1.56(9)	1.557	Benzaldehyde, 2-nitro- (C7H5NO3)	151	T	[11]
1.68(8)	1.691	Benzaldehyde, 4-nitro- (C7H5NO3)	151	T	[5, 6, 11]
0.89(5)	0.911	Benzene, 1-methoxy-4-nitro- (C7H7NO3)	153	T	[6, 8]
1.04(10)	1.040	Benzene, 1-methoxy-3-nitro- (C7H7NO3)	153	T	[14]
0.92(7)	0.850	4-Cyanobenzoic acid methyl ester (C9H7NO2)	161	T	[5, 6]
0.74(4)	0.711	Benzene, 1,3,5-trimethyl-2-nitro- (C9H11NO2)	165	T	[5–8]

TABLE A5.2 (*Continued*)

EVAL	NIST	Name/Formula	MW	Mtd.	Reference
1.33(9)	1.327	3-Nitroacetophenone (C8H7NO3)	165	T	[11]
1.40(9)	1.396	Ethanone, 1-(2-nitrophenyl)- (C8H7NO3)	165	T	[11]
1.57(9)	1.565	Acetophenone, 4′-nitro- (C8H7NO3)	165	T	[6, 11]
0.75(9)	0.750	Benzenamine, 3,5-dimethyl-4-nitro- (C8H10N2O2)	166	T	[6]
0.98(6)	0.984	Benzenamine, N,N-dimethyl-3-nitro- (C8H10N2O2)	166	E, T	[7]
1.65(10)	1.652	Benzene, 1,2-dinitro- (C6H4N2O4)	168	T	[7, 14]
1.66(10)	1.657	Benzene, 1,3-dinitro- (C6H4N2O4)	168	T	[7, 14]
2.00(9)	2.003	Benzene, 1,4-dinitro- (C6H4N2O4)	168	T	[7, 14]
1.16(9)	1.184	Naphthalene, 2-nitro- (C10H7NO2)	173	T	[16]
1.23(9)	1.227	Naphthalene, 1-nitro- (C10H7NO2)	173	T	[16]
1.36(9)	1.362	N-phenylmaleimide (C10H7NO2)	173	T	[15]
1.41(9)	1.430	2,6-diMe-4-CN-nitrobenzene (C9H8N2O2)	176	T	[5, 6]
1.38(9)	1.391	2,6-diMe-4-CHO-nitrobenzene (C9H9NO3)	179	T	[5, 6]
0.66(10)	0.663	Anisole, 3,5-dimethyl-4-nitro- (C9H11NO3)	181	CI	[4, 5]
1.23(9)	1.227	3-O2N-C6H4-COOCH3 (C8H7NO4)	181	T	[6]
1.48(9)	1.461	4-Nitrobenzoic acid methyl ester (C8H7NO4)	181	T	[6, 19]
1.47(5)	1.470	Benzene, 2-methyl-1,3-dinitro- (C7H6N2O4)	182	T	[7]
1.60(5)	1.600	Benzene, 1-methyl-2,4-dinitro- (C7H6N2O4)	182	P	[18]
1.77(5)	1.770	Benzene, 1-methyl-2,3-dinitro- (C7H6N2O4)	182	T	[18]
1.77(5)	1.770	Benzene, 4-methyl-1,2-dinitro- (C7H6N2O4)	182	T	[18]
1.03(10)	1.028	Naphthalene, 2-methyl-1-nitro- (C11H9NO2)	187	T	[16]
1.1(1)	2.168	p-t-Amyl-nitrobenzene (C11H15NO2)	193	CI	[21]
1.30(9)	1.309	2,6-diMe-4-COMe-nitrobenzene (C10H11NO3)	193	T	[5, 6]
2.05(9)	2.051	3-Nitrophthalic anhydride (C8H3NO5)	193	T	[15]
2.13(9)	2.129	4-Nitrophthalic anhydride (C8H3NO5)	193	T	[15]
2.16(10)	2.160	Benzonitrile, 3,5-dinitro- (C7H3N3O4)	193	T	[8]
1.66(9)	1.660	Benzene, 1,3-dimethyl-2,5-dinitro- (C8H8N2O4)	196	T	[5, 6]
1.07(10)	1.067	1,1′-Biphenyl, 2-nitro- (C12H9NO2)	199	T	[11]
1.12(10)	1.123	1,1′-Biphenyl, 3-nitro- (C12H9NO2)	199	T	[11]
1.20(10)	1.201	1,1′-Biphenyl, 4-nitro- (C12H9NO2)	199	T	[11]
1.10(9)	1.101	Naphthalene, 1-methoxy-4-nitro- (C11H9NO3)	203	T	[14]
1.26(9)	1.258	4-Cyanobenzophenone (C14H9NO)	207	T	[6]
1.19(9)	1.201	2,6-diMe-4-CO2Me-nitrobenzene (C10H11NO4)	209	T	[5, 6]
2.63(3)	2.628	1,3,5-Trinitrobenzene (C6H3N3O6)	213	M	[22]

TABLE A5.2 (*Continued*)

EVAL	NIST	Name/Formula	MW	Mtd.	Reference
1.77(10)	1.765	Naphthalene, 1,5-dinitro- (C10H6N2O4)	218	T	[16]
1.78(8)	1.782	Naphthalene, 1,3-dinitro- (C10H6N2O4)	218	T	[16]
1.15(9)	1.153	N-Phenylphthalic acid imide (C14H9NO2)	223	T	[15]
1.43(10)	1.431	Anthracene, 9-nitro- (C14H9NO2)	223	T	[16]
1.28(9)	1.284	3-Nitrobenzophenone (C13H9NO3)	227	T	[6]
1.57(9)	1.570	4-Nitrobenzophenone (C13H9NO3)	227	T	[6]

TABLE A5.3 CHONX Molecules by Value (in eV)

EVAL	NIST	Name/Formula	MW	Mtd.	Reference
0.87(9)	0.867	2,6-diMe-4-F-nitrobenzene ($C_8H_8FNO_2$)	169	T	[5, 6]
0.98(9)	1.005	2,6-diMe-4-Cl-nitrobenzene ($C_8H_8ClNO_2$)	185.5	T	[5, 6]
1.01(5)	1.010	4-Fluoro-2-methyl-nitrobenzene- (C7H6FNO$_2$)	155	T	[7]
1.09(5)	1.075	2-Fluoro-nitro-benzene ($C_6H_4FNO_2$)	141	T, E	[7, 10]
1.10(5)	1.119	4-Fluoro-nitro-benzene ($C_6H_4FNO_2$)	141	T, E	[6–8, 10]
1.13(5)	1.162	2-Chloro-nitrobenzene- ($C_6H_4ClNO_2$)	157.5	T, E	[6–8, 10]
1.16(10)	1.162	2-Bromo-nitrobenzene- ($C_6H_4BrNO_2$)	202	T	[8]
1.22(5)	1.236	3-Fluoro-nitro-benzene ($C_6H_4FNO_2$)	141	T	[6–8]
1.24(4)	1.258	4-Chloro-nitrobenzene- ($C_6H_4ClNO_2$)	157.5	T	[6–8]
1.27(5)	1.280	3-Chloro-nitrobenzene- ($C_6H_4ClNO_2$)	157.5	T	[6–8]
1.29(10)	1.292	4-Bromo-nitrobenzene- ($C_6H_4BrNO_2$)	202	T	[8]
1.29(10)	1.318	3-Bromo-nitrobenzene- ($C_6H_4BrNO_2$)	202	T	[8]
1.29(5)	1.292	1,2-Dichloro-nitrobenzene- ($C_6H_3C_{12}NO_2$)	192	T	[7]
1.33(10)	1.331	2-CF3-nitrobenzene ($C_7H_4F_3NO_2$)	191	T	[8]
1.41(5)	1.414	3-CF3-nitrobenzene (C_7H_4F3NO_2)	191	T	[6–8]
1.44(5)	1.444	3,4-Dichloro-nitrobenzene- ($C_6H_3C_{12}NO_2$)	192	T	[7]
1.49(9)	1.500	4-CF3-nitrobenzene ($C_7H_4F_3NO_2$)	191	T	[8]
1.5(1)	1.453	Pentafluoronitrobenzene- ($C_6F_5NO_2$)	213	T, E	[23]
1.53(8)	1.500	3,5-Dichloronitrobenzene ($C_6H_3C_{12}NO_2$)	192	T	[5, 6]
1.79(10)	1.787	3,5-Bis(CF3)nitrobenzene ($C_8H_3F_6NO_2$)	259	T	[7]
1.99(7)	1.987	3,5-Dinitrobenzotrifluoride ($C_7H_3F_3N_2O_4$)	236	T	[24]
2.0(1)	2.000	Trifluoronitrosomethane (CF_3NO)	99	Misc.	[25]

TABLE A5.4 CHONX Molecules by Molecular Weight (in eV)

EVAL	NIST	Name/Formula	MW	Mtd.	Reference
2.0(1)	2.000	Trifluoronitrosomethane (CF3NO)	99	Misc.	[25]
1.09(5)	1.075	2-Fluoro-nitro-benzene (C6H4FNO2)	141	T, E	[7, 10]
1.10(5)	1.119	4-Fluoro-nitro-benzene (C6H4FNO2)	141	T, E	[6–8, 10]
1.22(5)	1.236	3-Fluoro-nitro-benzene (C6H4FNO2)	141	T	[6–8]
1.01(5)	1.010	4-Fluoro-2-methyl-nitrobenzene- (C7H6FNO2)	155	T	[7]

TABLE A5.4 (*Continued*)

EVAL	NIST	Name/Formula	MW	Mtd.	Reference
1.13(5)	1.162	2-Chloro-nitrobenzene- (C6H4ClNO2)	157.5	T, E	[6–8, 10]
1.24(4)	1.258	4-Chloro-nitrobenzene- (C6H4ClNO2)	157.5	T	[6–8]
1.27(5)	1.280	3-Chloro-nitrobenzene- (C6H4ClNO2)	157.5	T	[6–8]
0.87(9)	0.867	2,6-diMe-4-F-nitrobenzene (C8H8FNO2)	169	T	[5, 6]
0.98(9)	1.005	2,6-diMe-4-Cl-nitrobenzene (C8H8ClNO2)	185.5	T	[5, 6]
1.33(10)	1.331	2-CF3-nitrobenzene (C7H4F3NO2)	191	T	[8]
1.41(5)	1.414	3-CF3-nitrobenzene (C7H4F3NO2)	191	T	[6–8]
1.49(9)	1.500	4-CF3-nitrobenzene (C7H4F3NO2)	191	T	[8]
1.29(5)	1.292	1,2-Dichloro-nitrobenzene- (C6H3Cl2NO2)	192	T	[7]
1.44(5)	1.444	3,4-Dichloro-nitrobenzene- (C6H3Cl2NO2)	192	T	[7]
1.53(8)	1.500	3,5-Dichloronitrobenzene (C6H3Cl2NO2)\	192	T	[5, 6]
1.16(10)	1.162	2-Bromo-nitrobenzene- (C6H4BrNO2)	202	T	[8]
1.29(10)	1.292	4-Bromo-nitrobenzene- (C6H4BrNO2)	202	T	[8]
1.29(10)	1.318	3-Bromo-nitrobenzene- (C6H4BrNO2)	202	T	[8]
1.5(1)	1.453	Pentafluoronitrobenzene- (C6F5NO2)	213	T, E	[23]
1.99(7)	1.987	3,5-Dinitrobenzotrifluoride (C7H3F3N2O4)	236	T	[24]
1.79(10)	1.787	3,5-Bis(CF3)nitrobenzene (C8H3F6NO2)	259	T	[7]

REFERENCES

1. Chen, E. C. M.; Welk, N.; Chen, E. S.; and Wentworth, W. E. *J. Phys. Chem. A* **1999**, 103, 9072–9079.
2. Grimsrud, E.; Caldwell, G.; and Kebarle, P. *J. Amer. Chem. Soc.* **1985**, 107, 4627.
3. Compton, R. N.; Reinhardt, P. W.; and Cooper, C. D. *J. Chem. Phys.* **1978**, 68, 4360.
4. Chen, G.; Cooks, R. G.; Corpuz, E.; and Scott, L. T. *J. Amer. Soc. Mass Spectrom.* **1996**, 7, 619.
5. Mishima, M.; Kang, C. H.; Huh, C.; Fujio, M.; and Tsuno, Y. *Chem. Lett.* **1993**, 885.
6. Huh, C.; Kang, C. H.; Lee, H. W.; Nakamura, H.; Mishima, M.; Tsuno, Y.; and Yamataka, H. *Bull. Chem. Soc. Japan* **1999**, 72, 1083–1091.
7. Fukuda, E. K. and McIver, R. T., Jr. *J. Amer. Chem. Soc.* **1985**, 107, 2291.
8. Chowdhury, S.; Heinis, T.; Grimsrud, E. P.; and Kebarle, P. *J. Phys. Chem.* **1986**, 90, 2747.
9. McDonald, R. N. and Chowdhury, A. K. *Tetrahedron* **1986**, 42, 6253.
10. Chen, E. C. M.; Chen, E. S.; Milligan, M. S.; Wentworth, W. E.; and Wiley, J. R. *J. Phys. Chem.* **1992**, 96, 2385.
11. Chowdhury, S.; Kishi, H.; Dillow, G. W.; and Kebarle, P. *Can. J. Chem.* **1989**, 67, 603.
12. Desfrancois, C.; Periquet, V.; Lyapustina, S. A.; Lippa, T. P.; Robinson, D. W.; Bowen, K. H.; Nonaka, H.; and Compton, J. *J. Chem. Phys.* **1999**, 111, 4569.
13. Chen, E. C. M.; Wiley, J. R.; Batten, C. F.; and Wentworth, W. E. *J. Phys. Chem.* **1994**, 98, 88.
14. Kebarle, P. and Chowhury, S. *Chem. Rev.* **1987**, 87, 513.
15. Paul, G. and Kebarle, P. *J. Amer. Chem. Soc.* **1989**, 111, 464.

16. Heinis, T.; Chowdhury, S.; and Kebarle, P. *Org. Mass Spectrom.* **1993**, 28, 358.
17. Chowdhury, S. and Kebarle, P. *J. Amer. Chem. Soc.* **1986**, 108, 5453.
18. Mock, R. S. and Grimsrud, E. P. *J. Amer. Chem. Soc.* **1989**, 111, 2861.
19. Mishima, M.; Huh, C.; Lee, H. W.; Nakamura, H.; Fujio, M.; and Tsuno, Y. **1995**, 36, 2265.
20. Bartmess, J. E. *J. Amer. Chem. Soc.* **1980**, 102, 2483.
21. Burinsky, D. J.; Fukuda, E. K.; and Campana, J. *J. Amer. Chem. Soc.* **1984**, 106, 2270.
22. Page, F. M. and Goode, G. C. *Negative Ions and the Magnetron.* New York: Wiley-Interscience, **1969**.
23. Dillow, G. W. and Kebarle, P. *J. Amer. Chem. Soc.* **1989**, 111, 5592.
24. Sharpe, P.; Alameddin, N. G.; and Richardson, D. E. *J. Amer. Chem. Soc.* **1994**, 116, 11098.
25. Harland, P. W. *Int. J. Mass Spectrom. Ion Phys.* **1977**, 25, 61.
26. National Institute of Standards and Technology (NIST). *Chemistry WebBook*, **2003**. Available at http://webbook.nist.gov.

TABLE A6.1 Bergman Dewar Set

0 Benzene	0+	eV	42 Tetrabenz[de,h,kl,rst]-pentaphene	1.70	eV
1 Naphthalene	0.16	eV	43 Tetrabenz[de,hi,op,st]pentacene	1.68	eV
2 Anthracene	0.68	eV	44 Pyrene	0.61	eV
3 Tetracene	1.08	eV	45 Benz[e]pyrene	0.56	eV
4 Pentacene	1.39	eV	46 Benz[a]pyrene	0.82	eV
5 Phenanthrene	0.30	eV	47 Dibenz[b,e]pyrene	1.10	eV
6 Benz[a]anthracene	0.72	eV	48 Dibenz[a,e]pyrene	0.88	eV
7 Benz[a]tetracene	1.04	eV	49 Dibenz[e,l]pyrene	0.59	eV
8 Benz[a]pentacene	1.38	eV	50 Dibenz[a,h]pyrene	1.16	eV
9 Triphenylene	0.29	eV	51 Dibenz[a,i]pyrene	1.06	eV
10 Dibenz[a,c]anthracene	0.69	eV	52 Naphtho[2,3,b]-pyrene	0.67	eV
11 Dibenz[a,c]naphthacene	1.02	eV	53 Naphtho[2,3,e]-pyrene	1.23	eV
12 Dibenz[a,c]pentacene	1.30	eV	54 Dinaphtho[a,h]pyrene	1.55	eV
13 Dibenz[a,h]anthracene	0.70	eV	55 Benz[x,y,z]heptaphene	1.28	eV
14 Dibenz[a,h]naphthacené	1.03	eV	56 Tribenz[a,e,1]pyrene	1.03	eV
15 Dibenz[a,h]pentacene	1.43	eV	57 Dibenz[cd,lm]perylene	1.25	eV
16 Chrysene	0.42	eV	58 Tetrabenz[a,cd,j,lm]-perylene	1.18	eV
17 Ben[b]chrysene	0.89	eV	59 Dibenz[def,nmno]-chrysene	1.14	eV
18 Dibenz[b,k]chrysene	1.05	eV	60 Pyranthene	1.32	eV
19 Benz[c]phenanthrene	0.58	eV	61 Naphthocenonaphthacene	1.65	eV
20 Dibenz[a,j]anthracene	0.66	eV	62 Fluoranthene	0.82	eV
21 Benz[g]chrysene	0.58	eV	63 Benz[a]fluoranthene	1.25	eV
22 Tribenz[a,c,f]-anthracene	0.61	eV	64 Dibenz[a,f]-fluoranthene	1.58	eV
23 Dibenz[pg,p]chrysene	0.79	eV	65 Benz[b]fluoranthene	0.95	eV
24 Benz[c]chrysene	0.60	eV	66 Benz[j]fluoranthene	1.01	eV
25 Dibenz[b,g]phenanthrene	0.78	eV	67 Benz[k]fluoranthene	0.89	eV
26 Picene	0.54	eV	68 Naphtho-[2,3,j]fluoranthene	1.17	eV
27 Pentaphene	0.75	eV	69 Naphtho-[2,3,k]fluoranthene	0.90	eV
28 Hexaphene	1.06	eV	70 s-Indenofluoranthene	1.33	eV
29 Benz[p]pentaphene	0.78	eV			

TABLE A6.1 (*Continued*)

30 Dibenz[c,m]pentaphene	0.76 eV	71 Benz-indenofluoranthene	1.73 eV
31 Naphtho[2,3-c]pentaphene	0.91 eV	72 Diindenonaphthacene	2.04 eV
32 Perylene	0.98 eV	73 Biphenyl	0.13 eV
33 Benz[a]perylene	1.24 eV	74 p-Terphenyl	0.27 eV
34 Dibenz[a,j]perylene	1.45 eV	75 p-Quarterphenyl	0.41 eV
35 Dibenz[a,n]perylene	1.28 eV	76 Dibenz[de,mn]-tetracene	1.44 eV
36 Benz[g,h.i]perylene	0.89 eV	77 Benzbenz[c]phenanthrene	0.58 eV
37 Dibenz[b,pqr]perylene	0.86 eV	78 Pentabenzpentaphene	1.90 eV
38 Benz[qr]naphthopentacene	1.03 eV	79 Hexacene	1.60 eV
39 Coronene	0.80 eV	80 Styrene	0.10 eV
40 Dibenz[fg,qr]-naphthacene	1.10 eV	81 Azulene	0.84 eV
41 Dibenz[bc,hl]-coronene	1.41 eV		

See Appendix II.

TABLE A6.2 Values Different from NIST Values

EVAL	NIST	Name/Formula	MW	Mtd.	Reference
		From Tables A2.1 and A2.3 CHX (in eV)			
0.10(5)	—	Styrene (C8H8)	104	E	[2]
0.16(2)	−0.200	Naphthalene (C10H8)	128	E	[1]
0.40(5)	0.890	Biphenylene (C12H8)	152	CI	[3]
0.42(4)	0.397	Chrysene (C18H12)	228	E	[1]
0.58(1)	0.545	Benzo[c]phenanthrene (C18H12)	228	E	[1]
0.60(5)	0.550	Anthracene, 1-methyl- (C15H12)	192	CI	[3]
0.61(2)	0.500	Pyrene (C16H10)	202	E	[1]
0.67(3)	0.591	Dibenz[a,j]anthracene (C22H14)	278	E	[1]
0.68(2)	0.530	Anthracene (C14H10)	178	E, P, T	[1, 4, 5]
0.69(3)	0.595	Dibenz[a,h]anthracene (C22H14)	278	E	[1]
0.69(3)	—	Dibenz[a,c]anthracene (C22H14)	278	E	[1]
0.72(1)	0.390	Benz[a]anthracene (C18H12)	228	E	[1]
0.80(10)	0.550	1,3,5,7-Cyclooctatetrene (C8H8)	104	E, P, T	[1, 6–8]
0.80(5)	0.403	Acenaphthylene (C12H8)	152	E	[1]
0.80(5)	0.470	Coronene (C24H12)	300	P, CI	[9, 10]
0.82(5)	0.630	Fluoranthene (C16H10)	202	E	[1]
0.84(5)	0.694	Azulene (C10H8)	128	E, P, T	[1, 11, 12]
0.2(1)	0.100	Ethene, 1,1-dichloro- (C2H2Cl2)	97	E	[1], this work
0.30(5)	0.094	Benzene, 1,2-dichloro- (C6H4Cl2)	147	E	[3], this work
0.46(5)	0.400	Trichloroethylene (C2HCl3)	131.5	E	[1, 4], this work
0.49(3)	0.340	Benzene, 1,3,5-trichloro- (C6H3Cl3)	181.5	E	[3], this work
0.70(5)	0.640	Tetrachloroethylene (C2Cl4)	166	E	[1, 4], this work
0.66(3)	0.450	Benzene, 1,2,4,5-tetrachloro- (C6H2Cl4)	216	E	[1, 3], this work
0.73(5)	0.434	Benzene, pentafluoro- (C6HF5)	168	E, T	[6, 7], this work
0.80(10)	0.720	1-Bromopyrene (C16H9Br)	281	E, T	[10]
0.86(2)	0.520	Benzene, hexafluoro- (C6F6)	186	E, T, P	[6, 7, 9, 11]

TABLE A6.2 (*Continued*)

EVAL	NIST	Name/Formula	MW	Mtd.	Reference
0.90(10)	0.610	9-Bromoanthracene (C14H9Br)	257	CI	[10]
0.90(5)	0.729	Benzene, pentachloro- (C6HCl5)	250.5	E, T	[12]
1.0(2)	0.824	1,1′-Biphenyl-decafluoro- (C12F10)	334	T	[6]
1.0(2)	0.859	Perfluorotoluene- (C7F8)	236	T	[6]
1.00(5)	0.815	Benzene, chloropentafluoro- (C6ClF5)	202.5	E, T	[6]
1.07(5)	1.062	Perfluoromethylcyclohexane (C7F14)	350	N, T	[14]
1.15(5)	0.915	Benzene, hexachloro- (C6Cl6)	285	E, T	[1, 12], this work
1.2(1)	0.110	Methyl iodide (CH3I)	142	A	[15, 16]
1.3(1)	1.149	Benzene, bromopentafluoro- (C6BrF5)	247	E	[6]
1.5(1)	1.414	Benzene, pentafluoroiodo- (C6F5I)	294	T	[6]
1.65(10)	0.622	Chloroform (CHCl3)	119.5	M, T	[11, 18]
1.8(2)	1.479	Ethane, hexachloro- (C2Cl6)	237	M	[11]
1.80(15)	0.911	Methane, bromotrifluoro- (CBrF3)	149	A	[16]
1.80(15)	1.101	Trichloromonofluoromethane (CCl3F)	137.5	A	[17, 18]
2.04(10)	0.805	Carbon tetrachloride (CCl4)	154	M, T	[11, 18]
2.20(20)	1.570	Methane, trifluoroiodo- (CF3I)	196	A	[19–21]
		From Table A3.1 CHNX (in eV)			
0.50(5)	0.450	1,3,5-Triazine (C3H3N3)	144	Ered	[7]
0.95(5)	0.012	Adenine (C5H5N5)	135	DB	[1–4]
2.95(10)	3.166	Tetracyanoethylene (C6N4)	128	M, T	[13, 19]
		From Tables A4.1 and A4.3 CHOX (in eV)			
0.338(2)	0.334	Acetophenone (C8H8O)	120	E	[4]
0.39(2)	0.373	Benzaldehyde, 4-methyl- (C8H8O)	120	E	[4]
0.457(5)	0.429	Benzaldehyde (C7H6O)	106	E	[4]
0.68(5)	0.620	Benzophenone (C13H10O)	1820	E, T	[1, 7–12]
0.85(5)	0.85	Ethanone, 1-(2-hydroxyphenyl)- (C8H8O2)	136	T	[11]
0.9(1)	0.823	Cinnamylaldehyde (C9H8O)	132	E	[1]
1.03(9)	0.971	M-phthalaldehyde (C8H6O2)	134	T	[9, 10]
1.07(9)	1.158	4-HC(O)-C6H4-COOCH3 (C9H8O3)	164	T	[9, 10]
1.30(9)	1.236	1,4-Benzenedicarboxaldehyde (C8H6O2)	134	T	[9, 10]
0.52(5)	0.395	Ethanone, 1-(4-fluorophenyl)- (C8H7FO)	138	E	[4]
0.49(3)	0.442	o-Fluoroacetophenone (C8H7FO)	138	E	[4]
0.57(5)	0.486	Benzaldehyde, 4-fluoro- (C7H5FO)	124	E	[4]
0.64(5)	0.585	Ethanone, 1-(4-chlorophenyl)- (C8H7ClO)	155	E, T	[1, 9–11]
0.67(5)	0.616	Acetophenone, 3′-chloro- (C8H7ClO)	155	E	[4]
0.70(5)	0.620	4-Fluorobenzophenone (C13H9FO)	200	T	[9, 10, 12]
0.66(4)	0.637	Benzaldehyde, 2-fluoro- (C7H5FO)	124	E	[4]

TABLE A6.2 (*Continued*)

EVAL	NIST	Name/Formula	MW	Mtd.	Reference
0.79(5)	0.642	2′-(Trifluoromethyl)acetophenone (C9H7F3O)	188	E	[4]
0.68(9)	0.659	Benzaldehyde, 4-chloro- (C7H5ClO)	141	T	[9, 10]
0.71(9)	0.668	Benzaldehyde, 3-chloro- (C7H5ClO)	141	T	[9, 10]
0.85(10)	0.815	Benzaldehyde, 3-(CF3)- (C8H5F3O)	174	T	[9, 10]
0.97(10)	0.941	p-CF3C6H4CHO (C8H5F3O)	174	T	[10]
1.03(9)	0.989	3,5-Dichlorobenzaldehyde (C7H4Cl2O)	175	T	[10]
1.15(9)	1.106	3,5-diCl-benzophenone (C13H8Cl2O)	251	T	[10]
1.29(9)	1.232	3,5-Bis(CF3)benzaldehyde (C9H4F6O)	242	T	[10]
1.3(1)	0.77	Di-MeTerephthalate tetraCl- (C10H6Cl4O4)	332	E	this work
2.1(15)	1.461	Fluoro-p-benzoquinone (C6H3FO2)	126	M	[15]
		From Table A5.1 CHON (in eV)			
0.56(5)	0.086	Cytosine (C4H5N3O)	111	text	this work
0.56(5)	0.230	Cytosine (C4H5N3O)	111	text	this work
0.79(5)	0.069	Thymine (C5H6N2O2)	126	text	this work
0.80(5)	0.086	Uracil (C4H4N2O2)	112	text	this work
0.90(5)	0.854	Benzene, 1,2-dimethyl-3-nitro- (C8H9NO2)	151	E, T	[7, 8], this work
0.92(7)	0.850	4-Cyanobenzoic acid methyl ester (C9H7NO2)	161	Misc.	[5, 6]
1.1(1)	2.168	p-t-amyl-nitrobenzene radical (C11H15NO2)	193	CI	[21]
1.51(5)	—	Guanine(C5H3N5O)	149	text	this work

TABLE A6.3 Unpublished or Updated Gas Phase Values not in NIST Tables

EVAL	Name/Formula	Mtd.	Table
0.00(20)	Pyridine (C_5H_5N)	E	12.01
0.13(5)	C_6H_5F	E	11.06
0.07((5)	$C_6H_5F^*$	E	11.06
0.10(5)	Styrene (C_8H_8)	E	10.06
0.14(5)	$1,2CH_3C_6H_4Cl$	E	11.07
0.17(10)	C_6H_5Cl	E	11.07
0.19(5)	p-$C_6H_4F_2^*$	E	11.06
0.20(5)	Benzyl acetate	E	10.10
0.20(5)	Ethyl acetate	E	10.10
0.21(5)	Acetic anhydride	E	10.10
0.25(5)	p-$C_6H_4F_2$	E	11.06
0.25(10)	$1,2CH_3C_6H_4Br$	E, CEC	11.10
0.45(15)	CF_3CHF_2	E	11.05
0.26(15)	$CF_3CHF_2^*$	E	11.05
0.30(10)	C_6H_5Br	E, CEC	11.10

TABLE A6.3 (*Continued*)

EVAL	Name/Formula	Mtd.	Table
0.32(10)	Dibenzofuran ($C_{12}H_8O$)	E	11.13
0.38(15)	CF_3CH_2F	E	11.05
0.38(5)	1,4$CF_3C_6H_4Cl$	E	11.07
0.40(5)	1,2,3,4-$C_6H_2F_4^*$	E	11.06
0.52(5)	1,2,3,4-$C_6H_2F_4$	E	11.06
0.40(5)	1,2$CF_3C_6H_4C_1$ (possible excited state)	E, CEC	11.10
0.42(10)	Xanthene ($C_{13}H_{10}O$)	E	11.13
0.50(15)	CF_3CHClF (possible excited state)	E, CEC	11.05
0.50(15)	Benzoquinoline ($C_{13}H_9N$)	E	12.01
0.50(10)	1,2FC_6H_4Br	E, CEC	11.10
0.50(15)	CHC_1F_2	E	11.05
0.50(5)	1,3,4-$FC_6H_3C_{12}$	E	11.07
0.55(5)	1,2,4$CF_3C_6H_3C_{12}$	E	11.07
0.13(10)	Cytosine*** ($C_4H_5N_3O$)	Est. PES	12.03
0.30(10)	Cytosine**($C_4H_5N_3O$)	Est. PES	12.03
0.45(10)	Cytosine* ($C_4H_5N_3O$)	Est. PES	12.03
0.60(10)	Cytosine (C4H3N3O)	Est. PES	12.03
0.60(15)	CF_3CHClF (possible excited state)	E	11.05
0.61(5)	Benzophenone-4-methoxy ($C_{14}H_{12}O_2$)	E	10.09
0.64(5)	Benzophenone-4-methyl ($C_{14}H_{12}O$)	E	10.09
0.64(5)	Benzophenone-4-ethyl ($C_{15}H_{14}O$)	E	10.09
0.69(3)	Dibenz[a,c]anthracene ($C_{22}H_{14}$)	E	10.06
0.70(10)	1,3$CF_3C_6H_4Br$	E, CEC	11.10
0.77(5)	m-hydroxyacetophenone	E	10.10
0.78(10)	Benzophenone-p-Cl ($C_{13}H_9OC_1$)	E	10.09
0.12(10)	Thymine***($C_5H_5N_2O_2$)	Est. PES	12.03
0.30(10)	Thymine**($C_5H_5N_2O_2$)	Est. PES	12.03
0.50(10)	Thymine* ($C_5H_5N_2O_2$)	Est. PES	12.03
0.79(10)	Thymine ($C_5H_5N_2O_2$)	Est. PES	12.03
0.15(10)	Uracil*** ($C_4H_3N_2O_2$)	Est. PES	12.03
0.30(10)	Uracil** ($C_4H_3N_2O_2$)	Est. PES	12.03
0.50(10)	Uracil* ($C_4H_3N_2O_2$)	Est. PES	12.03
0.80(10)	Uracil ($C_4H_3N_2O_2$)	Est. PES	12.03
0.8(1)	Ethyl trifluroacetate	E	10.10
0.90(10)	Benzophenone-p-Br ($C_{13}H_9OBr$)	E	10.09
1.02(5)	$CF_3C_5F_9$	NIMS	11.06
1.06(5)	$CF_3C_6F_{11}$	NIMS	11.06
1.09(5)	3-Cl,6-Me-nitrobenzene ($C_7H_6C_1NO_2$)	E	10.12
1.10(20)	Benzophenone-p-I ($C_{13}H_9O_1$)	E	10.09
1.30(15)	CH_2Cl_2	M	10.01
1.50(20)	Benzophenone-p-NO2 ($C_{13}H_9NO_3$)	E	10.09
2.06 (10)	CBr4	M	10.01

E = electron capture detector.
E, CEC = electron capture detector and CURES-EC.
M = magnetron.
NIMS = negative-ion mass spectrometry.
Analyze PES = assign peaks found in PES spectra.
* First excited state.
** Second excited state.
*** Third excited state.

TABLE A6.4 Values for Adenine, Guanine, Cytosine, Uracil, Thymine, and Their Hydrates (in eV)

EVAL	NIST	Name/Formula
1.51(5)	—	Guanine($C_5H_3N_5O$)
0.95(5)	0.012	Adenine ($C_5H_5N_5$)
0.80(5)	0.086	Uracil ($C_4H_4N_2O_2$)
1.00(5)	0.370	H_2O uracil radical anion ($C_4H_6N_2O_3-$)
1.20(5)	0.700	$(H_2O)_2$ uracil radical anion ($C_4H_8N_2O_4-$)
1.40(5)	0.940	$(H_2O)_3$ uracil radical anion ($C_4H_{10}N_2O_5-$)
0.79(5)	0.069	Thymine ($C_5H_6N_2O_2$)
0.99(5)	0.300	H_2O thymine radical anion ($C_5H_8N_2O_3-$)
1.19(5)	0.570	$(H_2O)_2$ thymine radical anion ($C_5H_{10}N_2O_4-$)
1.39(5)	0.820	$(H_2O)_3$ thymine radical anion ($C_5H_{12}N_2O_5-$)
1.59(5)	1.000	$(H_2O)_4$ thymine radical anion ($C_5H_{14}N_2O_6-$)
1.79(5)	1.170	$(H_2O)_5$ thymine radical anion ($C_5H_{16}N_2O_7-$)
0.56(5)	0.086	Cytosine ($C_4H_5N_3O$)
0.56(5)	0.230	Cytosine ($C_4H_5N_3O$)
0.76(5)	0.320	H_2O cytosine radical anion ($C_4H_7N_3O_2-$)
0.96(5)	0.490	$(H_2O)_2$ cytosine radical anion ($C_4H_9N_3O_3-$)
1.16(5)	0.690	$(H_2O)_3$ cytosine radical anion ($C_4H_{11}N_3O_4-$)
1.36(5)	0.850	$(H_2O)_4$ cytosine radical anion ($C_4H_{13}N_3O_5-$)
1.56(5)	1.040	$(H_2O)_5$ cytosine radical anion ($C_4H_{15}N_3O_6-$)

TABLE A6.5 Values for Charge Transfer Complex Acceptors not in NIST Tables

Selected	CT	Compound
1.55(5)	1.47(10)	Dimethylaminobenzoquinone
1.65(10)	1.65(10)	Tetracyanoxylene
1.66(5)	1.57(10)	2,5-Di-t-butylbenzoquinone
1.67(10)	1.67(10)	2,4-Dinitrotoluene
1.70(10)	1.70(10)	Tetrachlorophthalic anhydride
1.76(5)	1.69(10)	Methoxybenzoquinone
2.18(10)	2.18(10)	1,2,3-Trinitrobenzene
2.19(10)	2.19(10)	Pentacyanotoluene
2.19(5)	2.21(10)	Bromobenzoquinone
2.20(10)	2.20(10)	Iodobenzoquinone
2.20(5)	2.20(10)	2,3-dichloro-1,4-naphthoquinone
2.21(5)	2.29(10)	Acetylbenzoquinone
2.25(5)	2.35(10)	Di-bromo-dimethylbenzoquinone
2.30(10)	2.30(10)	Bromochloro-1,4-naphthoquinone
2.30(5)	2.22(10)	Cyanobenzoquinone
2.32(5)	2.38(10)	Trifluoromethylbenzoquinone
2.37(10)	2.37(10)	2,4,6-Trinitro-p-xylene
2.40(10)	2.40(10)	Trinitrofluorenone
2.41(10)	2.41(10)	Dibromo-1,4-naphthoquinone
2.52(10)	2.52(10)	2,4,6-Trinitroanisole
2.57(10)	2.57(10)	2,4,6-Trinitrotoluene

TABLE A6.5 (*Continued*)

Selected	CT	Compound
2.59(10)	2.59(10)	1,2,4-Trinitrobenzene
2.61(10)	2.61(10)	Di-bromobenzoquinone
2.63(10)	2.63(10)	Nitrobenzoquinone
2.64(10)	2.64(10)	2,4,6-Trinitrobenzoic acid
2.66(5)	2.66(10)	2,3-dicyano-1,4-naphthoquinone
2.67(10)	2.67(10)	o-tetrachlorobenzoquinone
2.78(10)	2.78(10)	2,3-dicyanobenzoquinone
2.78(5)	2.78(10)	p-tetrabromobenzoquinone
2.81(10)	2.81(10)	1,2,3,5-Tetranitrobenzene
2.83(10)	2.83(10)	p-tetraiodobenzoquinone
2.87(10)	2.87(10)	Tetracyanobenzoquinone

The values with the lower uncertainty are the average of the charge transfer complex value and the reduction potential value.

TABLE A6.6 Values for Chlorinated Hydrocarbons from Reduction Potentials and CURES-EC (in eV) [22 and this work]

Species	$E_a(E_{1/2})$	E_a(CEC)	Difference
$C_6H_5C_1$	0.17	0.10	0.07
2-$C_{10}H_7C_1$	0.30	0.35	−0.05
2,2′-$C_{12}H_8C_{12}$	0.37	0.60	−0.23
3-$C_{12}H_9C_1$	0.39	0.41	−0.02
2-$C_{12}H_9C_1$	0.40	0.38	0.02
4-$C_{12}H_9C_1$	0.44	0.40	0.04
2,4′-$C_{12}H_8C_{12}$	0.46	0.48	−0.02
1,2,4-$C_6H_3C_{13}$	0.47	0.49	−0.02
3,3′-$C_{12}H_8C_{12}$	0.47	0.66	−0.19
4,4′-$C_{12}H_8C_{12}$	0.50	0.66	−0.16
1,6-$C_{10}H_6C_{12}$	0.51	0.50	0.01
1,2,3-$C_6H_3C_{13}$	0.51	0.45	0.06
1,7-$C_{10}H_6C_{12}$	0.52	0.49	0.03
2,4-$C_{12}H_8C_{12}$	0.52	0.66	−0.14
2,3-$C_{10}H_6C_{12}$	0.54	0.50	0.04
2,3-$C_{12}H_8C_{12}$	0.54	0.61	−0.07
2,3-$C_{12}H_8C_{12}$	0.54	0.61	−0.07
1,5-$C_{10}H_6C_{12}$	0.55	0.51	0.04
1,3-$C_{10}H_6C_{12}$	0.56	0.52	0.04
1,4-$C_{10}H_6C_{12}$	0.56	0.51	0.05
2,5-$C_{12}H_8C_{12}$	0.56	0.66	−0.10
1,2-$C_{10}H_6C_{12}$	0.58	0.64	−0.06
2,6-$C_{10}H_6C_{12}$	0.58	0.64	−0.06
2,7-$C_{10}H_6C_{12}$	0.58	0.64	−0.06

TABLE A6.6 (*Continued*)

Species	$E_a(E_{1/2})$	E_a(CEC)	Difference
2,6,2′,6′-$C_{12}H_6C_{14}$	0.58	0.61	−0.03
3,5-$C_{12}H_8C_{12}$	0.60	0.65	−0.05
1,8-$C_{10}H_6C_{12}$	0.61	0.65	−0.04
3,4-$C_{12}H_8C_{12}$	0.63	0.66	−0.03
2,3,6-$C_{12}H_7C_{13}$	0.66	0.70	−0.04
1,2,3,5-$C_6H_2C_{14}$	0.68	0.70	−0.02
1,2,3,4-$C_6H_2C_{14}$	0.73	0.72	0.01
1,3,6-$C_{10}H_5C_{13}$	0.73	0.79	−0.06
1,3,7-$C_{10}H_5C_{13}$	0.73	0.78	−0.05
1,2,6-$C_{10}H_5C_{13}$	0.74	0.78	−0.04
1,4,6-$C_{10}H_5C_{13}$	0.74	0.77	−0.03
1,2,7-$C_{10}H_5C_{13}$	0.75	0.74	0.01
1,6,7-$C_{10}H_5C_{13}$	0.76	0.76	0.00
2,4,5-$C_{12}H_7C_{13}$	0.76	0.89	−0.13
1,2,6-$C_{10}H_5C_{13}$	0.78	0.75	0.03
1,3,5-$C_{10}H_5C_{13}$	0.78	0.77	0.01
1,2,4-$C_{10}H_5C_{13}$	0.80	0.75	0.05
2,5,2′,5′-$C_{12}H_6C_{14}$	0.80	0.80	0.00
1,2,3-$C_{10}H_5C_{13}$	0.81	0.74	0.07
1,3,8-$C_{10}H_5C_{13}$	0.82	0.77	0.05
1,4,5-$C_{10}H_5C_{13}$	0.82	0.75	0.07
2,3,5-$C_{12}H_7C_{13}$	0.82	0.83	−0.01
2′,5′,2,4,5-$C_{12}H_5C_{15}$	0.83	0.78	0.05
1,2,8-$C_{10}H_5C_{13}$	0.85	0.74	0.11
2′,4′6′,2,4,6-$C_{12}H_4C_{16}$	0.89	0.96	−0.07
3,4,5-$C_{12}H_7C_{13}$	0.90	0.86	0.04
2,3,5,6-$C_{12}H_6C_{14}$	0.91	0.91	0.00
2,3,4,6-$C_{12}H_6C_{14}$	0.92	0.88	0.04
3,4,3′,4′-$C_{12}H_6C_{14}$	0.94	1.05	−0.11
1,2,3,7-$C_{10}H_4C_{14}$	0.97	0.92	0.05
1,2,4,6-$C_{10}H_4C_{14}$	0.97	1.02	−0.05
1,3,5,7-$C_{10}H_4C_{14}$	0.97	1.03	−0.06
3,5,3′,5′-$C_{12}H_6C_{14}$	0.98	1.02	−0.04
1,4,6,7-$C_{10}H_4C_{14}$	0.99	1.02	−0.03
1,2,3,5-$C_{10}H_4C_{14}$	1.00	1.00	0.00
1,2,3,4-$C_{10}H_4C_{14}$	1.02	0.98	0.04
2,3,4,5-$C_{12}H_6C_{14}$	1.02	1.06	−0.04
2,3,4,5,7-$C_{12}H_5C_{15}$	1.03	1.03	0.00
1,3,5,8-$C_{12}H_6C_{14}$	1.04	1.02	0.02
2′,4′,5′,2,4,5-$C_{12}H_4C_{16}$	1.04	1.16	−0.12
1,4,5,8-$C_{10}H_4C_{14}$	1.07	1.00	0.07
1,2,3,5,7-$C_{10}H_3C_{15}$	1.07	1.29	−0.22
$C_{12}Cl_{10}$	1.39	1.57	−0.18
$C_{10}C_{18}$	1.57	1.63	−0.06

TABLE A6.7 Values for Biological Compounds from Reduction Potentials

Molecule	E_a (eV)	CEC	Table
Cyanopyridine $C_6H_5N_2$	1.05(5)	1.08	12.01
Hydroxyquinoline C_7H_8NO	0.85(5)	0.85	12.01
Purine	1.04(5)	1.05	12.06
Vitamin K	1.81(10)	1.83	12.06
1,2,4,5-Tetrazine	1.67(10)	1.67	12.06
Riboflavin	1.60(10)	1.66	12.06
1,2,4-Triazine	0.91(10)	0.89	12.06
Vitamin A acetate	0.90(10)	0.91	12.06
Vitamin A alcohol	0.75(10)	0.76	12.06
Quinoline	0.50(15)	0.45	12.06
Benzoquinoline	0.50(15)	0.55	12.01
Pyrazine	0.36(5)	0.38	12.06
Pyradazine	0.31(5)	0.33	12.06
Pyrimidine	0.20(5)	0.22	12.06
Butadiene	0.00(10)	0.00	12.05
Hexatriene	0.44(10)	0.41	12.05
Octatetrene	0.66(10)	0.64	12.05
Decapentaene	0.85(10)	0.80	12.05

Author Index

Abdoul-Carmine, H., 135, 136, 327, 328, 361
Abouaf, R., 166, 230
Achiba, Y., 191
Adams, N.G., 138
Aflatooni, K., 327
Alajajian, S.H., 138
Alameddin, N.G., 377
Albyns, K., 73, 294
Alge, E., 138
Allan, M.J., 137, 167
Allouche, A.R., 191
Almeida, D.P., 233
Altshuler, A.P., 294
Anbar, M., 327
Andersen, T., 45, 190, 352, 353
Andersson, K.T., 353
Ansdell, D.A., 191
Arnold, C.C., 354
Arnold, D.W., 192, 231
Arnold, S.T., 233, 354
Arshardi, 295
Aten, A., 360
Aubert-Frecon, M., 192
Austin, T.J., 328
Ayala, J.A., 44, 167, 230
Ayotte, P., 360
Ayres, P.W., 230
Azria, R., 167, 230

Babcock, L.M., 233
Babic, D.S., 230
Baeda, A.P.M., 102, 230
Baguenard, B., 192
Bailey, C.G., 362
Bailey, T.L., 102, 191, 231
Balling, P., 353
Bansal, K.M., 137
Barden, C.J., 134
Bartlett, R.J., 20, 263
Bartmess, J.E., 136, 377
Bass, A.D., 231
Bates, D.R., 231
Batista, V.S., 167, 230
Batten, C.F., 73, 101, 233, 263, 294, 359, 376
Bauschlicher, C.W., 230
Beauchamp, J.L., 328
Beck, D.R., 353
Becker, R.S., 43, 44, 74, 101, 265, 361, 370, 376
Belic, D.S., 230
Bengali, A.A., 354
Bengtsson, G., 328
Bennett, J.C., 328
Bennett, R.A., 102, 230
Berger, S., 370, 376
Bergman, I., 265
Berkout, V., 20, 101
Berkowitz, J., 136, 231
Berry, R.S., 135, 190
Berthold, H., 43, 327
Berzinsh, U., 353
Besler, B., 327
Biakowski, S.E., 294
Biernbaum, V.M., 232, 265
Bilodeau, R.C., 352, 353, 354
Blondel, C., 352, 353
Boesch, S.E., 134, 136
Bognar, J.A., 294, 359
Bordas, Ch., 191
Borden, W.T., 264
Boulas, P., 45, 265
Bowen, K.H., 135, 136, 231, 232, 263, 264, 327, 328, 353, 354, 376
Bowie, J.H., 328

The Electron Capture Detector and the Study of Reactions with Thermal Electrons
by E. C. M. Chen and E. S. D. Chen
ISBN 0-471-32622-4

Bradforth, S.E., 192, 231
Branscomb, L.M., 6, 44, 135, 230
Brauman, J.I., 135, 233, 231, 264, 370, 376
Breyer, F., 353
Briegleb, G., 21, 43, 167
Brinkman, E.A., 135, 362, 370, 376
Breyer, F., 353
Brodie, C.A., 352
Brown, S.T., 134
Brown, T., 45, 191
Bruning, F., 232
Brunot, A., 191
Buckman, S.J., 191
Burch, D.S., 102, 135
Burdon, J., 138
Burinsky, D.J., 136, 265, 377
Burrow, P.D., 20, 134
Burton, G.R., 354

Cacciani, P., 353
Cai, H., 45, 101
Cai, Z., 328
Calabrese, D., 191, 353
Caldwell, G., 263, 328, 376
Caledonia, G.E., 230
Calvin, M., 135
Campana, J.E., 136, 265, 377
Campbell, B.H., 294
Campbell, S., 328
Carles, S., 135, 136
Carman, H.S., 135, 136
Carpenter, D.L., 191, 353, 354
Carr, S.D., 45, 73, 135, 136, 166, 232
Carson R., 43
Carter, J.G., 20, 233
Casey, S.M., 73, 166, 354
Celotta, R.J., 102, 230
Cha, C.-Y., 354
Chandra, A.K., 328
Chantry, P.J., 136, 232
Chen, C.L., 136, 191
Chen, E.C.M., 6, 20, 45, 73, 74, 101, 102, 134, 135, 136, 137, 166, 167, 190, 230, 232, 233, 263, 265, 293, 294, 327, 328, 359, 361, 370, 376, 376
Chen, E.S., 43, 44, 73, 74, 101, 102, 134, 135, 136, 137, 138, 166, 167, 191, 232, 263, 293, 294, 327, 328, 359, 361, 376
Chen, G.D., 136, 166, 359, 376
Cheng, C-L., 354
Cherhnovsky, O., 191, 231, 353
Chew, F., 328
Chowdhury, S., 6, 20, 134, 233, 232, 235, 264, 265, 294, 360, 362, 376, 377
Christodoulides, A.A., 6, 20, 135, 166, 191, 230, 263, 294
Christophorou, L.G., 6, 20, 44, 45, 74, 135, 166, 191, 230, 233, 264, 264, 294, 359
Chu, Y., 232
Chupka, W.A., 136, 231
Chutjian, A., 138
Clark, C.W., 191
Clementi, E., 137
Clemons, C.A., 294
Coe, J.V., 232, 231, 354
Colson, A-O., 328
Comita, P.B., 231, 264, 370 376
Compton, R.N., 44, 102, 137, 138, 166, 233, 232, 230, 263, 264, 327, 360, 362, 370 376, 376
Conceicao, J., 191, 231
Cooks, R.G., 136, 166, 376
Cooper, C.D., 44, 135, 233, 263, 264, 294, 360, 362, 370, 376, 376
Corderman, R.R., 191
Corpuz, E., 136, 166, 376
Cosby, P.C., 232
Cottin, M., 191
Coulson, C.A., 167
Covington, A.M., 191, 353, 354
Cowles, D.C., 231
Cox, D.M., 231
Craycraft, M.J., 191, 231
Crick, F.H.C., 43
Crocker, L., 73, 135, 265
Crompton, R.W., 20, 137
Curl, R.F., 192
Curtis, L.A., 137

Dalgarno, A., 167
Datkos, P.G., 135, 233, 294, 359
Davico, G.E., 166, 370, 376
Davidson, J.A., 232
Davies, D.K., 136
Davis, F.J., 137, 263
Davis, V.T., 191, 353, 354
De Puy, C.H., 232, 265
De-Clercq, H.L., 327
DeDomenico, A., 263
Deinzer, M.L., 20, 101, 264
Delsart, C., 353
Delsart, C., 352
Deluze, M.S., 192
Deming, W.E., 44, 45
Denault, J.W., 136, 265
Denifi, G., 232
Desfrancois, C., 135, 136, 263, 327, 328, 361, 376

Dessent, C.E.H., 362
Dewar, M.J.S., 6, 44, 137, 166, 265
Dibeler, V.H., 231
Dick, J.P., 354
Dillard, J.G., 45, 134
Dillow, G.W., 137, 327, 359, 362, 370, 376, 376, 377
Ding, C.F., 327
Dirac, P., 137
Dispert, H., 264, 293, 359
Dojahn, J., 45
Donnart, M.H., 191
Donovan, R.J., 233
Dotan, I., 232
Doty, P.M., 191
Dressler, R., 137
Drzaic, P.S., 6, 20, 264
D'sa, E.D., 45, 73, 138, 294
Dukel'skil, V.M., 191
Duncan, M.A., 135, 166
Dunkin, D.B., 232
Dunning, F.B., 137, 232
Dussak, L., 45, 73
Duysing, W., 44

Eaton, J.G., 231, 354
Eberhardt, W., 354
Echt, O., 232
Edge, C.J., 353
Eguchi, D., 294
Eguchi, D., 137
Ehdaie, S., 45, 101, 361
Eliav, E., 191
Elizer, I., 167
Ellison, G.B., 6, 21, 134, 135, 166, 191, 230, 231, 232, 264, 352
Elmore, D.T., 328
Elving, P.J., 327
Engleking, P.C., 232, 231
Ervin, K.M., 73, 166, 231, 354
Ettre, L.S., 43
Eguchi, 359
Eyring, H., 167

Faeder, J., 167
Farragher, A.L., 231
Fehsenfeld, F.C., 137, 232
Feigerle, C.S., 191, 363
Feldman, D., 191, 232, 354
Feng, M.J., 328
Fenn, T., 354
Fenzlaff, H.-P., 137, 231
Ferguson, E.E., 136, 232
Fessenden, R.W., 137
Fine, J., 191
Fletcher, G.D., 352
Foldiak, G., 73, 135
Francois, J.P., 192
Franklin, J.L., 233, 232, 263
Freeman, R.R., 44, 230
Freidhoff, C.B., 232, 354
Frenking, G., 231
Frey, W.F., 264
Frey, P., 353
Friedhoff, C.B., 231
Fukuda, E.K., 136, 264, 370, 376, 377
Fujio, M., 362, 370, 376
Futrell, J.H., 232

Gadea, F.X., 230
Galbraith, J.M., 233
Gallup, G.A., 327
Gant, K.S., 137
Gantefor, G.F., 231, 354
Gareyev, R., 265
Garwan, M.A., 45, 73, 352
Gausa, M., 354
Gehrk, C.W., 328
George, R., 73, 101, 166, 232, 294
George, R.D., 20, 44, 137
Gerber, G., 354
Giessner-Prettre, G., 43, 327
Gilles, M.K., 233
Gilmore, F.R., 231
Giuffreda, M.G., 192
Glockler, G., 135
Golay, M.J.E., 43
Goldan, P.D., 136
Golden, D.M., 294
Goldfarb, F., 352
Goode, G.C., 6, 20, 73, 134, 167, 191, 233, 264, 293, 359, 362, 370 376
Goodman, L., 43
Gotchiguian, P., 191
Grabowski, J.J., 232
Grafton, A.K., 134
Greenblatt, B.J., 167, 230, 354
Grimsrud, E.P., 135, 232, 263, 264, 265, 294, 295, 327, 328, 359, 360, 376, 377
Gronert, S., 328
Groothuis, G., 294, 360
Guichon, G., 294
Gunther, E., 135, 362
Gustafsson, M., 353
Gutman, D., 136, 231
Gutsev, G.L., 232, 233
Gygax, R., 264
Hadad, G.N., 20, 137

Haeffler, G., 352, 353
Hahndorf, I., 327, 328
Hall, J.L., 102, 230
Hamill, W.H., 231, 328
Han, C.C., 44, 73
Hanstorp, G., 352, 353
Harland, P.W., 233, 377
Harris, F.M., 233
Hartman, H., 136
Haselbach, E., 134
Hashemi, R., 231
Hashmall, J.A., 45, 74, 265
Hauffe, B., 102, 138
Haugen, H K., 45, 190, 352, 353, 354
Heinis, T., 232, 264, 359, 370, 376, 377
Heinrich, N., 231
Helmy, E.M., 232
Hendricks, J.H., 135, 136, 327, 328, 354, 362
Henglein A., 44, 136
Heni, M., 231
Herbst, E., 232
Herman, Z., 353
Hernandez, N., 45, 370, 376
Herschbach, D.R., 2, 6, 45, 102, 166, 230, 264, 294, 360
Hildebrandt, G.F., 137
Hine, J., 167
Hiraoka, K., 137, 294, 359
Hirose, T., 232
Hirsch, W., 44, 102
Hirshfelder, J.O., 167
Hites, R.A., 6, 20, 294
Ho, J., 231, 232, 354
Hobbs, R.H., 231
Hopper, D.G., 233
Hoshino, K., 264, 294
Hotop, H., 44, 45, 352, 353
Hoyland, J.R., 43
Hrovat, D.A., 136, 264
Huang, J., 45, 101
Huang, Y., 328
Hubers, M.M., 233
Huels, M.A., 327
Hughes, B.M., 21, 102, 136, 230, 294, 359
Huh, C., 137, 265, 362, 370, 376, 377
Hunsicker, S., 231
Hunter, R., 328
Hurst, G.S., 44, 102, 230, 233
Hylleras, E., 43, 137

Ichikawa, T., 167
Iino, T., 137, 294, 359
Ilin, R.N., 353
Illenberger, E., 137, 138, 232, 327
Ingemann, S., 294, 360
Ishikawa, Y., 191

Jackson, R.L., 231
Jaffke, T., 138, 231
Jager, K., 263
James, A.T., 43
Arrold, C.C., 354
Johnson, M.A., 135, 136, 360, 362
Jones, P.L., 230, 232
Jones, R.O., 231
Jordan, K.D., 20, 134

Kadish, K.M., 45, 73
Kaldor, U., 191, 231
Kallenbach, N.R., 328
Kamenev, A.G., 191
Kang, C.H., 137, 265, 362, 370, 376, 377
Kao, L.W., 44, 265, 361, 370, 376
Kass, S.R., 166
Kato, A., 135, 166, 265
Kawamata, H., 192
Kay, J., 233
Kaya, K., 135, 166, 264
Kebarle, P., 6, 20, 134, 135, 136, 137, 232, 233, 263, 264, 265, 294, 295, 327, 359, 362, 370, 376, 377
Keiichi, T., 327
Keith, H., 44, 73
Kellert, F.G., 137
Kelley, J.A., 360
Kenttamaa, H., 328
Khvostenko, V.I., 191
Kidder, L.H., 231, 354
Kishi, H., 376
Kiyan, I., 352, 353
Kim, E.H., 232
Kim, J., 360
King, R.A., 233
Kitsopoulos, T.N., 192, 231, 354
Kleyn, A.W., 21, 45, 230, 264
Kline, L.E., 136
Klinkmlller, A.E., 352, 353
Klopcic, S.A., 354
Klots, C.E., 137, 263, 362
Knapp, A., 232
Knight, A.M., 135, 166
Knighton, W.B., 265, 294, 359
Knott, W.J., 264
Knox, J.H., 233
Koch, W., 231
Kocher, C.A., 264
Koepnick, M.C., 191, 353
Kohler, B.E., 230

Kohno, M., 191
Kondow, T., 362
Koshland, D.E., 327
Kozanecki, N., 6, 21, 73, 135, 166, 265
Kraus, K., 44, 135, 232
Krauss, M., 231
Kreisle, D., 232
Kristensen, K., 353
Kumada, T., 166
Kumagai, J., 166
Kuo, K., 328
Kuhn, 369
Kurepa, M.V., 230
Kvale, T.J., 191, 353

Lacmann, K., 102, 230, 264, 293, 360
Landau, A., 191
Langford, M.L., 233
Laramee, J.A., 20, 101, 264
Lardin, H.A., 166
Larson, D.J., 352, 353
Larson, J.W., 233
Lavrich, D.J., 231
Le Barbu, K., 264
Lecomte, F., 135, 136
Lee, C.K., 45, 361
Lee, G.H., 231
Lee, H.S., 265
Lee, H.W., 362, 370, 376, 377
Lee, L.C., 232
Lee, M.L., 328
Lee, T.J., 137
Leininger, M.L., 230, 327
Leopold, D.G., 353, 354
Lepine, F., 192
Lesk, A.M., 137
Levine, J., 102, 230
Levins, R.J., 369
Lewis, G.N., 7
Lewis, T.P., 327
Li, X., 327, 328, 354
Lifshitz, C., 21, 102, 136, 230, 232, 294, 359
Limero, T., 45, 73, 138, 294, 359
Lilly, A.C., 369
Lin S.N., 44, 101, 294
Lineberger, W.C., 6, 20, 73, 190, 230, 231, 232, 233, 264, 265, 352, 353, 354
Lippa, T.P., 135, 136, 353, 354, 376
Lipsky, S.R., 43
Litherland, A.E., 45, 190, 352
Liu, Y., 192
Ljungblad, U., 352
Lobo, R.F.M., 263
Locke, M.J., 328
Los, J., 233, 263
Lovelock, J.E., 6, 20
Luder, C., 192
Lutz, H.O., 354
Lyapustina, S.A., 135, 136, 263, 264, 327, 353, 354, 362, 376
Lykke, K.R., 137
Lyons, L.E., 44, 73, 136, 264

Mahadevan, P., 102, 138
Mahan B.H., 19, 44, 294
Makowski, J.A., 233
Mallion, R.B., 167
Manaa, M.R., 231, 354
Maneira, M.J.P., 264, 293
Mansour, N.B., 353
Mark, T.D., 138, 232
Marks, J., 6, 20, 370, 376
Martin, A.J.P., 20, 43
Marzluff, R.M., 328
Maslen, P.E., 167
Massey, H.S.W., 6, 20, 190, 231
Matejcik, S., 135, 138
Mathur, B.P., 264, 294
Mauk, P.H., 232
Mayer, J.E., 191, 135
Mazurkiewicz, P., 20, 101, 166
McCorkle, D.L., 135, 166, 263, 294
McCullum, K.J., 191
McDonald, R.N., 370, 376
McDowell, M.R.C., 166
McGeoch, M.W., 231
McHugh, K.M., 231, 232, 354
McIver, R.T., 6, 20, 135, 136, 264, 265, 294, 328, 370, 376
McMahon, T.B., 233
McMillen, J.F., 294
McNamee, P.E., 264, 294
Mead, R.D., 6, 20
Meinke, M., 231
MeiwesBroer, K.H., 192, 354
Michaelson, H.B., 191
Michels H.H., 44, 231
Miller, A.E.S., 137, 233
Miller, D.A., 327
Miller, S., 138
Miller, T.M., 137, 138, 294
Miller, W.H., 166, 230, 354
Milligan, M.S., 45, 263, 376
Mishima, M., 137, 265, 362, 370, 376, 377
Miyzaki, T., 166
Mizuno, T., 137, 294, 359
Mock, R.S., 135, 264, 377
Moore, H.W., 6, 20, 45, 136, 265, 294

Moravec, V.D., 354
Moriwaki, T., 191
Morris, R.A., 137, 233
Moseley, J.T., 232
Mothes, K.G., 137
Moutinho, A.M.C., 134, 230, 264, 293
Muccini, G.A., 44, 136
Muigg, D., 232
Muller, J.C., 191
Muller-Duysing, W., 135, 136
Mulliken, R.S., 6, 21, 167, 233
Murphy, R.F., 328
Murray, K.K., 137

Nadeau, M-J., 45, 190, 352
Naff, W.T., 44, 264
Naganuma, T., 264, 294
Nagao, S., 135, 166
Nagata, T., 232
Nakajima, A., 135, 166, 264, 294
Nakamura, H., 137, 370, 376, 377
Nakamura, Y., 135, 166, 265, 362
Nakao, K., 264
Nalley, S.J., 102, 230
Nandi, S., 6, 20, 134, 166, 191, 230, 264, 352
Negishi, Y., 135, 166, 264
Nazrenko, V.A., 362
Nelson, D.R., 137
Nenner, I., 45, 327, 362
Nesbet, R.K., 167
Neuert, H., 44, 102, 136, 138, 230, 232
Neumark, D.M., 136, 166, 232, 327, 328, 354
Nielsen, S.B., 360
Nguyen, M.T., 230, 231
Nibbering, N.N.M., 294, 360
Nilles, J.M., 191, 264, 353, 354
Nimlos, M.R., 232
Nonaka, H., 136, 376
Nordlander, P., 232
Norquist, P.L., 353
Novich, S.E., 232

O' Hair, R.J., 328
Oakes, J.M., 135, 231
Odegard, W., 45
Ogawa, Y., 138
Oldham, G., 328
O'Leary, B., 167
O'Neil, S.V., 166
Ono, F., 264, 294

Pack J.L., 6, 20, 102, 135, 230
Page, F.M., 6, 20, 73, 134, 167, 191, 231, 233, 264, 293, 359, 362, 370, 376, 377
Paidarova, I., 230
Paleev, V.I., 191
Palmer, L.D., 135, 136
Papanikolas.J.M., 167
Parenteau, L., 231
Parkes, D.A., 231
Parr, R.G., 230
Parson, R., 167
Paszek, A.P., 232
Patterson, T.A., 232
Paul, G., 370, 376
Pauling, L., 7, 167
Paulson, J.F., 137, 138, 294
Pearson, P.K., 232
Pegg, D.J., 352, 353
Pekeris, C.L., 137, 352
Pellerite, M.J., 231
Pentchev, P.N., 327, 328
Periquet, V., 135, 136, 376
Person, W.B., 43
Peters, H.L., 264
Petrunin, V.V., 353
Pettiettehall, C.L., 191, 231, 353
Phelps, V., 6, 20, 102, 135, 230
Picardi, G., 135
Pokhodenko, V.D., 362
Polak, M.L., 233, 354
Poole, C., 43
Pople, J.A., 137
Porterfield, W.W., 191
Pritchard, H.O., 6, 20, 102, 190, 191, 230
Pullman, A., 43, 327

Raaen, V.F., 137, 264, 362
Rains, L.J., 6, 20, 294
Ramaker, D.E., 231
Ramond, T.M., 73, 166, 370, 376
Rangell, J., 353
Rao, H., 45, 101
Reck, G.P., 232, 264, 360
Recknagel, E., 232
Redfern, P.C., 137, 294
Reed, D.R., 45, 166, 359
Reese, R.M., 231
Refaey, K.M.A., 231, 233
Reimann, C.W., 135
Reinhardt, P.W., 102, 135, 230, 232, 233, 263, 264, 294, 370, 376
Reid, C.J., 354
Renneboog, R., 328
Richardson, D.E., 377, 232
Richardson, J.H., 232

Richardson, N.A., 328
Richardson, P., 45, 359
Rienstra-Kiracofe, J.C., 6, 6, 20, 45, 166, 191, 264
Ristau, W., 44, 264, 370, 376
Robertson, B.H., 328
Robinson, D.W., 135, 136, 376
Robinson, J.M., 45, 101
Rodgers, M.T., 328,
Rolla, L., 135
Romer, B.C., 370, 376
Rosner, S.D., 230
Roth, E.W., 232, 264, 360
Ruoff, R.S., 45, 73
Rzepa, H.S., 45, 137

Sakharov, V.I., 353
Sambe, H., 231
Sanche, L., 231, 327
Sandstrom, J., 353
Sane, N., 6, 21, 73, 135, 137, 166, 265, 327, 361
Sarkas, H.W., 231, 354
Sauer, M.H.M., 327
Sauers, I., 264, 294
Saykally, R., 192, 265
Schaefer, H.F., 6, 20, 134, 166, 191, 232, 233, 264, 327, 352
Schaefer, O., 362
Schwartz, R.L., 370, 376
Scheer, M., 353, 354
Scheidt, J., 73, 102, 136, 230, 232, 264, 327, 370, 376
Schermann, J.P., 135, 136, 361
Schiff, H.I., 136
Schindler, R.N., 137
Schweinler, H.C., 370, 376
Schlag, E.W., 136, 264
Schlier, R.E., 231
Schmeltekopf, A.L., 136,
Schultz, G.J., 45, 166, 362
Schulz, A., 327, 328
Schulze, S., 6, 21, 73, 135, 137, 166, 191, 327, 361
Schwartz, R.L., 73, 166
Scott, L.T., 136, 166
Scottt, S.L., 370, 376
Seely, D.G., 191, 363
Seidel, C.A.M., 327
Senn, G., 232
Serenkov, I.T., 353
Serxner, D., 231
Sevilla, M.D., 327
Sheer, M.D., 191, 352, 353
Shimamori, H., 38, 191
Shuie, L.R., 45, 101, 233
Siegel, M.W., 102, 230
Skogerboe, K., 327
Slager, V.L., 354
Smalley, R.E., 192, 231, 363
Smith, D., 20, 138
Smith, G.P., 232
Smith, H., 135
Smith, K.A., 137
Smith, L.J., 45, 137, 327
Smith, S.J., 102, 135
Snodgrass, J.T., 232, 231, 354
Soep, B., 167, 230
Sokolov, V.M., 191
Sousa, S.R., 294
Spanel, P.I., 20, 135
Spyrou, S.M., 264, 294
Squires, R.R., 166, 264
Srawley, R.A., 353
St.John, G.A., 327, 328
Stadler, J., 328
Stanecke, P.O., 360
Stearns, S.D., 45, 101
Stebbings, R.F., 137
Steelhammer, J.C., 44, 73, 328, 359
Steenken, S., 328
Steinfelder, K., 6, 20
Stemmler, E.A., 6, 20, 135, 265
Stephenson, L.M., 232
Stevens, A.E., 6, 20
Stockdale, J.A.D., 44, 102, 230, 263
Streit, G.E., 232, 232
Streitweiser, A.S., 6, 21, 137, 167
Sugden, T.M., 135, 137
Sugioka, T., 264, 294
Sun, K., 45, 101
Sunagawa, T., 138
Sutton, J., 191
Sutton, P., 135
Suzuki, S., 191
Szent-Györgi, A., 7, 43

Tachikawa, H., 233
Taguwa, T., 166, 264
Takao, K., 137, 294, 359
Talley, L., 6, 21, 73, 135
Tang, S.Y., 232, 264, 360
Tatsumi, Y., 138, 265
Taylor, H.S., 167
Taylor, K.J., 191, 231, 353
Taylor, T.R., 167, 230, 354
Tellet-Billy, D., 166, 230
Thogresen, J., 352, 353

Thompson, G.P., 7, 20
Thompson, J.J., 7, 20
Thompson J.S., 191, 353, 354
Thynne, J.C.J., 233
Tiernan, T.O., 21, 102, 136, 233, 232, 230, 231, 294, 359
Tittel, F.K., 192
Tommasen, H., 44, 135
Trainham, R., 353
Trainham, R., 352
Travers, M.J., 231
Trinajatic, N., 45, 74
Tschumper, G.S., 6, 20, 134, 166, 191, 264, 265, 352
Tsuda, S., 328
Tsukuda, T., 232, 362
Tsuno,Y., 137, 362, 370, 376, 377
Tummler, R., 6, 20
Tung, R., 44, 73, 265, 293, 294

Uchimaru, T., 328

Valli, C., 352
Van de Wiel, H.J., 44, 135
Van den Berg, H.W., 328
Van Doren, J.M., 232
Van Orden, A., 192
Vandoren, J.M., 233
Viggiano, A.A., 137, 233
Villaalta, P.W., 354
Vogt, D., 102, 138
Volstad, J.D., 353
Von.Ardenne, M., 6, 20
Vorburger, T.V., 232

Wahl, A.C., 233
Wang, L.S., 328
Wang, T.B., 73, 135, 354
Wang, X.B., 328, 354
Wanzek, K.P., 136
Warren, L.J., 44, 73
Watanabe, K., 264, 294
Watson, J.D., 43
Weddle, G.H., 360
Weigman, U., 264, 294
Weik, F., 231
Weinkauf, R., 73, 102, 136, 230, 232, 264, 327, 370, 376
Weis, A., 137
Welk, N., 45, 102, 191, 232, 263, 376
Wenthold, P.G., 136, 166
Wentworth, W.E., 7, 20, 43, 44, 45, 46, 73, 74, 101, 102, 134, 135, 136, 137, 138, 167, 190, 191, 230, 232, 233, 264, 265, 293, 294, 327, 328, 352, 359, 361, 369, 370, 376
Wesolowski, S.S., 327
Westerdahl, G., 328
Wheeler, J.C., 190, 352
Wheeler, R.A., 134, 190
Wheeler, R.C., 231
Wiberg, W.B., 327
Wiley, J.R., 45, 73, 135, 263, 294, 327, 359
Williams, J.K., 166
Williams, J.M., 231
Williams, W.W., 191, 353
Wojnarvits, L., 73, 135
Wolcott, M.H., 328
Wong, S.F., 232
Woo, S.B., 232
Wright, L.A., 231
Wu, H., 354
Wu, R.L.C., 233, 232
Wu, W., 28

Xu, C.S., 232, 354
Xu, S.K., 264, 265
Xu, S.J., 191, 265, 353, 354

Yamabe, S., 137, 294, 359
Yamataka, H., 137, 362, 370, 376
Yang, S., 191
Yang, S., 231
Young, C.E., 20, 44
Younkin, J.M., 45, 137, 327

Zandberg, E.Y., 191
Zanni, M.T., 167, 354
Zeegers-Huyskens, T., 328
Zeforov, Y.V., 190
Zhang, Oi., 45, 101, 327, 328
Zhang, Q-L., 191
Zhang, Y., 191, 294
Zhao, X-L., 190, 352
Zlatkis A., 43, 45, 361, 369

Subject Index

Ab initio calculations, 125–132, 194, 226, 305
Accuracy, 39, 168
Activation energies, 11, 200–218, 269–274, 293, 315, 318, 332
Albyns, K., 59–64, 43
Alkali Metal Beam (AMB), 121–124, 163, 221, 238, 240–241, 268, 274–275, 331
Anion Hydrate Photoelectron
Spectroscopy (Ah-PES), 110–116, 298, 302
Appearance Potential (AP), 93–96, 241
Atmospheric Pressure Negative Ion Mass Spectrometry (API-NIMS), 37–40, 47–50, 93–96, 289–290
Avogaadro, A., 5
Ayala, J., 38

Bear, G., 42
Becker, R. S., 26, 28
Bergman, M., 5
Bergman-Dewar Structures, 336-339
Boltzman, L, 5
Bond Dissociation Energy, 59–64, 90, 93–96, 194, 196, 220, 268–272, 279, 293, 311–315, 332
Bond Order, 194–196, 207–209, 214 268
Born, M, 24
Briegleb, G., 31, 161
Brown, T., 43
Browne, T., 4

Cai, H., 42
Carbon Clusters, 1, 2, 180–185
Carr, S., 43
Carson, R., 27
Charge Densities, 39
Charge transfer spectra, 17, 67–71, 276 319–320, 332, 382
Chen, E. C. M., xiii, 31
Chen, E. S., 41
Cobb, M., 42
Compton, R. N., 161
Collisional Ionization, 116–121, 148, 243, 263
CURES-EC, Ch IX-XII, 1–4, 17, 40, 140–150, 169, 184–185, 189–190, 266, 274, 278–293, 331, 335

D(m), Dc(m) *see* MPEC
Dalagarno, A., 154
Darvish, P., 43
De Columb, C-A., 4
DeBroglie wavelength, 14, 32, 251, 272, 332
DeBtroglie, L., 14
Deming, W. E., 36, 103
Density Functional, 125–132, 185, 298, 305–311, 318
Deoxyribonucleic Acid (DNA), 296–299 318–325, 382
DNA Radiation Damage, 320–326
Dewar, M. J. S., 33, 141
Dipole bound state, 8, 110–116, 215, 303–305, 320, 332
Dillard
Dirac, P. A. M., 125
Dissociation Energy, 8, 18, 59–64, 267
Dissociative capture *see* Thermal electron attachment
Dojahn, J., 42, 43
Donor Acceptor complexes, 1–4, 16, 67–71, 257–261, 299, 320
D'sa E. D., 40

The Electron Capture Detector and the Study of Reactions with Thermal Electrons
by E. C. M. Chen and E. S. D. Chen
ISBN 0-471-32622-4

Du Fay, C.-F, 4
Dussack, L., 43

EDEA, 12, 18, 59–64, 140, 151–160, 210, 218, 226–227, 235, 251, 267, 280–287, 293, 311, 332
Ehdaie, S., 42
Electron affinity, Ch IV-XII, 8, 51–56, 85–96, 349–383, 332
  Acridine, 309–310
  Aromatic Hydrocarbons, 30, 244
  Adenine, Guanine, Thymine, Cytosine (AGCUT), 41, 296–305
  Adiabatic, 8, 299, 304, 309, 322, 326, 331
  Alkyl halides, 59–64, 266–267, 274
  Amino Acids, 1–4, 298, 316–317
  Anthracene, 11, 30, 55, 129–134, 140–149, 248
  Atoms, 168–178, 349–351
  Atomic clusters, 184–190
  Azide radical, 216, 229
  Bergman-Dewar Set, 27–33, 245
  Biological Molecules, 1–4, 40–42, 326
  Carbonyl Compounds, Aromatic, 51, 86
  Coronene, 113, 148–149
  CS_2, 89, 220, 222
  d block elements, 177
  Diatomic, . Triatomics, 54, 55, 193–194, 216, 228–229, 351–352
  Dioxins, 288–289
  Environmental Pollutants, Ch XI
  Excited State, 8, 268, 274
  Fluorocompounds, 272
  Geometry, 319
  Group IVA, 173
  Halogenated aliphatics, 8, 246, 357–360, 382–383
  Halogenated aromatics, 276–289, 292–293, 357–360, 382–383
  Halogens, 1–4, 171, 205
  HaloUracils, 307
  Hydrocarbons, 355–357
  Hydrogen Bonded Base Pairs, 318
  Isoelectronic Equivalence, 161
  Lead, 173, 196
  Linear Acenes, 51, 140
  Main group atoms, 168
  Naphthalene, 52, 242, 298
  Nitrobenzenes, 56, 241
  Nitrogen, 176
  NO, 41, 214, 241, 296
  Organic Molecules, Ch X, 31, 51, 355–387
  Organic Nitro compounds, 52, 253
  Oxygen, 41, 91, 112, 205–206, 241, 296
  Ozone, 216
  Pentafluorophenyl, 277
  Periodicity in atoms, 169, 174
  Phenanthrene, 51, 310
  Purines, Pyrimidines *see* AGCUT
  Substituent Effects, 261–263
  Sulfur fluorides, 56, 224–229
  Tetracene, 28–33, 244
  Transition Metals, 194
  Valence State, 8
  Vertical, 8, 141, 267, 280, 304, 335
  Vinyl Chloride, 275
  Work functions, 169
Electron Capture Detector (ECD)
  kinetic Model, Ch. IV, IX-XI, 47–63, 13, 164, 299, 310–312 316, 318, 332
  temperature Regions, 47–63, 335
Electron Impact Spectra, 15, 156, 218–224, 237, 267, 279, 297–313, 332
Electron Swarm, 15, 194, 333
Electron Transmission, 1–4.121–124, 298 304, 333
Electron Transport, 318
Electronegativities
Correlations, 1–4, 140
    Mulliken, 161, 168, 178–179, 183, 185
    Pauling, 178–179
    Periodicity, 180
    Rochow, 178
    Work function, 181, 186
    Hydrocarbons, 246
Elizer, I., 154
Endothermic Charge Transfer, 15, 331
Energy Level Diagrams, 305–306
Equilibrium Methods, 8–15, 105–110
Ettre. L. E., 22, 23
Evaluated Value (EvV), 333
Eyring, H., 154

Faraday, M., 4
Feynman, R, 139
Franck Condon region, 140, 151, 156–157, 206, 208, 220, 229
Franklin. B., 4
Freeman, R. R., 37

Galvani, L, 4
Gas Phase Acidities (GPA), 15–16, 140, 298, 302, 309–312, 315, 333
Geiger, J., 5
George, R., 34
George, R., 43

George, R. D., 34
Gilbert, W., 4
Glossary, 331–336
Golay, M. C. E., 23
Goodwin(Shell Laboratory), 4
Gremaud, G., 42

Haber, F., 24
Half Wave Reduction Potentials, 16, 28–40, 140, 165, 234, 246, 249–250, 263, 267 283, 284, 298–302, 307, 309, 332
Hartree Fock Equation, 17, 125–132, 142
Helias, N., 43
Hernandez, N., 40
Herschbach, D. R., 18, 40, 41, 197–198, 201
Herz, R. H., 5
Hirschfelder, J. O., 154
History Electron, Ch I
History Wentworth, Ch III
HYPERCHEM, 1–4, 17, 40, 139, 141, 146–147, 166, 224, 299, 324

Internuclear Distance, 18, 37–40, 150–161, 198–202, 210
Ion Pair, 15
Ionization Potential, 13, 64–66, 246, 333
  AGCUT, 299
  Periodicity in atoms, 179–180
  Role in charge transfer, 67–71, 320
  Calculations, 141, 161
  Atoms, 178–184
Isoelectronic Principle, 140, 151–156, 209

Keith, H., 34
Klein, (Shell Laboratory), 27
Koopman's Theorem, 1–4, 28, 140–150
Kozanecki, N., 43
Krebs, H., 27
Krick, F. N. C., 27

Least Squares Procedure, 1–4, 35, 40–41, 85–91, 96–101, 254, 339–349
Lee, C. K., 40
Lesk, A. M., 127
Lewis, G. N., 5
Limero, T., 40
Lin, S. N., 33
Lipsky, S. R., 23, 25
Lovelock, J. E., xiii, 1, 23, 24
Lowest Unoccupied Molecular Orbital (LUMO), 1–4, 28, 140–150, 144–146, 305, 334

Lyons, L. E., 25–27, 31, 32

M(m), Mc(m) *see* MPEC
Magnetron, 13, 28, 105–110, 224–226, 238–240, 263, 268, 274–276, 334
Mahan, B. H., 33
Mandubushi, J., 42
Marchand. A., xiii
Martin A. J. P., 22–25
Massey, H. S. W., 24
Maxwell, J. C., 5
McAfee, K. B., 25
McIver, R. T., 117
Melton. C. E., 39
Mendonca, S., 43
Mercury Pool Electrode, 146–147
Michaels, H. H., 37
Milligan, M. S., 43
Millikan, R., 5
Minus Delta Delta G (mDDG) *see* Solution Energy Difference
Moore, H. W., 117
Morris, G. C., 32
Morse Parameters, 18, 154, 156, 158, 312, 332, 334, 335
Morse Potential Energy Curves (MPEC), 18, 40
  Sulfur Fluorides, 224–225
  Alkyl Halides, 267–276
  Antibonding, 237 268–269
  Benzene, 157–161
  Bonding, 237, 268–269
  CO_2, COS, CS_2, 221–223
  Cytosine, Thymine, 305 311 313–314
  Dipole Bound, 268
  Group VI diatomic, 205
  Group VII diatomics, 197
  H_2, 153–156
  Herschbach Ionic (HIMPEC), 1–4, 18, 139, 150–153, 157, 166, 198–202, 205–206, 227 311, 315, 332, 333
  Homonuclear Diatomics, 229
  I_2, 156
  Naphthalene, 157–161
  Nitrobenzene, Nitromethane, 235–236, 263
  Organic molecules, Ch X
  Triatomics, 193
Polarization *see* MPEC Dipole bound
Morse, P. M., 9, 18, 35, 38
Mulliken, R. S., 1, 5, 17, 18, 26, 75
Multiconfiguration Configuration Interaction (MCCI), 139, 141, 143, 310, 334

Negative Ion Mass Spectrometry (NIMS), 40–42, 47–50, 105–110, 226, 235–240, 263–266 276, 284, 287, 291, 298, 310–311, 316, 326, 334
Nesbet, R. K., 154
Nondissociative capture *see* thermal electron attachment
Nucleic Acids, *see* DNA Nucleic acids

Ohm, G. S., 4
Orbitals *see* MCCI

Pariser Parr Pople, 125–128, 142–143
Page, F. M., 162
Parr, R. G., 195–205
Partition Function Ratio, Q, 13, 28–33, 50–59, 333, 334
Pauling, L., 5
Periodicity parameter (P), 181–190
Person, W. B., 195–205
Photodetachment (PD), 15, 110–116, 216, 241–243, 292, 333, 334
Photoelectron Spectroscopy (PES), 15, 110–117, 140–142, 148, 164, 171–173, 193, 198, 200–202, 206 208–221, 226, 237, 334
Photon methods, 15, 110–118, 171, 238, 241–245, 263, 276, 298–303, 326
Pi Stack, 162, 318–319, 325
Planck, M., 5
Precision, 31, 35–37, 93–101, 334
Precision and Accuracy Plots, 31, 103–124, 168, 334
Pre-exponential term, Ai, 8–15, 331
Pritchard, H. O., 2

QQQ, 2
Quantum Mechanical Calculations
AM1, 17, 40, 142–150, 164, 298, 310, 315, 319, 321, 331
Biological Molecules, 41, 327
Electron Affinity Calculations, 141
MNDO Parametric Method-3 (PM3), 17, 40, 298, 310, 315
Modified Intermediate Neglect of
Differential OverlapMINDO, 17, 40
Modified Neglect of Differential
Overlap (MNDO), 17, 40, 315
Semi-empirical, 17, 40, 140, 305, 335

Ranatunga, R., 42
Random uncertainties Atoms, 168 171
Rains, L. J., 117
Rao, Lei., 43
READS-TCT, 139–140, 147–150, 165, 169, 189, 193, 223, 243–244 323
Rearrangement Energy, 38, 214, 333
Reduction Potentials *see* Half Wave Reduction Potentials
Reed, W. R., 43
Riboflavin, 309
Richardson, P., 43
Ristau, W., 33
Robinson, J. M., 42
Rydberg Electron Transfer, 121–124, 221

Sane, S., 43
Schultza, S., 43
Seebeck, T. J., 4
Shen, S., 43
Shewhart, W. A., 103
Shuie, L. R., 40
Solution Energy Difference, Ch. X, 16, 41, 334
Spin Densities
Adenine (−), 323
AT:AT:AT stack, 325
CS2, COS, 224
GC enol, 323
GC keto, 323–324
GC:GC:GC stack, 324
Guanine (−), 321
Steelhammer, J. C., xiii, 34
Stockdale, J. A. D., 34
Stoney, G. J., 5
Streitweiser, A. S., 27
Substitution and Replacement Rules, 1–4, 139, 161, 165, 238, 26, 278–284, 289 299–300, 307
Sun, K., 42
Swatlowski, R., 43
Szent-Gyorgi, A., 28, 47

Taylor, H. S., 154
Temperature Dependence *see* ECD
Thales of Miletus, 4
Thermal Charge Transfer (TCT), 15, 116–121, 142, 164, 216, 221, 234, 237, 242–243 247, 249, 251–257, 263, 268, 274–278, 282–283, 293 298, 308–310, 316, 320, 326
Thermal Electron Attachment (TEA), 14, 270, 273, 317, 332
Thompson, J. J., xiii
Thomson, G. P., 5
Thomson, J. J., 5, 15
Timelines, 31, 103–124, 168, 335
Townsend, J. S., 5

Tswett, M. S., 23–25
Tung, R. T., 34

Unimolecular Dissociation, 59–64, 218 267

Volta, A., 4

Wang, L. M., 33
Warren, L. J., 32
Watanesk, S., 43
Watson, T. J., 27
Watson-Crick base pairs *see* DNA
Watts, (Shell Laboratory), 27
Wentworth, W. E., xiii-xvi, 1, 23, 28, 31, 35, 339
White, B., 42
Wiley, J. W., 33, 42
Williams, J. K., 154

Young, C. E., 33

Zhang, L., 43
Zhang, O., 43
Zlatkis, A., 22– 26, 31